Trends in Interfacial Electrochemistry

NATO ASI Series

Advanced Science Institutes Series

A series presenting the results of activities sponsored by the NATO Science Committee, which aims at the dissemination of advanced scientific and technological knowledge, with a view to strengthening links between scientific communities.

The series is published by an international board of publishers in conjunction with the NATO Scientific Affairs Division

A	Life Sciences	Plenum Publishing Corporation
B	Physics	London and New York
C	Mathematical and Physical Sciences	D. Reidel Publishing Company Dordrecht, Boston, Lancaster and Tokyo
D	Behavioural and Social Sciences	Martinus Nijhoff Publishers
E	Engineering and Materials Sciences	The Hague, Boston and Lancaster
F	Computer and Systems Sciences	Springer-Verlag
G	Ecological Sciences	Berlin, Heidelberg, New York and Tokyo

Series C: Mathematical and Physical Sciences Vol. 179

Trends in Interfacial Electrochemistry

edited by

A. Fernando Silva
Department of Chemistry, Faculty of Sciences,
University of Porto, Porto, Portugal

D. Reidel Publishing Company

Dordrecht / Boston / Lancaster / Tokyo

Published in cooperation with NATO Scientific Affairs Division

Proceedings of the NATO Advanced Study Institute on
Trends in Interfacial Electrochemistry
Viana do Castelo, Portugal
2-13 July, 1984

Library of Congress Cataloging in Publication Data

NATO Advanced Study Institute on Trends in Interfacial Electrochemistry (1984 : Viana do Castelo, Portugal)
Trends in interfacial electrochemistry.

(NATO ASI series. Series C, Mathematical and physical sciences; vol. 179)
"Proceedings of the NATO Advanced Study Institute on Trends in Interfacial Electrochemistry, Viana do Castelo, Portugal, 2–13 July 1984"—
"Published in cooperation with NATO Scientific Affairs Division."

Includes index.
1. Electrodes—Congresses. 2. Electrolytes—Congresses. 3. Electric double layer—Congresses. 4. Surface chemistry—Congresses. I. Silva, A. Fernando (António Fernando), 1949– . II. North Atlantic Treaty Organization. Scientific Affairs Division. III. Title. IV. Series: NATO ASI series. Series C, Mathematical and physical sciences; vol. 179.
QD571.N286 1984 541.3'7 86–10112
ISBN 90–277–2271–4

Published by D. Reidel Publishing Company
P.O. Box 17, 3300 AA Dordrecht, Holland

Sold and distributed in the U.S.A. and Canada
by Kluwer Academic Publishers,
101 Philip Drive, Assinippi Park, Norwell, MA 02061, U.S.A.

In all other countries, sold and distributed
by Kluwer Academic Publishers Group,
P.O. Box 322, 3300 AH Dordrecht, Holland

D. Reidel Publishing Company is a member of the Kluwer Academic Publishers Group

Printed in The Netherlands

CONTENTS

PREFACE

This book presents an up to date review of many aspects of Interfacial Electrochemistry and points direction of future developments.

Traditional routes for the study of the electrochemical interface are reviewed, critically discussed and the available experimental data is analysed. Complementary information is presented as obtained from the sucessful application of the various in-situ reflectance spectroscopies.

The use of single crystal face electrodes to study the electrochemical interface is emphasized with particular relevance to the technique to prepare clean surfaces. Some relevant results obtained for single crystal face electrodes are presented. This book presents also the techniques to study other interfaces such as the ionic-solution, immiscible liquid-liquid and gas-solid interfaces. The information gained is put in parallel to the solid-electrolyte solution interface.

More specific aspects of the electrochemical interface are covered with chapters on phase-transitions occuring on 2D adsorbed layers and on electrochemical kinetics where the dependence of the electrochemical rate parameters upon the solvent structure and electrode material is emphasized.

The final chapters reviews of the present state of solvent models interfaces and the recently developed theories of the electrochemical interfaces in which an association is made of models for the metal with the models for the electrolyte solution.

This book is a result of a NATO A.S.I. on "Trends in Interfacial Electrochemistry", held in Viana do Castelo from 2 to 13th July 1984 which brought together experts of the above mentioned aspects of Interfacial Electrochemistry.

I would like to thank all the lecturers for their cooperation and care in preparing the manuscripts of their lectures, for the quality of the oral presentations and for their enthusiasm and support during the Institute. My sincere thanks are also due to Prof. G. Hills, Prof. R. Parsons and Prof. S. Trasatti for all the help, invaluable advice and encouragement given since the begining of the organization of the Institute to the end of it.

I would like to thank most sincerely to NATO Scientific Affairs Division for the generous support of the A.S.I..

I also thank Calouste Gulbenkian Foundation (Lisbon) and Instituto Nacional de Investigação Científica for their financial support.

Finally I would like to express my gratitude to all participants in the Institute for their enthusiasm and active participation. It is our hope that this book may be useful for the development of Interfacial Electrochemistry and for their future careers.

University of Porto, Portugal
November, 1985

António Fernando Sousa da Silva

COMPONENTS OF THE ELECTRODE POTENTIAL. CONCEPTS AND PROBLEMS

Sergio Trasatti

Department of Physical Chemistry and Electrochemistry, University of Milan, Via Venezian 21, 20133 Milan, Italy.

ABSTRACT

This paper gives an introductory account of quantities, concepts and problems commonly encountered when delaing with electrode/solution interfaces. After some description of the structural features of the free surface of a metal and a polar liquid, equations are derived and discussed for the metal/polar liquid interface. In particular, attention is focussed on (i) the dependence of the potential of zero charge on the electron work function, (ii) the contact (Volta) potential difference, (iii) the temperature coefficient of the electrode potential, (iv) the effect of the heterogeneity of the surface, (v) the relevance of adsorption of water from the gas phase to the electrochemical situation, (vi) the shift of the electrode potential upon adsorption, (vii) the permittivity of the interfacial layer of solvent molecules, and (viii) the concept of "electrochemical" work function.

1. INTRODUCTION

The aim of this lecture is to discuss concepts and quantities which very often recurr in other papers of this series. An electrode is customarily thought of as an ensemble of an electronic conductor (most frequently a metal) and an ionic conductor (electrolyte solution, solid electrolyte, molten salt). Other solid/liquid interfaces are also relevant to electrochemistry; for example that between an ionic solid (oxide, halide, etc.) and an electrolyte solution. The mechanisms of charging are however very different in the two cases. At metal/solution interfaces the surface charge is associated with accumulation or depletion of electrons

A. F. Silva (ed.), Trends in Interfacial Electrochemistry, 1–24.

on the metal side and ions on the solution side. In the case of ionic solids the interface becomes charged as a consequence of some ion exchange between the solution and the solid surface. Different cases are further to be distinguished if the ion exchanged is a real component of the solid phase or is simply present in its surface layer. Accordingly, the description of the interfacial region is based on parameters which may differ conceptually, qualitatively and quantitatively for the various situations.

This lecture will be focussed on the metal/electrolyte solution interface which is the basic system on which the majority of experimental studies at electrodes have been carried out and to which most of the models thus far proposed have been specifically applied. Metal/solution interfaces may be either non-polarizable or polarizable depending on whether or not they are permeable by charged species [1,2]. While structural aspects are in fact common, the thermodynamic approach differs [3] and for the latter group of interfaces is less readily associated with physical models.

The most specific quantity characterizing an electrode is its potential. Its value depends on a number of factors [4], including primarily the state of charge of the interface. While it is not easy to measure directly the electrode potential at a given value of charge, the potential at which the charge on the metal vanishes can be straightforwardly obtained by a number of experimental techniques [5]. This *potential of zero charge*, $E_{\sigma=0}$, is not simply a quantity. It is in fact as typical a property of the given metal/solution interface as the electron work function is of a metal surface [6]. Accordingly, the potential of zero charge depends on the structural details of the electrode surface and is the most direct and simplest information we can obtain experimentally about the structure of the electrode/solution interface. The problem is that such information is not open but must be appropriately decoded.

The system considered here consists of a pure metal in contact with a pure solvent. The presence of ions is in principle conceptually unnecessary to discuss the structural parameters of uncharged surfaces. Where necessary, we shall however assume that ions in solution are not specifically adsorbed so that their concentration has no detectable effect on the experimental parameters. Although the charge on the metal and on the solution surface regions is usually considered as being smeared out uniformly on both sides of the ideal interface, recent experiments [7,8] on adsorption from the gas phase have shown that the localization of the ionic charge should be taken into account when translating experimental parameters into structural models. While discreteness of charge has been considered in electrochemical kinetics [9,10] and adsorption isotherm [11], is still not well clear how to introduce this effect into interfacial models in a plane parallel to the metal surface. Therefore, in this introductory lecture, no explicit attempt will be made to account for this.

While only one potential of zero charge exists for a polarizable interface, the potential of zero *free* charge is distinguished from potentials of zero *total* charge for non-polarizable interfaces [12]. Only the former is related to the structural details of the interface; the latter are associated with the thermodynamics of exchange of common components between the two phases [3]. Therefore, only polarizable interfaces will be explicitly considered in the treatment which follows.

The metal/solution interface has been for long depicted on the basis of the so-called Gouy-Chapman-Stern-Grahame model [13-15] according to which the interphase is split into two regions, the compact layer and the diffuse layer, where ions and solvent molecules are treated as behaving in a totally different and independent way. It has been recently pointed out [16-18] that this *a priori* separation needs not be introduced in double layer models. While this is in principle true since the interphase is a three-dimensional system where lateral interactions do not take place only in planes parallel to the ideal interfacial plane, it must be stressed that such a separation is still very useful for the elaboration of the experimental data of adsorption studies, since the error implicit in the original Gouy-Chapman theory for the diffuse distribution of ions around an electrode is quantitatively small in the charge range usually explored experimentally [19]. On the other hand, when discussing potentials of zero charge only, there is no need to deal with diffuse layer problems since this vanishes as the charge on the metal is zero.

2. ROLE OF THE ELECTRONS AT THE METAL SURFACE

The metal has for long been considered in double layer problems as an ideally smooth, perfectly conducting wall whose interaction with the environment takes place through electron exchanges and charge image effects; in other words, the role of the solid phase has very often been underestimated. In fact, even though the structural details of the surface are not considered explicitly, the metal is to be regarded as responsive to electrical polarizations and chemical interactions which can modify the intrinsic properties of its surface.

While models used in the theory of metals may be more complex, the structure of the metal surface is presently taken into account in double layer problems essentially on the basis of the "jellium" model [20,21] where the ion cores are treated as a uniform positive background terminating at the ideally smooth surface, and electrons are distributed smoothly across the ideal boundary thus protruting towards the exterior and giving rise to a surface dipolar layer. The associated potential drop is denoted [2,22] by χ^M (surface potential of the metal) (Fig. 1). It is to be stressed that the simple "jellium" model accounts neither for the finite size of the ions cores at the surface nor for the relaxation of the me

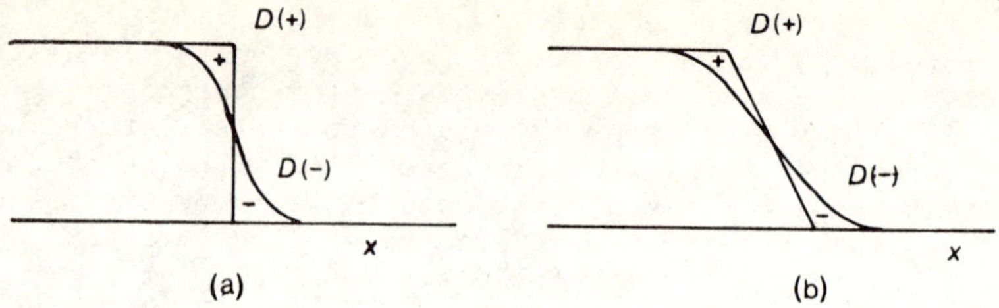

Fig. 1 - Onset of the surface potential of the metal according to the "jellium" model (a). $D(+)$ is the density of the positive charge of the ion cores, $D(-)$ is the density of the electronic charge. (b) "Modified" jellium to take into account the roughness of the surface at the atomic level.

tal lattice. While the length of electron escape from the metal may be of the order of 0.1-0.2 nm [23], the relaxation of the lattice may involve several atomic layers decreasing exponentially towards the interior of the solid [24]. The relevance of this effect to double layer problems may however be neglected compared to that of surface electrons. Because of all these reasons, the properties of different faces of the same metal cannot be treated by means of the simple "jellium" model. Although the introduction of ion core discreteness effects at the surface of a metal into models is difficult, this specific aspect is however of primary experimental importance and is to be taken always into account when working with single crystal faces.

As particles of another phase approach the metal surface, electrons located there will experience a different potential energy. They will react to the new condition[1] by adopting a new distribution across the surface layer. The new potential drop is usually denoted [2] by $g^M(\text{dip})$ to emphasize that it does not correspond to the situation pertaining to the free metal surface in a vacuum. Therefore:

$$g^M(\text{dip}) = \chi^M + \delta\chi^M \tag{2.1}$$

where $\delta\chi^M$ can be regarded as a perturbation term.

The energy of electrons at the Fermi level of an uncharged isolated metal is measured by the real potential of the electrons, α_e^M, which includes chemical and surface contributions [1,4]:

$$\alpha_e^M = \mu_e^M - F\chi^M \tag{2.2}$$

The chemical potential, μ_e^M, includes all short-range interactions keeping the electrons bound to the interior of the metal phase. It can only be calculated [6] with the aid of the electronic theory of metals.

The most typical experimental quantity of metals is the elec tron work function, Φ^M, measuring the minimum work to extract electrons from the Fermi level of the metal [26,27]. It follows readily that:

$$\Phi^M = -\alpha_e^M = F\chi^M - \mu_e^M \tag{2.3}$$

As a metal comes in contact with another phase, say a polar liquid, the surface potential is modified as shown in eqn.(2.1) and so is the work function. From eqns.(2.1) to (2.3) it is clear that the change in the work function coincides with $\delta\chi^M$ since μ_e^M is a bulk property and is not modified by surface events.

3. STRUCTURING OF SOLVENT MOLECULES AT INTERFACES

Molecules of a liquid polar phase (*i.e.* a solvent) at a newly created surface are subjected to unbalanced forces. As a consequence, they adopt an orientation which may differ from the average one in the bulk [28]. The layer where the structure of the liquid is disturbed, is the surface region and a net electric potential drop may arise there as a result of some preferential orientation of the molecular permanent dipoles [29]. This is the surface potential of the solution, χ^S [2].

If particles of another phase approach the liquid surface, the polar molecules will react to the new condition[2] by adopting new configurations and distributions to minimize energy. As a result of the interaction between the two phases, the surface potential of the solvent will be modified and will be denoted by $g^S(\text{dip})$. It follows readily that:

$$g^S(\text{dip}) = \chi^S + \delta\chi^S \tag{3.1}$$

where $\delta\chi^S$ is a perturbation term measuring the effect of the formation of the interface.

The thickness of the surface region usually depends on the strength of the interactions between the two phases. More specifically, the penetration of the perturbation into the liquid phase may be different depending on the specific property being considered to observe the surface [30]. For homogeneous metals, such as liquid Hg, experimental results suggest [31] that only the first solvent layer is probably affected by short-range interactions. In the case of high-energy surfaces such as silica etc. several monomolecular layers of the liquid may be subjected to modifications especially for hydrogen-bonded solvents [32]. In principle, there are no reasons to expect that modifications should be restricted to the first monomolecular layer, but it is only a matter of simplification if it is very often done so. Recent models however do not pose a priori such a constraint [16-19,25,33].

4. BASIC EQUATION OF THE POTENTIAL OF ZERO CHARGE

As the potential of an electrode M is measured against a reference electrode, the resulting quantity is in fact the electric potential difference between the terminals of the cell (Fig. 2) [4,34,35]:

$$M|S|R|M' \tag{4.1}$$

where R is the reference electrode, S the solution phase and M' differs from M for the electrical state only. M is assumed to be of the same material as that constituting the measuring instrument. The measured quantity E^M is known as the electrode potential of M relative to R and can be written as the sum of the electric potential drop across all the interfaces in cell (4.1):

$$E^M = (\phi^M - \phi^S) + (\phi^S - \phi^R) + (\phi^R - \phi^{M'}) \tag{4.2}$$

Since both R and M are metallic phases, an electronic equilibrium is established at the R/M' interface. Therefore:

$$\tilde{\mu}_e^R = \tilde{\mu}_e^{M'} \tag{4.3}$$

where $\tilde{\mu}_e$ are electrochemical potentials of electrons in the respective phases. On the basis of the definition of electrochemical potential [1]:

$$\tilde{\mu}_i^\alpha = \mu_i^\alpha + z_i F\phi^\alpha \tag{4.4}$$

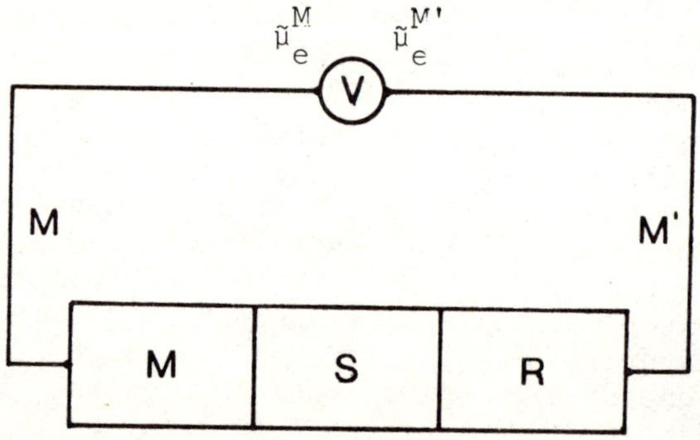

Fig. 2 - Sketch of the set-up to measure electrode potentials. M is the metal of the electrode under study, S the solution and R the metal of the reference electrode. The measuring instrument and the cables to the cell are assumed to be made of metal M. M' differs from M by the electrical state only.

where μ_i^α is the chemical potential of species i in phase α, ϕ^α is the inner electric potential of phase α and z_i is the charge number of species i, it follows from eqn.(4.3) that:

$$(\mu_e^R - \mu_e^M)/F = (\phi^R - \phi^{M'}) \qquad (4.5)$$

where the prime has been dropped from M in μ_e^M since the chemical state of M and M' is the same. From eqns.(4.2) and (4.5), it results:

$$E^M = (\phi^M - \phi^S) + (\phi^S - \phi^R) + (\mu_e^R - \mu_e^M)/F \qquad (4.6)$$

Since E^M depends only on the nature of M for a constant reference electrode, it is possible to write [34,36]:

$$E^M = (\phi^M - \phi^S) - \mu_e^M/F - E_T^R \qquad (4.7)$$

where E_T^R depends only on the nature of the reference electrode (and of the temperature). This quantity has been discussed in other papers [34,35] and is of no particular importance here since attention is focussed on the properties of the M/S interface. Equation (4.7) shows that the "electrode potential", as obtained in practice, does not measure the *electric potential drop* between the bulk of the metal and the bulk of the solution, but it is a more complex quantity which includes both electrical and chemical contributions. This is because E^M results from the difference in electronic energy between dissimilar phases. Equation (4.7) has been obtained by thermodynamic means only. E^M can however be split into several components which are not amenable to independent experimental determination [1,37,38]. Therefore, in order to relate E^M to the structural aspects of the M/S interface, it is necessary to introduce some model assumption.

The electric potential drop across a charged M/S boundary may be regarded as determined by both free charges and dipoles:

$$(\phi^M - \phi^S) = g_S^M(\text{ion}) + g_S^M(\text{dip}) \qquad (4.8)$$

This splitting rests on the separation of induced and permanent dipole effects[3]. It may often be somewhat artificial since as long as permanent dipoles respond to the electric field, part of the potential drop due to these dipoles turns out to be in fact attributed to induced dipoles. The separation is however useful from a modellistic point of view and is rigorous at least under the conditions when the free charge on the metal phase vanishes ($\sigma^M = 0$). Then, $g_S^M(\text{ion}) = 0$ and the measured potential becomes, from eqn. (4.7):

$$E_{\sigma=0}^M = g_S^M(\text{dip})_0 - \mu_e^M/F - E_T^R \qquad (4.9)$$

where $E_{\sigma=0}$ is now the *potential of zero charge*.

The term $g_S^M(\text{dip})_0$ accounts for the potential drop across the M/S boundary associated with the electron distribution at the surface of the metal and for that due to any layer of preferentially oriented solvent dipoles in the solution surface region (Fig. 3). These two components are in principle of difficult separation since interpenetration of the two dipole layers may take place. The separation is however useful from a modellistic point of view and is implicit in the fact that two distinct approaches are often used to calculate the particle distributions in the two surface regions supplemented by an account of the interactions occurring between the two phases.

Equation (4.9) can be rewritten in the following form:

$$E^M_{\sigma=0} = g^M(\text{dip})_0 - g^S(\text{dip})_0 - \mu_e^M/F - E_T^R \tag{4.10}$$

On the basis of the definition of electron work function (2.3) and of the significance of $g^M(\text{dip})$ (2.1), eqn.(4.10) becomes:

$$E^M_{\sigma=0} = \Phi^M/F + \delta\chi_0^M - g^S(\text{dip})_0 - E_T^R \tag{4.11}$$

This equation relates a typical electrochemical quantity, the electrode potential, to a typical physical quantity, the electron work function and accounts for perturbation terms which originate from the contact between the two phases. It is the equation from which any attempts to estimate the components of the electrode potential of zero charge usually start [41].

The sole value of $E_{\sigma=0}$ gives no structural information be-

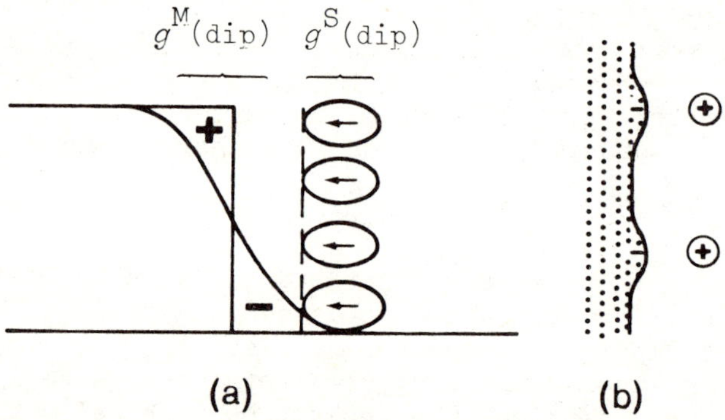

Fig. 3 - (a) Sketch to show the dipolar contribution to the electrode potential. The tail of the electron distribution at the metal surface "penetrates" the liquid phase where the preferential orientation of solvent molecules has been emphasized. (b) Sketch to illustrate that the localized ionic charges in solution are expected to be balanced by localized electronic charges in the metal, where the charge density is thus not smeared out uniformly.

cause it is a relative quantity. If Φ is independently known, no definite information is still obtained. The only way to obtain information which does not depend on the reference electrode is to correlate $E_{\sigma=0}$ and Φ for different metals. The plot can thus give information [38] on the sum of the two terms $[\delta\chi_0^M - g^S(\text{dip})_0]$. There is no way to obtain direct information on only one of the two terms separately and a given metal must always be chosen as a reference.

This approach has been used for long [42] and is the most direct one to gain insight into the structure of the M/S interface [38,41,43]. Its main drawback is however that Φ is generally measured on a surface whose state may not correspond to that of the metal during the measurement of $E_{\sigma=0}$.

5. CONTACT (VOLTA) POTENTIAL DIFFERENCE AT METAL/SOLUTION INTERFACES

The contact (Volta) potential difference between two phases is a quantity which is amenable to experimental determination [1]. It gives information which is related to the electrode potential but does not involve chemical terms (the electrode work function) [37].

An electrode is actually a three phase system since it includes the (inert) atmosphere of the environment (Fig. 4). As a metal is brought in contact with a liquid polar phase and a polarizable interface is created, a redistribution of charges and dipoles takes place at the M/S interface and at the M/air and solution/air interfaces so that an electrostatic equilibrium is reached. The amount of charge needed to reach the equilibrium conditions is usually compositionally negligible. Irrespective of the nature

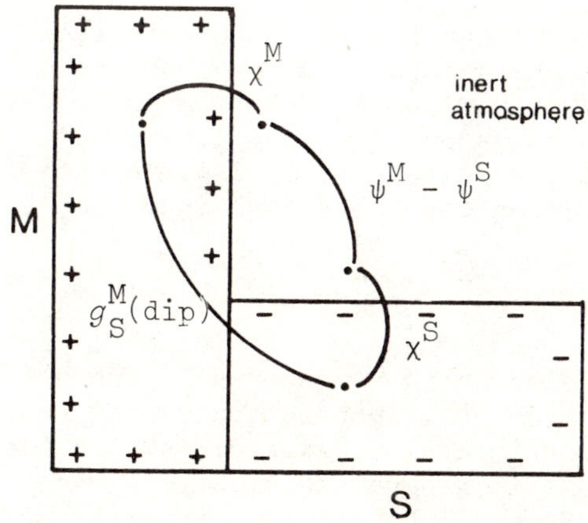

Fig. 4 - Picture of the charge and electric potential distribution for a metal/solution system at the potential of zero charge (no free charge is present at the M/S boundary).

of the mechanism by which surface charging may take place, it is possible in principle to define a Volta potential difference also between a metal and a pure solvent.

Under equilibrium conditions it must be (Fig. 4):

$$(\phi^M - \phi^S) = (\psi^M - \psi^S) + (\chi^M - \chi^S) \tag{5.1}$$

since the electric potential differences does not depend on the route. Therefore:

$$(\psi^M - \psi^S) = (\phi^M - \phi^S) - (\chi^M - \chi^S) \tag{5.2}$$

where ψ is the outer (Volta) electric potential [1]. If $(\phi^M - \phi^S)$ is written in terms of eqn.(4.8), it results:

$$(\psi^M - \psi^S) = g^M_S(\text{ion}) + g^M_S(\text{dip}) - (\chi^M - \chi^S) \tag{5.3}$$

Equation (5.3) shows that the contact potential difference depends on the state of charge of the interface; owing to the dipolar term, this dependence may be complex. The last term, $(\chi^M - \chi^S)$, is a constant as long as the charges accumulated at the free surfaces of the phases do not disturb the dipole distributions there.

At the potential of zero charge $g^M_S(\text{ion}) = 0$, but the Volta potential difference is non-zero if there are structural differences among the various interfaces. Thus, for $\sigma = 0$:

$$(\psi^M - \psi^S)_{\sigma=0} = g^M(\text{dip})_0 - \chi^M + \chi^S - g^S(\text{dip})_0 \tag{5.4}$$

Recalling eqns.(2.1) and (3.1), it results [44]:

$$(\psi^M - \psi^S)_{\sigma=0} = \delta\chi^M_0 - \delta\chi^S_0 \tag{5.5}$$

Equation (5.5) shows that the Volta potential difference at the potential of zero charge measures how much the distributions of dipoles at the M/S interface differ from those at the free surfaces of the separate phases.

In practice, the Volta potential difference is measured by means of a cell with a gap containing inert gas [45-47]:

$$\text{M}|\text{gas}|\text{S}|\text{R}|\text{M}' \tag{5.6}$$

M may be a liquid metal (Hg) or a solid metal. In the former case, a Hg jet is used along a tube whose walls are wet by the flowing solution. If a solid metal is used the system becomes a static one. Available experimental results differ for the two arrangements. Problems are in general associated with organic impurities at the static surface of the solution and with possible modifications of the surface state of the metal electrode due to solvent vapour and inert gas adsorption. Once a standard value for the Volta potential difference is obtained in cell (5.6) by extrapola tion to zero concentration of the electrolyte to which R is reversible, the Volta potential difference for the condition of zero charge is readily obtained by the equation:

$$(\psi^M - \psi^S)^\circ - (\psi^M - \psi^S)_{\sigma=0} = E^\circ - E^M_{\sigma=0} \tag{5.7}$$

where E° is the standard potential for the reversible electrode in cell (5.6).

6. TEMPERATURE COEFFICIENT

The effect of temperature is readily obtained from eqn.(4.11) (superscript M is now dropped):

$$dE_{\sigma=0}/dT = (d\Phi/dT)F + d(\delta\chi_0)/dT - dg^S(\text{dip})_0/dT - dE^R_T/dT \tag{6.1}$$

Equation (6.1) shows that the experimental temperature coefficient of the potential of zero charge contains four terms which are either not measurable or can be estimated with difficulty. Models which relate $dE_{\sigma=0}/dT$ to the solvent behaviour only may be inadequate. The last term on the right-hand side of eqn.(6.1) represents the temperature coefficient of the reference electrode and requires the knowledge of the absolute entropy of single ions. $d\Phi/dT$ can be measured independently [48] but this is not the case for $d(\delta\chi_0)/dT$. Therefore, treating the temperature coefficient of $E_{\sigma=0}$ only in terms of the behaviour of a layer of solvent dipoles may be highly misleading.

If χ^S is added to and substracted from eqn.(4.11), this becomes:

$$E_{\sigma=0} = \Phi/F + (\psi^M - \psi^S)_{\sigma=0} - E^R_K \tag{6.2}$$

where E_K is the so-called absolute potential of the given reference electrode and is related to E_T by $E_K = E_T + \chi^S$. The physical meaning of these electrode potentials has been discussed in a recent publication to which the reader is referred [35]. Equation (6.2), besides illustrating the relationship between electrode potential and metal/solution contact potential difference, shows that if the temperature coefficient is corrected simply by the temperature coefficient of Φ, the temperature coefficient of the Volta potential difference is obtained, which is in fact an experimentally determinable quantity. Therefore, no direct information on the behaviour of the solvent at the interface can be obtained, unless extrathermodynamic assumptions are introduced. Thus, in principle, even the sign of $dE_{\sigma=0}/dT$ is not indicative of the behaviour of the solvent.

7. SURFACE HETEROGENEITY

Polycrystalline surfaces are very often used with solid electrodes. The metal surface can thus be regarded as consisting of patches of single crystal faces. Equation (4.11) applies in fact to each of the patches but, as a consequence of the surface heterogeneity and of the equipotentiality of the metal surface, the condition of zero charge cannot be fulfilled everywhere.

Let us suppose to have a metal surface exposing two different crystal faces of equal surface area (Fig. 5). Let Φ^A be the work function of face A and Φ^B the work function of face B. Since the two grains are in contact, they are in electronic equilibrium. Therefore:

$$\tilde{\mu}_e^A = \tilde{\mu}_e^B \tag{7.1}$$

which is achieved at expenses of some transfer of electrons from the crystallite with the lower work function to the other one. In the absence of an external field, for the electroneutrality condition:

$$\sigma^A = - \sigma^B \tag{7.2}$$

This means that the so-called potential of zero charge of polycrystalline surfaces does not correspond locally to the conditions of uncharged metal.

The applicability of eqn.(4.11) to polycrystalline surfaces depends on whether or not the various quantities are averaged in the same way over the whole surface. This turns out to depend in fact on the particular property and the experimental method used to measure it.

Thermodynamically, an average work function can be defined for a polycrystalline surface [4,6,27]:

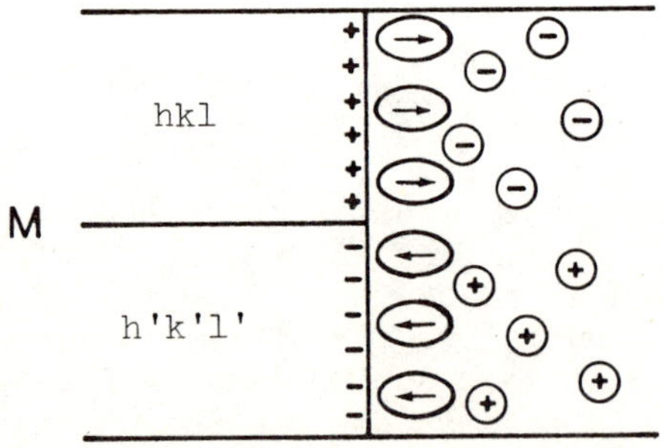

Fig. 5 - Sketch to emphasize the difference in free charge and dipole contributions at the two crystal faces (hkl) and (h'k'l') of a polycrystalline metal M at the potential of zero charge, *i.e.* $\sigma(h'k'l') = - \sigma(hkl)$.

$$\langle\Phi\rangle = \sum_i \Phi_i \theta_i \tag{7.3}$$

where Φ_i is the work function of patch (face) i and θ_i is the surface fraction occupied by this face on the actual sample. The "true" value of $\langle\Phi\rangle$ can be in fact obtained only by the contact potential difference measurement which is a thermodynamic experimental approach. Other techniques such as photoelectric, thermoionic and field emission measurements, privilege some specific emitting spots on the metal surface [27]. Although this effect may be small, it should be considered case by case.

The potential of zero charge is mostly obtained by observing the condition of maximum diffusiveness of the ionic atmosphere around the electrode surface, for instance from the minimum of the experimental capacitance. While it has been recognized for long that the heterogeneity of the surface gives less sharp results [49,50], it has recently been shown [51-53] quantitatively that because of the asymmetrical behaviour of double layer properties around the potential of zero charge, ionic atmosphere diffusiveness effects for the different patches on a electrode surface turn out not to be averaged so simply as in the case of the electrode work function. This makes $E_{\sigma=0}$ for polycrystalline surfaces a questionable quantity which should be handled with caution (Fig. 6).

It follows from the previous discussion that values of $E_{\sigma=0}$

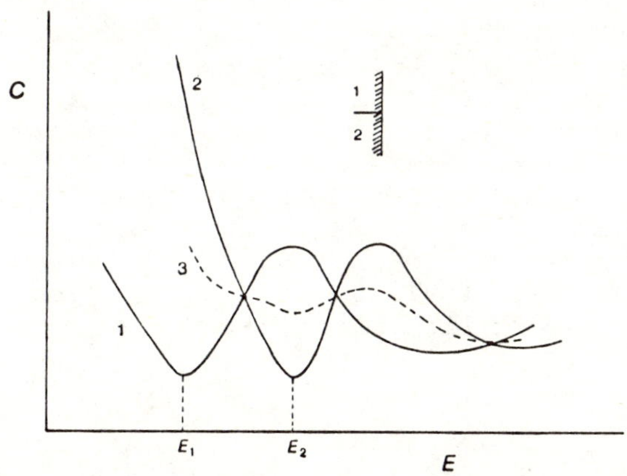

Fig. 6 - Polycrystalline metal surface composed of two crystal faces 1 and 2. Curves 1 and 2 are the capacitance-potential relationships for the corresponding single crystal face electrodes. E_1 and E_2 are the respective potential of zero charge. It is seen that in this specific case the potential of zero charge of the polycrystalline sample practically coincides with that of face 2. Curve 3 is the capacitance of the polycrystalline surface calculated from $(C_1 + C_2)/2$.

and Φ are strictly definable only for single crystal faces. Equation (4.11) can thus be rewritten as follows:

$$E_{\sigma=0}(\mathrm{hkl}) = \Phi(\mathrm{hkl})/F + \delta\chi_0(\mathrm{hkl}) - g^{\mathrm{S}}(\mathrm{dip})_0(\mathrm{hkl}) - E_{\mathrm{T}}^{\mathrm{R}} \quad (7.4)$$

where (hkl) are Miller's indexes. Since the distribution of atoms in the surface plane differs for different crystal orientation, the electron profile across the phase boundary also changes. As a consequence of the different solid/liquid particle interactions the dipolar terms on the right-hand side of eqn.(7.4) turn out to be face specific. Therefore, different crystal faces of the same metal have different $E_{\sigma=0}$ not only because of different Φ values but also because solvent molecules may adopt different preferential orientations at the surface. The role of $\delta\chi_0$ is still to be investigated in this specific case.

It is to be stressed that while the uniform positive background in the "jellium" model never comes in contact with solvent molecules, as the surface of real metals are considered the picture becomes dramatically more complex. Valleys and peaks can be distinguished on the atomic scale in the cross section of a real surface (Fig. 7) [4,54]. Since the electronic profile does not follows the atomic profile (smoothing effects, cf. section 2) [55], the valleys are filled by electrons while the peaks are deprived of electrons. Therefore, peaks are mainly positively charged and the interaction with solvent molecules may be qualitatively and quantitatively very different from that in the valleys. An increased complexity is found in stepped surfaces which represent actually the transition from mono- to polycrystalline surfaces [56].

The surface structure of a single crystal face can be checked by UHV techniques, specifically low-energy electron diffration (LEED) [57] and its cleanliness controlled by Auger electron spectroscopy (AES) [58]. As the face comes in contact with a solution, some reconstruction can be brought about by interactions with the liquid phase which may modify the bond strength of surface atoms with their neighbourings [59]. Such a reconstruction can also take place if electrochemical cleaning procedures are adopted leading

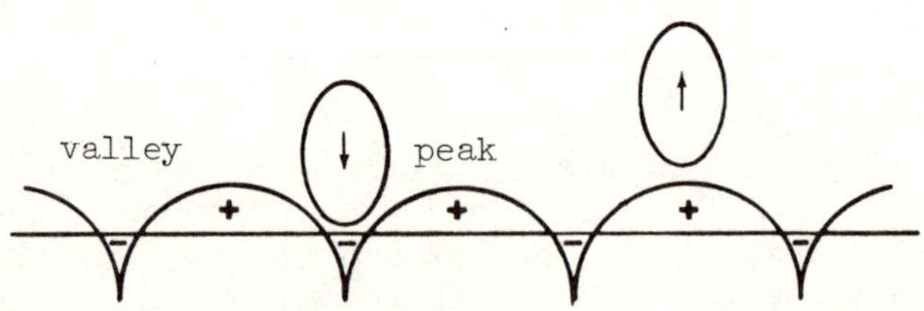

Fig. 7 - Sketch showing the profile of a real surface where ion cores are distinguished. The possible different orientations of dipolar molecules at different sites are emphasized.

to surface phase formation and successive removal [60-62]. In this case the final surface may be related only nominally to the original crystal face. In particular, the electron work function of a crystal face used as an electrode might differ from that expected for the same face in a vacuum. This might make any correlation of $E_{\sigma=0}$ and Φ difficult to interpret unambiguously.

The degree of heterogeneity of a metal surface is determined by the looseness of its surface atoms. This is very qualitatively measured by the melting point [30,63], more specifically by the lattice cohesion and the atomic mass which govern the trend of atoms to autodiffuse [64]. More quantitatively, the variation of Φ from face to face may give a straightforward idea of the degree of heterogeneity of a polycrystalline surface of the given metal [65].

8. SURFACE ORIENTATION OF SOLVENT MOLECULES

If a definite value of $g^{\mathrm{S}}(\mathrm{dip})$ is possibly estimated, this may be related to the orientation of surface dipoles by means of the Helmholtz formula [1]

$$g^{\mathrm{S}}(\mathrm{dip}) = \frac{p_{\perp}N}{\varepsilon} \qquad (8.1)$$

where N is the surface density of solvent molecules assumed to be in a monolayer and $p_{\perp}$ is the average normal component of the molecular dipole.

N is usually calculated from the bulk density of the solvent. This is in principle an inappropriate approach. Depending on the structuring action of the solid surface the local properties of the liquid may be very different [66]. Moreover, the solvent molecule population at the interface may depend on the atomic structure of the crystal face and vary accordingly. In this case any conclusion concerning the orientation of solvent molecules may turn out to be inadequate. Finally, the solvent molecules affected by the interactions with the solid phase may be not only those in the first monolayer but also in the successive layers.

ε is the permittivity of the dipole layer. Calculations are often been carried out by using for ε the value pertaining to the limiting static dielectric constant of the solvent in the high frequency region [31]. This may be again inappropriate. The value of ε essentially determines the derived magnitude of $p_{\perp}$ and therefore the resulting molecular picture of the interface. Physicists usually adopt the permittivity of the free space [67]. This implies that large specific orientations are not needed to explain large potential drops.

Information on the behaviour of solvent molecules at metal/solution interfaces is often derived [30,41] from studies of adsorption of the solvent from the vapour [7,8,68]. As a rule, the work function of a metal changes upon adsorption of polar or polarizable molecules [6]. The magnitude of this effect is not straightforwardly attributable to $g^S(\text{dip})$. It includes several other effects which can be tentatively listed as follows:

i) The surface electron distribution is disturbed ($\delta\chi^M$);

ii) Solvent molecules in the gas phase possess different permanent dipole moment from the liquid state;

iii) The structure of the solvent layer at electrodes depends on the M/S interaction but also on the interaction with the bulk liquid;

iv) Lateral interactions in thin layers may differ from those in thick layers;

v) There is a "surface potential" at the thin layer/vapour interface which is not present at electrodes and should be accounted for in quantitative calculations.

For all reasons above any translation of gas phase data to the electrochemical situation is not unambiguous. However, such data can give a qualitative indication of the sign of the main terms in eqn.(4.11).

9. ADSORPTION POTENTIAL SHIFT

Adsorption of foreign species at electrodes is very often used as a probe to obtain information on the state of the solvent at the interface [69]. The electrostatic effect of specific ionic adsorption on the potential of zero charge usually overrides any possible change in $g^S(\text{dip})$. It is therefore more convenient to use neutral molecules, essentially organic compounds.

Adsorption is believed to occur by displacement of solvent molecules close to the metal surface which are replaced by the adsorbate molecules[70]. At $\theta = 0$, the potential of zero charge is given by eqn.(4.11). As the surface is saturated with the adsorbate B ($\theta = 1$), the potential of zero charge can be written as

$$E'_{\sigma=0} = \Phi/F + \delta\chi'_0 - g^B(\text{dip})_0 - E^R_T \qquad (9.1)$$

Comparison with the situation at $\theta = 0$ gives:

$$E'_{\sigma=0} - E_{\sigma=0} = (\delta\chi'_0 - \delta\chi_0) + g^S(\text{dip})_0 - g^B(\text{dip})_0 \qquad (9.2)$$

Equation (9.2) shows that the value of $g^S(\text{dip})_0$ can be derived only if $g^B(\text{dip})_0 = 0$ and $\delta\chi'_0 = \delta\chi_0$.

While the constancy in $\delta\chi_0$ upon adsorption may be not far from reality because it essentially depends on electron gas-electron cloud interactions, the condition of apolarity of the organic molecule is very difficult to meet. This is especially true for aqueous solutions where reasonable solubility of organic compounds is achieved only for molecules with hydrophilic polar heads. Moreover, in principle, the polarity of a molecule as measured in the gas phase or in apolar liquids may differ from that at the interface because of distorsions and polarizations resulting from the interactions with the metal surface.

While eqn.(9.2) is valid only for $\theta = 1$, a condition difficult to reach experimentally, the value of $\Delta E_{\sigma=0}$ is often estimated by means of extrapolations to $\theta = 1$. This procedure is very delicate and the result is often very misleading. The variation of $E_{\sigma=0}$ with θ may be linear or non-linear [71] depending on the lateral interactions of the dipoles [72]. If the replacement of a solvent molecule at the electrode surface does not involve any disturbance of the neighbouring particles (this might be defined as a "regular" behaviour), then eqn.(9.2) can be written:

$$E^{\theta}_{\sigma=0} - E_{\sigma=0} = (\delta\chi'_0 - \delta\chi_0)^{\theta} + [g^S(\mathrm{dip})_0 - g^B(\mathrm{dip})_0]^{\theta} \qquad (9.3)$$

and the potential shift is a linear function of θ. If lateral interactions are involved, other terms which do not appear in eqn. (4.11) become in fact operative (Fig. 8).

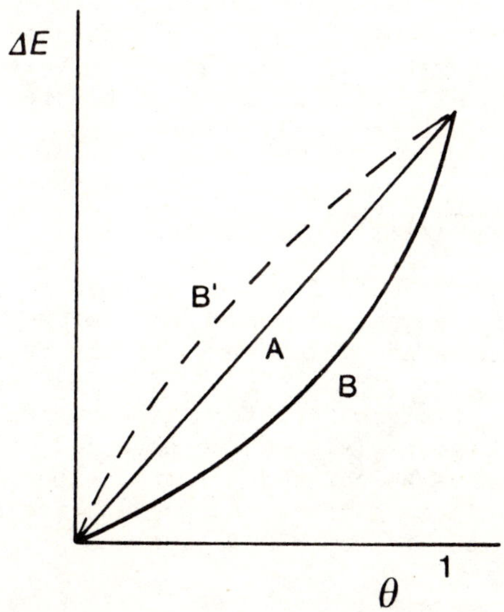

Fig. 8 - Potential shift as a function of adsorbate coverage. Curve A: "regular" behaviour (no lateral interactions between particles on the surface); Curve B: Typical behaviour observed at metal/solution interfaces; Curve C: Behaviour observed in some cases at the air/solution interface.

An aspect of difficult treatment is the nature of the boundary between the adsorbate layer and the bulk of the solution. Solvent molecules are now in contact with an organic layer and the kind of interaction is expected to differ substantially from that with the bare surface of the metal. The layers of solvent molecules in the immediate proximity of the adsorbate might exhibit some preferential orientation which is not explicitly accounted for in eqn.(9.2) and this adds some additional ambiguity to the physical interpretation of the results.

The comparison of the adsorption of a given molecule at the air/solution and at the metal/solution interface is a convenient way of obtaining some information on the role of the metal surface [73,74]. At the air/solution interface the potential shift is simply

$$\Delta\chi' = \chi^S - g^B(\mathrm{dip})_{air} \tag{9.4}$$

If $g^B(\mathrm{dip})_{air} = g^B(\mathrm{dip})_0$, the two adsorption potential shifts may enable some estimate of $\delta\chi^S$ to be made. In fact, from eqns.(9.2) and (9.4):

$$(E'_{\sigma=0} - E_{\sigma=0}) - \Delta\chi' = \Delta\delta\chi_0 + \delta\chi^S + g^B(\mathrm{dip})_{air} - g^B(\mathrm{dip})_0 \tag{9.5}$$

This however still includes the electronic term. On the other hand $g^B(\mathrm{dip})_{air}$ may differ from $g^B(\mathrm{dip})_0$ as a consequence of different interactions with the metal [75,76]. Therefore, the interpretation of adsorption potential shifts is always subjected to a number of assumptions which cannot be easily checked.

10. INTERFACIAL PERMITTIVITY

Although this lecture is focussed on the components of the potential of zero charge, it is convenient to dwell on a few other properties of the interface which are strictly related to the situation at $\sigma = 0$.

The value of $E_{\sigma=0}$ gives no information on the permittivity of the interfacial region. This can be obtained by measuring the interfacial capacitance which is determined experimentally from the rate of change of the surface charge with the electrode potential. Therefore, at the potential of zero charge:

$$C_{\sigma=0} = \mathrm{d}\sigma/\mathrm{d}E \tag{10.1}$$

Since the charge around $E_{\sigma=0}$ is necessarily involved in the definition of $C_{\sigma=0}$, the electrode potential should be better written by combining eqns.(4.3), (4.7) and (4.8). Thus:

$$E = \Phi/F + \delta\chi - g^S(\mathrm{dip}) + g^M_S(\mathrm{ion}) - E^R_T \tag{10.2}$$

where all terms, except Φ and E_T, are dependent on charge. If equation (10.2) is now differentiated with respect to charge:

$$dE/d\sigma = d(\delta\chi)/d\sigma - dg^S(\text{dip})/d\sigma + dg^M_S(\text{ion})/d\sigma \qquad (10.3)$$

from which, at $\sigma = 0$:

$$1/C_{\sigma=0} = 1/C^M - 1/C_{\text{dip}} + 1/C_{\text{ion}} \qquad (10.4)$$

The first term accounts for the polarizability of the metal electrons at the surface, the second term for the ability of interfacial solvent dipoles to reorient under the effect of electric field, and the third term for the induction effect of the electric field on the solvent molecules at constant dipole orientation. The last term includes also the contribution from the polarizability of the ionic atmosphere around the electrode (so-called diffuse layer). Since $1/C_{\text{ion}}$ is mainly determined by the bulk polarizability, the total interfacial capacitance is essentially determined by $1/C^M$ and $1/C_{\text{dip}}$.

The separation between these two terms cannot be achieved experimentally and it is in principle difficult to say which of the two is more important in determining the permittivity of the interface. Models have been proposed that either disregard [77,78] or emphasize the importance of $1/C^M$ [79,80]. Both are reasonably able to reproduce the main features of the interfacial capacitance and its dependence on the nature of the metal. In both cases, however, a more complex structure than a simple ensemble of dipoles must be introduced for the interfacial solvent [81-83], which testifies to the importance of lateral interactions in the behaviour of particles in the interfacial region.

11. ELECTROCHEMICAL WORK FUNCTION

It has been mentioned that the value of Φ for an electrode is actually unknown because relaxation and reconstruction may take place at the metal surface upon contact with the liquid phase. It is thus of interest to devise a method to derive the value of Φ from potential values. This quantity has been termed the "electrochemical work function" [6,41].

Since E is in fact a relative quantity, no absolute value of Φ can be obtained from data for a single metal. Let us write eqn. (10.2) for metal 1:

$$E_1 = \Phi_1/F + \delta\chi_1 - g^S(\text{dip})_1 + g^S_M(\text{ion})_1 - E^R_T \qquad (11.1)$$

and for metal 2 at the same so strong electric field (the same high value of σ) that the solvent orientation can be assumed to reach saturation:

$$E_2 = \Phi_2/F + \delta\chi_2 - g^{\mathrm{S}}(\mathrm{dip})_2 + g^{\mathrm{M}}_{\mathrm{S}}(\mathrm{ion})_2 - E^{\mathrm{R}}_{\mathrm{T}} \qquad (11.2)$$

Comparison of E_1 with E_2 gives:

$$E_1 - E_2 = (\Phi_1 - \Phi_2)/F + \Delta^1_2\delta\chi - \Delta^1_2 g^{\mathrm{S}}(\mathrm{dip})_\sigma + \Delta^1_2 g^{\mathrm{M}}_{\mathrm{S}}(\mathrm{ion}) \qquad (11.3)$$

If a specific condition is found for which the last three terms are close to zero, then the difference in potential is close to the difference in work function. Equation (11.3) shows why the difference in potential of zero charge cannot be taken as measuring the difference in work function. It ensues from section 7 that physical significance can be attached to eqn.(11.3) only for homogeneous polycrystalline surfaces or for single crystal faces.

CONCLUSIONS

The electrode potential is determined not only by the electric potential drop across the metal/solution interface but also by the electronic energy in the metal. Therefore, all parameters of the interface which are derived from the electrode potential contain a contribution from the metal and a contribution from the polar liquid (solvent). This makes it difficult to describe the properties of the interface in molecular terms since any experimental separation between the two contributions involves a number of extra-thermodynamic assumptions which cannot be easily checked.

Nevertheless, it has been suggested how to derive some estimate of the *total* effect due to the formation of the interface from a number of experimental parameters. This kind of information is especially needed as a function of the nature of the metal, of the temperature, of the crystal face for a given metal and of the nature of the solvent. Combination of all these pieces of information is expected to lead to gain better insight into the factors responsible for the properties of the interface. Although a trend to a dichotomy (metal properties or solvent properties) is still apparent in the literature, this is not justified. It is necessary to bear in mind in any approach that it should not result in a model where the details of one of the two phases are neglected.

Acknowledgement. Financial support to this work by the Italian National Research Council (CNR, Rome) is gratefully acknowledged.

NOTES

1 This may be related to interactions with the electron clouds of solvent molecules and to polarization in the field of the solvent permanent dipoles [25].

2 This will include possible interaction with surface orbitals, with the surface electrons and with the field of the metal surface dipolar layer.

3 It is of interest to discuss eqn.(4.8) for interfaces other than those including a metallic conductor. If M is a semiconductor (*e.g.* Si or Ge), g_S^M(ion) is further separable into two terms because of the electric potential drop in the solid phase across the space charge region which is physically distinguishable from the electric potential drop in the solution double layer [39]. If M is an ionic solid, irrespective of the main features of its conductivity, the separation implied in eqn.(4.8) is conceptually much more difficult [32]. With oxides, for instance, undissociated surface OH groups play the role of adsorbed water molecules. However, ionized OH groups are actually the source of the surface charge and the conceptual approach to the components of the interfacial electric potential drop must be somewhat more flexible. At the point of zero charge the splitting of dipolar and free charge contribution is, like in the case of metals, physically meaningful. The origin of $\delta\chi^M$ is however different and the dipolar layer in solution cannot be treated in terms of a simple ensemble of dipoles, since it includes chemisorbed, hydrogen-bonded and physisorbed solvent molecules whose spatial distribution can only be understood in chemical terms [32,40].

REFERENCES

[1] Parsons, R. in Modern Aspects of Electrochemistry (Bockris, J.O'M. and Conway, B.E., Eds.), Vol. 1, Butterworth, London, 1954, p. 103.

[2] Trasatti, S. and Parsons, R. 1983, Pure Appl. Chem. 55, p.1251.

[3] Parsons, R. in Comprehensive Treatise of Electrochemistry (Bockris, J.O'M. Conway, B.E. and Yeager, E., Eds.), Vol. 1, Plenum Press, New York, 1980, p. 1.

[4] Trasatti, S. Comprehensive Treatise of Electrochemistry (Bockris, J.O'M. Conway, B.E. and Yeager, E., Eds.), Vol. 1, Plenum Press, New York, 1980, p. 45.

[5] Frumkin, A.N. Petrii, O.A. and Damaskin, B.B. in Comprehensive Treatise of Electrochemistry (Bockris, J.O'M. Coway, B.E. and Yeager, E., Eds.), Vol. 1, Plenum Press, New York, 1980, p.221.

[6] Trasatti, S. Adavances in Electrochemistry and Electrochemical Engineering (Gerischer, H. and Delahay, P., Eds.), Vol. 10, Wiley Interscience, New York, 1976, p. 213.

[7] Bange, K. Grider, D. and Sass, J.K. 1983, Surf. Sci. 126 p.437.

[8] Grider, D.E. Bange, K. and Sass, J.K. 1983, J. Electrochem. Soc. 130 p. 246.

[9] Fawcett, W.R. 1981, Can. J. Chem. 59 p. 1844.

[10] Krylov, V.S. 1978, Elektrokhimiya 14 p. 1315.

[11] Levine, S. and Fawcett, R. 1979, J. Electroanal. Chem. 99 p. 265.

[12] Frumkin, A.N. and Petrii, O.A. 1975, Electrochim. Acta 20, p. 347.
[13] Grahame, D.C. 1947, Chem. Rev. 41, p. 441.
[14] Reeves, R. in Comprehensive Treatise of Electrochemistry (Bockris, J.O'M. Conway, B.E. and Yeager, E., Eds.), Vol. 1, Plenum Press, New York, 1980, p. 83.
[15] Delahay, P. Double Layer and Electrode Kinetics, Interscience, New York, 1966.
[16] Henderson, D. and Blum, L. 1982, J. Electroanal. Chem. 132, p. 1.
[17] Badiali, J.P. Rosinberg, M.L. Vericat, F. and Blum, L. 1983, J. Electroanal. Chem. 158, p. 253.
[18] Carnie, S.L. and Chan, D.Y.C. 1982, Adv. Colloid Interface Sci. 16, p. 81.
[19] Henderson, H. 1983, Progr. Surf. Sci. 13, p. 1047.
[20] Lang, N.D. 1969, Solid State Commun. 7, p. 1047.
[21] Appelbaum, J.A. and Hamann, D.R. 1976, Rev. Mod. Phys. 48, p. 479.
[22] IUPAC Commission on Electrochemistry, 1974, Pure Appl. Chem. 37, p. 499.
[23] Horiuti, J. and Toya, T. in Solid State Surface Science (Green, M., Ed.), Vol. 1, Marcel Dekker, New York, 1969.
[24] Barnett, R.N. Landman, U. and Cleveland, C.L. 1983, Phys. Rev. B 28, p. 1685.
[25] Schmickler, W. 1983, Chem. Phys. Lett. 99, p. 135.
[26] Hass, G.A. and Thomas, R.E. in Measurements of Physical Properties (Passaglia, E., Ed.), Part I, Interscience, New York, 1972.
[27] Rivière, J.C. in Solid State Surface Science (Green, M., Ed.), Vol. 1, Marcel Dekker, New York, 1969.
[28] Rowlinson, J.S. 1978, Chem. Soc. Rev. 7, p. 329.
[29] Randles, J.E.B. 1977, Phys. Chem. Liq. 7, p. 107.
[30] Trasatti, S. 1983, Electrochim. Acta 28, p. 1083.
[31] Trasatti, S. in Modern Aspects of Electrochemistry (Coway, B.E. and Bockris, J.O'M., Eds.), Vol. 13, Plenum Press, New York, 1979, p. 81.
[32] Daghetti, A. Lodi, G. and Trasatti, S. 1983, Mater. Chem. Phys. 8, p. 1.
[33] Henderson, H. 1981, Kinam 3A, p. 87.
[34] Trasatti, S. 1974, J. Electroanal. Chem. 52, p. 313.
[35] Trasatti, S. 1982, J. Electroanal. Chem. 139, p. 1.
[36] Bockris, J.O'M. and Argade, S.D. 1968, J. Chem. Phys. 49, p. 5133.
[37] Trasatti, S. 1974, J. Chem. Soc. Faraday Trans. I 70, p. 1752.
[38] Trasatti, S. 1983, J. Electroanal. Chem. 150, p. 1.
[39] Pleskov, Yu.V. in Comprehensive Treatise of Electrochemistry (Bockris, J.O'M. Conway, B.E. and Yeager, E., Eds.), Vol. 1, Plenum Press, New York, 1980, p. 291.
[40] Furlong, D.N. Yates D.E. and Healy, in Electrodes of Conductive Metallic Oxides (Trasatti, S., Ed.), Part B, Elsevier, Amsterdam, 1981, p. 367.

[41] Trasatti, S. 1971, J. Electroanal. Chem. 33, p. 351.
[42] Frumkin, A.N. and Gorodetzkaya, A. 1928, Z. Phys. Chem. 136, p. 451.
[43] Trasatti, S. 1980, Colloids Surf. 1, p. 173.
[44] Frumkin, A.N. 1960, Electrochim. Acta 2, p. 351.
[45] Randles, J.E.B. 1956, Trans. Faraday Soc. 52, p. 1537.
[46] Gomer, R. and Tryson, G. 1977, J. Chem. Phys. 66, p. 4413.
[47] Farrell, J.R. and McTigue, P. 1982, J. Electroanal. Chem. 139, p. 37.
[48] Hölzl, J. and Schulte, F.K. 1979, Spring Tracts Mod. Phys. 85, p. 1.
[49] Frumkin, A.N. 1967, J. Res. Inst. Catal., Hokkaido Univ. 15, p. 61.
[50] Valette, G. and Hamelin, A. 1973, J. Electroanal. Chem. 45, p. 301.
[51] Grigoryev, N.B. 1976, Dokl. Akad. Nauk SSSR 229, p. 647.
[52] Bagotskaya, I.A., Damaskin, B.B. and Levi, M.D. 1980, J. Electroanal. Chem. 115, p. 189.
[53] Vorotyntsev, M.A. 1981, J. Electroanal. Chem. 123, p. 379.
[54] Herring, C. and Nichols, M.H. 1949, Rev. Mod. Phys. 21, p. 185.
[55] Smoluchowski, R. 1941, Phys. Rev. 60, p. 661.
[56] Lecoeur, J. Andro, J. and Parsons, R. 1982, Surf. Sci. 114, p. 320.
[57] Marcus, P.M. and Jona, F. 1982, Appl. Surf. Sci. 11/12, p.20.
[58] Grant, J.T. 1982, Appl. Surf. Sci. 13, p. 35.
[59] Shih, H.D. Jona, F. Jepsen, D.W. and Marcus, P.M. 1981, Phys. Rev. Lett. 46, p. 731.
[60] Tilak, B.V. Conway, B.E. and Angerstein-Kozlowska, H. 1973, J. Electroanal. Chem. 48, p. 1.
[61] Scortichini, C.L. and Reilley, C.N. 1982, J. Electroanal. Chem. 139, p. 233.
[62] Hamelin, A. 1982, J. Electroanal. Chem. 142, p. 299.
[63] Valette, G. 1982, J. Electroanal. Chem. 139, p. 285.
[64] Reed-Hill, R.E. Physical Metallurgy Principles, Van Nostrand Reinhold, New York, 1964, p. 171.
[65] Bouwman, R. and Sachtler, W.M.H. 1971, Surf. Sci. 24, p. 350.
[66] Drost-Hansen, W. 1978, Physi. Chem. Liq. 7, p. 243.
[67] Macdonald, J.R. and Barlow, C.A. 1963, J. Chem. Phys. 39, p. 412.
[68] Heras, J.M. and Viscido, L. 1980, Appl. Surf. Sci. 4, p. 238.
[69] Trasatti, S. 1974, J. Electroanal. Chem. 53, p. 335.
[70] Bockris, J.O'M. Devanathan, M.A. and Müller, K. 1963, Proc. Roy. Soc. 274A, p. 55.
[71] Frumkin, A.N. Damaskin, B.B. and Survila, A.A. 1968, J. Electroanal. Chem. 16, p. 493.
[72] Daghetti, A. Trasatti, S. Zagorska, I. and Kowzorowski, Z. 1981, J. Electroanal. Chem. 129, p. 253.
[73] Frumkin, A. and Damaskin, B. 1967, Pure Appl. Chem. 15, p. 263.

[74] Daghetti, A. Trasatti, S. Zagorska, I. and Koczorowski, Z. 1981, J. Electroanal. Chem. 129, p. 253.
[75] Borkowska, Z. 1983, J. Electroanal. Chem. 146, p. 385.
[76] Mingozzi, S. Daghetti, A. Trasatti, S. Zagorska, I. and Koczorowski, Z. 1983, J. Chem. Soc. Faraday Trans. I 79, p.2801.
[77] Damaskin, B.B. and Frumkin, A.N. 1974, Electrochim. Acta 19, p. 173.
[78] Parsons, R. 1975, J. Electroanal. Chem. 59, p. 229.
[79] Badiali, J.P. Rosinberg, M.L. and Goodisman, J. 1983, J. Electroanal. Chem. 143, p. 73.
[80] Vorotyntsev, M.A. Izotov, V.Yu. Kornyshev, A.A. and Schmickler, W. 1983, Elektrokhimiya 19, p. 295.
[81] Guidelli, R. 1981, J. Electroanal. Chem. 123, p. 59.
[82] Parsons, R. and Reeves, R.M. 1981, J. Electroanal. Chem. 123, p. 141.
[83] Fawcett, W.R. Levine, S. de Nobriga, R.M. and McDonald, A.C. 1980, J. Electroanal. Chem. 111, p. 163.

POTENTIALS OF ZERO CHARGE. WHAT THEY SUGGEST ABOUT THE STRUCTURE OF THE INTERFACIAL REGION

Sergio Trasatti

Department of Physical Chemistry and Electrochemistry, University of Milan, Via Venezian 21, 20133 Milan, Italy.

ABSTRACT

Plots of potential of zero charge vs work function are shown for polycrystalline metals and single crystal faces of Au, Ag and Cu in aqueous solutions, dimethylsulfoxide and acetonitrile. The resulting picture is discussed in terms of contributions to the electrode potential from surface electrons on the metal side and from oriented solvent dipoles on the solution side of the electrode interface. Quantitative estimates for the various interfacial parameters are made for the specific case of Hg and are compared to calculated values. Results are also reported and discussed for the temperature coefficient of the potential of zero charge, the Volta potential difference, the work function change upon water adsorption from the gas phase and the differential capacitance of the electrode interface.

1. INTRODUCTION

In the previous lecture [1], the relationships between the structural parameters of the electrode/solution interface and some properties of the two separate phases have been derived and discussed in detail from a theoretical point of view. The purpose of the present lecture is to analyse the available experimental evidence in the light of the above relationships to obtain some indications about the possible actual structure of the electrode/solution interface in dependence on the nature of the metal and of the solvent. The final aim is to find out the factors which are responsible for the structural features of the electrochemical interface.

The experimental quantities which will be analysed are: (i)

A. F. Silva (ed.), Trends in Interfacial Electrochemistry, 25–48.

the potential of zero charge; (ii) the electron work function; (iii) their temperature coefficients; (iv) the contact (Volta) potential difference; (v) the shift of the potential and the work function upon adsorption; (vi) the capacitance at the potential of zero charge. Where possible, experimental data will be compared to the predictions of models.

2. POTENTIAL OF ZERO CHARGE-WORK FUNCTION RELATIONSHIP

This relationship has been discussed several times in the literature [2-6]. Although adjustments have proved to be necessary for some of the metals [7], the general picture suggested in 1971 by the present author [4] has not changed appreciably during about 14 years. Before analysing the $E_{\sigma=0}$ vs Φ plot, it is convenient to spend a few words on the selection of the experimental parameters. When the experimental field offers a number of data showing a not negligible scatter, some selection of the available results is unavoidable in order to be able to use a well definite value. "Selection" is admittedly a word implying the possibility of a subjective point of view. This is to be ruled out for most of the potentials of zero charge for which the reproducibility and the experimental accuracy are presently very high. It is however possible with work functions especially for metals with large surface heterogeneity.

2.1. Potentials of Zero Charge

Available experimental data for *sp*- and *d*-metals in aqueous solutions are reported in Table 1 [4,7,8]. *sp*-Metals are usually *soft* from a structural point of view. They have low melting points, the surface atoms are intrinsically mobile and their distribution is thus not really frozen [9]. As a consequence, the surface is basically homogeneous. Liquid metals such as Hg, Ga (above 32 °C), In-Ga, Tl-Ga belong obviously to this group. They actually give the most reliable experimental data and are thus used (specifically Hg) as reference systems.

The situation with *d*-metals is dramatically different. They either do not show any or show a very narrow potential region where the surface is really bare and in intimate contact with the pure solvent [10-12]. They often adsorb (or even absorb) hydrogen and oxygen in aqueous solutions. Moreover, polycrystalline surfaces have been mostly used to determine double layer parameters. Their values of $E_{\sigma=0}$ are thus subjected to the criticism related to the heterogeneity of the surface. In addition, the potential of zero *free* charge must be distinguished from potentials of zero *total* charge which are mostly obtained with the usual experimental techniques [10]. Nevertheless, some of the available values are reproducible and accurate and they can be used to obtain some semi-quantitative information on the structure of the electrode interface.

Values for single crystal faces are available for a number of

TABLE 1

Potentials of Zero Charge and Work Functions of Polycrystalline Metals

sp-Metals[(a)]	$E_{\sigma=0}$/V(SHE)	Φ/eV[(b)]
Hg (liquid)	- 0.192±0.001	4.50±0.02
Tl(Ga) (liquid alloy)	- 0.71±0.01	4.10±0.05
Bi	- 0.39	4.35
Sn	- 0.38	4.40
Pb	- 0.60	4.16
Cd	- 0.75	4.11
In	- 0.65	4.15
In(Ga) (liquid alloy)	- 0.67	4.15
Ga (liquid)	- 0.69	4.30
d-Metals[(c)]		
Pd	0.10	5.12
Rh	- 0.01	4.98
Ni	- 0.33	4.80
Co	- 0.43	4.70
Fe	- 0.70	4.35
Nb	(- 0.79)	4.20
Ta	(- 0.85)	4.15
Ti	(- 1.05)	4.10

(a) From Ref. [4,7,8]
(b) From Ref. [4,7,15-18]
(c) From Ref. [4,7,10,12]

metals and are summarized in Table 2 [8,13]. Significant heterogeneity is shown only by Cu, Ag and Au which thus constitute a specific group (*sd*-metals). The surface of Zn is also appreciably heterogeneous, but its potentials of zero charge can only be obtained by extrapolation because they are more positive than the dissolution potential [14].

2.2. Electron Work Functions

It is impossible to discuss in this paper all the aspects of the selection of this parameter. The reader is referred to previous papers [4,15-18]. Briefly, the work functions of the polycrystalline surfaces of *d*-metals have been derived from plots of Φ vs the sintering temperature of evaporated films; the rationale behind

TABLE 2

Potentials of Zero Charge and Surface Heterogeneity[a] of Single Crystal Faces of Metals

Metal	Face	$E_{\sigma=0}$/V(SHE)		$\Delta\Phi$/eV	
Au	(111)	0.57±0.02	[13]	0	[20]
	(100)	0.38		- 0.11	
	(110)	0.19±0.01		- 0.28	
Ag	(111)	- 0.459±0.002	[13]	0	[19]
	(100)	- 0.609		- 0.06 to - 0.24	
	(110)	- 0.734±0.005		- 0.19 to - 0.32	
Cu	(111)	- 0.33±0.05	[22]	0	[20,59]
	(100)	- 0.46		- 0.11	
	(110)	- 0.63		- 0.22	
Zn	(0001)	- 0.77(±0.05)	[14]	-	
	$(10\bar{1}0)$	- 0.85			
	$(11\bar{2}0)$	- 0.87			
Bi	(011)	- 0.36±0.01	[70]	-	
	$(2\bar{1}\bar{1})$	- 0.33			
	(111)	- 0.41			
Pb	(111)	- 0.62±0.01	[71]	-	
	(100)	- 0.59			
	(110)	- 0.58			
Sn	(001)	- 0.36±0.01	[72]	-	
	$(\bar{1}10)$	- 0.36			
	(110)	- 0.37			

(a) The surface heterogeneity is expressed in terms of the change in work function from face to face.

this approach is described specifically in two previous chapters [17,18]. The selection of Φ for single crystal faces of Ag has been illustrated in detail in another paper [19], the same is the case for the Φ values of single crystal faces of Au and Cu [20]. For the *sp*-metals, the experimental work functions as a rule agree closely with the *electrochemical work functions* derived from capacitance data [2] (see specifically the discussion in one of the previous papers [18]). The values of Φ used in this work are summarized in Table 1 and 2.

2.3. $E_{\sigma=0}$-Φ Plot for Aqueous Solutions

Fig. 1 shows the graphical relationship for polycrystalline electrodes in aqueous solutions. Hg is taken as a reference metal. Metals can be seen to split into two main groups: *sp*- and *d*-metals. The solid straight line drawn through the point of mercury is of unit slope. According to the equation relating $E_{\sigma=0}$ and Φ

$$E_{\sigma=0} = \Phi/F + [\delta\chi_0^M - g^{H_2O}(\mathrm{dip})_0] + \mathrm{const} = \Phi/F + X + \mathrm{const} \quad (1)$$

where the constant term depends on the nature of the reference electrode only, the solid straight line represents the place where all metals with the same value of the term in brackets as for Hg should be located. As a matter of fact, most of the *sp*-metals gather around a straight line of slope different from one. Ga falls far from this line. It is useful to anticipate that Ga lies close to the expected position for polycrystalline Ag [5].

The fact that the slope of the dashed line grouping the *sp*-metals is different from one suggests that the quantity X in eqn. (1) is metal dependent within this group. More specifically, this

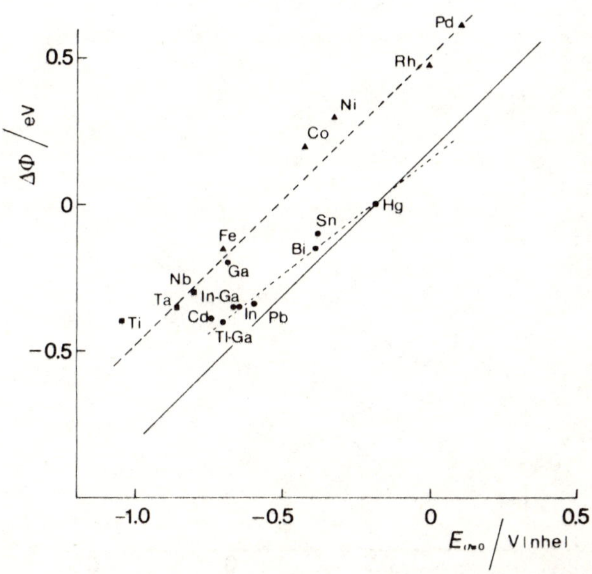

Fig. 1 - Plot of the potential of zero charge of polycrystalline metal electrodes in aqueous solutions *vs* the relative value of the work function (Hg is the reference metal). (——) Line of unit slope. (●) *sp*-Metals; (■,▲) *d*-Metals. Data of higher (▲) and lower (■) reliability are distinguished.

term introduces a more negative contribution to the electrode potential compared to Hg. Were the value of this term for a given *sp*-metal the same as for Hg, that metal would lie on the solid straight line.

d-Metals too gather around a straight line whose slope is possible one. For the same work function, *d*-metals show a much more negative value of X in eqn.(1) than the corresponding *sp*-metal. This indicates that the contribution of X to the electrode potential is even more negative and, moreover, it appears to be essentially metal independent.

Fig. 1 provides only relative values of the quantity X. Its value for a given metal relative to Hg corresponds to the distance, on the potential axis, of the point for that metal from the solid straight line. The magnitude of X depends on both the electron distribution at the metal surface and the orientation of solvent molecules on the solution side. No experimental separation can be obtained on a thermodynamic basis, but some definite information can be derived by means of extra-thermodynamic assumptions.

2.4. $E_{\sigma=0}$ vs Φ Plot for Non-aqueous Solvents

Fig. 2 shows the available experimental data [5]. A fairly complete picture is possible only for acetonitrile (ACN) and partly for dimethylsulfoxide (DMSO). The pattern appears to resemble that for aqueous solutions. In particular, transition metals fall

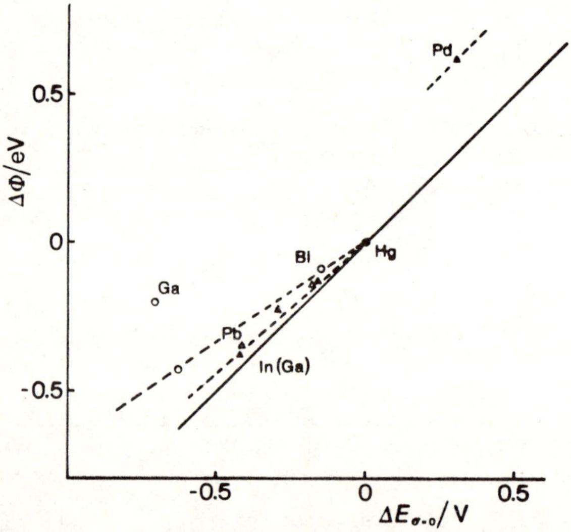

Fig. 2 - Plot of the relative value of the potential of zero charge of polycrystalline *sp*-metal electrodes in various non-aqueous solvents *vs* the relative value of the work function (Hg is the reference metal). (△) Methanol; (▲) Acetonitrile; (■) Dimethylsulfoxide.

far away from *sp*-metals whereas *sp*-metals show a possibly linear dependence of $E_{\sigma=0}$ on Φ with a slope lower than one. This suggests that the same factors are probably responsible for the structural features of the electrode/non-aqueous solution interface, and that the metals can apparently be arranged in a sequence which does not change significantly with the nature of the solvent. In particular, note the position of the point for Ga in DMSO and ACN. Like in water, this metal falls far from the group of the other *sp*-metals, but not together with the *d*-metals.

2.5. $E_{\sigma=0}$ vs Φ Plot for Single Crystal Faces

Fig. 3 includes only the data for Ag, Au and Cu. The value of $E_{\sigma=0}$ for the last metal are not as reliable as for the first two because they have been estimated indirectly from adsorption data [22]. The work functions for the other metals for which $E_{\sigma=0}$ values have been measured are not available.

It is remarkable that the three main faces of each metal parallel the behaviour of the main group of *sp*-metals in Fig. 1. In particular, the points lie on straight lines whose slope is always

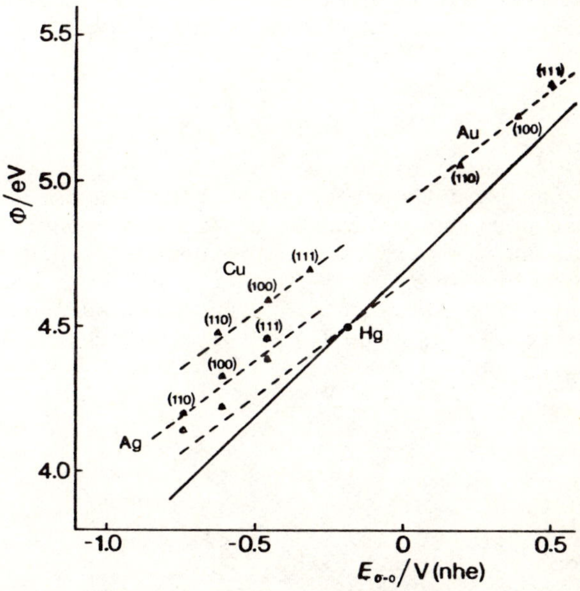

Fig. 3 - Plot of the potential of zero charge in aqueous solutions *vs* the work function of single crystal faces of *sd*-metals. Hg is also shown for reference. (——) Line of unit slope. The dashed line through the point of Hg is taken from Fig. 1 and gathers all *sp*-metals but Ga. Different symbols for Ag refer to differently determined work function values [19].

different (lower) from one. This indicates that the quantity X in eqn.(1) is face specific. It is also interesting that the points for all the faces lie on the left of the solid straight line of unit slope drawn through the point of Hg. This indicates that the quantity X gives a negative contribution to the electrode potential compared to Hg, in any case.

3. THE CASE OF Hg

The work function of Hg is known with an accuracy of *ca* ±20 meV and its potential of zero charge can be measured to ±1 mV and even better. In addition, a number of other experimental parameters are available for this electrode which makes it possible to try some quantitative calculations to separate $\delta\chi_0$ from $g^{H_2O}(\text{dip})_0$. In particular, the Volta potential difference between Hg and an aqueous solution of not specifically adsorbed electrolytes has been found to be - 0.25 V [23]. Since [1]:

$$(\psi^M - \psi^S) = \delta\chi_0^M - \delta\chi_0^S \tag{2}$$

it ensues that the contact between Hg and water provokes variations in the surface potentials of the two phases which amount to - 0.25 V. Thus:

$$\delta\chi_0^{Hg} - \delta\chi_0^{H_2O} = -\ 0.25\ \text{V} \tag{3}$$

No value of $\delta\chi_0^{Hg}$ can be estimated experimentally, but some estimation of $\delta\chi_0^{H_2O}$ can be attempted. It has been found [24] that the potential shift upon adsorption of a series of monofunctional aliphatic alcohols at the air/solution interface is higher by about 0.06 V than the corresponding zero charge potential shift at the Hg/solution interface. If such difference is assumed to measure the difference between χ^{H_2O} and $g^{H_2O}(\text{dip})_0$ at Hg, it results:

$$\chi^{H_2O} - g^{H_2O}(\text{dip})_0 = 0.06\ \text{V} \tag{4}$$

Recalling [1] that $g^{H_2O}(\text{dip})_0 = \chi^{H_2O} + \delta\chi_0^{H_2O}$, it follows that:

$$\delta\chi_0^{H_2O} = -\ 0.06\ \text{V} \tag{5}$$

Therefore, from eqns.(3) and (5):

$$\delta\chi_0^{Hg} = -\ 0.31\ \text{V} \tag{6}$$

Equation (6) suggests that as the liquid phase comes in contact with Hg, the tail of the electron distribution at the surface of the metal is pushed back so that χ^M decreases. This conclusion agrees with calculations of some authors [25] but contrasts with those of others according to whom electrons are pulled out as the metal comes in contact with the solvent [26]. In order to obtain a positive value of $\delta\chi_0^M$ from eqn.(3), it is necessary to assume that $\delta\chi_0^{H_2O} > 0.25$, a value difficult to rationalize in the light of current ideas.

The value of $g^{H_2O}(\text{dip})_0$ cannot be measured directly. In principle, it could be determined from adsorption potential shifts [27] but an organic substance with zero net normal dipole moment has not yet been found. Attempts have been made to estimate χ^{H_2O} [28, 29]. Depending on the approach values between 0.08 and 0.13 V have been suggested but as low values as 0.03-0.05 V have also been proposed [23,30]. Adsorption potential shifts at the air/solution interface suggest that χ^{H_2O} is probably around 0.1 V [29]. If the range of values 0.08-0.13 V is assumed, from eqn.(4) it results that:

$$g^{H_2O}(\text{dip})_0 \text{ (at Hg)} = 0.02\text{-}0.07 \text{ V} \tag{7}$$

The sign of this equation suggests that water molecules are preferentially oriented with the negative end of the dipole towards the electrode; therefore, some interaction between the oxygen atom and the metal surface takes place. It is worth stressing that this conclusion involves two extra-thermodynamic assumptions regarding the values of $\delta\chi_0^{H_2O}$ and χ^{H_2O}.

3.1. Comparison with Calculations

It is worthwhile at this point to compare the values of the parameters $\delta\chi_0^{Hg}$ and $g^{H_2O}(\text{dip})_0$ for Hg as derived here with calculations based on models. Badiali *et al.* [25] have calculated for $\delta\chi_0^{Hg}$ a value in the range - 0.21 to - 0.24 V. Both sign and magnitude are in good agreement with experiments. Schmickler [26] has found that the field of solvent dipoles is such that it tends to pull the electrons out of the metal. Therefore, he obtains $\delta\chi_0^{Hg} \simeq$ 0.04 V and $g^{H_2O}(\text{dip})_0 \simeq 0.27$ V; thus, $\delta\chi_0^M - g^S(\text{dip})_0 = -0.23$ V, and the predicted Volta potential difference at the Hg/water interface is *ca* - 0.10 V compared to the measured - 0.25 V.

The value of $g^{H_2O}(\text{dip})_0$ at Hg has been several times calculated by means of models in which only the features of the solvent have been taken into consideration. These models [29,31-38] predict values for the solvent dipole potential in the range 0-0.04 V, somewhat lower than the value found here, but with the same sign. It is to be stressed that the calculations have been based on a curve fitting procedure making use of capacity data. Actually, these models do not neglect $\delta\chi^M$ but rather its electric field dependence. For this reason, the value of the interfacial potential drop derived from this models do not include the electronic contribution and can be safely treated as $g^S(\text{dip})$.

The estimation of $g^S(\text{dip})_0$ for Hg in non-aqueous solvents is much more difficult because the experimental potential of zero charge includes the unknown liquid junction potential between the (usu ally) aqueous reference electrode and the non-aqueous solution. However, Volta potential differences for Hg and Bi in contact with methanol, ethanol, DMSO, acetone and dimethylformamide have been reported [39]. There are however not enough data to try to disentangle $\delta\chi_0^M$ and $\delta\chi_0^S$ for these systems.

3.2. Temperature Coefficient of the Potential of Zero Charge

The temperature coefficient of $E_{\sigma=0}$ for Hg has been measured by Randles and Whiteley and found to be 0.57 mV K^{-1} [40]. Since the temperature coefficient of the reference electrode has been taken into account in this value, it results:

$$(\mathrm{d}\Phi/\mathrm{d}T)/F + \mathrm{d}(\delta\chi_0^M)/\mathrm{d}T - \mathrm{d}g^{H_2O}(\mathrm{dip})_0/\mathrm{d}T = 0.57 \text{ mV K}^{-1} \qquad (8)$$

Even though the temperature coefficient of $\delta\chi_0^M$ is neglected to a first approximation, the experimental value still differs from the temperature coefficient of the dipole potential drop. There is only one (unpublished) experimental result for Hg which gives for $\mathrm{d}\Phi/\mathrm{d}T$ a positive value (0.4 meV K^{-1}) [41]. From eqn.(8) the correct dipolar term is thus

$$\mathrm{d}g^{H_2O}(\mathrm{dip})_0/\mathrm{d}T \text{ (at Hg)} = -0.17 \text{ mV K}^{-1} \qquad (9)$$

One molecular model for water at electrodes predicts a positive value for this coefficient [35]. Another model gives - 0.35 mV K^{-1}, close to that in eqn.(9) [32]. On the other hand if $\mathrm{d}\Phi/\mathrm{d}T$ for Hg is estimated by making use of the concept of *electrochemical work function* [42], a negative value of - 0.66 meV K^{-1} is obtained which is in agreement with the value of *ca* - 0.4 meV K^{-1} derived by McTigue and Farrell from the temperature coefficient of the Volta potential difference [43]. The temperature coefficient of $g^{H_2O}(\mathrm{dip})_0$ at Hg may thus be somewhat higher, for instance - 1.3 mV K^{-1} [44]. A value close to this estimate has been obtained by Dutkiewicz and Lamperski (0.85 mV K^{-1}), although the value of $g^{H_2O}(\mathrm{dip})_0$ for Hg is far too negative from their model, *i.e. ca* - 0.65 V [45].

The situation is thus unclear in this specific field and further experimental investigations are needed.

4. OTHER *sp*-METALS

The value (relative to Hg) of the quantity $X = \delta\chi_0^M - g^{H_2O}(\mathrm{dip})_0$ in eqn.(1) can be estimated in Fig. 1 from the horizontal distance between the point for the given metal and the solid straight line. The derived values are given in Table 3. In the same Table the values of $\delta\chi_0^M$ calculated by Badiali *et al.* [25] are also given. These increase much faster than X does, which indicates that the calculations tend probably to overestimate the role of the surface electrons [46]. In fact, if the calculated values of $\delta\chi_0^M$ are sub-

TABLE 3

Components of the Potential of Zero Charge in Aqueous Solution

Metal	$n \times 10^3$/a.u.	$\Delta[\delta\chi_0^M - g^S(\text{dip})_0]$/V		$\Delta(\delta\chi)$/V	
		(a)	(b)	(c)	(d)
Hg	12.7	0	0	0	0
Bi	-	- 0.05	-	-	-
Pb	19.7	- 0.07	- 0.093	- 0.50	0.06
Sn	17.4	- 0.09	- 0.064	-	0.04
Tl(Ga)	15.4	- 0.12	- 0.037	-	0.02
In(Ga)	17.1	- 0.13	- 0.060	- 0.47	0.035
Cd	13.8	- 0.17	- 0.015	- 0.32	0.01
Ga	22.3	- 0.30	- 0.123	- 0.62	0.11

(a) This work, from $E_{\sigma=0} - \Phi$ plots.
(b,d) Calculated by Schmickler [26].
(c) Calculated by Badiali *et al.* [25].

tracted from the experimental values of X, the value of $g^S(\text{dip})_0$ results to be increasingly negative as X increases, which indicates that water molecules are preferentially oriented with the hydrogen atoms towards the solid at the metals with the lower work function. This is difficult to rationalize from a chemical point of view and is hard to accept. On the other hand, Schmickler's treatment [26] gives much lower values of $\delta\chi_0^M$ but with the opposite sign, although the metals are arranged in a sequence which parallels the experimental one and the calculated values for X have the correct sign.

Some ideas based on chemistry can help at this point rationalize the experimental picture. If water at Hg is preferentially oriented with the oxygen atom towards the metal, it is reasonable to assume that the interaction is governed by the *affinity* of the surface for oxygen. Since the heat of adsorption of oxygen from the gas phase on *sp*-metals is not available and this quantity is linearly related to the heat of oxide formation in the case of *d*-metals [47], it may be assumed that the affinity of the metal surface for oxygen (water molecules) is measured by the enthalpy of formation of the oxide MO, ΔH_f° [48]:

$$\text{M(solid)} + \frac{1}{2}\,\text{O}_2\text{(gas)} = \text{MO(solid)} \qquad (10)$$

It is worth stressing that the *affinity* of the metal for oxygen takes into account all kinds of energy involved in bond formation, thus including terms related to the redistribution of electrons and to the crystal field stabilization energy at the surface. The quantity ΔH_f^o is thus to be properly correlated to the quantity X rather than to the single term $g^S(\text{dip})$.

Fig. 4 shows a plot of $[\delta\chi_0^M - g^S(\text{dip})_0]$ vs ΔH_f^o [49] for aqueous solutions. It can be seen that the value of X becomes more negative as the affinity of the surface for oxygen increases. This indicates that probably both $\delta\chi_0^M$ and $g^S(\text{dip})_0$ increase (the former becomes more negative, the latter more positive). Water molecules are thus expected to adopt a stronger preferential orientation the higher the affinity of the metal surface for oxygen.

It is difficult and probably unnecessary to try to separate $\delta\chi_0^M$ from $g^S(\text{dip})_0$ for all metals. According to the *jellium* model [26], the value of $\delta\chi_0^M$ should increase with the electron density. If the quantity X is plotted as a function of the electron density, the experimental data are not arranged in a rational sequence. They only show a broad trend to co-vary (Fig. 5). This may indicate that the electronic term does not prevail in governing the interfacial potential but both contributions are equally important.

4.1. The Concept of Hydrophilicity

The term hydrophilicity means *affinity for water* and has been introduced [50] into the field of the double layer at electrodes to express the ability of a metal surface to bind water molecules. Quantitatively, the concept of hydrophilicity may depend on the property chosen to measure it. As a general rule, a surface is hydrophilic if it is wetted by water. This means that a drop of water on the surface spreads out and the contact angle is zero. Wa-

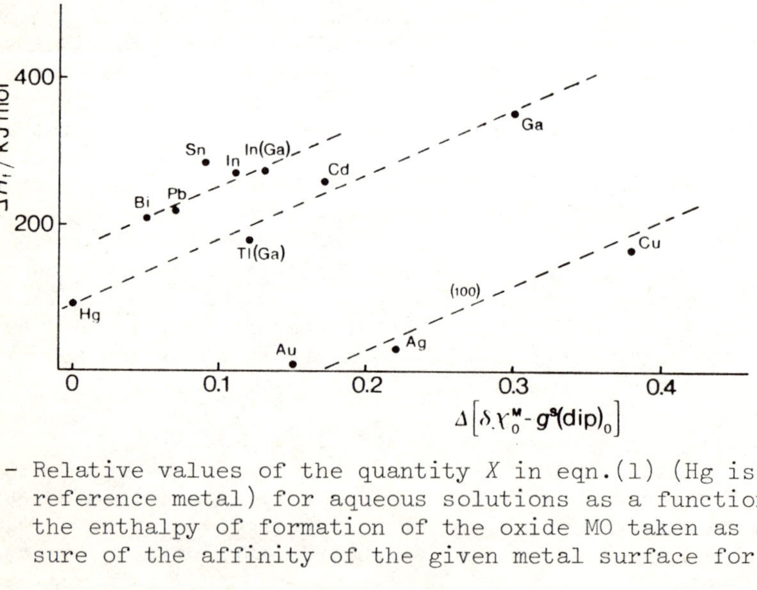

Fig. 4 - Relative values of the quantity X in eqn.(1) (Hg is the reference metal) for aqueous solutions as a function of the enthalpy of formation of the oxide MO taken as a measure of the affinity of the given metal surface for water.

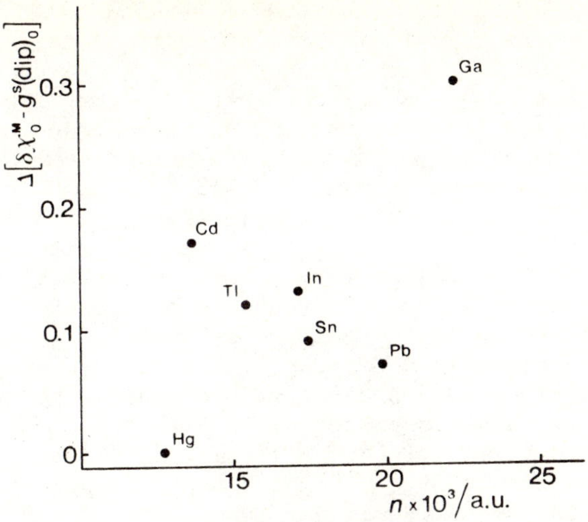

Fig. 5 - Plot of the quantity X in eqn.(1) as a function of the electron density of *sp*-metals.

ter does not wet a hydrophobic surface, the drop retains its spherical shape and the contact angle is 180°.

In terms of electrode potential, the hydrophilicity may be expressed by the electric potential drop across any preferentially oriented solvent dipole layer. Since the free surface of water possesses an intrinsic surface potential, a metal may be defined as hydrophilic if, after contact with the solution, $g^{H_2O}(dip)_0 > \chi^{H_2O}$ and hydrophobic if the reverse is true. In other words, hydrophilic metals possess structure-breaking surfaces and hydrophobic metals exhibit structure-making surfaces towards water [9]. However, this definition does not prove to be a general one. Fig. 4 shows that two metals may exhibit the same value of $g^{H_2O}(dip)_0$ but different affinity for oxygen. Therefore, hydrophilicity may be better expressed in terms of adsorption energy and not of electric potential, although $g^S(dip)$ can in fact be used as a parameter measuring the hydrophilicity in a relative sense within a homogeneous group of metals. This aspect will become clearer as the behaviour of *d*-metals will be examined.

5. *sp*-METALS IN NON-AQUEOUS SOLVENTS

The quantity X in eqn.(1) can be derived from Fig. 2 also for non-aqueous solvents. Although only a few experimental data are available, regularities following the same pattern as for aqueous solutions can be observed. In the case of methanol or DMSO, the affinity of the surface for oxygen can still be used to organize the experimental data since these solvents becomes adsorbed with the oxygen atom towards the metal surface. In the case of ACN, this parameter can also be used if it is assumed that the interaction of the metal surface with the nitrogen atom (negative end of the dipole) governs the mode and the strength of adsorption. In this respect, it is worth stressing that metals have been found to exhibit a scale of reactivity towards different molecules (H_2, O_2, CO_2, N_2, CO) which does not appear [17,51] to depend on the nature of the adsorbate (provided it is molecularly simple). Therefore, the reactivity scale (affinity) of *sp*-metals for nitrogen (ACN) is expected to run parallel to (or to vary in the same direction as) that for oxygen. ACN has been found to be oriented with the CN group towards the metal in the pure solvent [52]. The orientation may be understandably different if adsorption takes place from aqueous solutions [53] because of hydrophilic-hydrophobic interactions with the aqueous phase.

Fig. 6 shows a plot of $[\delta\chi_0^M - g^S(\mathrm{dip})_0]$ (=X) as a function of ΔH_f^o for DMSO and ACN. It can be seen that the former quantity, like for water, is probably a linear function of the chemical parameter. For DMSO, the metals are likely to split into two groups resembling the behaviour in water. The splitting is not apparent for ACN, but in this case the magnitude of X is so small that it is difficult to reach a definite conclusion.

It is to be stressed that, for a given metal, the value of X increases in the sequence ACN < W < DMSO. While the higher values in DMSO can be understood in terms of its higher permanent dipole than that of water, the situation in ACN probably reflects the dif

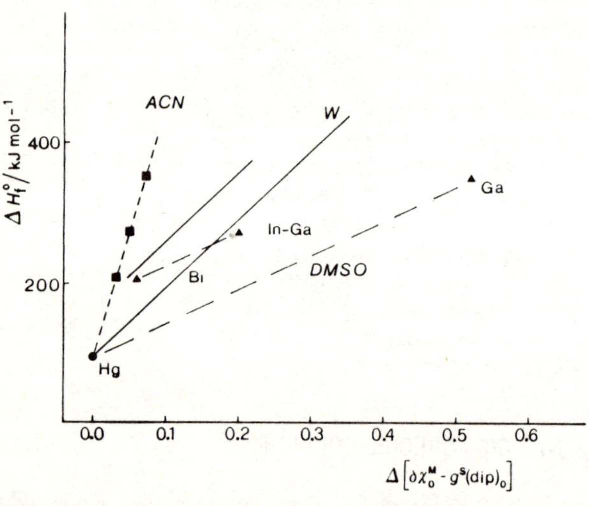

Fig. 6 - Plot of the quantity X in eqn.(1) for polycrystalline *sp*-metals in non-aqueous solvents *vs* the enthalpy of formation of the oxide MO taken as a measure of the affinity of the surface for the electronegative end of the solvent molecules.

ficulty for this molecule to adopt a preferential orientation owing to its asymmetric structure.

6. TRANSITION (*d*-)METALS

As mentioned above, the situation of transition metals is still not completely settled because polycrystalline surfaces are as a rule used. However, from a qualitative point of view the indications of the experimental data are rather unambiguous.

Fig. 7 shows the plot of the quantity *X* in eqn.(1) as a function of the affinity of the transition metals for oxygen as measured by the experimental heat of adsorption of O_2 from the gas phase. The plot shows that *X* is high, negative and, to a first approximation, constant for all *d*-metals. Although $\delta\chi_0^M$ values may be in principle quite considerable for these multielectron metals [26], there is ample independent experimental evidence that water molecules are chemisorbed through the oxygen atom [9]. The interaction may involve the partly empty *d*-bands of the metal and is sometime so strong that water molecules may break into OH and H on the surface. Unless a close compensation of $\delta\chi_0^M$ and $g^{H_2O}(\text{dip})_0$ occurs, the results in Fig. 7 suggest that water molecules are strongly oriented (relatively to *sp*-metals) when in contact with *d*-metals. The interaction is apparently strong enough as to bring the orientation to its maximum value (compatibly with steric effects and bond distorsion) for all the metals in this group. Therefore, $g^S(\text{dip})_0$ appears to be independent of the affinity of the metal surface for oxygen.

Transition metals appear to occupy the same position relatively to *sp*-metals also in ACN [5]. Figs. 2 and 7 show that $g^S(\text{dip})_0$ is higher also in this non-aqueous solvent. However, the constancy

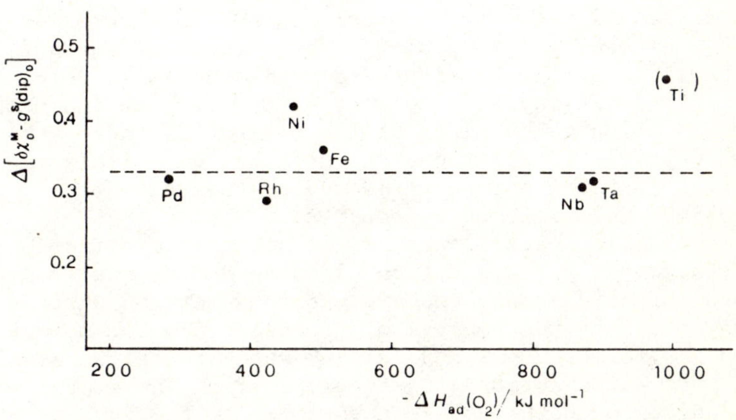

Fig. 7 - The quantity *X* in eqn.(1) for *d*-metals in aqueous solution as a function of the heat of adsorption of O_2 from the gas phase.

of this parameter within the group of *d*-metals is not verified [54] and it is probable that the difference in chemical interaction with the organic molecule is operative in this case also for *d*-metals. There is clear spectroscopic evidence that ACN is bound through the CN group to, for instance, platinum electrodes [55].

7. SINGLE CRYSTAL FACES. *sd*-METALS

The picture for these metals is rather debated [6,19,56,57]. The major point of disagreement is the value of Φ to be used in the plot in Fig. 3. It is obvious that different (even opposite) conclusions can be drawn from the same set of electrochemical data.

The points for Au, Ag and Cu in Fig. 3 are always on the left of Hg. This suggests that $[\delta\chi_0^M - g^S(\text{dip})_0]$ is more negative on these metals. In terms of Fig. 4, this would imply a higher affinity of the metal surface for water (oxygen). This can be easily seen to be really the case with Cu, but discrepancies are found with Au and Ag. In particular, the value of ΔH_f° for these metals is lower than for Hg [49]. Therefore, in Fig. 4 this group of metals fall far from the main group of *sp*-metals.

It is not possible to use different values of the affinity for water for the various faces of a metal, although this is very likely to be the case.

The reason is that ΔH_f° is a bulk property and cannot discriminate between the various crystal orientations. Therefore, to a first approximation, the tabulated ΔH_f° values have been assigned to the (100) face of each metal. If X for the (100) face is plotted against ΔH_f° in Fig. 4, regularities are again observed within this group of metals with $g^S(\text{dip})_0$ increasing as ΔH_f° increases [20]. However, very high values of X are in this case associated with very low values of ΔH_f°.

The explanation for this outcome is not unambiguous. It may be that $\delta\chi_0^M$ exceeds $g^S(\text{dip})_0$ with these structured surfaces because water molecules can be accomodated in the valleys where electrons accumulate [58]. It may also be that strong interaction promoting strong orientation takes place at the peaks [59] where the surface atoms are deprived of electrons and unfilled *d*-orbitals might protrude. These facts cannot be reflected by the value of ΔH_f° quantitatively since this is a bulk property.

There is evidence [19] that O_2 is strongly chemisorbed on Ag at $\theta \sim 0$ and water spreads out on clean Ag and Au surfaces. On the other hand, organic substances are adsorbed more on Ag than on Hg which suggests that the removal of preadsorbed water molecules is easier on the former metal [9]. At any rate, the picture in Fig. 3 suggests that the quantity X is face specific. Therefore, either $\delta\chi_0^M$ or $g^S(\text{dip})_0$ or both must be a function of the surface atom distribution. Fig. 3 indicates that the quantity X decreases in the sequence (110) > (100) > (111) for all *sd*-metals. This evidence is questioned by other authors [6] on the basis of differently estimated values of Φ. There is strong need for further experimental work, especially the determination of Φ for the same samples used to measure the potential of zero charge. The present evidence however suggests that the components of the electrode potential are very probably face specific and follow the same pattern for metals with the same crystal structure.

8. ADSORPTION OF WATER

While it is difficult to disentangle adsorbate contributions from water effects in adsorption potential shifts at electrodes, water adsorption from the gas phase may give some more direct indications on the structuring of the solvent at the interface with metals.

Experimental results show that [60-62], unless water molecules are broken with oxidation of the metal surface, the work function of metals is markedly decreased by the adsorption of water molecules. To a first approximation it is possible to write:

$$\Delta\Phi/F = \delta\chi_0^M - g^S(\text{dip})_0 \tag{11}$$

Therefore, the decrease in work function should account for both the electronic term and the dipolar term in the electrode potential. A major difficulty in comparing $\Delta\Phi$ with electrochemical data is that only values relative to Hg (or another metal) are obtained in electrochemistry while $\Delta\Phi$ is, in this context, an absolute quantity. Quantitatively, the agreement is only approximate.

Most adsorption studies have been carried out on transition metals for which, on average, $\Delta\Phi \simeq -1$ eV. Fig. 7 shows that for d-metals:

$$\Delta[\delta\chi_0^M - g^{H_2O}(\text{dip})_0] \simeq -0.33 \text{ V} \tag{12}$$

with respect to Hg. Since it has been estimated in section 3 that $\delta\chi^{Hg} = -0.31$ V and $g^{H_2O}(\text{dip}) = 0.02$ to 0.07 V, then

$$[\delta\chi_0^{Hg} - g^{H_2O}(\text{dip})_0] \text{ (at Hg)} \simeq -0.33 \text{ to } -0.38 \text{ V} \tag{13}$$

From eqn.(12) and (13) it results:

$$[\delta\chi_0^M - g^{H_2O}(\text{dip})_0] \text{ (at } d\text{-metals)} \sim -0.65 \text{ to } -0.7 \text{ V} \tag{14}$$

The experimentally observed decrease in Φ is thus larger than the value expected from electrochemical data. Reasons for this may involve the role played by the bulk liquid phase which lacks in $\Delta\Phi$ measurements, and the possibility of some partial charge transfer if the metal is electrically floating (as in $\Delta\Phi$ studies) and not under potential control. Calculations show that electron transfer from water to the metal may play a role [63]. In fact, the importance of the density of electronic states in the metal has been pointed out on the basis of spectroscopic measurements [64]. It is to be mentioned that the work function of metals decreases also upon adsorption of rare gases which do not possess permanent dipoles [65].

Fig. 8 shows a typical dependence of $\Delta\Phi$ on the surface concentration of water molecules [60]. It is seen that Φ drops linearly with θ only at very low coverages. As θ increases the rate of change of Φ with θ decreases monotonically. This can be interpreted in terms of decreasing normal component of the permanent dipole of the adsorbing molecule [62]. This is presumably related to lateral interactions and may be due either to dipole-dipole depolarization effects or to clustering of water molecules with formation of hydrogen bonds. In any case, the final average preferential orientation of water molecules is very small. It may correspond to some molecules bonded to the metal surface through one of the lone pairs of oxygen and hydrogen-bonded to other molecules in the first monolayer and in the next layer [29,48]. A water molecule directly bonded to a metal surface in the possible position of maximum orientation is depicted in Fig. 9. Not all molecules in the monolayer can be in the same orientation because of the constraint of the hydrogen-bonds. The intrinsic structure of bulk water plays thus a primary role in determining the electrical parameters of the interface.

Adsorption studies from the gas phase suggest that metal-water interactions are probably weak [66] and not very different from metal to metal. This is in fact in line with the electrochemical

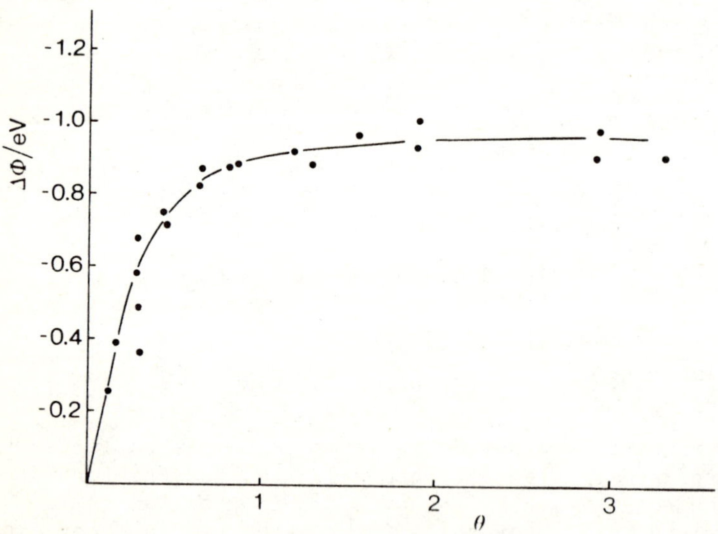

Fig. 8 - Typical work function drop upon adsorption of water from the gas phase onto a metal (Fe) surface [60]. θ is the surface coverage with adsorbed water molecules.

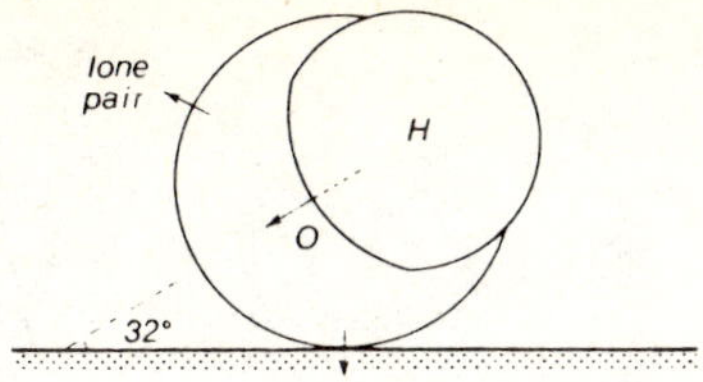

Fig. 9 - Sketch of a water molecule adsorbed in the supposed position of maximum orientation of the permanent dipole.

evidence, since the adsorption bond of water on *sp*-metals has been estimated [67] to vary by only a few kJ mol^{-1} from Hg to Ga. This indicates that the concept of hydrophilicity has always a relative sense. The adsorption bond is always very weak but comparatively stronger on Ga (hydrophilic) than on Hg (hydrophobic). This result also points out that *minor energy changes* are required to produce *major potential drop variations*. In this respect, studies in the gas phase, suggesting weak adsorption of water, and electrochemical data, showing large variations in surface potential, are not in contrast.

9. PERMITTIVITY OF THE INTERFACE

Differential capacitance (C) measurements are available for a number of *sp*-metals and a few solvents [7,29,68]. Results for *d*-metals are very scanty and are difficult to interpret in terms of a simple model.

Fig. 10 shows that the capacitance of the electrode/solution interface varies with the nature of the metal following the same sequence as the hydrophilicity. In fact, C^i data are arranged in a rational order by the affinity of the surface for oxygen [7]. The nature of the solvent affects the quantitative value of C^i but not its qualitative trend.

This result is not unexpected and can be understood on a qualitative basis. If water molecules are strongly oriented because of strong chemical interaction, a large variation of the orientation is to be expected as an electric field is superimposed because of the modification in the interaction strength. As a result, the molecules will be strongly reoriented by the field and this will result in a high capacitance. Therefore, high capacitances are expected to be associated with large values of $[\delta\chi_0^M - g^S(\mathrm{dip})_0]$.

On the other hand, also the polarizability of the electrons at the metal surface may play a role. Calculations have shown that the *metal* capacity increases with the electron density [69]. As a matter of fact, this component shows the same trend as the experimental capacitance. Table 4 shows a comparison between the experimental C^i and the calculated C_M. However, the polarizability of water probably prevails over that of the surface electrons. This

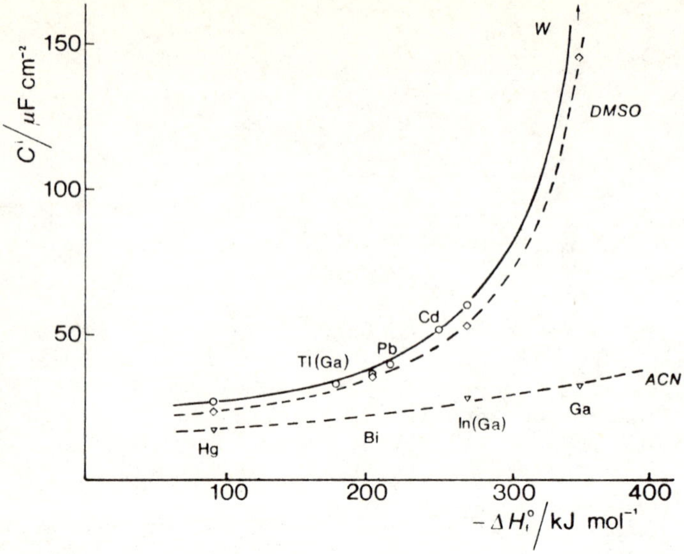

Fig. 10 - Inner layer capacity of metal/solution interfaces for different solvents as a function of the enthalpy of formation of the oxide MO. W: water; DMSO: dimethylsulfoxide; ACN: acetonitrile.

TABLE 4

Inner Layer Differential Capacity at the Potential of Zero Charge in Aqueous Solution

Metal	$C^i/\mu F\ cm^{-2}$ [29]	$C_M/\mu F\ cm^{-2}$	$C_{dip}/\mu F\ cm^{-2}$
Hg	28	31.2	- 18.1
Tl(Ga)	33.5	-	-
Bi	36	-	-
Sn	39	-	-
Pb	40	127.5	- 24.0
Cd	52	55.3	- 17.3
In(Ga)	60	0	- 23.7
Ga	250 [73]	- 89.3	- 22.9

C^i is the inner layer capacity from experiments. C_M is the *metal surface* capacity calculated by Badiali *et al.* [25]. C_{dip} is the capacity associated with the reorientation of water molecules obtained by combining C^i and C_M.

is presumably supported by the strong dependence of C^i on the nature of the solvent, although the complex dependence of C_M on the features of the solvent molecules makes it difficult to predict the magnitude of the metal role in this case. The very low value of C^i for ACN is understandable in terms of the difficult reorientation of this asymmetrical molecule in the electric field. The change of C^i from Hg to Ga points to the small effect of the metal surface too. The high values of C^i in DMSO compared to those in H_2O (higher than the ratio between the bulk dielectric constants) points to the large polarizability of DMSO molecule in the interface reflecting the different bulk structure of this solvent. More detailed discussions of the permittivity of the electrode/solution interface have been presented in previous papers [5,7,29].

CONCLUSIONS

Analysis of potentials of zero charge leads to a classification of metals in terms of the structure of the electrode/solution interface. The modifications in the structure of the surface regions brought about by the contact between the metal and the solvent are higher for *d*-metals than for *sp*-metals. In the former group the interfacial parameters are probably metal independent, whereas they change regularly with the nature of the solid phase in the latter group. Results for single crystal faces of a given metal show that the interfacial structure is definitely face specific.

Comparison of electrochemical and gas phase data suggests that water molecules are associated also at the interface so that the intrinsic structure of the solvent still plays a major role in the surface region. Chemical interactions with the metal surface are as a rule relatively weak but small energetic changes produce large variations in electrical parameters. In a sense, it is possible to state that the electrode potential is a quantity whose components reflect and amplify chemical events occurring at the interface.

While it is not evident from experiments, models emphasize the role of the surface electrons of the metal in determining the electrical parameters of the interface. Although no direct measurement of this effect is possible, some estimation can be attempted for the Hg electrode. Conclusions are that the role of the metal surface cannot be neglected or disregarded but it should probably not be overemphasized. The contribution is more substantial for the value of the electrode potential than for the capacitance.

The most recent models describe the metal in terms of the *jellium* model and treat the solution as an ensemble of hard spheres with point charges and point dipoles. Although the predictions of the models are encouraging and substantially correct, *chemistry* is still lacking. Metal-water interactions are essentially expressed in terms of electrostatic factors (electron-electron, electron-dipole, dipole-dipole, etc.). Quantum chemical calculations should probably be extended to these models.

The *jellium* model is probably adequate for homogeneous metals, but inadequate for single crystal faces and transition metals. Studies of water adsorption from the gas phase are available only for transition metals and few *sd*-metals. These should be extended to *sp*-metals for which a large amount of electrochemical data is available and to which models can be applied. On the other hand, data for single crystal faces are still insufficient. Investigations should be extended to non-aqueous solvents and studies of adsorption of suitable organic substances should be increased. The situation with transition metal is still underdeveloped in the field of the electrode/solution interface.

Acknowledgement. Financial support by the Italian National Research Council (CNR, Rome) is gratefully acknowledged.

REFERENCES

[1] Trasatti, S. in Trends in Interfacial Electrochemistry. NATO Advanced Study Institute, Reidel, Dordrecht, 1984, p.
[2] Frumkin, A.N. 1965, Sven. Kem. Tidskr. 77, p. 300.
[3] Parsons, R. in Bockris, J.O'M. (Ed.), Modern Aspects of Electrochemistry, Vol. 1, Butterworth, London, 1954.
[4] Trasatti, S. 1971, J. Electroanal. Chem. 33, p. 351.
[5] Trasatti, S. 1980, Colloids Surf. 1, p. 173.
[6] Valette, G. 1982, J. Electroanal. Chem. 139, p. 285.
[7] Trasatti, S. 1981, J. Electroanal. Chem. 123, p. 121.
[8] Frumkin, A.N. Potentials of Zero Charge, Nauka, Moscow, 1979.
[9] Trasatti, S. 1983, Electrochim. Acta 28, p. 1083.
[10] Frumkin, A.N. and Petrii, O.A. 1975, Electrochim. Acta 20, p. 347.
[11] Angerstein-Kozlowska, H. Conway, B.E. Barnett, B. and Mozota, J. 1979, J. Electroanal. Chem. 100, p. 417.
[12] Lazarova, E.M. 1978, Elektrokhimiya 14, p. 1300.
[13] Hamelin, A. Vitanov, T. Sevastyanov, E. and Popov, A. 1983, J. Electroanal. Chem. 145, p. 225.
[14] Batrakov, V.V. Damaskin, B.B. and Ipatov, Yu.P. 1974, Elektrokhimiya 10, p. 144.
[15] Trasatti, S. 1971, Chim. Ind.(Milan) 53, p. 559.
[16] Trasatti, S. 1972, J. Chem. Soc. Faraday Trans. I 68, p. 229.
[17] Trasatti, S. in Gerischer, H. and Tobias, C.W. (Eds.), Advances in Electrochemistry and Electrochemical Engineering, Vol. 10, Wiley-Interscience, New York, 1976, p. 213.
[18] Trasatti, S. in Bockris, J.O'M. Conway, B.E. and Yeager, E. (Eds.), Comprehensive Treatise of Electrochemistry, Vol. 1, Plenum Press, New York, 1980, p. 45.
[19] Trasatti, S. 1984, J. Electroanal. Chem., in press.
[20] Trasatti, S. in Chemistry and Physics of Electrocatalysis,

The electrochemical Society, Pennington, 1983, in press.
[21] Frumkin, A. Damaskin, B. Bagotskaya, I. and Grigoryev, N. 1974, Electrochim. Acta 19, p. 75.
[22] Hennig, H. and Batrakov, V.V. 1979, Elektrokhimiya 15, p. 1833.
[23] Farrell, J.R. and McTigue, P. 1982, J. Electroanal. Chem. 139, p. 37.
[24] Damaskin, B.B. Survila, A.A. and Rybalka, L.E. 1967, Elektrokhimiya 3, p. 146.
[25] Badiali, J.P. Rosinberg, M.L. and Goodisman, J. 1981, J. Electroanal. Chem. 130, p. 31.
[26] Schmickler, W. 1983, Chem. Phys. Lett. 99, p. 135.
[27] Trasatti, S. 1974, J. Chem. Soc. Faraday Trans. I 70, p. 1752.
[28] Randles, J.E.B. 1977, Phys. Chem. Liq. 7, p. 107.
[29] Trasatti, S. in Conway, B.E. and Bockris, J.O'M. (Eds.), Modern Aspects of Electrochemistry, Vol. 13 Plenum Press, New York, 1979, p. 81.
[30] Gomer, R. and Tryson, G. 1977, J. Chem. Phys. 66, p. 4413.
[31] Fawcett, W.R. 1979, Isr. J. Chem. 18, p. 3.
[32] Fawcett, W.R. Levine, S. de Nobriga, R.M. and McDonald, A.C. 1980, J. Electroanal. Chem. 111, p. 163.
[33] Guidelli, R. 1980; 1981, J. Electroanal. Chem. 110; 123, p. 205; p. 59.
[34] Damaskin, B.B. and Frumkin, A.N. 1974, Electrochim. Acta 19, p. 173.
[35] Parsons, R. 1975, J. Electroanal. Chem. 59, p. 229.
[36] Damaskin, B.B. Palm, U.V. and Salve, M.A. 1976, Elektrokhimiya 12, p. 232.
[37] Damaskin, B.B. 1977, J. Electroanal. Chem. 75, p. 359.
Bockris, J.O'M. and Habit, M.A. 1977, Electrochim. Acta 22, p. 41.
[38] Damaskin, B.B. and Shlepakov, A.V. 1982, Elektrokhimiya 18, p. 752.
[39] Damaskin, B.B. and Kaganovich, R.I. 1977, Elektrokhimiya 13, p. 293.
[40] Randles, J.E.B. and Whiteley, K.S. 1956, Trans. Faraday Soc. 52, p. 1509.
[41] Hölzl, J. and Schulte, F.K. 1979, Springer Tracts Mod. Phys. 85, p. 1.
[42] Trasatti, S. 1978, Appl. Surf. Sci. 1, p. 341.
[43] Farrell, J.R. and McTigue, P. 1984, J. Electroanal. Chem. 163, p. 129.
[44] Trasatti, S. 1977, J. Electroanal. Chem. 82, p. 391.
[45] Dutkiewicz, E. and Lamperski, S. 1984, J. Electroanal. Chem. 162, p. 13.
[46] Trasatti, S. 1983, J. Electroanal. Chem. 150, p. 1.
[47] Bond, G.C. 1970, R.I.C. Rev. 3, p. 1.
[48] Trasatti, S. 1975, J. Chim. Phys. 72, p. 561.
[49] Samsonov, G.V. The Oxide Handbook, I.F.I./Plenum, New York, 1973.
[50] Frumkin, A. Damaskin, B. Grigoryev, N. and Bagotskaya, I.

1974, Electrochim. Acta 19, p. 69.
[51] Trasatti, S. in O'Grady, W.E. Ross, P.N. and Will, F.G. (Eds.) Electrocatalysis, The Electrochemical Society, Pennington, 1981, p. 73.
[52] Bagotskaya, I.A. 1980, Elektrokhimiya 16, p. 216.
[53] De Battisti, A. and Trasatti, S. 1973, J. Electroanal. Chem. 48, p. 213.
[54] Petrii, O.A. Khomchenko, I.G. and Zelinskii, A.G. 1979, Elektrokhimiya 15, p. 400.
[55] Pons, S. 1983, J. Electroanal. Chem. 150, p. 495.
[56] Trasatti, S. 1982, J. Electroanal. Chem. 138, p. 449.
[57] Bagotskaya, I.A. and Shlepakov, A.V.S. 1980, Elektrokhimiya 16, p. 565.
[58] Valette, G. 1981, J. Electroanal. Chem. 122, p. 285.
[59] Lecoeur, J. Andro, J. and Parsons, R. 1982, Surf. Sci. 114, p. 320.
[60] Heras, J.M. and Viscido, L. 1980, Appl. Surf. Sci. 4, p. 238.
[61] Heras, J.M. Papp, H. and Speiss, W. 1982, Surf. Sci. 117, p. 590.
[62] Bange, K. Grider, D. and Sass, J.K. 1983, Surf. Sci. 126, p. 437.
[63] Kuznetsov, A. Reinhold, J. and Lorenz, W. 1984, J. Electroanal. Chem. 164, p. 167.
[64] Kolotyrkin, Ya.M. Lazorenko-Manevich, R.M. Plotnikov, V.G. and Sokolova, L.A. 1977, Elektrokhimiya 13, p. 695.
[65] Nieuwenhuys, B.E. Van Aardenne, O.G. and Sachtler, W.M.H. 1974, Chem. Phys. 5, p. 418.
[66] Kretzschmar, K. Sass, J.K. Bradshow, A.M. and Holloway, S. 1982, Surf. Sci. 115, p. 183.
[67] Trasatti, S. 1975, J. Electroanal. Chem. 65, p. 815.
[68] Shlepakov, A.V. and Bagotskaya, I.A. 1982, Elektrokhimiya 18, p. 26.
[69] Badiali, J.P. Rosinberg, M.L. and Goodisman, J. 1983; 1983, J. Electroanal. Chem. 143; 150, p. 73; p. 25.
[70] Palm, U.V. Pyarnoya, M.P. and Salve, M.A. 1977, Elektrokhimiya 13, p. 873.
[71] Khmelevaya, L.P. Chizhov, A.V. and Damaskin, B.B. 1978, Elektrokhimiya 14, p. 1304.
[72] Khmelevaya, L.P. and Damaskin, B.B. 1981, Elektrokhimiya 17, p. 1721.
[73] Bagotskaya, I.A. and Shlepakov, A.V. 1982, Elektrokhimiya 18, p. 462.

TEMPERATURE EFFECTS OF INTERFACIAL PROPERTIES

A. Fernando Silva

Departamento de Química - Faculdade de Ciências
4000 PORTO - PORTUGAL

Thermodynamic relations for the electrochemical interface are presented. Emphasis is laid on temperature coefficients of the differential capacity and of the interfacial tension. Procedures to evaluate the entropy of formation of the interphase are outlined. The entropy of formation of the Hg/aqueous solution interface containing adsorbing ions or organic molecules as a function of the interfacial composition is presented and discussed.

The electrochemical interface, specially the metal-solution interface, have been extensively studied varying the chemical composition of the system. This produced a large amount of thermodynamic data consistent with elementary theories of adsorption. Somewhat surprisingly much less work have been done involving other thermodynamic variables, namely the temperature, in spite of the valuable information it is believed it can be obtained to elucidate the fundamental structure of the electrochemical interface.

I - THERMODYNAMIC RELATIONS

A number of formal descriptions have been given of the temperature coefficients of the electrical and compositional state of the metal-solution interface [1, 2, 3, 4]. The Gibbs equation for the equilibrium interfacial free energy per unit area can be written as

$$d\gamma = -\Gamma_s dT + \Gamma_v dP + \sum_i \Gamma_i^{\sigma} d\mu_i - \sigma d\varepsilon_{\pm} \qquad (1)$$

where $\varepsilon_{\pm}$ is the difference of electrochemical potentials between one charged species in each of the adjoining phases and the other

A. F. Silva (ed.), Trends in Interfacial Electrochemistry, 49–70.

terms have their usual significance. This generalized potential can always be expressed in terms of the applied potential and a combination of the chemical potentials of the reversible electrode components.

$$\varepsilon_{\pm} = E_{\pm} + K/F \qquad (2)$$

where $E_{\pm}$ is the potential of the ideally polarized electrode measured with respect to an electrode reversible to a cation (+) or anion (-) in the solution under study, K is a constant sum of free energy terms related to the type of reference electrode employed. For instance for Na(Hg)/Na electrode $K = \mu_{Na}/F$, whereas for a calomel electrode $K = (\mu_{Hg_2Cl_2} - \mu_{Hg})/F$.

The important term in (1) when analysing the temperature dependence of interfacial properties is the excess entropy, and it is evidently defined as

$$\Gamma_s = -\left(\frac{\partial\gamma}{\partial T}\right)_{P,\mu_i,\varepsilon_{\pm}} \qquad (3)$$

where γ is simply the interfacial tension but where, because both μ and $\varepsilon_{\pm}$ are not necessarily independent of temperature even though the composition remains unchanged, care must given to the exact evaluation of the partial derivative.

For an interfacial zone, σ, bounded by two homogeneous phases α, β the exccess entropy can be further defined as

$$\Gamma_s = S^{\sigma} - \sum_j \Gamma_j \bar{S}_j^{\alpha} - \sum_i \Gamma_i \bar{S}_i^{\beta} - \sum_{j-1} \Gamma_{j(M)} \bar{S}_j^{\alpha} - \sum_{i-1} \Gamma_{i(w)} S_i^{\beta} \qquad (4)$$

where j and i refer to all components in phases α and β respectively, Γ_j and $\Gamma_{j(M)}$, Γ_i and $\Gamma_{i(w)}$ are respectively, the surface excess and relative surface excess.

M - reference component in the metallic phase
w - reference component in the solution phase
S^{σ} - is the total entropy per unit area of the interface
$\bar{S}$ - are partial molar entropies

It is clear that this quantity is very complex representing the sum of entropies of the excess components and the entropies of the other reference components which may be different from the corresponding bulk phases.

Γ_s cannot be regarded as the entropy of formation of the interphase. The entropy of formation of the interphase, S*, is defined as the difference in entropy of the components of the interphase when they are forming part of it, and when they are present in the bulk of the adjoining phases, i.e.

$$S^* = \sum_j \Gamma_j(\bar{S}_j^{\sigma} - \bar{S}_j^{\alpha}) - \sum_i \Gamma_i(\bar{S}_i^{\sigma} - \bar{S}_i^{\beta}) \qquad (5)$$

where the brackted terms represent the difference in the partial molar entropy of the component in the interphase, σ, compared to its value in the bulk phases α and β. If ions are present at the interface then a non-thermodynamic assumption intervenes in that single ion entropies are not known.

The complete route to S* is a slightly complex one and needs to be firmly established, especially in the presence of several adsorbed components.

The relation between S* and Γ_s is

$$S^* = \Gamma_s - \sum_j \Gamma_j \bar{S}_j^\beta \tag{6}$$

thus, for a system of a single univalent electrolyte eq. (6) becomes

$$S^* = \Gamma_s - \Gamma_- \bar{S}_-^\beta - \Gamma_+ \bar{S}_+^\beta \tag{7}$$

or, because of the electroneutrality condition

$$S^* = \Gamma_s - (\Gamma_+ + \frac{\sigma}{F})\ \bar{S}_-^\beta + \Gamma_+ \bar{S}_+^\beta = \Gamma_s - \Gamma_\mp\ \bar{S}_{salt}^\beta\ \pm \frac{\sigma}{F}\ \bar{S}_\pm^\beta \tag{8}$$

The $\bar{S}_\pm^\beta$ terms are the separate partial molar ionic entrotropies and $\bar{S}_{salt}^\beta$ is the readily determinable partial molar entropy of the salt.

The total relative excess entropy is thus redefined by re-arrangement of eq. (8) as

$$\Gamma_s = S^* + \Gamma_\mp\ \bar{S}_{salt}\ \pm \frac{\sigma}{F}\ \bar{S}_\pm \tag{9}$$

and this can be substituted into the alternative form of eq. (1), namely at constant pressure

$$- d\xi_\pm = \Gamma_s dT + \sum_i \Gamma_i d\mu_i - \varepsilon_\pm d\sigma \tag{10}$$

where ξ is the constant charge function (= γ + σE). Thus, at constant composition,

$$- d\xi_\pm = (S^* + \Gamma_\mp\ \bar{S}_{salt}\ \mp \frac{\sigma}{F}\ \bar{S}_\pm)\ dT - \ \varepsilon_\pm d\sigma + \Gamma_\mp d\mu_{salt} \tag{11}$$

and because at constant composition

$$d\mu_{salt} = -\bar{S}_{salt}\ dT \tag{12}$$

eq. (11) simplifies to

$$- d\xi_{\pm} = (S^*_{\mp} + \frac{\sigma}{F}\bar{S}_{\pm})\, dT - \varepsilon_{\pm} d\sigma \qquad (13)$$

Because $\varepsilon_{\pm}$ is a total differential, cross-differentiation leads to

$$\frac{\partial}{\partial\sigma}(S^*_{\mp} + \frac{\sigma}{F}\bar{S}_{\pm})_{m,T} = -(\frac{\partial\varepsilon_{\pm}}{\partial T})_{m,\sigma} \qquad (14)$$

and, therefore

$$(\frac{\partial S^*}{\partial\sigma})_{m,T} = -(\frac{\partial\varepsilon_{\pm}}{\partial T})_{m,\sigma} \pm \frac{\bar{S}_{\pm}}{F} \qquad (15)$$

The temperature coefficient of the electrochemical potential is related to the temperature coefficient of the potential, $E_{\pm}$ of the metallic interphase whith respect to a reversible reference electrode. Changes in temperature of the cell must take account of the entropies of the reference components not directly linked to the compositional dependence of the potential. In general terms,

$$(\frac{\partial\varepsilon_{\pm}}{\partial T})_{m,\sigma} = (\frac{\partial E_{\pm}}{\partial T})_{m,\sigma} + \frac{1}{F}\frac{\partial K}{\partial T} \qquad (16)$$

and, since

$$(\frac{\partial E_{\pm}}{\partial T})_{m,\sigma} = (\frac{\partial E_{\pm}}{\partial T})_{m,\sigma=0} + \int_0^{\sigma}\left(\frac{\partial(1/C)}{\partial T}\right)_{m,\sigma} d\sigma \qquad (17)$$

eq. (15) can be rewritten to include (16) and (17) and then integrated to give, for a metal-solution interface, hereafter designated respectively by M and S

$$S^*_{\sigma} - S^*_{\sigma_R} = -\int_{\sigma_R}^{\sigma}\left[\int_0^{\sigma}\left(\frac{\partial(1/C)}{\partial T}\right)_{m,\sigma} d\sigma + (\frac{\partial E_{\pm}}{\partial T})_{m,\sigma=0}\, d\sigma\right] - \left[\frac{1}{F}\frac{\partial K}{\partial T} \mp \frac{\bar{S}_{\pm}}{F}\right](\sigma - \sigma_R) \qquad (18)$$

where R refers to any reference state of the system and which may be, but not necessarily, thar of $\sigma = 0$. Eq. (18) establishes the complete route for the evaluation of S^* using diferential capacitance data and knowing the temperature coefficient of the potential of zero charge for the system. The value of the entropy of formation at the uncharged metal can be directly obtained from the temperature coefficient of the interfacial tension.

For electrocapillary measurements the use of the applied potential as the electrical variable is more convenient. In this case S^* is readily obtainable from the temperature coefficient of the interfacial tension using an isothermal cell

$$S^* = -\left(\frac{\partial\gamma}{\partial T}\right)_{E_\pm,\mu} - \frac{\sigma}{F}\left(\frac{\partial K}{\partial T} \mp \bar{S}_\pm\right) \tag{19}$$

II - THE DIFFUSE LAYER ENTROPY OF FORMATION

The potential drop across the metal-solution interface is given by (4)

$$\Delta_S^M\phi = E_\pm + (K \pm \mu_\pm)/F \tag{20}$$

where $\mu_\pm$ is the chemical potential of the species to which the reference electrode is reversible.

Differentiating this expression with respect to temperature at constant molatily

$$\left(\frac{\partial\Delta_S^M\phi}{\partial T}\right)_{P,m,\sigma} = \left(\frac{\partial E_\pm}{\partial T}\right)_{P,m,\sigma} + \frac{1}{F}\left(\frac{\partial K}{\partial T} \mp \bar{S}_\pm\right) \tag{21}$$

it gives the temperature coefficient of the potential drop across the interface between phase M and S.

Comparing eq. (21) with eq. (15) it results

$$\left(\frac{\partial S^*}{\partial\sigma}\right)_{m,T} = -\left(\frac{\partial\Delta_S^M\phi}{\partial T}\right)_{m,\sigma} \tag{22}$$

Under the usual assumptions made in the separation of the double layer capacitance, into two contributions: the inner layer and diffuse layer capacities, $\Delta_S^M\phi$ may be expressed as:

$$\Delta_S^M\phi = \Delta\phi_{inner} + \Delta\phi_{diffuse} \tag{23}$$

or, after differentiating

$$\left(\frac{\partial\Delta_S^M\phi}{\partial T}\right) = \left(\frac{\partial\Delta\phi_{inner}}{\partial T}\right) + \left(\frac{\partial\Delta\phi_{diffuse}}{\partial T}\right) \tag{24}$$

Combining eq. (22) with eq. (24) it results

$$S^* = S^*_{inner} + S^*_{diffuse} \tag{25}$$

meaning that the separation of the entropy of formation in two contributions as described by eq. (25) involves the same assumptions made in the separation of the double layer capacitance into the contributions from the inner and diffuse layers [5].

The second term on the r.h.s. of eq. (24) may be calculated using the Gouy-Chapman theory. Expressions for the calculation of the entropy of formation of the diffuse layer have been given by Harrison et al [2] and Hurwitz et al [3].

$$S^*_{diffuse} = - RT\left(\frac{2RT\varepsilon m\rho}{\pi Z^2F^2}\right)\left\{2\pi\cosh^{-1}(1+\theta^2)^{1/2} - \left[3 + \left(\frac{\partial\ln\varepsilon\rho}{\partial\ln T}\right)_{m,p}\right]\left((1+\theta^2)^{1/2} - 1\right)\right\} \qquad (27)$$

Where $\theta = \sigma_{OHP}/2(2RT\varepsilon m\rho/\pi)^{1/2}$

m - molality
ρ - density

Table 1 contains the values calculated by Hurwitz et al. [3]

$(\sigma + z_iF\Gamma_i)$ (μCcm^{-2})	$-S^*_{diff} \times (10^9\ Jk^{-1}\ mol^{-1}\ cm^{-2})$			
J = (mole kg^{-1})	1	0.1	0.01	0.001
2	0.036	0.109	0.298	0.604
4	0.138	0.415	0.940	1.636
6	0.315	0.855	1.757	2.844
8	0.554	1.409	2.697	4.178
10	0.838	2.039	3.721	5.608
12	1.191	2.739	4.807	7.303
14	1.573	3.498	5.948	8.632

Table 1: Values of the entropy of formation of the diffuse layer (ref. 3); J-ionic force

The contribution of $S^*_{diffuse}$ to the entropy of formation of the interface is, according to the Gouy-Chapman model, symmetrical with respect to the charge at the OHP and does not take into account the individual characteristics of the ions because it considers them as point charges.

It must emphasized that, when capacitance data are used in the evaluation of S^*_{inner} for a non-adsorbing electrolyte, the diffuse layer correction can be made by substracting the inverse of the diffuse layer capacitance, at each temperature, from the inverse of the total capacitance and the resulting $1/C_{inner}$ used directly in the calculation.

III - THE ENTROPY OF FORMATION OF THE Hg/AQUEOUS SOLUTION INTERFACE

i - Non-adsorbing electrolytes

Sodium fluoride have been considered as a non-adsorbing electrolyte at the Hg electrode. Values of S* for the interface Hg/NaF aq. were obtained by Harrison, Randles and Schiffrin [2] using differential capacitante and interfacial tension data. Hills and Hsieh [6] calculated values of S* for the same system using differential capacitance data. The agreement between both sets of data is excellent and the dependence of S* on the interface charge density is depicted in fig. 1.

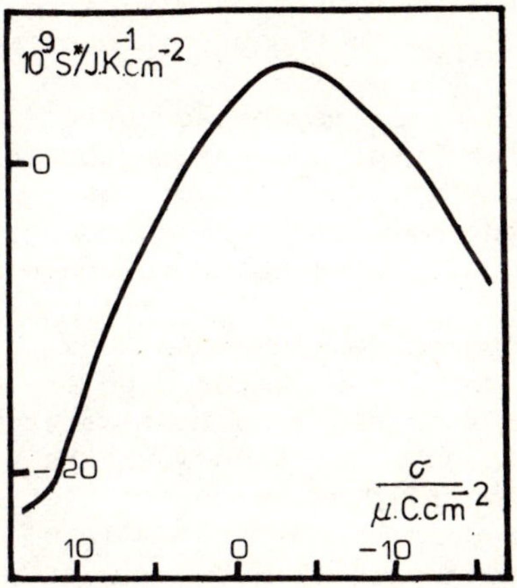

Figure 1 - The dependence of the entropy of formation, (S*) on σ for the Hg/0.1 M NaF interface at 298 k.

The striking aspect of this curve is the quasi parabolic nature of the dependence of S* on σ with a maximum at negative values of the interfacial charge density ($\sigma = -4 \mu C cm^{-2}$). Since, in the absence of adsorption

$$S^*_{inner} = S^{\sigma} - \sum_i \Gamma_i^{\sigma} \bar{S}_i^{\sigma} - \Gamma_{H_2O} \bar{S}^{\sigma}_{H_2O} - S^*_{diffuse} \tag{28}$$

or

$$S^*_{inner} = \sum_i \Gamma_i^{\sigma} (\bar{S}_i^{\sigma} - \bar{S}_i^{\alpha}) + \Gamma^{\sigma}_{H_2O} (\bar{S}^{\sigma}_{H_2O} - \bar{S}^{\beta}_{H_2O}) \tag{29}$$

Variations in S^*_{inner} with the interfacial charge density can be brought about by changes in the amounts $\Gamma^{\sigma}_{H_2O}$, Γ^{σ}_i in the interface and by changes of the mean value of the entropy of the components there. It is difficult to decide to which of the terms appearing in eq. (29) the variation of S^*_{inner} with σ is attributable. The components arising from the Hg electrode will be mercury ions and electrons. Their contribution with respect to charge will certainly not be symmetrical because at strong positive charge polarization the distribution of mercury ions in the interphase will be higher than at strong negative charge polarization. Nevertheless, a positive contribution of the mettalic components to S^*_{inner} would be expected. This is at variance with the observed decrease of the entropy of formation in the anodic branch of the curve which suggests that the charge dependence of the entropy of formation of the inner part of the double laver may be attributable to the other component present in the interphase, i.e. the solvent-water.

Further support to this proposition will be given later by the dependence upon σ of S^*_{inner} in the presence of adsorbing agents.

The charge dependence of S^* may be accounted for either by a)changes in the amount of solvent present in the inner layer which would be a minimum at $\sigma = - 4\mu Ccm^{-2}$ b) by changes in the entropy of the adsorbed solvent molecules due to changes in the orientation of the dipole moment of the water molecule with the electrical field.

To account for the charge dependence of S^*_{inner} solely by the variation in the number density of water dipoles in the interphase, changes in the order 50% of the amount of water would be required [2]. This would be in contradiction with the results obtained by Bewick and Robinson [7], using optical techniques, who estimated that the number density of water molecules is very insensitive to polarization, changing 4% between $- 4\mu Ccm^{-2}$ and $+ 8\mu Ccm^{-2}$. Then the observed entropy variations with σ may be attributed mainly to the change in the orientational distribution of water dipoles in the inner layer due to electrical field. The value of S^*_{inner} at $\sigma = 0$ indicates that the dipole of water have a preferential orientation at the uncharged metal that being with the negative poles pointing to the electrode. The decrease of S^*_{inner} with increase in polarization reflects the effect of an increasing electrical field upon dipole orientation whereby an increasing number of dipoles becomes more and more aligned with the electrical field therefore loosing degrees of freedom.

At high polarization the rate of variation of S^*_{inner} with σ decreases significantly which may reflect the increase in dielectric saturation.

To compare the state of interfacial water with bulk water, the absolute value of S^*_{inner} at the maximum is needed. This can be obtained using the temperature coefficient of γ at the PZC and if the metallic contribution is known. Such contribution is difficult to evaluate and the assumption usually made is that the contribution from the metal is charge independent and that it is given by the entropy of formation of pure mercury surface ($= - 20.9 \times 10^{-9}$ J K^{-1} cm^{-2}) [9]. It results that S^*_{inner} at the maximum is $- 6.0 \times 10^{-9}$ J K^{-1} cm^{-2} which indicates that the configuration of water at the state of maximum orientation disorder at the interphase still is more ordered than that of the bulk [10].

ii - Systems containing adsorbing species

The dependence of S* on σ in the presence of several adsorbing species have been estimated [1, 2, 7, 10] but the attempts to relate the data to molecular modeles were not successful because the entropy of formation of the inner layer was not presented in terms of its composition. To overcome the difficulty two quasi-thermodynamic treatments were proposed by Charles et al [10], for ionic systems at constant ionic strenght, and by Hills and Silva [11] for organic containing systems.

a) Adsorbing electrolytes

Figs. 2, 3 and 4 show experimental differential capacitance vs charge curves for three systems in different conditions. The influence of the adsorbed anion on the entropy of formation is shwon in figs. 5, 6 and 7 for systems with a single salt and with two salts at constant ionic strength. It is clear that the presence of an adsorbed ion in the interphase causes a decrease in S*. At constant charge density and composition in solution, S* shows a tendency to decrease with increasing adsorbability of the ionic species.

The presence of adsorbing ions in the interphase does not affect the general shape of the dependence of S* on σ reinforcing the idea that the common property to those systems is the entropic contribution of the major component in the interphase, the solvent, and its variations with the electric field.

The adsorption process involves the displacement of water from the inner layer as "free" water, as the hydration shell of ions or both. On the basis that the partial molar entropy of water in the interphase is smaller than that in the bulk, such displacement will give rise to positive contributions to S*. However if the only entropic consequence of the adsorption of ions was that resulting from the displacement of molecules, S* would increase with Γ_1^σ. From fig. 6 it is clear that the reverse is the observed, indicating a negative entropy change upon adsorption. In order to be able to evaluate the dependence of the entropy of formation as a function of the inner layer composition, Charles et al [10] assumed that the inner layer constituted a surface phase equivalent to a monomolecular layer the thickness of which is assumed to vary between the dimension of a water molecule and that of an hydrated ion.

The entropy of formation of this surface layer is defi-

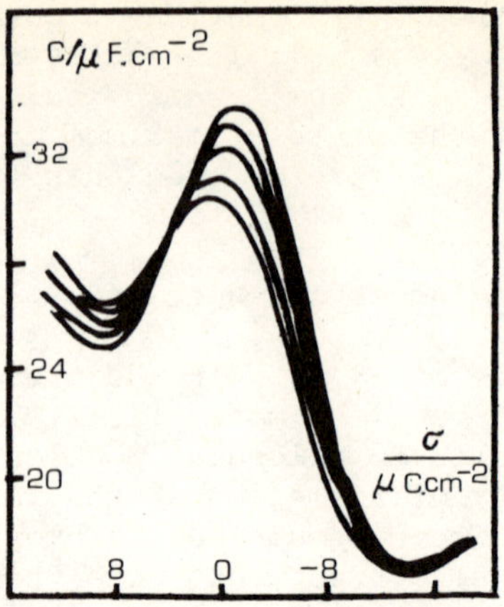

Figure 2 - Differential capacity - charge curves for Hg in contact with 1 M KNO_3 solution at different temperatures; from top to bottom: 278, 283, 288, 298, 308, 318 K.

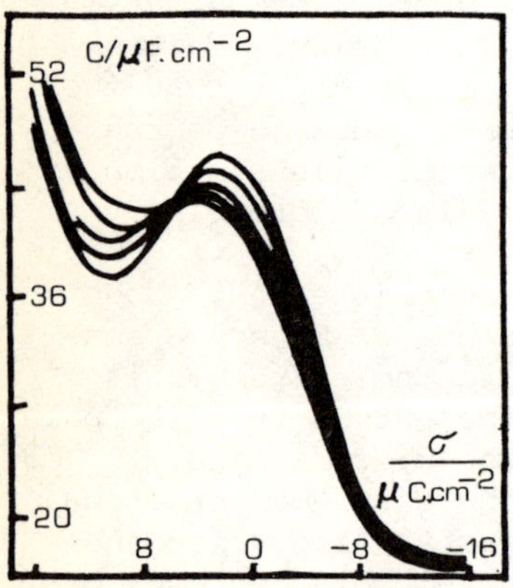

Figure 3 - Differential capacity - charge curves for Hg in contact with 1 M KCl solution at different temperatures; from top to bottom: 288, 298, 308, 318, 328 K.

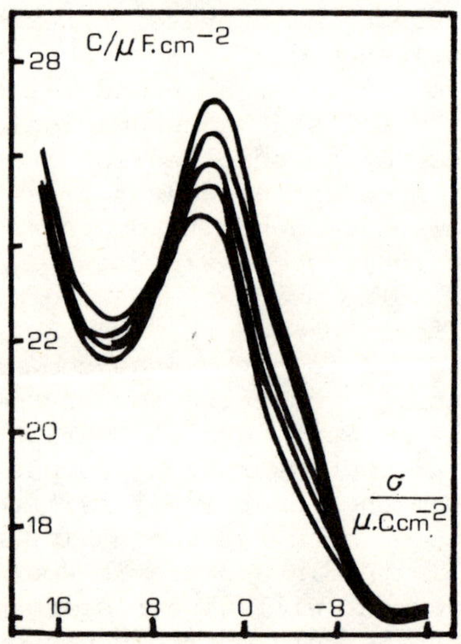

Figure 4 - Differential capacity - charge curves for Hg in contact with 0.1 M KPF_6 [2] solution at different temperatures; from top to bottom: 278.2, 287.6, 297.8, 307.3, 318.0 K.

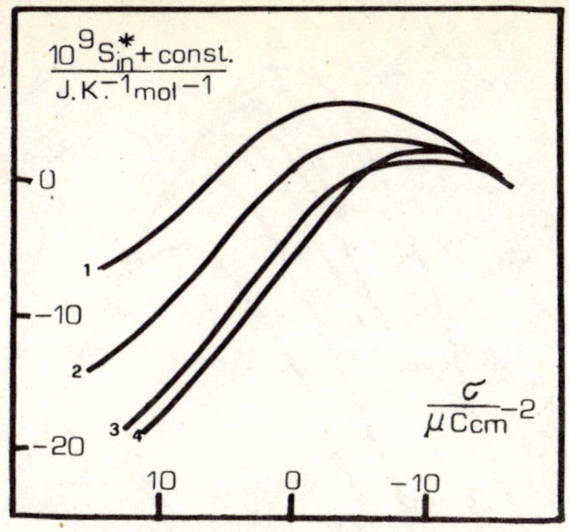

Figure 5 - Entropy of formation as a function of σ in the presence of different salts - 1 - NaF, 2 - KNO_3, 3 - KCl, 4 - KI.

ned as given in eq. (4). Using a system constituted by two salts with a common cation, at constant ionic strength, J, the entropy of adsorption of the solvent can be obtained through

$$\frac{\bar{S}^{\beta}_{H_2O} - \bar{S}^{\sigma}_{H_2O}}{a_{H_2O}} = \frac{\Delta S_{H_2O}}{a_{H_2O}} = -\left(\frac{\partial \varepsilon}{\partial T}\right)_{P,J,\sigma,\Gamma_{ic}} + \sigma \frac{\partial E_{ref}}{\partial T} + S^{*}_{diffuse} \qquad (30)$$

where

$$\frac{\partial E_{ref}}{\partial T} = \frac{1}{F}\frac{\partial K}{\partial T} + \frac{\bar{S}_{\pm}}{F} \; ; \qquad \Gamma_{ic} = \frac{1}{RT}\left(\frac{\partial \xi}{\partial \ln m_{AC}}\right)_{\Gamma,P,J,\sigma}$$

and m_{AC} is the molality of salt AC, and a_{H_2O} is the area occupied by a water molecule in the surface layer.

Figs. 8, 9 and 10 show the adsorption entropies of water

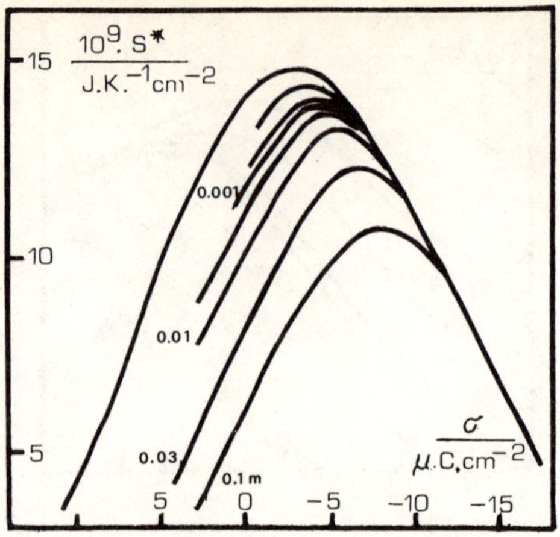

Figure 6 - Entropy of formation at 298 K of the double layer at Hg/NaF-NaI solutions at constant ionic strength

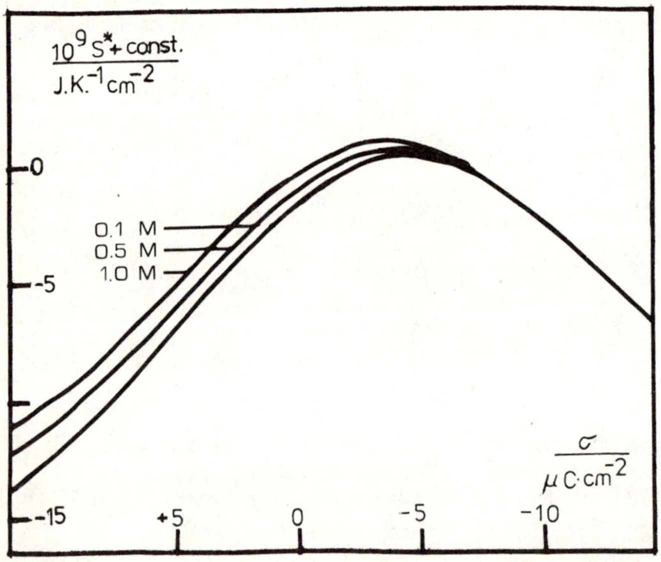

Figure 7 - Entropy of formation at 298 K of the double layer at Hg/KCl aqueous solutions.

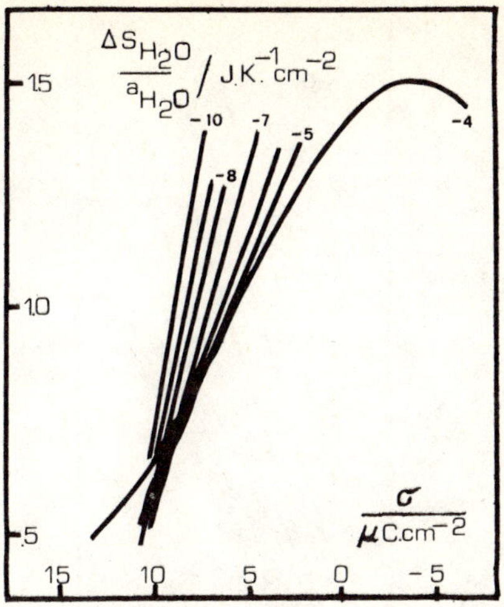

Figure 8 - Entropy of adsorption of water as a function of σ, at 298 K, for different amounts of Cl^- adsorbed [10] expressed in $\mu C cm^{-2}$.

in the inner layer calculated using eq. (30) in the presence of Cl^-, Br^- and I^-, as a function of σ and at constant amount of ions adsorbed. The dependence of the adsorption entropies of water on σ is pratically linear with slopes that are independent of the amounts adsorbed for the case of I^-. This means that in the presence of I^- the entropy of adsorption of water is linearly dependent of the amount of I^- adsorbed at constant value of the metal charge density. This was interpreted [10] as if the effect of the strongly adsorbing I^- ion on the interfacial solvent, at low coverages, does not go beyond the first circular sheet (hydration shell) of solvent molecules [10]. Therefore the presence of I^- in the interphase, at low values of coverage, does not affect the entropic properties of the remaining solvent. For the other ions, Br^- and Cl^-, the dependence of the entropy of adsorption of water on the charge density suggests that the interfacial solvent is more affected by their presence than in the case of I^-. These hypothesis are consistent with the proposed model for the adsorption of those ions, described by a virial isotherm with a charge independent virial coefficient in the case of I^-, while for Br^- and Cl^- the virial coefficients are charge dependent [10, 12].

The conclusion is that the effect of ion adsorption is to decrease the entropy of formation of the inner layer (Fig. 6). The overriding factor for the decrease is the partial molar entro

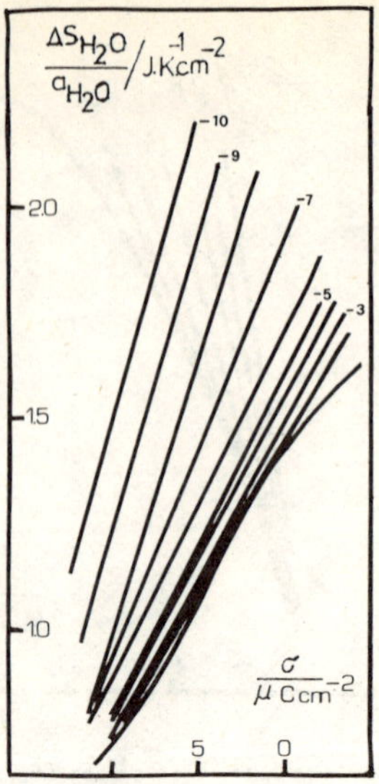

Figure 9 - Entropy of adsorption of water as a function of σ, at 298 K, for different amounts of Br^- adsorbed [10] expressed in μCcm^{-2}.

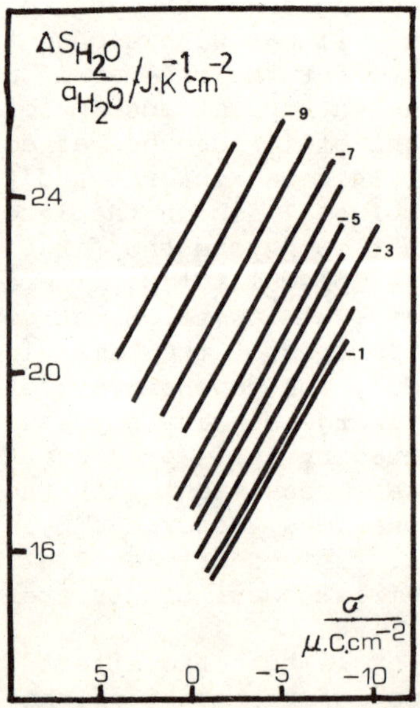

Figure 10 - Entropy of adsorption of water as a function of σ, at 298 K, for different amounts of I^- adsorbed [10] expressed in μCcm^{-2}.

py of the adsorbed ions and the properties of the remaining solvent molecules are less affected the strongest the adsorbability of the ion.

b) Adsorbing neutral species

The influence exerted by the presence of strong adsorbing organic molecules on the entropy of formation and, as a consequence on the interfacial solvent was investigated by Hills and Silva [11] using mixed solutions of thiourea and KCl.

The analysis of the temperature coefficient of the differential capacitance in order to obtain the entropy of formation required an extension of the thermodynamic treatment indicated earlier. It is a straightforward matter to included appropriate terms for the presence of an organic molecule (here denoted by O) into eq. (11), such that

$$- d\xi_{\pm} = (S^* \mp \frac{\sigma}{F}\bar{S}_{\pm} + \Gamma_{\mp}\bar{S}_{salt} + \Gamma_O\bar{S}_O)dT + \Gamma_{\mp}d\mu_{salt} + \Gamma_O d\mu_O - \varepsilon_{\pm}d\sigma \tag{32}$$

but, at constant composition

$$d\mu_{salt} = - \bar{S}_{salt}\,dT \quad \text{and} \quad d\mu_O = - \bar{S}_O dT \tag{33}$$

After substitution for $d\mu_{salt}$ and $d\mu_O$ in eq. (32), it gives simply

$$- (d\xi)_{m_{salt},m_O} = (S^* \mp \frac{\sigma}{F}\bar{S}_{\pm})\,dT - \varepsilon_{\pm}d\sigma \tag{34}$$

which is identical to eq. (13), except for the conditon that the concentration of the organic solute is also constant. The same sequence of equations then follows to give again eq. (18) which contains only observable quantities and $\bar{S}_{\pm}$ but no term directly related to the organic solute. The influence of the organic substance is of course manifested in the experimental values of the temperature coefficients of the differential capacitance and of the potential of zero charge determined at constant composition but dependent on the concentration and surface excess of the organic solute. It is possible to define a residual entropy $\Delta S^* = S^*_O - S^*_b$ which allows the determination of the effect of adsorbed organic molecules on the entropy of formation of the interphase and hence on the configuration of the solvent in the interphase.

Combining eqs. (18) and (34) it results

$$\frac{\partial}{\partial\sigma}(S_O^* - S_b^*) = -\left(\frac{\partial E_{\pm,O}}{\partial T}\right)_{\sigma=0} - \int_{\sigma=0}^{\sigma}\left(\frac{\partial(1/C_O)}{\partial T}\right)_{\sigma,m_O} + \frac{\bar{S}_{\pm,O}}{F} +$$

$$+ \left(\frac{\partial E_{\pm,b}}{\partial T}\right)_{\sigma=0} + \int_{\sigma=0}^{\sigma}\left(\frac{\partial(1/C_b)}{\partial T}\right)_{\sigma,m_b} - \frac{\bar{S}_{\pm,b}}{F} \qquad (35)$$

If the reasonable assumptions are then made (i) that the degree of ionic adsorption is independent of the concentration of the organic solute; (ii) that $\bar{S}_{\pm}$ is also independent of the presence of the organic substance, the difference $\bar{S}_{\pm,O} - \bar{S}_{\pm,b}$ can be equated to zero and the integration of eq. (35) then gives $S_O^* - S_b^*$ irrespective of the nature of the base electrolyte or the reference electrode, i.e. determined solely by the experimental quantities C and $E_{\pm}$ and the result integral is

$$S_O^* - S_b^* = \int_{\sigma_R}^{\sigma}\left[\int_{\sigma=0}^{\sigma}\left(\frac{\partial(1/C_b)}{\partial T}\right)_{m_b,\sigma} d\sigma - \int_{\sigma=0}^{\sigma}\left(\frac{\partial(1/C_O)}{\partial T}\right)_{m_O,\sigma} d\sigma\right] +$$

$$+ \left[\left(\frac{\partial E_{\pm,b}}{\partial T}\right)_{m_b,\sigma=0} - \left(\frac{\partial E_{\pm,O}}{\partial T}\right)_{m_O,\sigma=0}\right] d\sigma \qquad (36)$$

Fig. 11 shows a typical set of capacitante curves as a function of temperature for one composition namely 0.04 molar thiourea in 0.1 M KCl. The first aspect to notice is the temperature invariant point, characteristic of many if not all the capacitance curves in aqueous solutions; this will be referred later.

Fig. 12 shows values of S* as a function of σ but at constant molality of thiourea. The derived relationships shown as full lines are all quasi parabolic in nature and resemble those obtanied for ionic systems like I^- (fig. 6). The important feature for all these curves is again that each shows a maximum but the height and the position on the charge axis are systematically dependent on the composition of the solution and, therefore of the interphase, i.e. on the values of m_{TU}.

Values of $\Delta S^*(= S_{TU}^* - S_b^*)$ for a range of thiourea solutions are shown in fig. 13 as a function of interfacial charge density. The falling curves reflect the change in the interfacial composition. The fall in ΔS* is predominantly due to a change in the relative surface excess of thiourea because the effect of specific adsorption of chloride ion is already taken into account in the evaluation of ΔS*. Further reinforcement of this proposition is obtained from fig. 14 which shows the dependence of ΔS* on Γ_{TU} for different concentrations of thiourea. The lines stem from a commom origin at $\Gamma_{TU} = 0$. The slopes are essentially the same and certainly no significant dependence on m_{TU} was observed.

According to the definiton of S* early given ΔS* is also given by

$$\Delta S^* = (\sum_i \Gamma_i^\sigma(\bar{S}_i^\sigma - \bar{S}_i^b)) - (\sum_i \Gamma_i^\sigma(\bar{S}_i^\sigma - \bar{S}_i^b)_R) \tag{37}$$

where R denotes a convenient reference state in this case when $m_{TU} = 0$. Since the surface excess is closely related to the surface concentration it follows from eq. (37) that

$$\frac{\partial \Delta S^*}{\partial \Gamma_{TU}^\sigma} = \frac{\partial S^*}{\partial \Gamma_{TU}^\sigma} \cong \frac{\partial S^*}{\partial \Gamma_{TU(H_2O)}} = -(\bar{S}_{TU}^\sigma - \bar{S}_{TU}^b) + n(\bar{S}_{H_2O}^\sigma - \bar{S}_{H_2O}^b) \tag{38}$$

where n is the number of water molecules displaced or replaced by adsorption or desorption of a thiourea molecule.

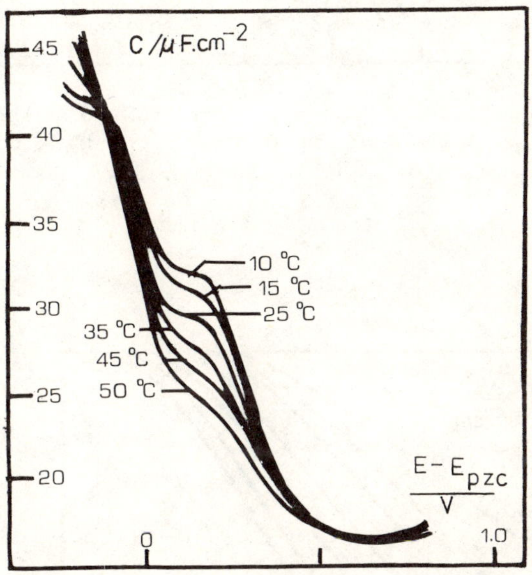

Figure 11 - Differential capacity of the interface between Hg and 0.04 M thiourea in 0.1 M KCl as a function of temperature and rational potential.

The linearity of the plots in fig. 14 suggests (i) that the partial molar entropy of adsorption of thiourea is not significantly dependent on its surface coverage; (ii) that the predominant term in eq. (38) is $(\bar{S}_{TU}^\sigma - \bar{S}_{TU}^b)$ which in fact is large and negative . Deconvoluting from the parabolic lines in fig. 12 the contribution due to the varying Γ_{TU}, thus leaving the resi-

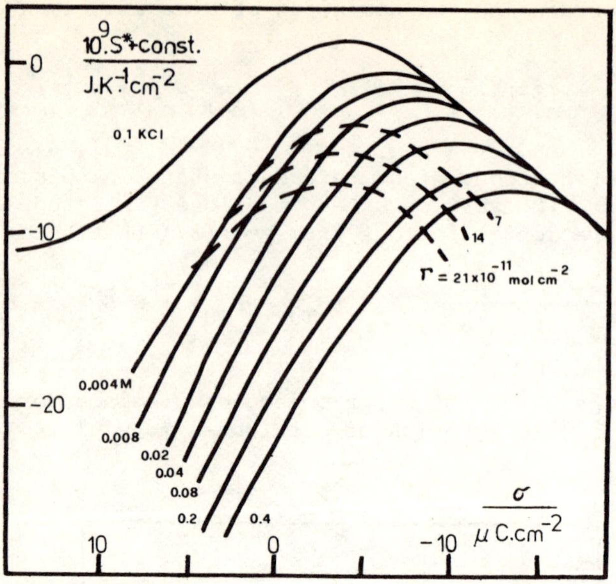

Figure 12 - S* (full lines) and S* - Γ_{TU} ($\partial\Delta S^*/\partial\Gamma_{TU}$) (dashed lines) as a function of σ and of the bulk concentration of thiourea at 298 K.

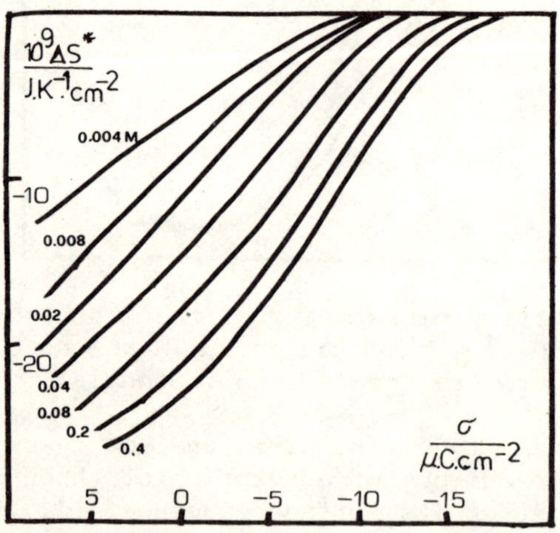

Figure 13 - ΔS* as a function of surface charge density for seven concentrations of thiourea.

dual contribution due to the solvent, the new relationships obtained are shown as dashed lines in the same figure. These parabolic curves corresponding to a variation with interfacial charge density of the entropy of formation at constant surface excess of thiourea are remarkable in that their maxima occur at the same value of σ as for solutions containing no thiourea i.e. when $m_{TU} = 0$. The charge density of zero preferred solvent orientation is therefore, as before, $- 3 \pm 1\ \mu C\ cm^{-2}$. It is clearly evident that the entropic nature of water in the interphase essentially in what it depends on the charge density is independent of its molecular number density at least up to moderate coverages of thiourea. This is a result similar to that obtained with the adsorption of iodide which suggests that the presence in the interphase of strongly adsorbing agents does no affect the electrical properties of the remaining solvent.

IV - TEMPERATURE INVARIANT POINT

As indicated earlier, an important feature existed in all differential capacitance curves for ionic and organic solutes. It is the presence of a temperature invariant point when C is displayed as a function of the electrode charge density. This isosbestic point is always placed on the anodic side of the potential or charge of the assumed zero preferred orientation of the solvent molecules. In organic solvents, isosbestic points are also observed [14, 15]. Their position on the charge scale varies from solvent to solvent and may occur on either side of the point of zero charge.

Such isosbestic points are not well defined on a normal differential capacitance - potential plot. The lack of definition arises because the potential scale itself is temperature dependent for reasons indicated in (I). If, however, at each temperature the electric potential is re-expressed on a rational scale then the convergence of the intersections also occurs (see fig. 11), confirming that one is dealing with a single electrical state of the systems at each temperature, the result of which is independent of whether it is expressed as charge or rational potential. The temperature invariant capacitante simply corresponds to an electrical state of the system no longer temperature dependent.

V - TEMPERATURE COEFFICIENT OF THE POTENTIAL DROP ACROSS THE INTERPHASE AND SOLVENT ORIENTATIONS

It is in the nature of thermodynamics that

$$\left(\frac{\partial C}{\partial T}\right)_{\sigma,m} = 0 = \left(\frac{\partial (1/C)}{\partial T}\right)_{\sigma,m} \tag{39}$$

and

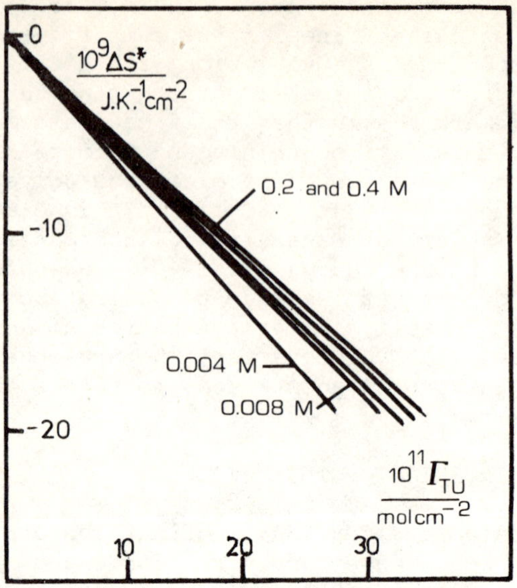

Figure 14 $-\Delta S^*$ as a function of Γ_{TU} for different concentrations of thiourea in 0.1 M KCl at 298 K.

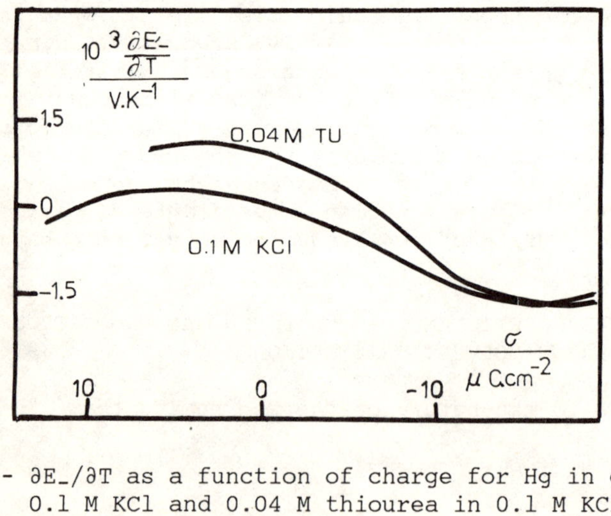

Figure 15 - $\partial E_-/\partial T$ as a function of charge for Hg in contact with 0.1 M KCl and 0.04 M thiourea in 0.1 M KCl.

$$1/C = \left(\frac{\partial E_{\pm}}{\partial \sigma}\right)_m \tag{40}$$

and that

$$\left(\frac{\partial(1/C)}{\partial T}\right)_{\sigma,m} = \frac{\partial}{\partial T}\left(\frac{\partial E_{\pm}}{\partial \sigma}\right) = \frac{\partial}{\partial \sigma}\left(\frac{\partial E_{\pm}}{\partial T}\right) \tag{41}$$

A typical plot of $\partial E_-/\partial T$ vs σ is shown in fig. 14. As expected the temperature invariant point is now defined by the maximum in $\partial E_-/\partial T$ or as

$$\frac{\partial \Delta_S^M \phi}{\partial T} = \frac{\partial E_-}{\partial T} - \frac{\bar{S}_-}{F} - \frac{1}{F}\frac{\partial k}{\partial T} \tag{42}$$

The temperature invariant point corresponds to a maximum in the temperature derivative of the potential drop across the metal--solution interphase. The potential of zero preferred solvent orientation (when S* is a maximum) coincides with the value $(\partial \Delta_S^M \phi/\partial T) = 0$.

At far cathodic values of E_- or σ, $\partial E_-/\partial T$ passes through a minimum indicating another isosbestic point discernible on the experimental differential capacitance curves.

The general symmetry of $\partial E_-/\partial T$ vs σ about the potential or charge of zero preferred solvent orientation will no doubt be explained in terms of changing interfacial composition and changing degree of solvent orientation upon a varying electrical field.

At zero charge the temperature derivative of $\Delta_S^M \phi$ has been used [16, 17] to assess the degree of solvent orientation at the uncharged electrode. A positive value of $\partial(\Delta_S^M \phi)/dT$ has been taken as indicate that the solvent molecules are preferentially oriented with the negative end of the dipoles towards the metal surface. This position is in agreement with the dependence of S* on σ and, as a consequence, with the dependence of $\partial E_-/\partial T$ whith σ. Criticisms have been raised [18] to the validity of using the value of $\partial(\Delta_S^M \phi)/\partial T$ at $\sigma = 0$, in the absence of specific adsorption, to assess solvent orientation due to the contribution of the unknown temperature coefficients $\delta\chi_S$ and $\delta\chi_M$ to $\partial(\Delta_S^M \phi)/\partial T$ at $\sigma = 0$. However the general symmetry of $\partial(\Delta_S^M \phi)/\partial T$ vs σ and of S* vs σ are indicative that qualitatively, its use to assess the average preferential orientation of the solvent is valid. Finnaly, the dependence of $\partial E_-/\partial T$ on σ depicted in fig. 15 also indicates the dependence of S* on σ is non parabolic but there exist two inflexion points corresponding to zero value of the derivative which indicates the existence of other contributions to S* other than those considered in the simple model used so far.

ACKOWLEDGEMENTS

The finantial support of Gulbenkian Foundation (Lisbon) and INIC (Lisbon) is acknowledge.

REFERENCES

1. G.J. Hills and R. Payne, Trans. Faraday Soc., 61 (1965) 326.
2. J.A. Harrison, J.E.B. Randles and D.J. Schiffrin, J.Electroanal. Chem. (48 (1973) 359.
3. H.D. Hurwitz and C.V. D'Alkaine, J. Electroanal. Chem., 42 (1973) 77.
4. R. Parsons, Can. J. Chem., 37 (1959) 308.
5. D.C. Grahame, J. Am. Chem. Soc., 79 (1957) 2093.
6. G.J. Hills and S. Hsieh, J. Electroanal. Chem., 58 (1975) 289.
7. A. Bewick and J. Robinson, J. Electroanal. Chem. 71 (1976) 131.
8. Nguyen Huo Cuong, C.V. D'Alkaine, A. Jenard and H.D. Hurwitz, J. Electroanal. Chem., 51 (1974) 377.
9. M.L. Wilkinson, Chem. Rev., 72 (1972) 575.
10. M.H. Charles, M. Mulenga, A. Jenard and H.D. Hurwitz, Can. J. Chem., 59 (1981) 2053.
11. G.J. Hills and F. Silva, Can. J. Chem., 59 (1981) 1853.
12. M.I. Sosa, D. Posadas and A.J. Arvia, J. Electroanal. Chem. 137 (1982) 307.
13. F. Silva, Ph. D. Thesis, Southampton, 1981.
14. Z. Borkowska and W.R, Fawcett, Can. J. Chem., 59 (1981) 710
15. Z. Borkowska, R.M. de Nobriga and W.R. Fawcett, J. Electroanal. Chem., 124 (1981) 263.
16. J.E.B. Randles and K.S. Whitley, Trans. Faraday Soc. 52 (1956) 1509.
17. W. Paik, T. Andersen and H. Eyring, J. Phys. Chem. 71 (1967) 1981.
18. S. Trasatti, This volume.

ELECTROCHEMISTRY ON WELL-DEFINED SOLID METAL ELECTRODES

Roger Parsons*

Laboratoire d'Electrochimie Interfaciale du CNRS -
1, Place Aristide Briand
92190 Meudon
France

INTRODUCTION

In this paper I shall try to convince the reader that it is possible to carry out electrochemical measurements under conditions where the composition and structure of the interface can be known with reasonable precision. The present situation is the culmination of many years of development of technique which finds its parallel in other branches of surface science. Electrochemical reactions like those of heterogeneous catalysis and other interfacial processes occur in a very small region of space often only a few atomic diameter thick. Thus, changes in the environment can be caused by very small amounts of adventitious impurities, or by changes in the structure resulting from the movement of samll numbers of atoms. The difficulty of controlling these properties of the solid-liquid interface led electrochemists in the past to concentrate on the use of mercury electrodes where the measurements could be made soon after the formation of the interface before contamination occurred. The use of a liquid-liquid interface assured a repeatable interfacial structure.

Comparable results heve been obtained on solid metals during the last few decades by using empirical developed pretreatment cycles, but these left the electrode in a state which was usually unknown, even if quite precisely repeatable.

* Present Address: Department of Chemistry
The University
Southampton SO9 5NH

A. F. Silva (ed.), Trends in Interfacial Electrochemistry, 71–81.

In the last few years the development of in-situ as well as ex-situ studies of electrode surfaces has shown that it is possible to obtain electrodes in a well-defined state and this has dramatically enlarged the possibilities for the precise understanding and control of electrode reactions. I shall illustrate this with emphasis on the work of my colleagues in Meudon but several laboratories have made important contributions to this development.

DEMONSTRATION OF ELECTRODE CLEANLINESS

It has been known for some years that cyclic voltammograms of whole metal electrodes are very sensitive to traces of impurities and, for example, Conway et al [1] have suggested that a particular form of the voltammogram for Pt in aqueous H_2SO_4 indicates that the interface is clean even though there was no direct proof. The type of direct proof which is now possible is illustrated by the work of Clavilier and Chauvineau [2] on spherical polycrystalline gold electrodes. They carried out parallel electrochemical and UHV (ultra high vacuum) experiments on two electrodes which were pretreated identically. When this pretreatment consisted of an argon glow discharge, the electrochemical experiment gave a cyclic voltammogram which was stable and reproducible from the first cycle immediately after contact with the solution. The parallel UHV experiment gave an Auger spectrum with the minimum amount of impurity. The sensitivity of the two types of experiment was demonstrated by a deliberate contamination of the electrodes by a brief exposure to the laboratory atmosphere. The Auger spectrum indicated the presence of about a third of a monolayer of a carbonaceous material while the voltammogram was significantly distorted on the first cycle [3] with suppression of the initial oxidation current of the gold and an enhanced current at higher potentials attributable to the oxidation of the carbonaceous material. The voltammogram on the second cycle returns almost to that on the clean electrode, while it indicates a completely clean electrode after 3-4 cycles.

Similar experiments were done with platinum electrodes [4] which would be expected to be much more sensitive to their environment through gold because of their greater adsorption ability. It was found that they could be cleaned effectively by making use of their catalytic function in a hydrogen-oxygen flame and these protected during transfer by a drop of ultrapure water--which is of course the product of the combustion. Uncontamined electrodes prepared in this way and brought into contact with the electrolyte under potentiostatic control showed a voltammogram on the first cycle of the type described by Conway et al [1]. Thus, it can be demonstrated that this does in fact correspond to a clean electrode surface.

ELECTRODES HAVING A WELL-DEFINED SURFACE STRUCTURE

Although a considerable body of work now exists for other metals, I shall restrict the discussion here essentially to platinum electrodes.

The characteristic voltammogram for polycrystalline platinum already referred to has two principal peaks in the hydrogen region. Since they correspond to an adsorption process which is fast they can be used to obtain thermodynamic information about this adsorption. A simple adsorption of a single species on a homogeneous surface would be expected to yield a single peak. Hence there have been speculation about the origin of the multiple peaks since they were first observed [5]. The existence of different crystallographic sites was a probable explanation but a great sensitivity to the nature of the electrolyte was also observed. [6, 7]. The first attempt to check the structural explanation was made by Will [8] who measured voltammograms on platinum single crystals oriented to present the (100), (111) and (110) faces and sealed in glass leaving only one of these faces exposed in each case.

His results show that the peak at more negative potentials (weakly adsorbed hydrogen) was larger for the (111) face while that after the positive potentials (strongly adsorbed hydrogen) was larger for the (100) face while there was no marked difference from polycrystalline platinum with the (110) face. However the total amounts of hidrogen adsorbed (obtained from the total charge under the peaks) was about twice that expected for a smooth surface of the given orientation. These results could be obtained only after preliminary electrochemical cleaning by cycling over 1.5 V for a hundred or so cycles. This suggests that the surfaces were very strongly perturbed and later work has confirmed this. Consequently, while Will's experiments suggest the existence of a structural effect, the precise nature of his surfaces is quite uncertain. Later work [9] seemed to suggest the absence of a structural effect, but it seem likely that this also suffered from the difficulty of maintaining surfaces in a satisfactory condition in the electrochemical cell.

In the last few years, considerable effort has been exerted to use the ex-situ techniques of surface structural characterization, notably LEED (low energy electron diffraction) in order to ensure that the electrochemistry was being done on a surface of known geometry. The direct transfer of a sample from an electrochemical cell to a UHV chamber or vice versa without alteration or contamination is extremely difficult. It requires careful design of the experiment and observation of phenomena at all stages. The first published results for the (100) and (111) faces of Pt by Hubbard et al [10] set a very high standard, but even these were obtained only after some treatment at each stage: a few electrochemical cycles or a slight argon ion bombardment. This leads to a slight uncertanty about structural changes.

Nevertheless, these results showed clearly that Will's suggestion was correct in the assignment of the two main hydrogen peaks of the voltammogram for polycrystalline Pt.

Other UHV/electrochemical experiments published about the same time gave less clear cut results [11, 12, 13] often with apparent coverages significantly less than unity, notably on the (111) surface, peaks with little fine structure which were modified by the first few cycles. It seems likely that some of the lack of agreement amongst these workers arises from the difficulty of transfering samples without structural change or contamination.

An alternative method of study to the direct transfer method was developed from the method used by Clavilier for the small spherical electrodes described above. This uses similar small Pt spheres which are single crystals (2-3 mm diameter) oriented and cut to expose the desired plane. Contact with the electrolyte can then be made by touching the surface with this plane and forming a meniscus, so that only the desired plane is in contact with the electrolyte [14, 15]. This has the advantage of avoiding any masking technique and permitting high temperature pretreatment of the electrode followed by very rapid transfer to the electrolyte. The latter can be done with immediate potentiostatic control so that any electrochemical process ocurring can be followed from the first contact of the electrode and electrolyte. It is found, in fact, that such transfer yields voltammograms stable from the first cycle, confirming that no contamination of the electrode is detectable. Parallel experiments, in which the electrode is introduced into an UHV apparatus, confirm that the structure is maintained up to the first atomic layer of the metal [15, 16].

Use of this technique has enable a variety of new observations of great interest in both electrochemistry and surface science. Some of these (which will be described in the next section) have been controversial. However the experimental results have been confirmed by several other groups using the same technique as well as quite recently by Ross [17] using a carefully designed direct transfer technique qith a thin layer cell in a UHV apparatus. Consequently there now seems little doubt that the relatively simple method developed by Clavilier yields precise and reliable data.

SOME RESULTS FOR LOW INDEX FACES OF PLATINUM

Of the results obtained using Clavilier's technique, those which have caused the most discussion were obtained using the (111) face [15]. A crystal annealed at high temperature, quenched with pure water and brought into contact with 0.5 M H_2SO_4 at a potential of 0.6 V RHE (with respect to a reversible hydrogen electrode in the same solution) shows a reduction transient which can be identified with oxygen adsorbed from the atmosphere. Cycling in the region 0.05 to 0.6 V RHE shows a broad region of current well above that expected for the double layer charging and stretching from 0.05 to nearly 0.5 V RHE. The oxidation current on the positive sweep is virtually identical at each point to the reduction current on the negative sweep and this symmetry is maintained to sweep speeds of up to 50 V S^{-1}. It therefore corresponds to a fast process. The charge corresponding to this region is 255 μ C cm^{-2} which will include a double layer contribution which is probably quite small. The charge calculated for an ideal (111) plane assuming one electron transferred per Pt atom is 243 μ C cm^{-2}. Extension of the potential sweep to 1.2 V (RHE) showed no other features above what might be expected from the double layer charging. This should be contrasted with the behaviour of polycrystalline Pt where the currents attributed to hydrogen adsorption do not extend above 0.3 V and those attributed to oxygen adsorption begin already at 0.8 V. This unusual voltammogram is stable for many cycles between 0.05 and 1.2 V (RHE). I shall call it type C. However, further extension of the potential sweep to 1.5 V leads in the first cycle to an oxidation charge of 550 μ C cm^{-2} and an equal reduction charge in a peak centered at 0.72 V (RHE) having a form very like the reduction peak of adsorbed oxide on polycrystalline Pt. At the same time the form of the current at potentials negative of 0.5 V is substantially modified although the total charge involved is unchanged. The modification can be discribed by saying that the most positive third of the change disappears and is replaced by a highly peaked region with the peak potential at 0.12 V. The voltammogram is now stable for many cycles and has almost exactly the characteristics of that published by Hubbard et al [10] for this orientation. (Type H voltammogram).

Deliberate contamination of the electrode before contact with the solution frequently yields voltammograms with rather featureless hydrogen regions reminiscent of those in refs [11] and [12]. The oxygen adsorption is similarly featureless with the higher currents delayed to the more positive potentials. Some 15 cycles to 1.5 V are required to reach a stable voltammogram which is then a type H. It appears that the decontamination process is much slower than the process in which a type C voltammogram is converted into a type H. Thus the former cannot be achieved with a contaminated electrode.

The anion present in the solution has a marked effect on the form of a type C voltammogram [18]. If 1 M perchloric acid is used in place of 0.5 M sulphuric acid the region of current between 0.05 and 0.5 V splits into two unequal parts. The larger corresponding to two thirds of the charge and at the more negative potentials is unaffected by the change of acid while the more positive one third moves to more positive potentials becoming located between 0.6 and 0.82 V (RHE). This voltammogram has the characteristics described above when the upper limit at the potential sweep is 0.9 V: i.e. the anodic and cathodic currents are equal at each potential up to 50 V s^{-1} and the voltammogram is stable. In addition, variation of the acid concentration over two orders of magnitude leaves the position of both regions unchanged on the RHE scale i.e. they both move following the hydrogen electrode. Addition of progressive amounts of sulphuric acid causes the more positive one third to move progressively towards the more negative two thirds.

Extension of the upper limit of potential cycling to 1.15 V yields a sharp anodic peak at 1.06 V with a charge corresponding to 130 $\mu C\ cm^{-2}$. At the same time the cathodic peak at 0.8 V is enlarged by the same amount of charge. Nevertheless the voltammogram remains stable and the change to a type H is achieved only by an extension to higher potentials and larger charges.

Most of these observations have been confirmed, notably by Reilley and Scortichini [19] and by Motoo and his coworkers [20]. Recently a careful LEED study by Wagner and Ross [17] has indicated that the change from voltammograms of type C to type H is associated with the disappearance of long range order and the formation of a disordered stepped surface. This confirms earlier suggestions by Clavilier that a structural change was involved and is consistent with Motoo's observations [20] of the effect of oxygen on the annealing process.

Disagreements still remain about the interpretation of the voltammogram. It seems to be generally accepted that oxidation currents above 0.8 V (RHE) are associated with the deposition of adsorbed oxygen species. It then seems necessary to accept that in the presence of sulphuric acid (HSO_4^- ions) this deposition is strongly inhibited up to 1.2 V.

It seems unlikely that this is due to contamination since the oxidation up to 1.5 V yields a very precise charge balance with the associated reduction peak. So it must be concluded that oxide species adsorbs weakly on this type of surface. Once they are forced to adsorb and a monolayer (550 $\mu C\ cm^{-2}$) is formed and removed the surface structure changes and the oxygen species adsorb in the normal way. This surface reconstruction is progressive; if less than a monolayer is adsorbed it is less complete. On the other hand in the absence of adsorbed anions ($HClO_4$) there is a threshold of oxygen adsorption (130 $\mu C\ cm^{-2}$) before reconstruction is initiated. This may be correlated with the observation in UHV that adsorption of oxygen up to a 2 x 2 ovlerlayer does not cause reconstruction of a (1 x 1) Pt (111) surface [21]. Under electrochemical conditions this behaviour seems to be maintained in the presence of water and weakly adsorbed anions but considerably modified by the more strongly adsorbed anion.

If the above is accepted then it is evident that the region of the voltammogram at more negative potentials is a very sensitive indication of surface structure. On polycrystalline platinum it has been accepted for many years that this region corresponds to the adsorption of atomic hydrogen and this explanation is not in dispute for the H-type voltammogram.

This may be extended to the two thirds of the charge unaffected by HSO_4^- adsorption without difficulty, but the one third at more positive potentials has not been generally accepted as due to H adsorption. This is primarily because the idea of such strongly bound hydrogen is unusual. Nevertheless the rapid nature of the process involved and the pH dependence point strongly to the identification of this charge with hydrogen and the transfer of precisely this charge to the peak in the H type voltammogram is more understandable in terms of hydrogen adsorption. The pH dependence could be interpreted if the adsorbed species were strongly bound oxide or hydroxide but the rapid adsorption is then unlikely and OH adsorption at 0.4 V is at least as improbable as H adsorption at 0.8 V. An explanation in terms of anion adsorption or water reorientation leads to difficulties with the amount of charge as well as the pH dependence. Hence in the light of the evidence available at present the existence of unusual adsorption state of hydrogen seems the most probable explanation, though it lacks a physical basis in terms of the bonding to the particular surface used. A study of hydrogen adsorption on Pt (111) surfaces annealed at high temperature would be particularly interesting.

The behaviour of surface with the (100) orientation is not less interesting although less spectacular in that no such unusual hydrogen adsorption states are observed. The voltammograms show usually two peaks in the hydrogen region at + 0.27 and + 0.37 V (RHE). The total charge in these peaks if integration is continued up to + 0.15 V, where there is a minimum in the current, is $205 \pm 5\ \mu C\ cm^{-2}$. This is close to that calculated for an ideal (1 x 1) Pt (100) plane which is $209\ \mu C\ cm^{-2}$. However, the relative height of the two peaks depends on the pretreatment before electrolyte contact as well as in the electrochemical cell [16]. A series of experiments in which the electrode was cooled in different atmospheres and quenched with pure water in equilibrium with different atmosphere at different temperatures showed that the electrode adsorbed different amounts of oxygen before contact with the electrolyte. This could be demonstrated by contacting at a potential of + 0.9 V where no transient current was observed. On the first negative sweep a cathodic peak at 0.785 V could be observed which was attributed to the reduction of thermally adsorbed oxygen. The largest peak obtained corresponded to a charge of $210\ \mu C\ cm^{-2}$ which could be related to the 2 x 2 overlayer [22]. This could be reduced to a very small amount by cooling the electrode in hydrogen and quenching in water satured in hydrogen. In the first case the peak at 0.37 V was largely suppressed at the expense of the peak at 0.27 V on the negative sweep whereas in the second case the peak at 0.37 V was much enhanced and that at 0.27 V much diminished. Similar effects could be produced by electrochemical treatment in the hydrogen or oxygen regions. Thus the electrode in the first case above showed a slightly different peak distribution on the following positive sweep as the peak at 0.37 V began to reappear, presumably as a result of the adsorption of hydrogen to saturation. Cycling between 0.08 V and 0.55 (i.e. without oxygen adsorption) gradually produced a peak distribution like that observed in the second case above. The process could be reversed by cycling into the oxygen region. It seems that adsorbed oxygen stabilizes a structure in which the more weakly hydrogen adsorbing sites predominate but the nature of these structures is not yet known. The fact that these changes are reversible under electrochemical conditions is particularly interesting. It contrasts with the behaviour of the (111) plane where the reverse process from voltammograms of type H to these of type C has only been achieved by high temperature annealing.

Electrodes of this orientation prepared by high temperature annealing show apparently simple behaviour [15] with a sharp peak at 0.12 V which with its shoulders corresponds to 220 $\mu C\ cm^{-2}$. The oxygen adsorption begins close to 0.8 V and has a structure somewhat different shape in detail from that of polycrystalline electrodes. Subsequent cycles show only subtle changes in the peak structure compared with those of the other low index planes. In fact, the agreement with other published results [12] [13] for this orientation is good, indicating that this plane is less sensitive to the details of its preparation.

OXIDATION REACTIONS ON SINGLE CRYSTAL PLATINUM

The oxidation of small organic molecules such as methanol and formic acid has been extensively studied on the platinum metals, but the results have been somewhat conflicting. This is partly due to the complexity of the reactions which involve the formation of a by-product that blocks the surface as well as the overall oxidation to form CO_2. Thus at least two or perhaps three adsorbed species are concerned in the reaction. Consequently, the use of polycrystalline surfaces has led to a complexity in the system which makes understanding difficult, if not impossible. The use of single crystal faces in the study of these reactions, therefore represents a major step forward in the attempt to understand the kinetics and to design electrocatalysts.

The first published work on formic acid oxidation [23] shows remarkably little effect of crystal orientation, but this has not been confirmed by later work of the same authors [24] as well as work using the technique of electrode preparation described above. The latter showed, in fact, remarkable differences of behaviour of the three low index planes on the first sweep. To avoid modification of the surface structure as far as possible the sweep was limited to an upper potential limit before oxygen adsorption began. The simplest behaviour is that of the (111) plane where a single broad peak extends from 0.2 to 0.8 V (RHE) and is almost the same on the positive and negative sweeps. However the peak current is rather small being about 1.5 mA cm^{-2}. This indicates a rather low catalytic activity for this plane, which has the advantage that it does not lead to the formation of the blocking intermediate. This is indicated by the close similarity of the positive and negative sweeps in the first cycle as well as the fact that continued cycling causes very little change in the voltammogram.

The other two low index planes show much greater activity. The (100) plane is completely blocked in the positive sweep, no current appearing until 0.8 V (RHE). Reversal of the sweep at 0.85 V however leads to a large anodic peak on the negative sweep whose maximum is nearly 10 times that on the (111) plane. Evidently the blocking intermediate is formed rapidly as the electrode contacts the solution. The high activity of this plane results also in the formulation of blocking species which cause a progressive decrease in the peak current on continued cycling. The behaviour of the (110) surface is somewhat similar except that the surface blocking is removed quite sharply on the first positive sweep at 0.66 V to produce a high peak at 0.72 V. After reversal of the sweep the anodic peak is lower and broader. Again continued cycling causes a progressive decay of the currents. As a result of this progressive poisoning of the (100) and (110) planes and the absence of this on the (111), the latter is the best operating catalyst for this reaction in the long term.

The behaviour of the low index planes of Pt in the electrooxidation of methanol is somewhat similar although there are specific differences [26] notably that it is only the (110) plane which shows extensive poisoning and high initial activity and also that there is a greater sensitivity to the anion as indicated by differences in perchloric and sulphuric acids.

A major problem in the study of these reactions is the identity of the blocking intermediate. Attempts have been made to characterize this by determining the charge required to oxidize the adsorbed species. The number of CO_2 molecules produced etc.. These have been carried out using polycrystalline or even platinized platinum electrode and have given somewhat contradictory results. The general view has been in favour of a mixture of species with a COH radical adsorbed on three sites predominatly. More recently direct evidence has been obtained using modulated infra-red spectroscopy [27] which shows that the predominant species is CO adsorbed either on a single site or (to a less extent) in bridged from both in the oxidation of formic acid and of methanol. A predominance of single site adsorbed CO would require an oxidation charge corresponding to 2 electrons per site. This is larger than the charge observed in most previous work but the discrepancy may arise from two causes: the ill-defined nature of the electrodes used or the different time scales of the experiment. Hence it is of great interest to carry out experimentals on well defined surfaces at time scales comparable to the usual electrochemical experiments.

This problem has been solved quite recently in the laboratories at Meudon by Clavilier and Sun. Electrodes prepared as described above are brought in contact with the electrolyte containing the reacting species. They are than transferred to a cell containing the base electrolyte. This transfer is done with the electrode protected by a drop of the electrolyte containing the fuel. After contact with the pure electrolyte under potentiostatic control, time is allowed for the unadsorbed fuel to diffuse away.

A negative sweep is then used to investigate the hydrogen adsorption to determine the fraction of the surface blocked by adsorbed species and the latter is oxidized on the subsequent positive sweep. The form of the oxidation peak is independent of the excursion into the hydrogen region. The area of the peak corresponds closely to a charge of 2 electrons per site in agreement with the proposed single bonded CO. Experiments with CO adsorbed directly yield somewhat similar oxidation peaks although the detailed form depends on the coverage. This demonstrates that CO evidently exists not only under the long term conditions of the early IR experiments. It is notable that the voltammograms on the single crystal have sharp peaks with well-defined fine structure in great contrast to the board unstructered peaks obtained in the dipping experiments with high area platinum electrodes [28]. The full explanation of the nature of the adsorbed intermediate(s) remains to be given as well as of the kinetics of these "simple" oxidation reactions. Nevertheless the possibility of progress is much greater with the use of single crystal electrodes.

CONCLUSION

The work summarized in this paper represents the threshold of a new method of approaching experimental electrochemistry in which electrode reactions can be carried out under well-defined conditions both with respect to composition and structure. The avoidance of empirical procedures used in the past gives the possibility for great progress in fundamental understanding of electrode reactions and the design of specific selective electrocatalysts.

REFERENCES

1. B.E. Conway, H. Angerstein-Koslowska, W.B.A. Sharp and E.E. Criddle, Anal. Chem., 45 (1973) 1331.

2. J. Clavilier and J.P. Chauvineau, J. Electroanal. Chem., 97 (1979) 199.
3. R. Parsons, Croat. Chem. Acta. 53 (1980) 133.
4. J. Clavilier and J. P. Chauvineau, J. Electroanal. Chem., 100 (1979) 461.
5. A. Eucken and B. Weblus,, Z. Elektrochemie, 55 (1951) 144
6. T. Loucka and J. Weber, J. Electroanal. Chem., 26 (1970) App 1.
7. B.E. Conway, H. Angerstein - Kozlowska and F.C. Ho, J. Vac. Sci. Technol. 14 (1977) 351.
8. F.G. Will, J. Electrochem. Soc., 112 (1965) 451.
9. V.S. Bagotsky, Yu. B. Vassiliev and I.I. Pyshnograeva, Electrochim. Acta, 16 (1971) 2141.
10. A.T. Hubbard, R.P. Ishikawa and J. Katekaru, J. Electroanal. Chem., 86 (1978) 271.
11. E. Yeager, W.E. O'Grady, M.Y.C. Woo and P.L. Hagens, J. Electrochem. Soc., 125 (1978) 348.
12. K. Yamamoto, D. Kolb, R. Kotz and G. Lempfühl, J. Electroanal Chem., 96 (1979) 233.
13. P.N. Ross, J. Electrochem. Soc., 126 (1979) 67
14. D. Dickertman, F.D. Koppitz and J.W. Schultze, Electrochim. Acta, 21 (1976) 967.
15. J. Clavilier, R. Faure, G. Guinet and R. Durand, J. Electroanal.Chem., 107 (1980) 205.
16. J. Clavilier, D. Armand and B.L. Wu, J. Electroanal. Chem., 135 (1982) 159.
17. F.T. Wagner and P.N. Ross, J. Electroanal. Chem., 150 (1983) 141.
18. J. Clavilier, J. Electroanal. Chem., 107 (1980) 211.
19. C.L. Scortichini and C.N. Reilley, J. Electroanal. Chem., 139 (1982) 233.
20. S. Motoo and N. Furuya, J. Electroanal. Chem., 172 (1984) 339.
21. J.L. Gland, Surf. Sci., 93 (1980) 487
22. B. Lang, P. Legave and G. Maire, Surf. Sci., 47 (1975) 89.
23. R.R. Adzic, W.E. O'Grady and S. Srinivasan, Surf. Sci., 94 (1980) L191.
24. R.R. Adzic, A.V. Tripkovic and W.E. O'Grady, Electrochemical Society Meeting, May 1981.
25. J. Clavilier, R. Parsons, R. Durand, C. Lamy and J.M. Leger, J. Electroanal. Chem., 124 (1981) 321.
26. J. Clavilier, C. Lamy and J.M.Leger, J. Electroanal. Chem., 125 (1981) 249.
27. B. Beden, A. Bewick, K. Kunimatsu and C. Lang, J. Electroanal. Chem., 142 (1982) 345.
28. A. Capon and R. Parsons, J. Electroanal. Chem., 45 (1973) 205.

TRENDS IN EXPERIMENTAL ELECTROCHEMISTRY ON sp AND sd METAL SINGLE CRYSTAL ELECTRODES

A. HAMELIN

Laboratoire d'Electrochimie Interfaciale du C.N.R.S.
1, Place Aristide Briand, 92195 Meudon Principal Cedex, France.

It was shown that some electrochemical results on solid metal electrodes are more easily understood when obtained with crystal faces, which have a controlled, well-determined, atomic arrangement, than when obtained with polycrystalline electrodes which have an uncontrolled distribution of atoms at their surface. Experimental difficulties to obtain and maintain a chemically clean and physically well-defined interphase will be described. A discussion of experimental results for different faces of sp and sd metals for : equilibrium potential, double layer and electrochemical reactions will be given.

I. INTRODUCTION

This communication is restricted to the case of sp and sd metals (Sn, Pb, Bi, Ag, Au, Cu) and the behaviour of their faces in aqueous solutions. Some simple models of the metal surface can explain some of the results already obtained as long as the interphase is chemically clean and physically well-defined.

On the solution side, a level of impurities which does not interfere with double layer (dl) measurements at a mercury drop (life time : 4 seconds), could completely prevent observations of significant current-potential curves (i(E)) or differential capacity-potential curves (C(E)) at a solid metal electrode which will stay several hours in the same solution. Not only the water, the salts and the galssware must be kept clean but also the gas used to removed oxygen. This is not part of my lecture.

The metal side of the electrochemical interphase must also be rigorously controlled : For crystal faces not only the chemical

A. F. Silva (ed.), Trends in Interfacial Electrochemistry, 83–102.

state, but also the physical state of the top layers of atoms (layers : 0, 1, 2 at least). This is the object of these lectures. Each metal brings specific difficulties : e.g. one oxidizes in air another does not,.... Practical requirements which are satisfactory for one metal are not necessarily valid for another one.

In most cases, the one who tries to apply his elaborate theory to experimental results is not the one who has obtained the data, and only the latter knows how results were selected before publication. This is another difficulty.

II. STRUCTURE OF METAL SURFACES (1, 2).

The characteristic of a single crystal is the periodic nature of its atomic structure which can be related to a network of points in space : the lattice. The coordinates of a given point in a lattice are referred to the crystal axes. The axes form the edges of a parallelepiped called the unit cell which is the fundamental building block of the crystal.

Metals crystalize in the cubic system (body centered, face centered : Au, Ag, Cu, Pb), the hexagonal system (Zn, Cd), the rhombohedral system (Bi) the tetragonal system (Sn). Miller indices are universally used for notation of crystal faces (figure 1). All parallel planes have the same indices. A family of planes is those which are equivalent in the crystal : the six different faces of a cube for instance ; all these have the same atomic configuration. Any two non parallel planes intersect along a line, they are planes of a zone. A set of planes which meet along parallel lines are planes of a zone.

The angular relationships among crystal faces cannot be accurately displayed by perspective drawings but they can be precisely recorded on a stereographic projection. Each face of a crystal can be represented by its pole on a reference sphere, then a map of the sphere is made (fig.2). Stereographic projection of low index planes in a cubic crystal is given figure 3. The high level of symmetry of crystals of pure metals allows all types of planes to be represented on a single triangle : the unit projected stereographic triangle.

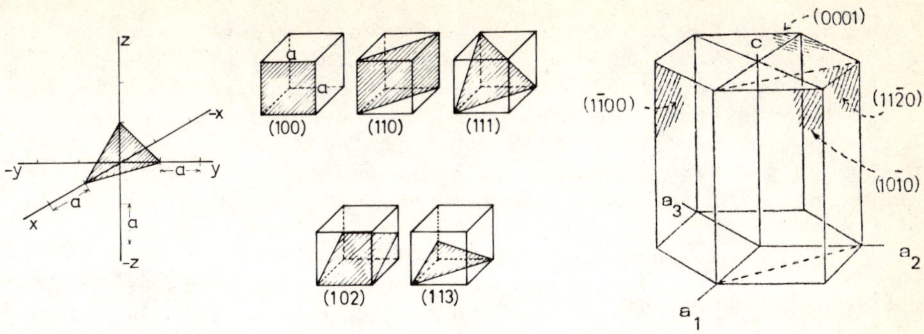

Figure 1. Miller indices for cubic and hexagonal systems.

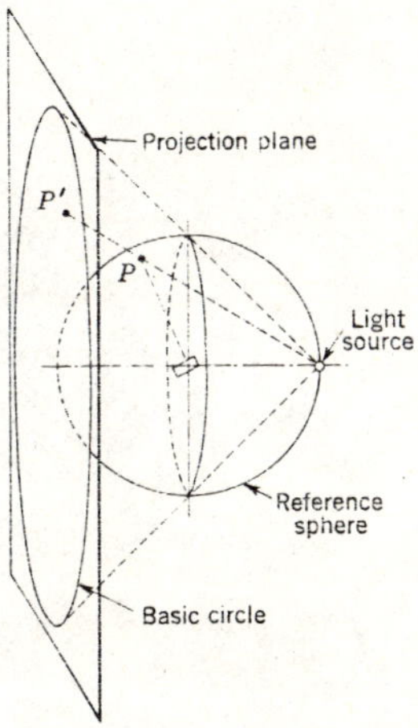

Figure 2. Stereographic projection (1).

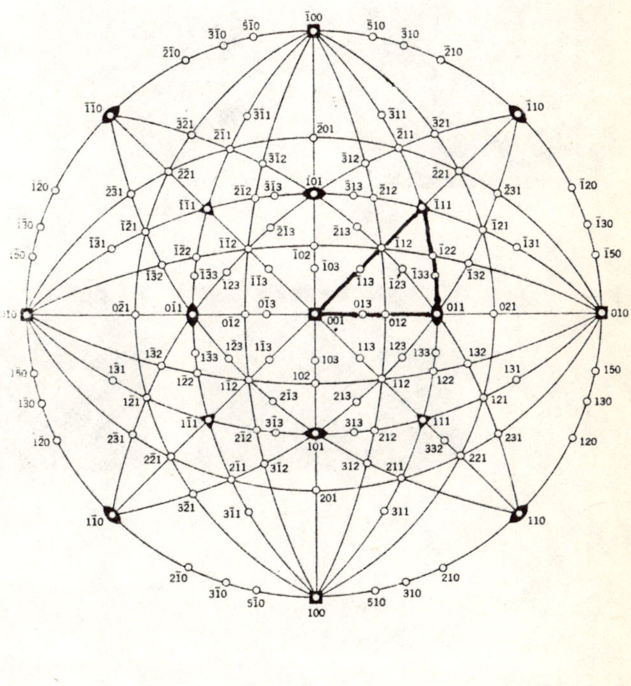

Figure 3. Standard (001) stereographic projection of poles and zones circles for cubic crystals (1).

Assuming that the atoms are hard spheres, any surface can be represented by ball models (diameter of the balls is the spacing between nearest-neighbour atoms). For the face centered cubic (fcc) system , on the three low index faces there is only one type of site where an atom can be added in the nearest neighbour position, they are called singular faces. Surfaces nearly parallel to a singular face are vicinal faces. The vicinal faces can be described by a composition of terraces of singular faces and monoatomic steps. This is the TLK model(terrace, ledge, kink) which is used to describe stepped faces (2). For instance the (221) face which is 15.79° from the (111) face on the zone from (111) to (110) is made of terraces (111) oriented and monoatomic steps denoted (111) because they have three fold symmetry like the (111) face (figure 4).

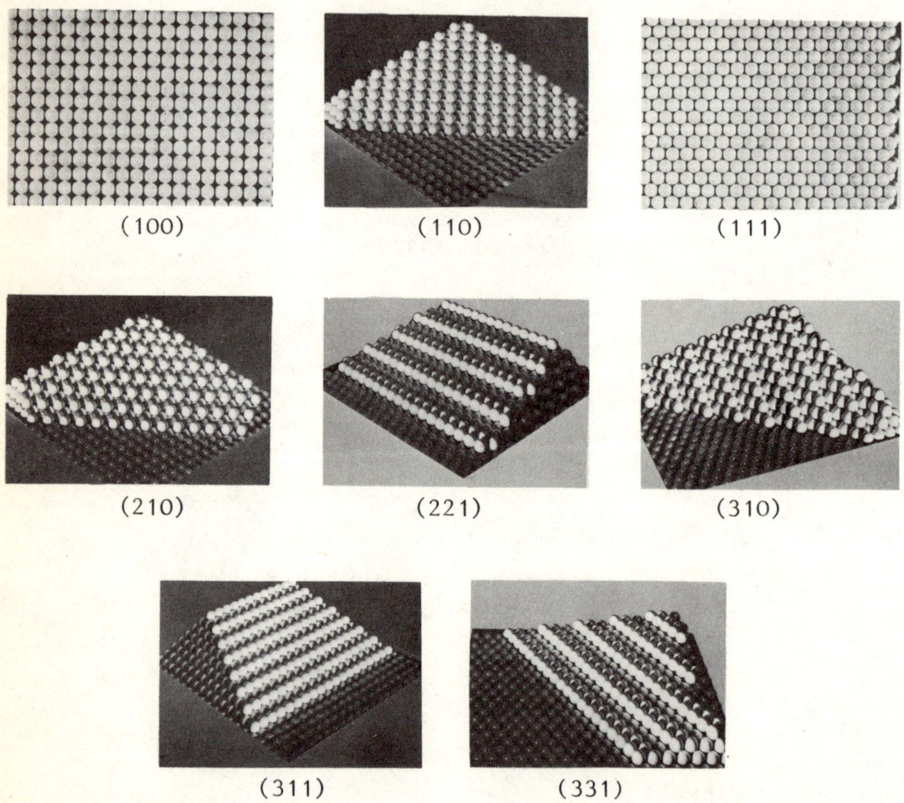

Figure 4. Ball models for 8 faces of the fcc system (2)

Accepting this TLK model, a convenient <u>notation</u> was proposed for stepped surfaces (Appendix of ref.3):

$n(h_t k_t l_t)-(h_s k_s l_s)$ where $(h_t k_t l_t)$ represents the crystallographic orientation of the terraces, $(h_s k_s l_s)$ represents the step face, n denotes the width of the terraces in number of atoms. For instance (221) is 4(111)-(111). Most of the faces which have been used by electrochemists are represented figure 5 with their Miller indices and their notation as stepped faces. Each of the three sides of the triangle seems to be made of two portions where there are systematic changes of the atomic arrangement of the face. The (311) and (210) faces appear as "turning points" on the (111)-(100) and (100)-(110) zones respectively. Are systematic changes of the experimental results in electrochemistry observed for systematic changes of the atomic distribution at the electrode surface? Yes to some extent. But for gold (as for platinum) <u>reconstruction</u> of the first outermost layer of atoms was observed for singular faces (in UHV):then the surface is not just a bulk termination denoted (1x1), Au(100) was approximated by a (20x5) unit cell, Au(110)by a (1x2) unit cell; monoatomic steps would be stabilized by repulsive forces.

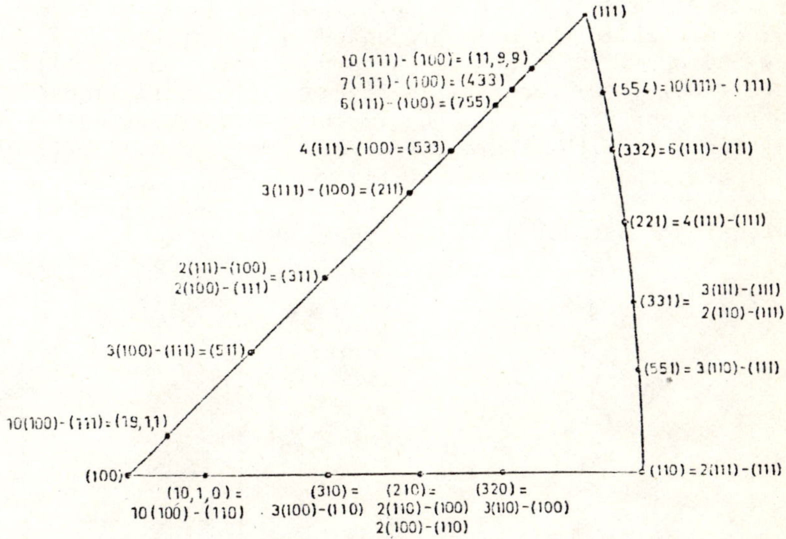

Figure 5. Unit projected stereographic triangle for the fcc structure accepting the notation of Lang et al.(3) : $n(h_t k_t l_t)-(h_s k_s l_s)$.

Besides this atomic aspect of the surface, the electronic aspect has to be considered. The positive ions are immersed in a "gas" of approximately uniform density of free electrons. At the surface the electron density dies off gradually, the electrons spill over the positive ions lattice : an electric double layer is created. Its dipole moment varies with the nature of the metal and with the crystallographic orientation (co) for a given metal. This is why the electronic work function for one metal is different on different faces, it varies by amounts which are sizable fractions of itself. Beside this "spreading" of the electrons perpendicularly to the surface, there is a "smoothing" of the electrons parallely to the surface because on actual surfaces the positive charge is not uniformly distributed. For steps at the surface, the contours of constant electronic density have ridges and valleys corresponding to the positive distribution, but they are smoother. The amount by which the "smoothing" decreases the double layer moment is greater the rougher (on an atomic scale) the surface. Therefore the rougher face has the lower electronic work function. All atomic irregularities lower the work function because they create point dipoles with their positive end outward.

Surface atoms have fewer neighbours than bulk atoms, hence contribute less to the binding energy of the crystal. Surface free energies were calculated assuming short range interactions between surface atoms. The countour lines for constant free energy for fcc metals in the unit projected stereographic triangle (4) show obviously that there is a turning point on the (111)-(100) and (100)-(110) zones as in figure 5. The variations of the surface free energy with co : the γ-plot, is represented figure 6. For this fcc system there are cusps for (100) and (111), (210) and (311) are other extrema as on figure 5.

Defects are neglected in this description.

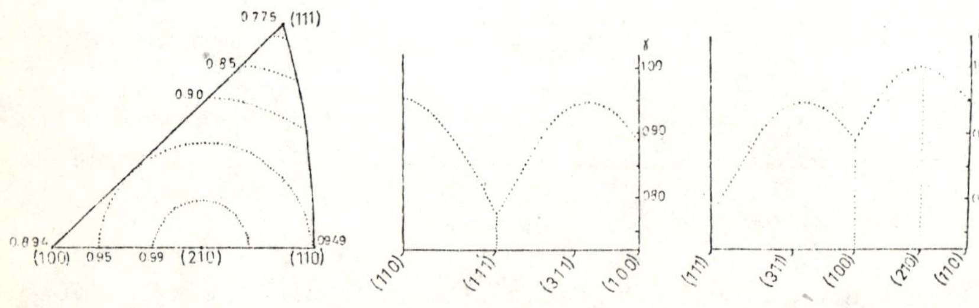

Figure 6. Iso-energy contour lines for the fcc system based on the nearest-broken-bond model (4) and corresponding γ-plots.

III. PREPARATION OF CRYSTAL FACE ELECTRODES (5)

Crystals are generally grown from the melt : in a crucible (Bridgman method) or by the crystal pulling method. They can also be obtained from metal vapour or from the metal electrolytic solution. The determination of the co of the obtained crystal is generally done by X-ray back-reflection Laüe method. The relected X-ray beams are governed by the distribution of atoms within the unit cell following Bragg's law and Laue's equations (1). The experimentater has just to use Greninger's chart for reading angles on back-reflection X-ray films. If the crystal is fixed on a goniometer head,by tilting it of proper angles any chosen co can be set parallel to the film. Then the crystal must be cut parallel to the film.

After cutting, the surface is left with a disturbed layer on top which is eliminated by mechanical polishing followed by electrochemical polishing. As only the face of interest has to be in contact with the solution,it has to be isolated. Three possibilities are shown on figure 6, each of which has its advantages. The "hanging electrolyte" (6) requires no isolating material, where used just after annealing in a torch and cooling in water (7) a non-contaminated interphase can be obtained.

Then a final surface preparation is necessary : either a slight hydrogen evolution (silver) or deposition of a monolayer of oxide which can be subsequently reduced (gold). This final surface preparation is at the same time a check of the quality of the electrochemical interphase.

Ex-situ check of the state of the electrode surface can be done in UHV but then there is a transfer problem. In-situ checks can be either the observation of dl parameters (potential of zero charge (pzc), shape of the contribution of the diffuse part of the dl on the C(E) curves, stability of the curves, comparison of i(E) and C(E) curves in the dl region,...) or other methods such as optical methods.

For gold the shape of the oxidation (and reduction) current peaks gives information on the chemical and physical states of the interphase.

IV. ELECTROCHEMICAL RESULTS

IV.1. Equilibrium potential

For two different crystals of one metal M having a different face hkl (and h'k'l') in contact with a solution of its salt M^+ the difference of inner potential is :

$$\varphi_{hkl} - \varphi_{h'k'l'} = \frac{\bar{\mu}_e^{hkl} - \bar{\mu}_e^{h'k'l'}}{|e|} = \frac{\bar{\mu}_{M^+}^{hkl} - \bar{\mu}_{M^+}^{h'k'l'}}{z|e|}$$

$\bar{\mu}_e$ electrochemical potential of the electron in M, $\bar{\mu}_{M^+}$ electrochemical potential of positive ions in M. In the two pieces of M, the chemical potential of the electron is the same. The difference of the chemical potentials of the positive ions in M is the difference of the sums of the energies of all the atoms of each piece of M :

$$n_{hkl}\,\mu_{M^+}^{hkl} - n_{h'k'l'}\,\mu_{M^+}^{h'k'l'} = \Delta(SdT + PdV + \gamma dA)$$

S,T,V,P entropy, temperature, volume, pressure of each crystal ; γ is the surface tension, A the area of the face. Only the last term differs from one piece of metal to the other because of the different face exposed ; as the number of atoms at the surface is very small, compared to the total number of atoms of each piece of M, it does not interfer (this will not be true for very thin films,for instance). Thus at equilibrium :

$$\varphi_{hkl} - \varphi_{h'k'l'} = \frac{\bar{\mu}_{M^+}^{hkl} - \bar{\mu}_{M^+}^{h'k'l'}}{z|e|} = \frac{\bar{\mu}_{M^+}^{s} - \bar{\mu}_{M^+}^{s}}{z|e|} = 0$$

$\bar{\mu}_{M^+}^{s}$ electrochemical potential of positive ions of M in solution.

This was observed experimentally (8).

IV.2. Potential of zero charge

When no charge transfer occurs across the interphase, only a change of the electrostatic charge on the metal (and correspondingly in solution) takes place. This can be observed in a more or less large region of potential according to the nature of the metal, the solvent, the ions in solution. It is the double layer dl region. It is nearly the same for all co of one metal.

When a metal is dipped into solution the charges developped at the surface are either positive or negative according the potential imposed to the metal electrode.

For one value of potential no charge exists on the metal surface, it is the potential of zero charge (pzc). As for other solid metal electrodes, for crystal faces the pzc can be determined by observation of the interference of the diffuse part of the dl on the C(E) curves in dilute solutions. The pzc varies with the co because it is linearly related to the electronic work function, which depends on the moment of the electric double layer at the metal surface (see II). The pzc may be in the dl region for all co, or for only some of them.

Pzc values are summarized in figure 7 as a function of the melting point of the metal, solid lines show to what extent the pzc varies with co. The co influences largely the pzc for high melting metals and has less influence for low melting metals (this is probably due to the mobility of the surface atoms at room temperature for low melting metals).

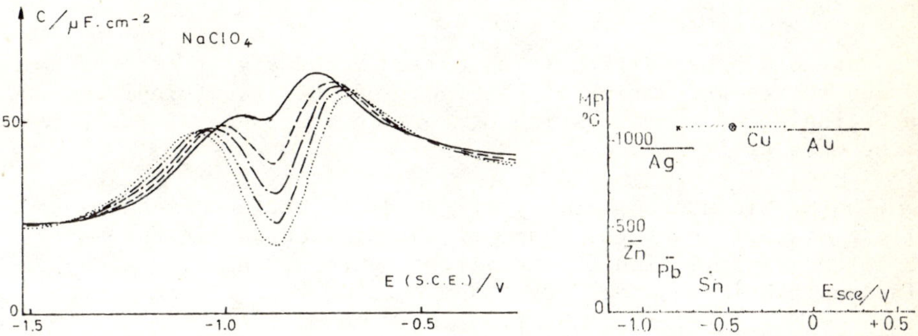

Figure 7. C(E) curves for Ag(100) in aqueous solutions of $NaClO_4$: 100, 40, 20, 10 and 5 mM (9). Anisotropy of the pzc of different metals as a function of their melting point ; xx : pzc of Cu(111) and (100) obtained recently by J. Lecoeur.

The surface energy varies with co (see II). For a large number of gold faces the pzc was determined ; it varies with co, the parallelism of the variations of the two parameters - one calculated, the other experimental - is very good (10). The "spreading" of the electrons and the "smoothing" by a lateral flow of electrons explain the decrease of the pzc from smooth to atomically stepped surfaces. Obviously the most densely packed face should have the most positive pzc. Any atomic irregularity will lower the pzc.

IV.3. Double layer on the solution side of the outer-Helmholtz-plane (5,10)

The variations of the capacity of the diffuse part of the dl with potential has the shape of an hyperbolic cosine (Gouy-Chapman theory for symmetrical electrolytes), the minimum of capacity corresponds to zero charge density on the electrode (figure 7).

For no (or slight) adsorption and dilute solution, the solution side of the oHp gives rise to a depression on the C(E) curves, deepening with dilution of the solution. This is clear as long as the charge is uniformly distributed on the surface of the solid electrode (as it is on mercury). This is not the case for polycrystalline electrodes of high melting metals.

For high index faces a number of steps and kinks are necessary for the average specified co, they alter the distribution of charge at the surface and,on the C(E), the part of the curves which reflects the contribution of the diffuse part of the dl. However the contribution of the dl on the solution side of the oHp is observable.

For gold other difficulties exist: Secondary effects probably due to reconstruction of the faces of the lowest indices, and the beginning of the oxidation of gold in the range of potentials of the more positive pzcs.

The atomic non-uniformity of the distribution of charge entails a distortion of the shape of the C(E) curves in the part due to the contribution of the diffuse part of the dl ; then blind application of the Gouy-Chapman theory would be a mistake.

IV.4. Double layer on the metal side of the outer Helmholtz plane (5,10)

In the case of no adsorption, on the metal side of the oHp there are only water dipoles. The inner layer capacity C^i varies with the charge density σ; according to the accepted model, their variations should be independent of the concentration of the solution. $C^i(\sigma)$ curves were published for Bi, Ag, Au (5,10) and lately for Au(210) in KBF_4 (figure 8). These curves describe the metal-water interactions.

Figure 8. $C_i(\sigma)$ curves for Au(210) in aqueous solutions of $NaBF_4$ (3 concentrations) from the work of A. Hamelin and Z. Borkowska.

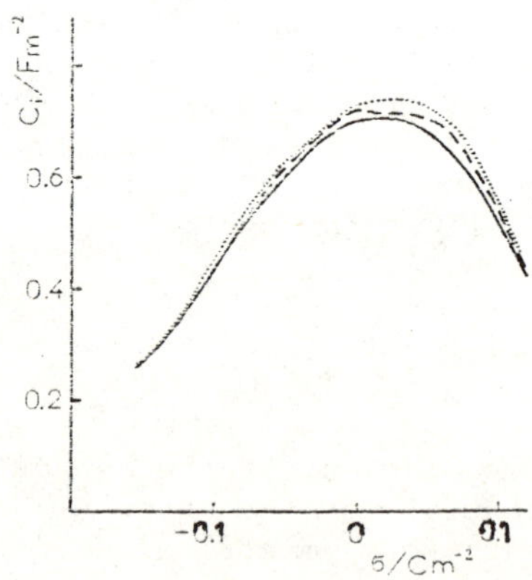

When there is specific adsorption (preadsorbed water displacement) these curves describe also the metal-ion interactions.

For adsorption on crystal faces the phenomenology is qualitatively what it is for other electrodes (5) : with increasing concentration of the anion in solution, on the C(E) curves progressive disappearance of the contribution of the diffuse part of the dl and negative shift of the adsorption capacity peaks are observed ; as it was for non-crystal electrodes.

The size of the ion influences the C(E) curves as for non-crystal electrodes : the bigger is the ion, the more it adsorbs (figure 9). The shift of potential is the same for all faces of one metal.

For one metal in a given medium what differs significantly from one face to the other, is the shape of the C(E) curves : quite distinctive patterns are observed experimentally (figure 9). One face has the same pattern in all solutions. Generally three capacity peaks are observed in the dl region, for some co they merge, then one peak can be only a shoulder on another peak. For silver in chloride, thermodynamic analysis was done for the (110) face : the most negative peak was shown to be possibly due to reorientation of water dipoles in the inner layer, the middle peak occurs in the range of half coverage and has been interpreted as being caused by the usual type of adsorption effect, the third peak could be due either to reorganisation of the adsorbate, or to reconstruction of the face or to charge transfer. For bromide or iodide on gold faces, adsorption is much more stronger;no effect of reorientation of the solvent can be observed. But for gold,secondary effects (12) complicate the results :
– for the (100) face, hysteresis of the C(E) curves (which diminishes when the negative limit of the explored range of potential is shifted positively) (figure 10) ;
– for the (111) face of gold, hysteresis of the C(E) curves (which is not influenced by the negative limit) ;
– for the (110) face an extra frequency dispersive peak.
Reconstruction of the surface was proposed as an explanation for these observations (12).

Understanding of ionic adsorption is easier when considering the influence of the co in a general way : to systematic changes of co correspond systematic changes of the pattern of the C(E) curves (figure 11) (13). The TLK model (see II) explains these systematic changes, this idea is in agreement with results obtained for underpotential deposition (UPD) (see IV-5).

When neutral molecules interact with the metal electrode the necessity to have an ionic base electrolyte in solution as well as the neutral substance complicates the phenomenon observed. The

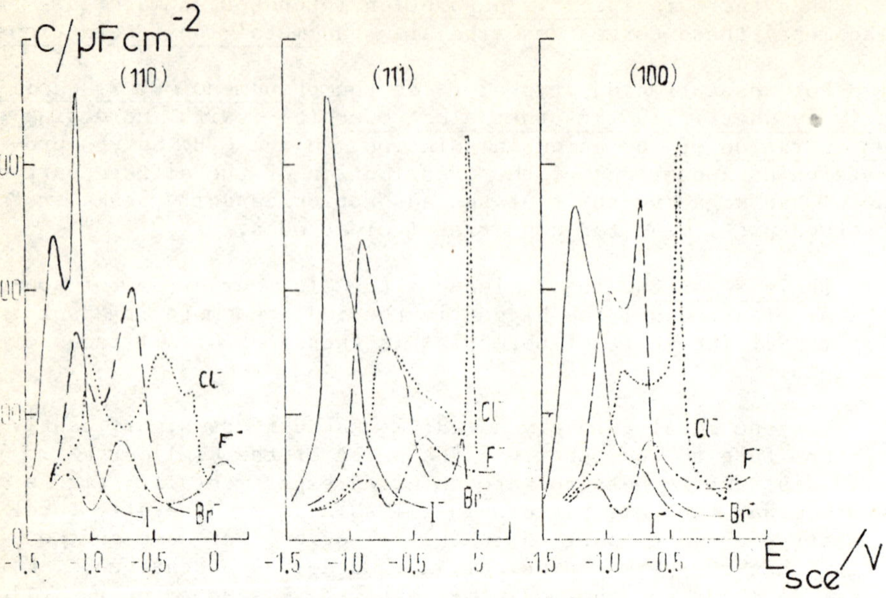

Figure 9. C(E) curves for Ag(111), (100) and (110) in fluoride, chloride, bromide and iodide aqueous solutions (11).

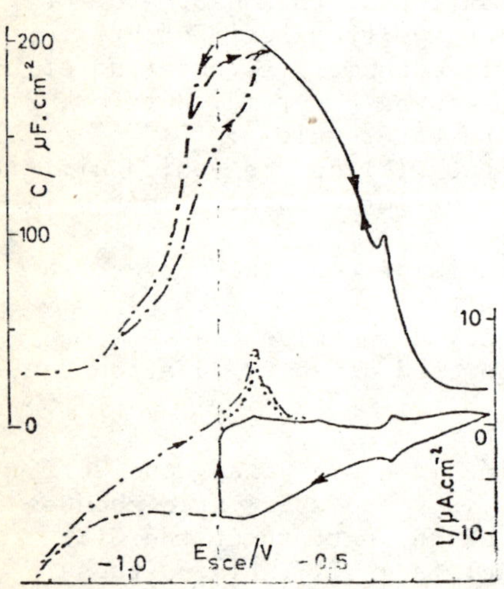

Figure 10. C(E) and i(E) curves for Au(100) in 0.1M KI for different negative limits of the range of potential (12).

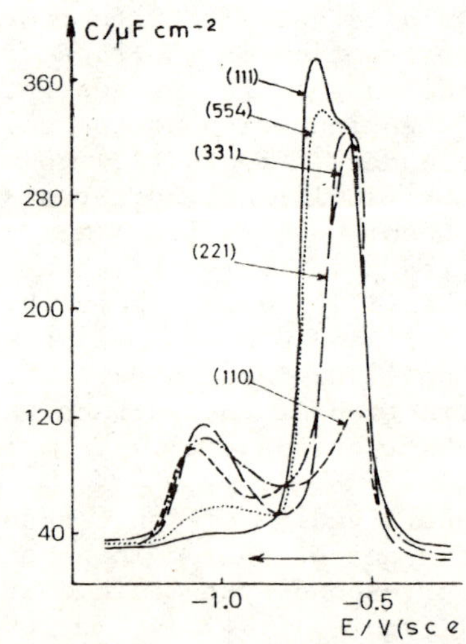

Figure 11. C(E) curves of five faces of the (111)-(110) zone in 100 mM KI (13).

anion of the base electrolyte may adsorb, but if the adsorption of the neutral substance occurs in a range of potential in which the charge due to adsorbed anions is very small, analysis of the phenomenon can be done (14). However for : temperature, concentration (figure 12), structure of the adsorbate, the phenomenology is qualitatively the same that for polycrystals. The C(E) pattern is close to what is expected. For polycrystalline electrode of high melting metals the pattern of the observed C(E) curves can be understood only from results obtained with faces of the same metal.

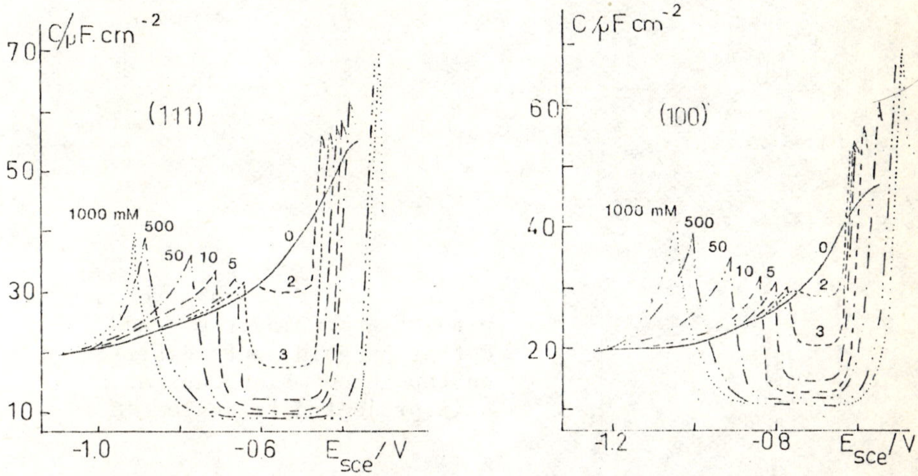

Figure 12. C(E) curves for adsorption of n-isobutanol in 50 mM Na_2SO_4 on Ag(111) and (100) (15).

IV.5. Underpotential deposition of a monolayer of a foreign metal on crystal faces

A general review of the phenomenon of UPD was published in 1978 (16). A pronounced influence of the substrate surface structure on the UPD of a foreign metal monolayer was already observed in 1974. When UPD occurs with complete charge transfer all along the UPD potential range and no alloy formation, the understanding of the i(E) curves (cyclic voltammogram) in this range is clear ; then the influence of the co of the substrate can be analysed. This is the case for electrochemical adsorption of a submonolayer or a monolayer of lead on gold faces (18,19).

For faces distributed on the three main zones of the unit projected stereographic triangle the electrochemical results are easily understood when divided in five groups. The (111), (100), (110), (210) and (311) faces display extreme electrochemical behaviours ; and in between these faces for systematic changes of the atomic arrangement at the gold surface, systematic changes of the i(E) curves are observed (figure 13). These experimental observations are explained by the TLK model of the substrate surfaces. Obviously the more negative peak is due to adsorption on steps, its amplitude increases from (111) to (110) and the other important peak is due to adsorption of lead on terraces of the gold substrates.

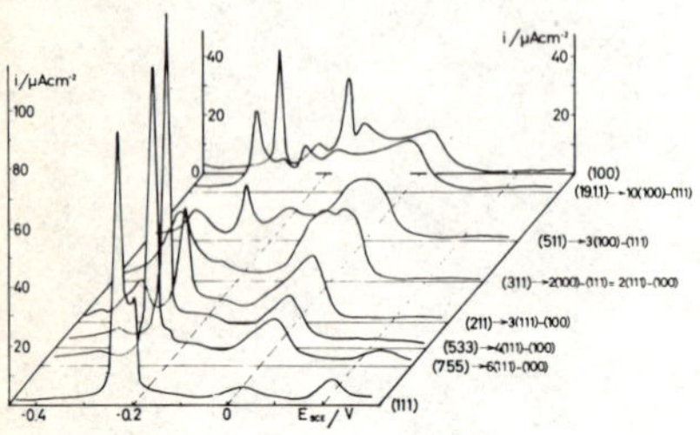

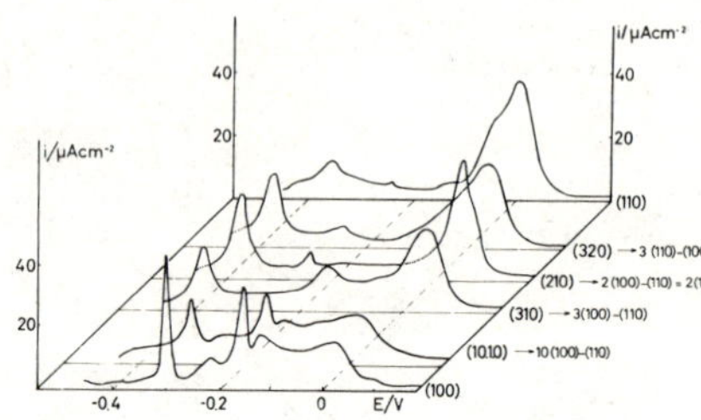

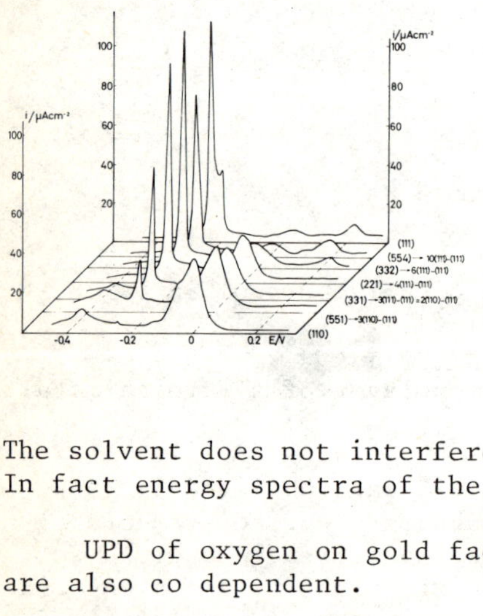

Figure 13. i(E) curves for stripping of a monolayer of lead from different gold faces distributed on the three main zones of the unit projected triangle (18,19).

The solvent does not interfere in this case of adsorption (19). In fact energy spectra of the faces are observed.

UPD of oxygen on gold faces, hydrogen on platinum faces,... are also co dependent.

Results are more understandable on faces that on polycrystalline metal surfaces which have a patchy uncontrolled density of charge at their surfaces.

IV.6. Electrode reaction kinetics on crystal faces.

The influence of the electrode material on the kinetics of electrode reactions was described (20). For metals two limiting types of reactions can be considered :

- the reactants form only the weakest of bonds with electrode surface, then there is no influence of the nature of the metal and -of course- no direct influence of the co of the metal surface, although the structure of the dl could affect the rate (MnO_4^{2-}/MnO_4^- , Fe^{2+}/Fe^{3+},...);

- the reactants or intermediates form strong bonds with the electrode surfaces (adsorption), then the nature of the metal has an effect; for the influence of the co several possibilities have to be considered.

First, let us consider two extreme cases: Either the reaction occurs in a range of potential in which the density of charge is highly negative for all faces of the metal (the "spreading" of the electron gas out of the lattice of positives cores masks the different distributions of atoms at the different faces (see II)) (range a on figure 14), or the reaction occurs in a range of potentials in which the density of charge is positive for all faces (range b on figure 14). In the first case (a) only the structure of the electrochemical dl (although maintained constant by an excess of supporting electrolyte) could explain different reaction rates on different faces, but structure of the solvent in the inner layer is nearly the same for all cos in this case. While in the second case (b) the structure of the surface will introduce differences.

Intermediates cases could also exist(for instance ranges c and d on figure 14).

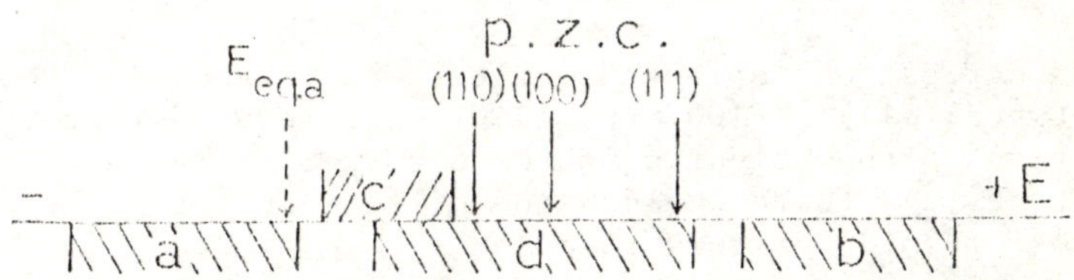

Figure 14. For fcc metals: positions of the pzcs on the potential axis and ranges of potential for four types of reactions : a, b, c and d.

The equilibrium (Nernst) potential of an electrochemical reaction is not co dependent (see IV.1). The pzc of a metal is co dependent especially for high melting metals (see IV.2), but it was demonstrated that -in case of metal electrodes without adsortion - in the reaction rate equation the difference of the pzcs is compensated by the difference of the electronic work functions (they are equal) (21). From these considerations it seems that for reactions of type (a) there will be no influence of the co of the metal surface on the reaction rate. For types b, c and d there could be an influence of the co on the observed reaction rate. Of course experimental results depend on the chemical and physical states of the interface (see I and II) but nowadays some experimentaters are able to have chemically clean and physically well-defined interfaces.

Many reactions were shown to be face dependent. Let us examine the case of gold faces . Recently results for reduction of solvated protons (hydrogen evolution) on gold faces were obtained in 1 M perchloric acid solution (22). For the (111), (110)(210), (311) and (100)faces,curves are represented figure 15 a,b.Slight differences are observed , are they significant?

Let us consider the curve calculated from experimental results of faces, for a model polycrystalline electrode of a fcc metal. It is assumed that only the five faces which have shown an extreme electrochemical behaviour exist at the surface and in the proportions they are in the stereographic projection of figure 3 ,i.e : 0.108 for (111), 0.162 for (110), 0.081 for (100), 0.324 for (311)and 0.324 for (210). This approach postulates that the reaction does not occur only on special sites (dislocations for instance). For reduction of solvated protons the calculated curve is compared to experimental results of a polycrystalline sample in figure 15 c. Is the difference between calculated and experimental curves due to grain boundaries or experimental uncertainties? In this set of experiments all crystal faces were electropolished and the polycrystalline sample were not , is this the explanation of the difference? On the other hand no roughness factor was taken into account (and a perfectly flat surface does not exist).

When correction for the dl effect was introduced, the difference between (111) , (100) and (110) were reduced (figure 15 d). As reduction of the solvated proton occurs in a range of potential of type (a), there would be no difference among the different faces of gold for this reaction.

On the contrary for the oxidation and the reduction of photogenerated atomic hydrogen in perchloric acid,differences larger than experimental uncertainties were observed between the curves of the (111) and (110) faces of gold, (figure 16)(23). These reactions occur in a range of potential of type c, when charge density is very negative for (111) and slightly negative for (110). This could explain the observed differences.

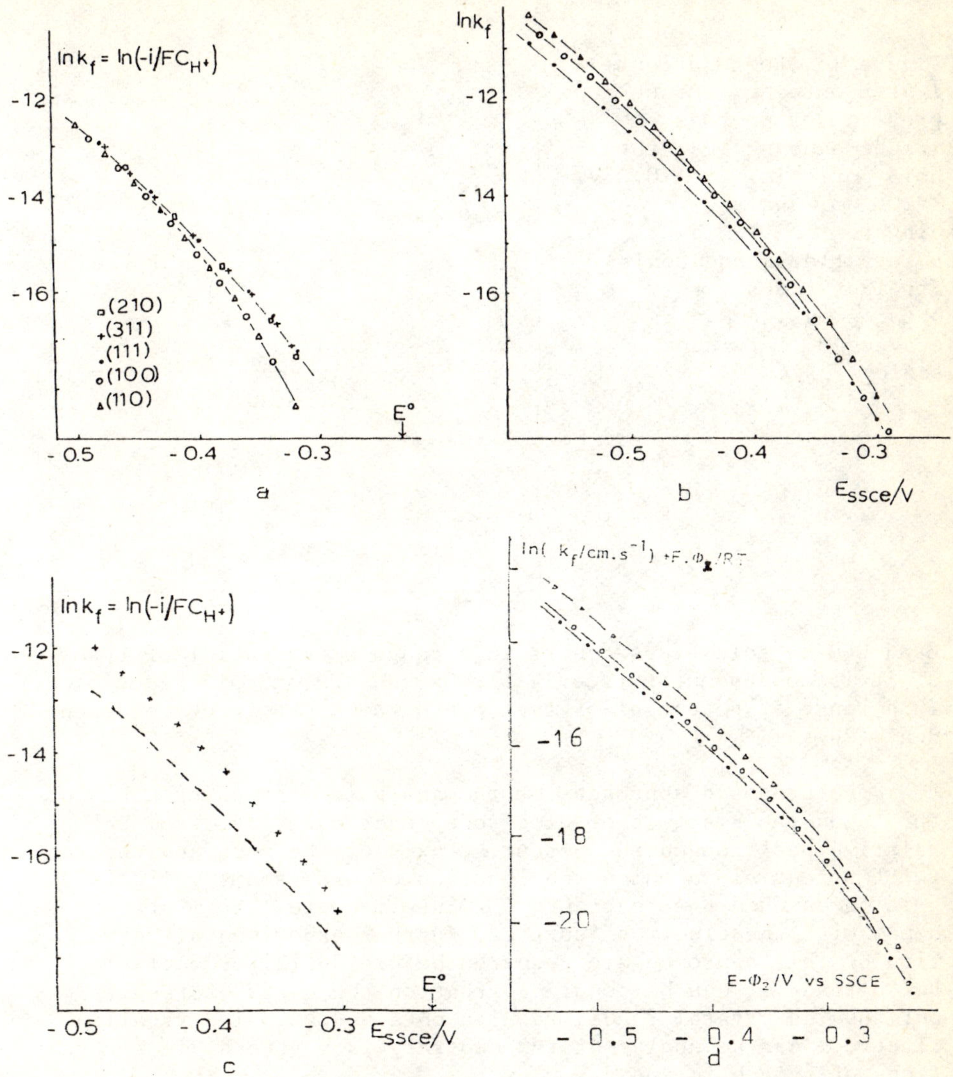

Figure 15. Logarithm of forward rate constant for reduction of solvated protons on gold faces: a)potentiostatic sweep measurements,b) potential step measurements, c) comparison of calculated and experimental curves for polycrystalline gold, d) curves corrected for the dl effect for (111), (100) and (110) faces of gold (22).

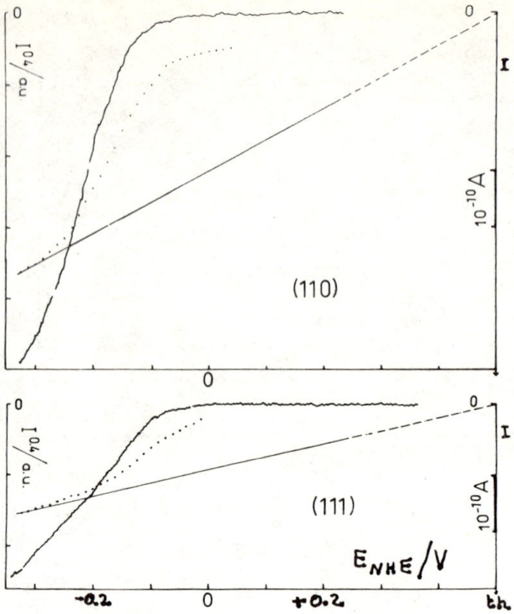

Figure 16. Photocurrent I vs E obtained in the presence of protons (right-hand scale), threshold potential (th)$\pm$0.03V, λ=300 nm, analysis according to the 5/2 law : points (left hand scale). Straight lines : results obtained using N_2O as scavenger (left-hand scale) (23).

Lead UPD on gold faces occurs in a range of potential of type d, it is co dependent (see IV-5). Oxygen UPD on gold faces occurs in a range of potential of type b and was shown to be co dependent (24).

The proposed approach for the analysis of the influence of the co on the kinetics of electrode reactions (i.e. from the relative position on the potential axis of the pzcs and the range of potential in which the reaction occurs) seems valid for gold faces when the reactants (or intermediates) form strong bonds with the electrode surface. Further experimental verification of this approach are required before definite conclusion. But already it can be concluded that on the basis of results obtained on crystal faces, results obtained on polycrystalline electrode can be analysed; the contrary is not possible for reactions of type b, c and d.

ACKNOWLEDGEMENTS.

The authors is indebted to Dr. Roger Parsons for corrections and fruitful discussions.

REFERENCES.

(1) Barrett, C.S. and Massalski, T.B., 1966,"Structure of Metals: Crystallograhic Methods, Principles and Data", Mc Graw-Hill, London.

(2) Nicholas, J.F. 1965,"An Atlas of Models of Crystal Surfaces", Gordon and Breach, Sci. Pub., New-York.

(3) Lang, B., Joyner, R.W. and Somorjai, G.A., 1972, Surf. Sci., 30, p. 454.

(4) Mac Kenzie, J.K., Moore, A.J.W. and Nicholas, J., 1962, J. Phys. Chem. Solids, 23, p. 185.

(5) Hamelin A., 1985, in Modern Aspects of Electrochemistry, Vol. 18, "Double Layer Properties at sp and sd Metal Single Crystal Electrodes" , to be published.

(6) Dickertmann, D.,Koppitz, F.D. and Schultze, J.W., 1976, Electrochimica Acta, 21, p. 967.

(7) Clavilier, J.,Faure, R., Guinet, G. and Durand, R., 1980, J. Electroanal. Chem. , 107, p. 205.

(8) Jenkins, L.H. and Bertocci, U., 1965, J. Electrochem. Soc., 112, p. 507.

(9) Valette, G., 1982, J. Electroanal. Chem., 138, p. 37.

(10) Hamelin, A., Vitanov, V., Sevastianov, E. and Popov, A.,1983, J. Electroanal. Chem. , 145, p. 225.

(11) Valette, G., Hamelin, A. and Parsons, R., 1978, Z. Phys. Chem. NF, 113, p.71.

(12) Hamelin, A., 1982, J. Electroanal. Chem., 142, p. 299.

(13) Hamelin, A. and Bellier,J.P.,1978, Surf. Sci., 78, p. 159.

(14) Lipkowski, J. Nguyen Van Huong, C., Hinnen, C., Parsons, R. and Chevalet , J., 1983, J. Electroanal. Chem., 143, p.375.

(15) Vitanov, T. and Popov, A., 1975, Trans. SAEST, 10, p.5.

(16) Kolb, D.M., 1978, in Adv. in Electrochemistry, Vol. 11, p. 125.

(17) Adzic, R., Yeager, E. and Cahan, B.D.,1974, J. Electrochem. Soc., 121 ,p. 474.

(18) Hamelin, A. 1984, J. Electroanal. Chem. ,165, p. 167.

(19) Hamelin, A. and Lipkowski, J., 1984, J. Electroanal. Chem.

(20) Parsons, R., 1964, Surf. Sci., 2, p. 418.

(21) Frumkin, A.N., 1965, Soviet Electrochem., 1, p. 341.

(22) Brug, G.J. , Sluyters-Rehbach, M., Sluyters, J.H. and Hamelin, A., 1985, J. Electroanal. Chem.,

(23) Hamelin, A., Picq, G. and Vennereau, P., 1983, J. Electroanal. Chem., 148, p. 61.

(24) Hamelin, A. and Sotto, M., 1970, C.R. Acad. Sci. Paris, Série C, 271, p. 609.

STRUCTURE OF THE DOUBLE LAYER AND RATES OF ION CROSSINGS AT "SINGLE" IMMISCIBLE LIQUID/LIQUID (L/L) INTERFACES

Jonathan D. Reid,[a] Owen R. Melroy,[b] William E. Bronner, Harold C. Hughes,[c] Petr Vanysek[d] and Richard P. Buck

Department of Chemistry, University of North Carolina
Chapel Hill, North Carolina 27514

[a]IBM Laboratories, Endicott, New York
[b]IBM Watson Laboratory, San Jose, California
[c]Western Australia Institute of Technology,
Hayman Road, Bently South, W. Australia 6102
[d]On Leave: J. Heyrovsky Institute, Prague, Czechoslov.

TABLE OF CONTENTS

A. F. Silva (ed.), Trends in Interfacial Electrochemistry, 103–185.

INTRODUCTION TO LIQUID/LIQUID (L/L) INTERFACE STUDIES

Separation of bulk and interfacial kinetic properties for two-interface, ion-transferring, ion-exchanging membrane systems is not theoretically possible in all situations. Consequently, detailed analyses of potential, field, and concentration profiles at single interfaces, such as aqueous electrolyte/ synthetic ion exchangers and aqueous electrolyte/immiscible organic electrolytes must be achieved by other methods. The obvious choice is use of thick phases such that electrodes can be placed within the phases on each side of the single interface under study. This option is not generally available for biological membranes, many polymer films, and synthetic ion exchange membranes. Use of immiscible L/L interface systems that mimic properties of membranes seems to open possibilities for determination of processes and electrochemical conditions at single interfaces.

In 1902, Nernst and Riesenfeld (1) began a study of ion crossings at immiscible L/L interfaces in the absence of supporting electrolytes. Guastalla pioneered the field in more modern times.(2-6) In the last 10 years this topic has been pursued in a fundamental way. Frumkin's last work (7-14) renewed interest, and research began anew in a few laboratories around the world: Gavach et. al. (Montpellier) (15-28), Koryta et. al. (Prague) (29-36), Samec et. al. (Prague) (37-48), Senda et. al. (Kyoto) (49-54), Buck et. al. (Chapel Hill) (55-67), Koczorowski et. al. (Warsaw) (69-75), Hundhammer (Addis Ababa) (76), Fujinaga et. al. (Tokai, Japan) (77-78), and Freiser (Tucson) (79-80). The electrochemical phenomena at the interface of two immiscible, ion-exchanging electrolyte solutions have been reviewed by Koryta et. al. (30-35) Most recent reviews are by Koryta (81) and Vanysek and Buck (82).

Structure of the double layer, (charge, potential, field) at L/L interfaces is a controversial topic, and the resolution is essential to correct interpretation of ion-crossing kinetics. For single or multiple partitioned salt systems, ($NR_4^+X^-$) at water-nitrobenzene interfaces for example, single-ion free energies in each phase determine equilibrium space charge and, in part, control potential differences, (pds). For nonequilibrium conditions, the overpotential may be entirely accounted for in space charge perturbations, or as pds in compact layers in each phase, or as combinations of both. The compact layers may be composed of solvent, adsorbed ions and adsorbed salts. The one- and two-ion transport problems at constant current have been treated for a variety of kinetic conditions, including reversible cases, with and without additional supporting electrolyte in each phase.(61-64) There is not, as yet, an analysis of responses to compact layers formed by salts that accumulate during current flow, e.g., compact layers that are a function of time. Chronoamperometric and cyclic voltammetric theory for redox processes at metal electrodes in supporting electrolytes has been used for L/L interfaces without modifications.(61) These theories apply only for a few, but practical L/L situations.

Experimental methods for studying rates of ion transport across L/L interfaces are timely for four reasons: (1) kinetic barriers (e.g., dependences on activities, interfacial pds, and temperature) for ion crossings need measurement to determine relationships that suggest rate-determining processes; (2) rates of interfacial ion crossings are presumed to reflect local quantities such as potential and concentrations in diffuse and compact layers at the interface. These structures are not presently known in great detail for L/L interfaces; (3) secondary processes such as involvement of homogeneous reactions during interfacial processes (e.g., rates of interfacial reactions between aqueous ions and oil-soluble translocators) may become directly measurable or calculable using conventional supported electrolyte transport theory of coupled processes; and (4) processes at L/L interfaces may resemble processes between electrolytes and cell or tissue components.

In this paper we present theory and experiments for the equilibrium, non-polarizable reversible interfaces and for completely blocked, polarizable interfaces. As examples of equilibrium experiments, electrocapillary curves of reversibly-partitioned salts, tetra-n-alkylammonium bromides (methyl through butyl) are presented.(65) Analysis of charge and capacitance is used to demonstrate the absence of large dipolar, compact layers and corresponding potential component (the chi or χ-potential). Results of Senda et. al. (54) on the electrocapillary behavior of the nearly blocked LiCl (water) TBATPB (nitrobenzene interface), TBA = tetrabutylammonium, TPB = tetraphenylborate are considerably more precise. Their data suggest a residual dipole potential at the pzc of about 20 mV. Samec et. al. (44) estimated the same quantity at 42 mV.

In the second part, small ac perturbation methods applied to the equilibrium interface provide direct measurement of the resistive and capacitive character of the interface. The results are quite different for a polarizable, blocked interface (66) such as LiCl(aqueous), TPAsTPB (nitrobenzene), TPAs = tetraphenylarsonium compared with a reversible, unblocked interface.(67) This work leads directly to large amplitude perturbation methods in Part 3. Although some potentiostatic results will be mentioned, the bulk of accumulated data have been obtained by constant current and cyclic voltammetric applications. Finally, some results of chemically modified interfaces using adsorbed proteins will be mentioned.

The question of electrochemical reversibility or irreversibility of ion transfers can be addressed theoretically. However, experimentally it is not clear whether the majority of ion transfers studied to date, mainly tetraalkylammonium ions, are reversibly or quasi reversibly transferred from water to nitrobenzene. It is possible to obtain rate constants and transfer coefficients using small voltage step applications (polarization method) at the foot of the wave. However, the results do not agree very well with the same parameters obtained using large amplitude methods, particularly constant current, in which both interfacial kinetics and mass transport are involved. Defects in experimental cell design and formation of resistive films at the interface during current flow (68) may be involved.

Equilibria at Electrified L/L Interfaces: Neutrals, Single Salts and Two Salts

The local equilibrium condition for neutral solvent species at constant temperature and pressure is equality of chemical potentials at the interface. The thermodynamic partition, or distribution coefficient is

$$K = \bar{a}/a = \exp[(\mu^o - \bar{\mu}^o)/RT] = \exp\{-\Delta G(w \rightarrow o)/RT\} \tag{1}$$

Bars indicate organic phase quantities and μ^o's and $\Delta G(w \rightarrow o)$ are equivalent symbols for partial molar free energies and the free energies of partitioning from water (w) to oil (o) phase, mutually saturated. For species of charge z, constancy of electrochemical potential is presumed to hold locally near the interface, and is usually applied to activities outside the space-charge region. The single ion partition coefficient is defined as

$$K_{\pm} = (\bar{a}/a)\exp[zf(\bar{\phi} - \phi)] = \exp[(\mu_{\pm}^o - \bar{\mu}_{\pm}^o)/RT] \tag{2}$$

where z is the ionic charge with its sign. This quantity is also designated as the medium effect coefficient in other literature.(83,84) In either phase, from equality of electrochemical potential at any point:

$$a(x) = a\exp[zf(\phi - \phi(x))] \tag{3}$$

$$\bar{a}(x) = \bar{a}\exp[zf(\bar{\phi} - \bar{\phi}(x))] \tag{4}$$

$$f = F/RT$$

When ions of both sign are partitioned at equilibrium, the usual salt partition coefficient is related to the single ion coefficients:

$$K_+K_- = \bar{a}_+\bar{a}_-/a_+a_- = K^2(\text{salt}) = \exp[(\mu_+^{o}+\mu_-^{o}) - (\bar{\mu}_+^{o}+\bar{\mu}^{o})]RT \tag{5}$$

whereas the interfacial potential for a single, reversibly partitioned ion is simply

$$\bar{\phi} - \phi = (1/zf)\ln(K_{\pm}a/\bar{a}) \tag{6}$$

When both ions partition, the potential takes the form:

$$\bar{\phi} - \phi = (1/zf)\ln\left[\frac{K_+a_+ + \bar{a}_-}{K_-a_- + \bar{a}_+}\right] \tag{7}$$

This equation simplifies, in the normal circumstance that a single salt is partitioned from and to electroneutral phases:

$$\bar{\phi} - \phi = (1/2zf)\ln\left[\frac{K_+\gamma_+\bar{\gamma}_+}{K_-\gamma_-/\bar{\gamma}_-}\right] \tag{8}$$

because $\bar{c}_+ = \bar{c}_-$ and $c_+ = c_-$, a result noted by Karpfen and Randles (85). This result is unusual in that the interfacial pd is independent of ion activities. When two ions of opposite charge are free to partition and to transport under an applied potential, the chemical system is not "simply adjustable" in the sense that individual ion activities on each side of an interface cannot be arbitarily selected from stoichiometric addition of salts. The initial condition is determined by the concentration of added salts after consideration of electroneutrality and partitioning equilibria. The result for single partitioning salts is that the pd cannot be controlled by salt activity adjustment.

In practical systems, there are both transferring and non-transferring (or blocked) species initially present. Blocked species have zero rate constants for interfacial transport and they do not partition or carry current across the interface. Typically, aqueous supporting electrolytes are comprised of blocked ions, e.g. LiF, while the substance under study, MX, can contain reversible, partially blocked, or kinetically-limited ions, e.g., a tetraalkylammonium bromide. In the organic phase, a very hydrophobic salt NY or MY is usually present to increase conductivity and to provide a common ion when required. A typical example is a tetraalkylammonium tetraphenylborate. In the reversible limit for four partitioning species, M, N, X, Y the interfacial pd is generally written

$$\bar{\phi}-\phi = (1/2f)\ln\left[\frac{K_M a_M/\bar{a}_M + K_N a_N/\bar{a}_N}{K_X a_X/\bar{a}_X + K_Y a_Y/\bar{a}_Y}\right] \tag{9}$$

The values of activities to be used are those calculated from equilibrium partitioning. But, when single salts MX and NY are used, one has:

$$\bar{\phi}-\phi = (1/2f)\ln\left[\frac{K_M a_M/\bar{\gamma}_M + K_N a_N/\bar{\gamma}N}{K_X a_X/\bar{\gamma}_S + K_Y a_Y/\bar{\gamma}_Y}\right] \tag{10}$$

There are then four single-ion partition coefficients, defined in eqn. (1) and two salt partition equilibrium constants, expressible in terms of products of single-ion coefficients. A more complete discussion of Galvani potential differences has been presented by Hung (86). These expressions are, in some sense derived from corrosion theory, "mean" or "mixed" potentials because they are determined by exchange of two or more charged species. It should be noted that the values computed from eqns. (7)-(10) must also be given by the individual equations for each partitioning and potential-determining ion:

$$\bar{\phi}-\phi = (1/f)\ln[K_+ a_+/\bar{a}_+] \tag{11}$$

and

$$\bar{\phi}-\phi = (1/f)\ln[\bar{a}_-/K_- a_-] \tag{12}$$

The Equilibrium Potential "Window" for Two Salt Systems with a Common Cation: MX, MY

Only three ions M^+, X^-, Y^-, are sufficient to duplicate the transition from two salt MX, MY equilibrium extraction to a liquid ion exchanger for M^+. One ion, Y^-, must change from reversible to nearly blocked behavior to effect this response change. The presence of one salt MY that is very much favored in the organic phase and a second salt MX that is favored in the aqueous phase <u>is necessary to develop a range of equilibrium interfacial pds</u>. As we have shown, the interfacial pd between water and an immiscible phase containing a single equilibrated simple salt, MX, is nearly constant and independent of salt activity. When the two salts are present a potential "window" is created that can be spanned by variation of aqueous (or organic) salt activities.

For a typical two phase, water/nitrobenzene system containing equilibrated salts MX and MY, some MX and MY will be found in each phase. The concentrations of each depend on the salt partition coefficients $K_M K_X$ and $K_M K_Y$. When M^+ is a short chain tetraalkylammonium ion TAA^+ and X^- = Cl^-, and Y = tetraphenylborate (TPB^-), MX is predominately in water while TAATPB is predominately in nitrobenzene. This result, determined experimentally, can now be predicted from tabulated values (51, 87-89) of K_i of ΔG (w → 0). For this example we have chosen $K_Y >> K_X$ and from the relation $\Delta G = -RT\ln K_i$, the transfer free energy for X is more positive than the value for Y.

The interfacial pd for an entirely equilibrated system containing constant [MY] = 10^{-2} M in nitrobenzene with variable added concentration of MX, can be calculated. For illustration, consider a system with $K_Y = 10^6$, $K_M = 10^2$ and $K_X = 10^{-2}$. The interfacial pd $\Delta\phi = \bar{\phi} - \phi$ is shown in Fig. 1.

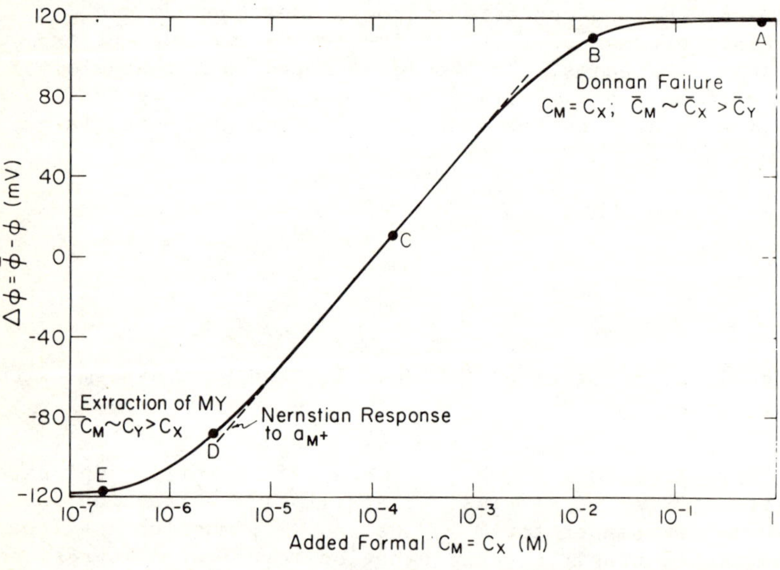

Figure 1. An example of the potential "window". Computation for two salts: MX mainly water soluble and MY mainly oil soluble. Formal concentration of MY = 10^{-2} mol/l; single-ion partition coefficients $K_y = 10^6$, $K_M = 10^2$, and $K_X = 10^{-2}$. The computed potential window is 336 mV. c_M and c_X and added concentrations, e.g., formal concentrations.

The potential "window" is determined by the limiting pds for excess MX (point A) and pure MY (point E). The limiting pd at point A is given by:

$$\Delta\phi = \frac{1}{2f}\ln\left[\frac{K_M \gamma_M \bar{\gamma}_X}{K_X \bar{\gamma}_M \gamma_X}\right] = 118 \text{ mV at } 25^\circ\text{C} \qquad (13)$$

while at point E

$$\Delta\phi = \frac{1}{2f}\ln\left[\frac{K_M \gamma_M \bar{\gamma}_Y}{K_Y \bar{\gamma}_M \gamma_Y}\right] = -118 \text{ mV at } 25^\circ\text{C} \qquad (14)$$

(activity coefficient values have been assumed to be unity). At point A, Donnan exclusion by Y is violated, while at infinite dilution of MX the interfacial pd is determined by extraction equilibrium of MY at point E.

In the midrange of the figure, point C, the response approaches Nernstian behavior that is most easily expressed by eqn. (11). In the transition regions, points B and D, the pd equations are quadratic:

$$\Delta\phi = \frac{1}{f}\ln\left[\frac{[\bar{Y}+(\bar{Y}^2+4K_M K_X a_M a_X/\bar{\gamma}_M \bar{\gamma}_X)^{1/2}]\bar{\gamma}_X}{2K_X a_X}\right] \qquad (15)$$

and for point D,

$$\Delta\phi = \frac{1}{f}\ln\left[\frac{K_M \gamma_M [a_M+(a_M^2+4\bar{Y}^2\bar{\gamma}_M \bar{\gamma}_Y/\gamma_M \gamma_Y K_M K_Y)^{1/2}]}{2Y\gamma_M}\right] \qquad (16)$$

where a_M is the added activity of MX. The potential "window" depends upon the ratios K_M/K_X and K_M/K_Y. The response along the activity axis depends on the extraction coefficient products $K_M K_X$ and $K_M K_Y$.

The range of potentials accessible at the water/nitrobenzene interface is estimated at 500 mV. Since the partition coefficients for OH^- and F^- are not known, but expected to be less than about 10^{-8} (for Cl^-), while that for TPB^- is ~10^6, a selectivity ratio range of perhaps 10^{16}–10^{17} leads to the 500 mV window. It is worth noting that partition coefficients and corresponding free energies of partitioning are the values for mutually saturated solvents. These differ from values for pure solvents (87).

EXPERIMENTAL METHODS

For L/L electrolyte interfaces, three types of cells could

be chosen for application of the Lippmann analysis. These are based on reversible anion, cation, and isolated reference electrodes. Variations of these occur throughout the L/L interface literature. We have chosen the former for reversible (unblocked) cells:

$$Cu^{I}|Ag^{I}|AgX|R^{+}X^{-}(aq.)\overset{x}{|}R^{+}B^{-}(org.)|AgB|Ag^{II}|Cu^{II} \quad \text{or}$$

$$Cu^{I}Ag^{I}|AgX|R^{+}X^{-}(aq.)\overset{x}{|}R^{+}B^{-}(org.)|R^{+}X^{-}(aq.)|AgX|Ag^{II}|Cu^{II}$$

The blocked interface under investigation is marked "x". The other interfaces are ionically or electronically reversible.

Unblocked (Reversible) Electrocapillary Measurements

Interfacial tensions of aqueous/nitrobenzene interfaces can be measured by the maximum bubble pressure method.(65) The cell is shown in Figure 2.

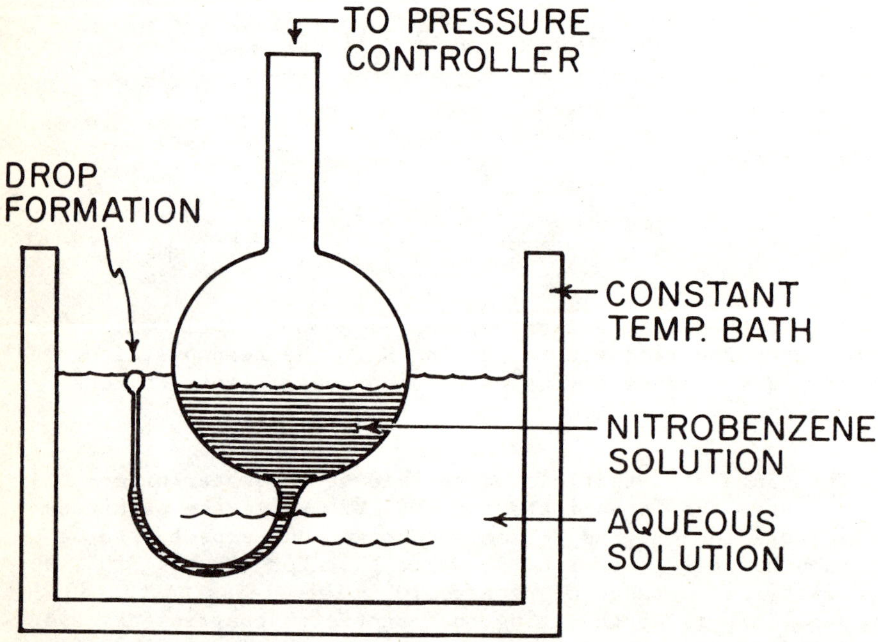

Figure 2. Cell used for measurement of nitrobenzene/water interfacial tensions.

Pressure differences are controlled and measured with the Ruska DDR 6000 pressure gauge with controller. The custom-made cell uses typically 2.0 ± 0.01 ml of the nitrobenzene solution such that the level in the reservoir section of the cell was approximately (within ±1 mm) level with the tip of the capillary section. The cell is submerged in a 50 ml beaker containing the aqueous solution such that the capillary tip is approximately 1 mm below the surface of the aqueous solution. A constant pressure ramp is generated using a custom-made potential ramp generator in place of the manual pressure control, supplied with the Ruska apparatus. The pressure is recorded when a drop of nitrobenzene becomes visible at the top of the capillary tube. Upon reaching the maximum bubble pressure, bubbles rapidly form, grow, and detach from the capillary tip.

For unblocked, reversible interfaces at which an ion exchange process is occurring, the pd at the interface cannot be controlled by external application of a voltage, because current would flow. Consequently, the electrocapillary curve is constructed from a series of aqueous/nitrobenzene solution pairs, each with a known ratio of exchanging ion. Typically, the nitrobenzene solution is held at 0.01 mol/l TAATPB, TAA = tetraalkylammonium ion, and the aqueous solution concentration of TAACl or TAABr is varied in the presence or absence of a supporting electrolyte such as NaCl or NaBr. From the calculated interfacial pds, the electrocapillary plot is constructed and derived parameters, surface charge density and capacitance, are calculated.

Blocked Electrode Electrocapillary Measurements

Conventional electrocapillary curves for blocked interfaces, such as the well-studied aqueous electrolyte/mercury interfaces, are obtained with dropping, pendant and sessile drop electrodes. Presumably, blocked L/L interfaces, for example aqueous LiCl/nitrobenzene containing TBATPB or TPAsTPB (TPAs = tetraphenylarsonium), can be studied by the same methods. Indeed, in two important papers by Kakiuchi and Senda (53,54), the ascending aqueous electrolyte "dropping" electrode has been successfully used. Their apparatus is shown in Figure 3 and is based on the cell

$$\text{Cu}|\text{Ag}|\text{AgCl}|\text{LiCl(aq.)}\overset{X}{|}\text{TBATPB(org.)}|\text{TBACl(aq.)}|\text{AgCl}|\text{Ag}|\text{Cu}$$

$$0.1\ \text{mol/l} \qquad 0.1\ \text{mol/l} \qquad 0.1\ \text{mol/l}$$

"x" indicates the polarizable interface. The second organic/aqueous interface is nearly reversible and contributes little non-faradaic current to the measurement. The L/L interface under study is not ideally blocked. It is only partially blocked, but sufficient for the experiment; less than 1μA of current flows over about 200 mV DC bias. The faradaic current is only precisely zero at the equilibrium potential set by the small, but finite solubility of TBATPB in aqueous LiCl.

Blocked (Irreversible) and Unblocked (Reversible) Interface Impedance Measurements

Admittances of the aqueous/nitrobenzene systems have been measured using an impedance bridge designed by Mathis and Buck.(90) This cross-correlation bridge makes possible accurate

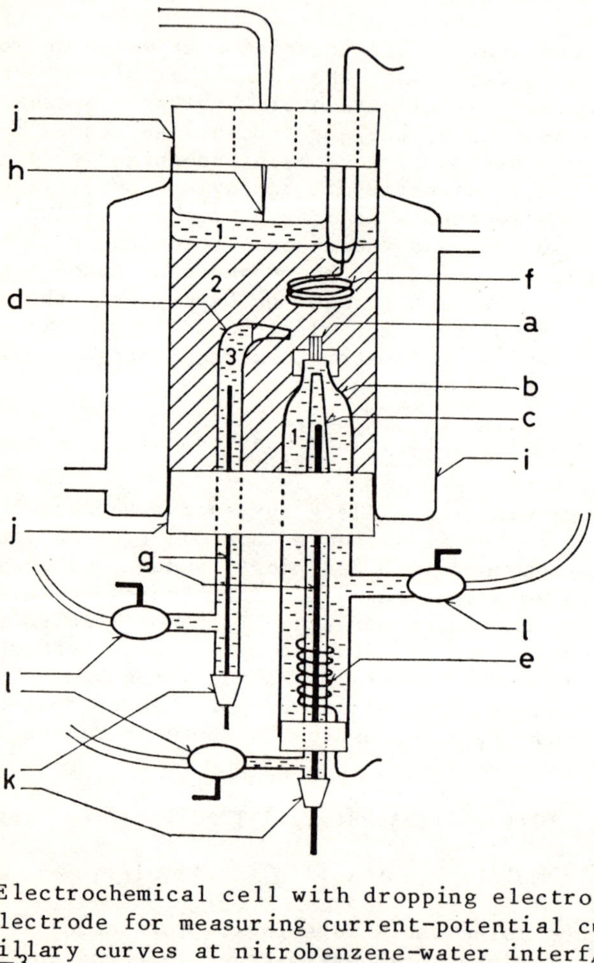

Figure 3. Electrochemical cell with dropping electrolyte solution electrode for measuring current-potential curves and electrocapillary curves at nitrobenzene-water interface. (1) 0.1 mol dm^{-3} LiCl aqueous solution; (2) 0.1 mol dm^{-3} TBATPB nitrobenzene solution; (3) 0.1 mol dm^{-3} TBACl aqueous solution; (a) dropping electrolyte solution electrode tip; (b) glass tube; (c and d) reference electrodes with Luggin capillaries; (e and f) auxiliary electrodes; (g) Ag/AgCl electrodes; (h) fine glass

capillary; (i) water-jacketted cell; (j) silicone rubber stopper; (k) silicone rubber cap; (l) stop cock.

admittance measurements from 0.5 Hz to 50 kHz.(66,67) Modifications now include use of a DC offset that permits the interface to be externally polarized while retaining the capability of measuring the admittance. Normally admittances are measured at the zero DC current, equilibrium pd. Commercial instruments made by PAR, Hewlett-Packard and Solartron perform these functions very well and over a wider frequency range.

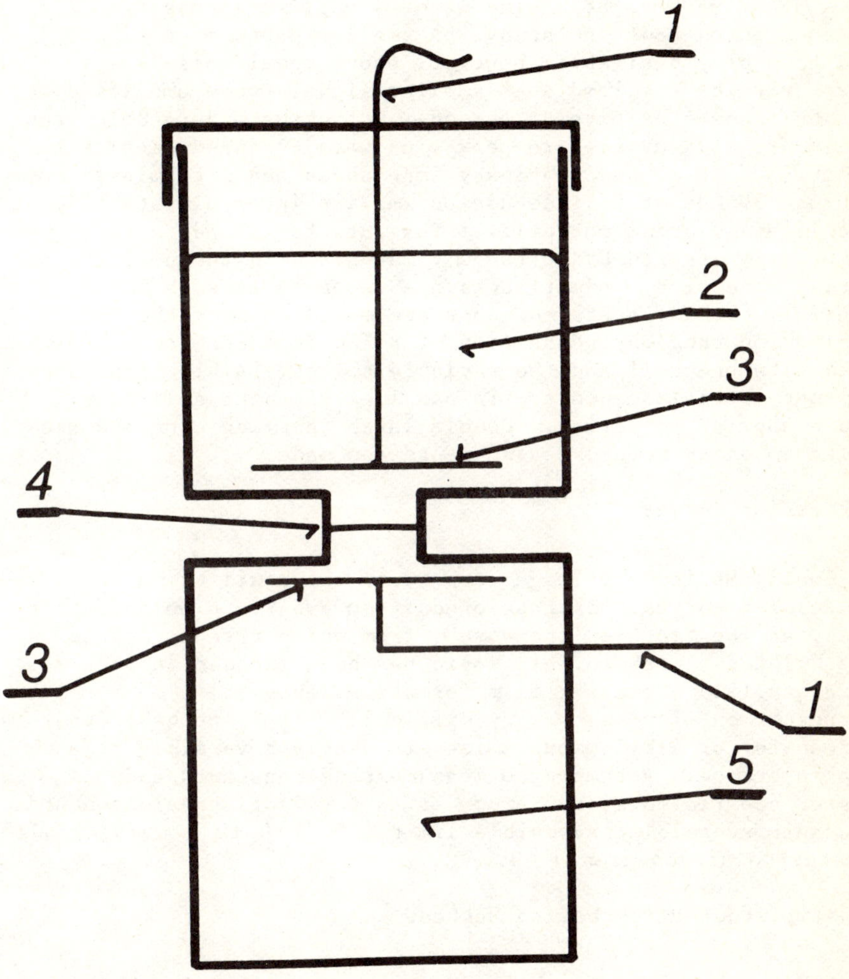

Figure 4. Side view of the circular cell. 1 - Pt connecting wires, 2 - aqueous phase, 3 - Ag/AgCl flag electrodes, 4 - interface, 5 - nitrobenzene phase.

The cell is based on Figure 4. The nitrobenzene phase is added to the cell such that the aqueous/nitrobenzene interface appears within the narrow region between the two reservoirs. The aqueous phase is then added to the cell immersing the upper electrode. The nitrobenzene/aqueous interfacial area of 0.5 ± 0.05 cm^2. Ag/AgCl flag electrodes with Pt wire connections are used in both phases. The large (4 cm^2) surface area and the resulting high capacitance and low current density are essential to elimination of electrode contributions to the measured admittance values.

A 35 mV peak-to-peak sine wave is applied across the interface throughout the study. A smaller perturbing signal would have been desirable; however, poor signal/noise characteristics resulted when smaller signals were amplified prior to phase-sensitive detection. Using the modern Solartron instrument, a 14 mV peak-to-peak sine wave produced excellent results.(67) Real and imaginary impedances are determined from 0.5 Hz to 50 kHz at half decade or smaller intervals starting at the cell equilibrium potential. The interfacial pd is then sequentially stepped by 20 to 50mV intervals in the positive or negative direction. Admittances are recorded at each DC potential. Steady state currents are measured over the polarization range by connection of a low impedance ammeter in series with the cell and the variable source. A high impedance voltmeter is again connected in parallel with the cell to measure applied potential. Double layer charging currents are allowed to decay before measurements are made.

Polarization Curves

Small amplitude voltage steps from the equilibrium pd produces I-t curves. Extrapolated I(t=o) vs V are polarization curves, at the "foot-of-the-wave", from which rate constants can be calculated. Work on this topic has been pursued in Buck's laboratory.(91) However, film formation makes the interpretation of results very difficult. It is probable that the transfer of alkylammonium ions is nearly reversible at a clean interface. Attempts to measure rate constants and transfer coefficients, by Samec, et. al. (44) give values near the quasi-reversible/reversible limit (10^{-1} to 10^{-3} cm/sec) and α (water to nitrobenzene) $\sim$0.3

Large Amplitude Perturbation Methods

In many works it has been shown that the interface between two immiscible liquids is, with proper experimental setup, and analog of the interface electrolyte/metal electrode (21-24, 28-35, 59-62), and that this interface can be studied by the large amplitude methods used in modern electrochemistry. Transport of ions across immiscible liquid-electrolyte interfaces, however, differs in several significant ways from the electrolyte/metal interfacial processes, as shown by Melroy et. al. (59): (1) diffusion-migration via Nernst-Planck equations, rather than Fick's laws must be considered because of the general low ion loadings in the organic phase; (2) interfacial transport of ions of opposite sign must be considered because of salt partition equilibria; (3) partition equilibrium is a non-linear ionic process that can lead to insoluble equations.

Experimentlly, the most extensively used techniques have been chronopotentiometry (16,18-24, 28,60-64)polarography (normal potentiostatic voltammetry) with electrolyte dropping electrodes (29,41,77-78,92), cyclic voltammetry (37-40), and impedance measurements (19,44,47,55,56,65-67,71). Polarographic experiments in the current-scan mode (galvanopotentiometry) have been recently reported.(79,80)

In contrast to modern electrochemical instrumentation for electrolyte/metal interface studies, the L/L interface requires four-electrode (rather than the conventional three-electrode) systems. In both cases, the investigator must control or measure the pd across the interface. For liquid interfaces, two sensor probes are placed close to, and on opposite sides of the interface. These electrodes should be small and may be either salt-bridge-isolated reference electrodes, typically Ag/AgCl, or simply shrouded platinum wires. They need to be placed close to the interface, and be small enough to avoid shielding an appreciable area of the interface. Two other large area "working" electrodes are positioned away from the interface, but in line with the interface so that applied electric field lines, or current pass perpendicularly through the interface. Instrumentation for potentiostatic, galvanostatic or pulse control are extensions of the more common equivalent three electrode devices. Likewise for voltage or current scan applications, four-electrode systems are used.(37) In principle, positive feedback can be used to correct for residual IR drop between the sensor or reference probes.(39) Voltages applied to the working electrodes automatically adjust to values such that the prode pd, or current, is controlled.

Chronopotentiometric and Voltammetric Static-Meniscus Cell Design

Cell designs derive from the classical U-tube experiments. Our preferred design is shown in Figure 5. The cell is constructed of Pyrex and consists of three compartments. Kontes

O-ring Teflon valves are opened for degassing and stirring with nitrogen, and for adjusting the level of the interface. The central compartment is the working area of the cell. This contains the four electrodes.

Chronopotentiometric and Voltammetric Sessile Drop and Droping Electrolyte Cell Designs

Analogous to the dropping mercury electrode, pendant drop and dropping nitrobenzene electrodes can be made.(93) However, it was found early (29) that ascending "drop" electrodes of aqueous electrolyte, formed in the nitrobenzene electrolyte were easier to fabricate and control. Improvements were made by Samec et. al. (41) and by Senda et. al.(53,54) The addition of a microsyringe by Yoshida and Freiser (79,80) permits attachment of the sessile drop.

EXPERIMENTAL RESULTS

Free Energies of Parition, Partition Coefficients and Standard Potentials

Free energies of solution of electrolytes in water and in nitrobenzene are obtained from solubility measurements.(87) Radiotracers, spectrophotometric, and gravimetric methods are used. The free energies of transfer are determined by combining the solubility results (as free energies of salt partitioning) to eliminate the pure component free energy. Free energies of salt partitioning that apply to solvent-saturated phases are determined by direct partitioning measurements between contacting phases followed by chemical analyses.

In the Parker/Popovych convention (83,84), the key partitioning compound is TPAsTPB whose single ion free energies are presumed to be equal, on the basis of similar size, charge and hydrophobic character. Thus.

$$\underline{\Delta}G(w\rightarrow o)/2 = \underline{\Delta}G(w\rightarrow o)_{TPAs\ and\ TPB} \tag{17}$$

By using a sequence of TPAs salts, the individual anion free energies of partitioning could be found from

$$\underline{\Delta}G(w\rightarrow o)_{X^-} = \underline{\Delta}G(w\rightarrow o)_{TPAsX} - \underline{\Delta}G(w\rightarrow o)_{TPAs^+} \tag{18}$$

and from a sequence of TPB^- salts, cation free energies could be found:

$$\underline{\Delta}G(w\rightarrow o)_{M^+} = \underline{\Delta}G(w\rightarrow o)_{MTPB} - \underline{\Delta}G(w\rightarrow o)_{TPB^-} \tag{19}$$

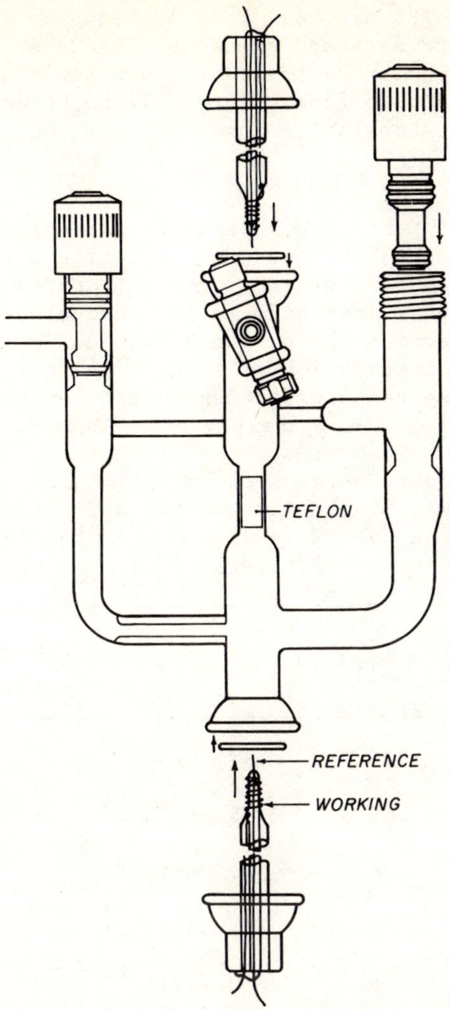

Figure 5. Cell for liquid/liquid transport studies, a piece of Teflon tubing, and a Teflon stopcock. There are two identical electrode housings (glass), each containing a reference and working electrode. The housing has two concentric tubes, the center serving to sheathe the electrode leads from each other. All the electrodes are platinum. The housings are connected to the central compartment by O-ring joints and held in place by joint clamps, assuring reproducible positioning. The compartment is narrowed in the interface region, giving a cross-sectional area of 0.32 cm^2 inside the Teflon tubing.

In fact, it is more convenient and accurate to determine the single ion coefficients for ClO_4^-, I^- and picrate and then compute cation coefficients from the partitioning of the various salts in addition to TPB^- salts. This overlapping gives internal checks on the data and shows up effects of ion pairing. The latter is expected for multivalent ion-containing salts and salts of least hydrophobic ions. Internal checks are provided for anions by using tetraalkylammonium salts as well as $TPAs^+$-salts.(87)

Free energies of partitioning and transfer differ on account of ion pairing differences and hydration differences in nitrobenzene. Small univalent ions remain hydrated, while the larger, more hydrophobic ions seem to be unhydrated. Koryta has sumarized direct determination of ion hydration in nitrobenzene.(34) A compilation of single ion free energies of partition, the corresponding partition coefficients and the standard potentials $\Delta\phi^o(w \rightarrow o)$ are given in Table I.

TABLE I. Standard Gibbs Energies of Partition From Water to Nitrobenzene, Single Ion Partition Coefficients (Nitrobenzene Over Water), and the Standard Electrical Potential Difference, $\phi_w - \phi_o$

	$\Delta G^o(w \rightarrow o)$ [kJ mol^{-1}]	$K_{o/w}$	$\Delta\phi^o(w \rightarrow o)$ [mV]	(Ref.)
Li^+	38.2	2.10×10^{-7}	395	88
Mg^{2+}	69.6	5.68×10^{-14}	361	94
Na^+	34.2	1.04×10^{-6}	354	88
Ca^{2+}	67.3	1.63×10^{-13}	349	95
Sr^{2+}	66.0	2.74×10^{-12}	342	94
H^+	32.5	2.02×10^{-6}	337	88
Ba^{2+}	61.7	1.55×10^{-11}	320	94
NH_4^+	26.8	2.02×10^{-5}	277	88
K^+	23.4	7.95×10^{-5}	242	88
Rb^+	19.4	3.99×10^{-4}	201	88
Cs^+	15.4	0.002	159	88
choline$^+$	11.3	0.0105	117	96

acetycholine$^+$	4.8	0.144	49	96
TMeA$^+$	3.4	0.254	35	88
TEA$^+$	- 5.7	9.97	- 59	88
TPA$^+$	-15.5	5.06×10^{2}	-160	88
TBA$^+$	-24.0	1.6×10^{4}	-248	88
TPAs$^+$	-35.9	1.95×10^{6}	-372	88
cryst. violet$^+$	-39.5	8.32×10^{6}	-410	97
dicarbollyl-cobaltate	-50.2	6.23×10^{8}	520	88
dipicrylaminate	-39.4	7.99×10^{6}	407	88
I_5^-	-38.8	6.27×10^{6}	401	88
TPB$^-$	-35.9	1.95×10^{6}	372	88
I_3^-	-23.4	1.26×10^{4}	242	88
octoate	- 8.5	30.8	89	97
picrate	- 4.6	6.4	47	88
dodecylsulphate	4.1	0.19	- 43	97
SCN$^-$	5.8	9.64×10^{-2}	- 61	97
ClO_4^-	8.0	3.97×10^{-2}	- 83	88
I^-	18.8	5.09×10^{-4}	-195	88
NO_3^-	24.4	6.6×10^{-4}	-253	98
Br^-	28.4	1.06×10^{-5}	-295	88
Cl^-	31.4	3.15×10^{-6}	-324	88
F^-	44.0	1.96×10^{-8}	-454	76

The latter quantity is calculated from

$$\Delta G^{o}(w\rightarrow o) = nF\Delta\phi^{o}(w\rightarrow o) \qquad (20)$$

Hydration numbers are compiled in Table II and intercomparisons of partition and transfer values are given in Table III.

TABLE II. Hydration Numbers of Ions in Nitrobenzene Saturated with Water

Ion	Hydration Number
Li^+	6.5
H^+	5.5
Ag^+	4.4
Na^+	3.9
K^+	1.5
Tl^+	1.0
Rb^+	0.8
Cs^+	0.5
Et_4N^+	0.0
ClO_4^-	0.4
NO_3^-	1.4
TPB^-	0.0

Ref. (34)

TABLE III. Comparison between observed and calculated free energies of transfer and free energies of partition based on $\Delta G_t^o(Ph_4As^+) = \Delta G_t^o(Ph_4B^-)$

Ion	ΔG_t^o Obs[a]	ΔG_t^o Calc[b]	ΔG_p^{oc}	ΔG_p^{od}
K^+	5.01	5.3	5.6	–

Rb^+	4.60	4.8	4.6	—
Cs^+	4.25	4.2	4.1	—
Me_4N^+	0.96	3.1	0.8	—
Et_4N^+	- 1.14	0.6	-1.4	—
Pr_4N^+	- 3.92	- 2.4	—	—
Br_4N^+	—	- 5.4	-5.7	—
Ph_4P^+	- 8.92	—	-8.6	—
Ph_4As^+	- 8.62	—	-8.6	—
Cl^-	10.49	11.9	7.1	10.2
Br^-	8.61	8.3	—	8.2
ClO_4^-	2.35	2.5	1.9	2.4
Pi^-	- 0.81	0.8	-1.1	—
Ph_4B^-	- 8.62	—	-8.6	—

[a]Ref (87). [b]By subtracting 1 kcal mol^{-1} for the cations and adding 1 kcal mol for the anions to single ion ΔG_t^o given in ref. (99) in order to have these values in accord with the Ph_4As^+/Ph_4B^- convention. [c]Ref. (88) and (99). [d]Ref. (89) and (99).

Electrochemical cells can also be used to determine these quantities.(69,74) However the cells invariably involve odd-numbers of interfaces and liquid junctions. The problems of junction potential corrections make these methods of second choice, but nevertheless useful for intercomparison of values.

Free energies of partitioning can also be determined from large amplitude methods. Senda et al. (51) used chronoamperometry and cyclic voltammetry to determine half-wave potentials for a collection of tetraalkylammonium ions. After corrections for diffusion coefficients free energies were computed and given in Table IV.

TABLE IV. The Reversible Half-Wave Potentials, $E^{r}_{1/2,j}$, Determined by Cyclic Voltammetry and Potential-Step Chronoamperometry, and the Standard Gibbs Free Energies of Transfer From Nitrobenzene to Water, $\Delta G^{\ominus,N\to W}_{tr,j}$, for Fifteen Alkylammonium Ions. $(R_1R_2R_3R_4)N^+$.

$(R_1R_2R_3R_4)N^+$				$E^{r}_{1/2,j}/V$		$\Delta G^{\ominus,N\to W}_{tr,j}$
R_1	R_2	R_3	R_4	cyclic voltammetry	potential-step chronoamperometry	$kJ\ mol^{-1}$
H	H	H	Et	0.45	---	-18.8
H	H	H	n-Pr	0.42	0.43	-15.9
H	H	H	n-Bu	0.40	0.40	-13.3
H	H	Me	Me	0.43	---	-16.6
H	H	Et	Et	0.39	0.39	-12.8
H	H	n-Pr	n-Pr	0.33	0.33	-7.3
H	Me	Me	Me	0.38	---	-11.9
H	Et	Et	Et	0.32	0.32	-5.6
H	n-Pr	n-Pr	n-Pr	0.23	---	2.4
Me	Me	Me	Me	0.32	0.32	-6.4
Me	Me	Me	Et	0.30	0.39	-3.9
Me	Me	Me	n-Pr	0.26	0.27	-0.6
Me	Me	Me	n-Bu	0.23	---	2.4

TABLE IV. (continued)

Et	Et	Et	Me	0.25	0.26	0.7
Et	Et	Et	Et	0.23	---	2.7

Ref. (51).

Thermodynamic and Electrostatic Analysis of the Interfacial Potential Distribution (Lippmann Equation Analysis)

Measurement of interfacial tensions between nitrobenzene containing a tetraalkylammonium tetraphenylborate (TAATPB) salt and an aqueous solution of the bromide salt of the same TAA cation allows determination of the interfacial charge as a function of cell potential. The latter is established by control of aqueous TAABr concentration at constant TAATPB. These experimentally derived charge values may then be compared with values calculated through Gouy-Chapman theory assuming arbitrary values for the χ potential drop across any compact layer. In this way the χ potential may be estimated at several Galvani potential differences.

Three types of cells can be chosen for application of the Lippmann analysis. We have completed the analysis for each type using reversible anion, cation, and isolated reference electrodes.(65) We have chosen the anion-reversible cell

$$Cu^{I}|Ag^{I}|AgBr|R^{+}Br^{-}(aq.)|R^{+}B^{-}(org.)|AgB|Ag^{II}|Cu^{II}$$

as noted above. $R^{+} = TAA^{+}$; $B^{-} = TPB^{-}$, $\underline{\Delta\phi}$ and $d\underline{\Delta\phi}$ are taken right-left e.g., organic-aqueous in this analysis. Differential interfacial tensions are given by:

$$-d\gamma = \Gamma_{R^{+}(H_2O)} d\mu_{RBr} + \bar{\Gamma}_{R^{+}(org.)} d\bar{\mu}_{RB} + \sigma_o d\underline{\Delta\phi}_{-} \tag{21}$$

Γ's are surface excesses (bar denotes the organic phase), and σ_o is the organic phase charge:

$$\sigma_o = F(\bar{\Gamma}_{R^{+}(org.)} - \bar{\Gamma}_{B^{-}(org.)}) \tag{22}$$

When the aqueous solution contains a supporting electrolyte with a common anion (Br^- in this study), eqn. (21) contains another term on the right side: $\Gamma_{Na^{+}(H_2O)} d\mu_{NaBr}$.

Since the cell under study contains a reversibly exchanging species R^+, (Br^- and TBP^- remained blocked) an additional equilibrium condition is:

$$d\tilde{\mu}_{R^{+}} = d\tilde{\bar{\mu}}_{R^{+}} \text{ or } d\mu_{RBr} - d\bar{\mu}_{RB} = d\tilde{\mu}_{Br^{-}} - d\tilde{\bar{\mu}}_{B^{-}} \tag{23}$$

By substitition and rearrangement of eqn. (23) in eqns. (21) and (22) new differential tension equations for reversible interfaces, corresponding to the anion-reversible reference electrode case, are obtained.

$$-d\gamma = (\sigma_o + F\Gamma_{R^+(H_2O)})d\underline{\Delta\phi}_- + (\bar{\Gamma}_{R^+(org.)} + \Gamma_{R^+(H_2O)})d\bar{\mu}_{RB} = (\sigma_o - F\bar{\Gamma}_{R^+(org.)})d\underline{\Delta\phi}_- + (\bar{\Gamma}_{R^+(org.)} + \Gamma_{R^+(H_2O)})d\mu_{RBr} \quad (24)$$

Because a constant organic phase ionic activity is used, from eqn. (24) the slope of the electrocapillary curve can be expressed:

$$-d\gamma/d\underline{\Delta\phi}_- = (\sigma_o + F\Gamma_{R^+(H_2O)}) = \sigma_o + \sigma_{+(aq.)} \quad (25)$$

Application of these equations to reversible liquid/liquid interface cells is considerably more complicated than applications to blocked cells discussed later. The variables $d\underline{\Delta\phi}_-$, $d\bar{\mu}_{RB}$ and $d\mu_{RBr}$ are not independent, and only one $\underline{\Delta\phi}_-$ value corresponds to zero current. Consequently, each $\underline{\Delta\phi}_-$ value is established by changing the chemical composition. In our work variation of RBr at constant RTPB is used. Plots of interfacial tension (or a proportional quantity) vs. $\underline{\Delta\phi}_-$ requires the calculation:

$$\underline{\Delta\phi}_- = \frac{RT}{F}\ln\left(\frac{a_R a_{Br}}{\bar{a}_R\,\bar{a}_{Br}}\right) = \frac{2RT}{F}\ln\left[\frac{(cf)}{(\bar{c}f)}\right] \quad (26)$$

where f's are activity coefficients. For each decade change in salt activity, a change of 118 mV in $\underline{\Delta\phi}_-$ occurs at 25°C.

Gros et al. (27) used only supported electrolytes in their measurements on these systems. The more general expression for σ_o applicable to both our work and that of Gros et al. (27) is:

$$(1/F)\sigma_o = \bar{\Gamma}_{R^+(org.)} - \bar{\Gamma}_{B^-(org.)} = -(\Gamma_{R^+(H_2O)} + \Gamma_{Na^+(H_2O)} - \Gamma_{Br^-(H_2O)}) \quad (27)$$

It should be noted that a given interfacial pd, established by $a_{R+}/\bar{a}_{R+}$ does not give the same charges for the aqueous (or organic) solution in the presence and absence of supporting electrolytes.

Equations for the available potential window have been stated above (eqns. 13 and 14). The two eqns. (13) and (14) are limiting expressions. A more general expression for the positive region response (from Nernstian response to the positive limit) was derived by assuming that the RBr concentration is large enough that any loss of bromide to the organic phase can be neglected, e.g. $(\overline{Br^-}) \ll (Br^-)$. The computed interfacial pds, $\Delta\phi$, are shown for the Nernstian region in Figure 6.

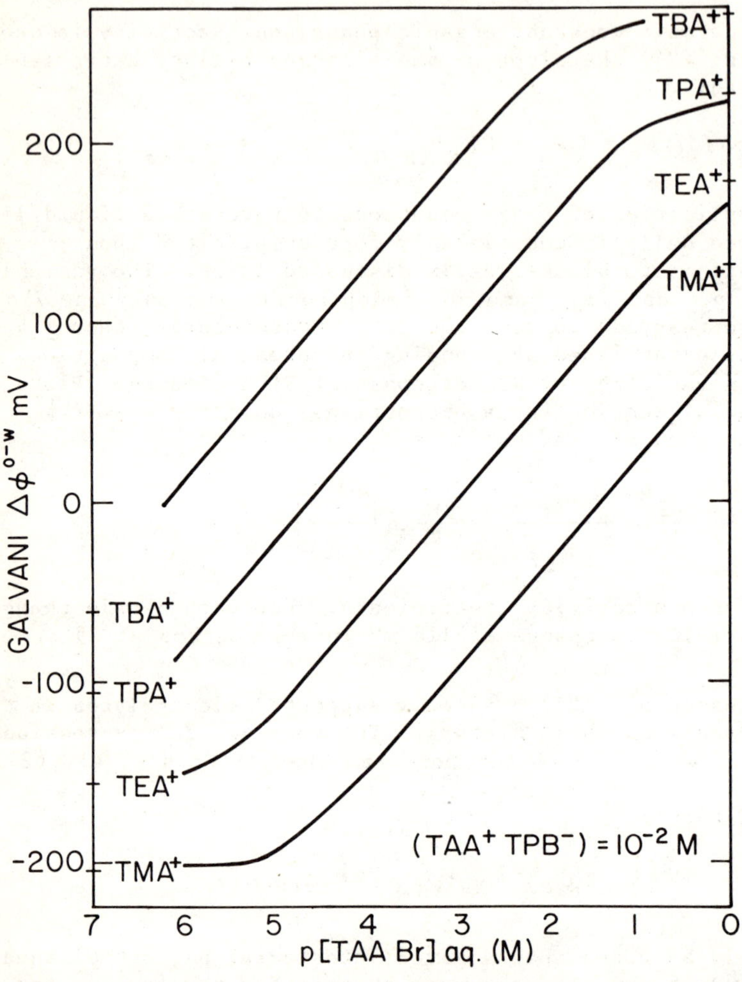

Figure 6. Calculated Galvani potential differences across the nirobenzene/water interface as a function of aqueous TAABr concentration. The concentration in nitrobenzene of the corresponding TAATPB is constant at 10^{-2} mol/l.

The potentials at the OHP's and the charge are calculated from eqns. (28) and (29)

$$\underline{\Delta}\phi_2^{w}(\phi_{bulk}=0) = \frac{2RT}{F}\sinh^{-1}\left[\frac{\epsilon_o \bar{c}}{\epsilon_w c}\right]^{1/2} \sinh\left[\frac{F}{2RT}(\underline{\Delta}\phi(o\rightarrow w)-\chi-\underline{\Delta}\phi_2^{w})\right] \quad (28)$$

$$\sigma_o = 2(2RT\epsilon_o\epsilon^o\bar{c})^{1/2}\sinh\left[\frac{F}{2RT}(\underline{\Delta}\phi(o\rightarrow w)-\chi-\underline{\Delta}\phi_2^{w})\right] \quad (29)$$

and the partial charge due to positive species in the aqueous solution is given by:

$$\sigma_+(H_2O) = (2RT\epsilon_w\epsilon^o c)^{1/2}[\exp(-F\underline{\Delta}\phi_2^{w}/RT)-1] \quad (30)$$

The first equation requires iterative solution. For the examples reported here; $\bar{c}_2$= [RTPB] = 0.01 mol/l; ϵ_o = 35.8; ϵ_w = 78.5; $\epsilon^o = 8.85 \times 10^{-12}\ C^2\ N^{-1}\ m^{-2}$ and c = [RBr] is varied. χ, the dipole potential is assumed to be zero in these calculations. In supporting 1:1 electrolytes, the aqueous concentration term is the sum of the bulk concentration of RBr and NaBr. Results of calculations of $\underline{\Delta}\phi_2^{w}$ and $\underline{\Delta}\phi_2^{o}$ are plotted in Figure 7 and 8.

Interfacial Tensions and Diffuse Layer Charges at a Reversible Interface

Figure 9 shows variation of interfacial tension as a function of aqueous TAABr concentration in the absence of supporting aqueous electrolyte. The electrocapillary curve maxima for TMABr, TEABr, and TPABr (tetramethyl, ethyl, propylammonium, respectively) occur at concentrations that agree well with those calculated from Gouy-Chapman-Verwey-Niessen theory assuming the chi potential is zero. Solid curves are drawn through the data points based on the Lippmann theory for reversible interfaces. Increases in tension at low concentrations follow from the significant additional term $\sigma_+ = F\Gamma_{R^+(H_2O)}$ occuring in eqn. (24). No maxima were observed in the TBATPB and THATPB (THA = tetrahexylammonium) electrocapillary curves. This result was anticipated because calculated maxima should occur at $10^{-6.2}$ mol/l and 10^{-9} mol/l, respectively.

In the absence of supporting aqueous electrolytes the interfacial tension curves reach plateau values at low TAABr concentrations. This behavior, which is different from the "inverted" negative cosh function predicted for electrolyte/metal interfaces, originates from two sources. One is the compensation of the charges σ_o and σ_+. The second is the approach to constant aqueous salt concentration at low dilutions of TAABr. The aqueous concentration becomes constant

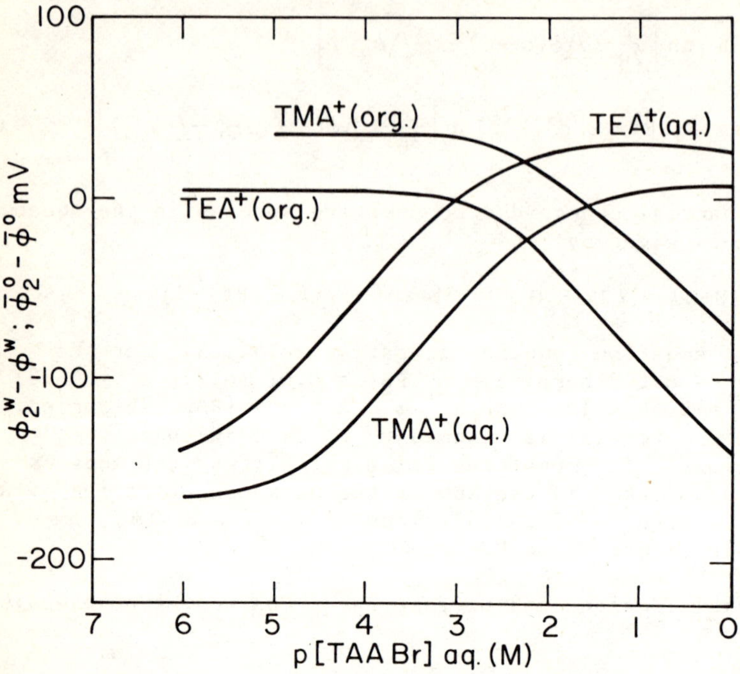

Figure 7. Diffuse layer potential drops in water and nitrobenze for varying aqueous concentrations of tetrapropylammonium and tetrabutylammonium bromides calculated assuming the chi potential is zero. The concentration of the coresponding TAATPB in nitrobenzene is 10^{-2} mol/l. No aqueous supporting electrolyte is present.

and determined by the equilibrium solubility of TAATPB in the aqueous phase. The charge must also reach a limiting non-zero value. This unusual behavior arises because the Galvani potential difference in unchanged by further dilution of TAABr. At low concentrations in one phase nearly all of the Galvani potential difference occurs in the diffuse layer of the dilute phase.

Figures 10 and 11 show variations of interfacial tensions as a function of TAABr concentration with 0.03 mol/l and 0.3 mol/l

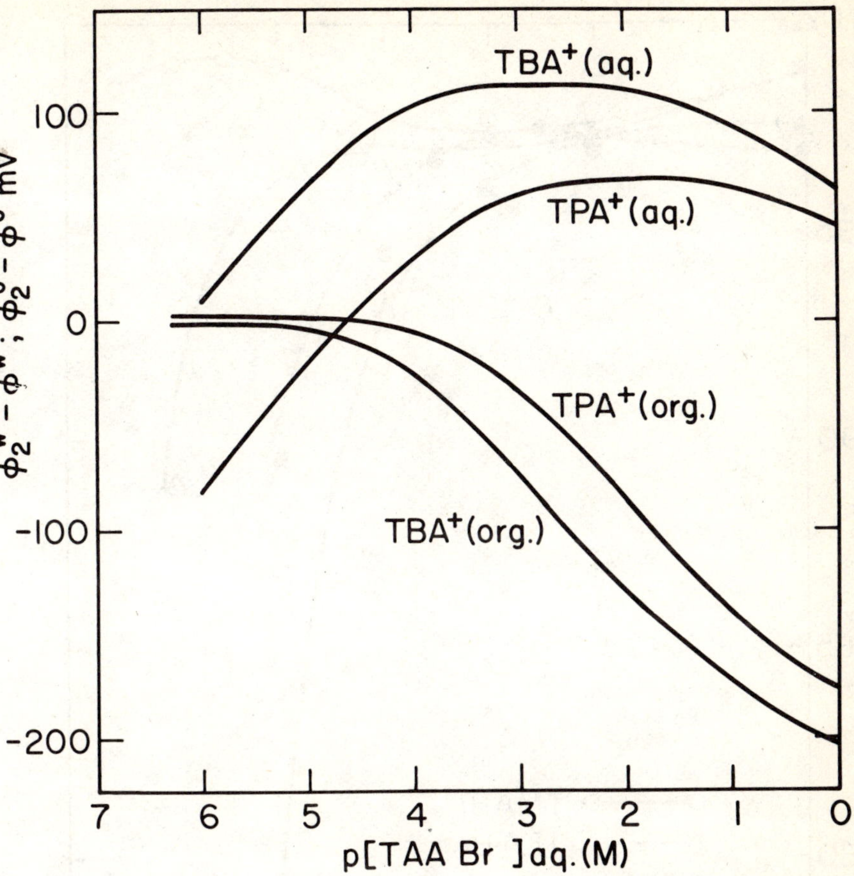

Figure 8. Diffuse layer potential drops in water and nitrobenzene for varying aqueous concentrations of tetramethylammonium and tetraethylammonium bromides calculated assuming the chi potential is zero. The concentration of the corresponding TAATPB in nitrobenzene is 10^{-2} mol/l. No aqueous supporting electrolyte is present.

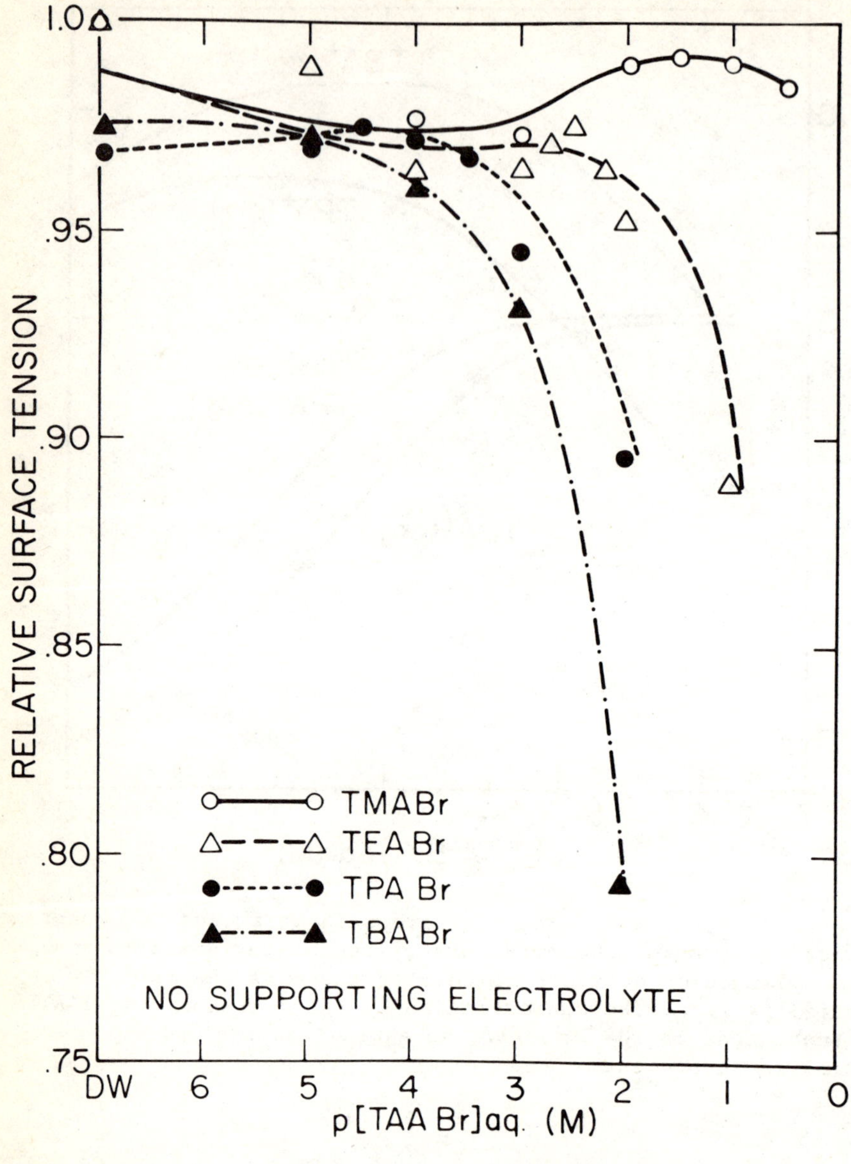

Figure 9. Relative interfacial tensions between 10^{-2} mol/l TAATPB in nitrobenzene and varying aqueous concentrations of corresponding TAABr salt. No supporting aqueous electrolyte.

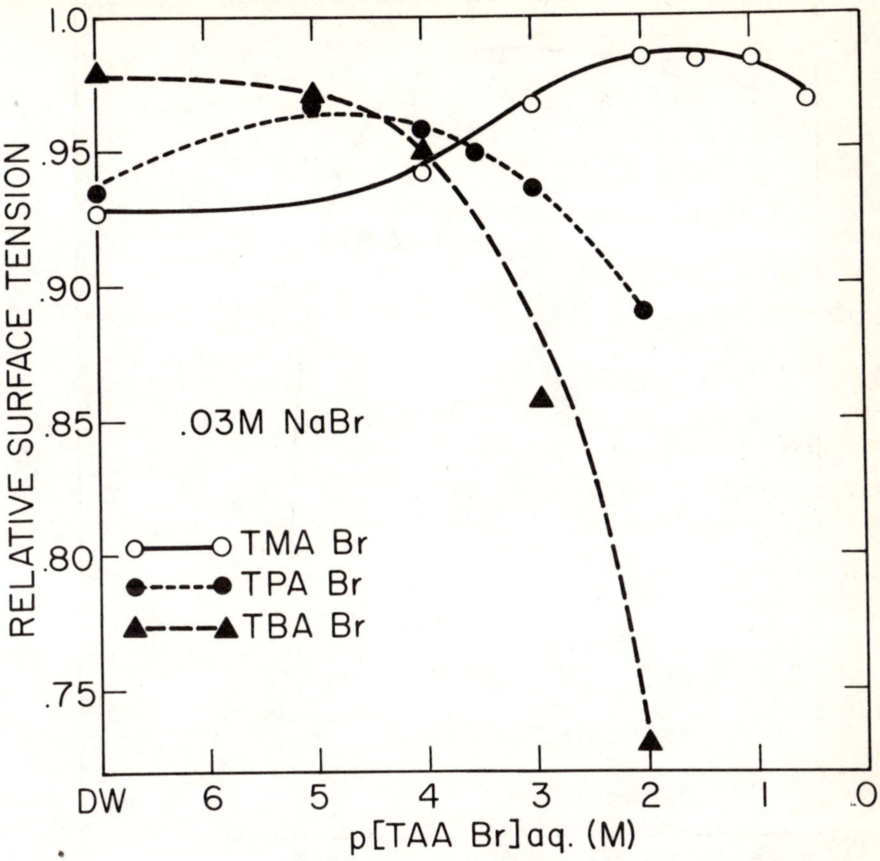

Figure 10. Relative interfacial tensions between 10^{-2} mol/l TAATPB in nitrobenzene and varying aqueous concentrations of corresponding TAABr salt. Supporting electrolyte: $3x10^{-2}$ mol/l NaBr.

NaBr as supporting electrolyte. The electrocapillary curve maxima agree well with calculated values, again assuming the χ potential is zero. The good agreement between electrocapillary maxima and calculated concentrations of zero charge (or zeros for the diffuse layer potentials) indicate that the χ potential is probably within a few mV of zero ($\pm$ 5 mV). This determination agrees with the conclusion of Gros et al. (27).

In the presence of 0.03 mol/l and 0.3 mol/l NaBr supporting electrolyte, interfacial tensions decrease as aqueous TAABr concentrations decrease beyond the pzc. This behavior

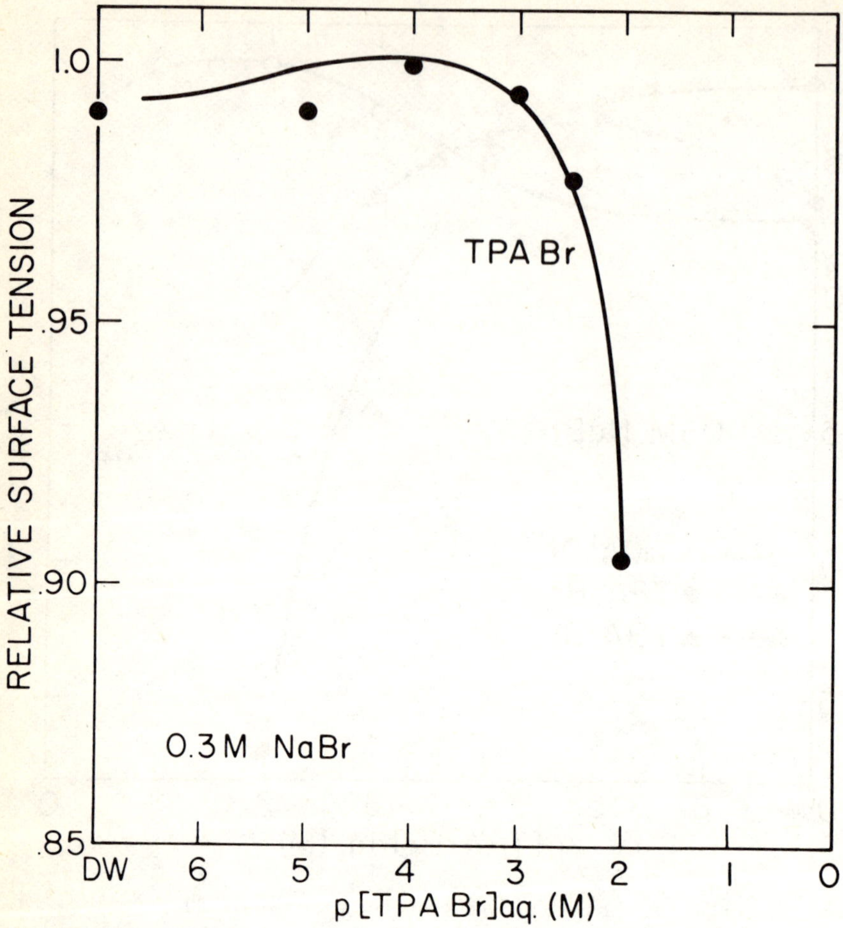

Figure 11. Relative intefacial tensions between 10^{-2} mol/l TPeATPB in nitrobenzene and varying aqueous concentrations of corresponding TABr salt. Supporting electrolyte: 0.3 mol/l NaBr.

continues until the Galvani potential difference reaches the low concentration limit determined by the equilibirum solubility of TAATPB in water. To convert the abscissa from log concentration to Galvani potential difference, using hypothetical anion-reversible probes, 118 mV were assigned to each decade of concentration change.

Figures 12-14 show the calculated quantity $\sigma_o + \sigma_+(H_2O)$ for aqueous solutions containing zero, 0.03 mol/l and 0.3 mol/l NaBr supporting electrolyte. The principal feature of the results is the low value of the surface excesses computed as charge. In contrast to metal/electrolyte interfaces where charges are frequently 10X larger, these interfaces show generally low charge densities. The reason for this

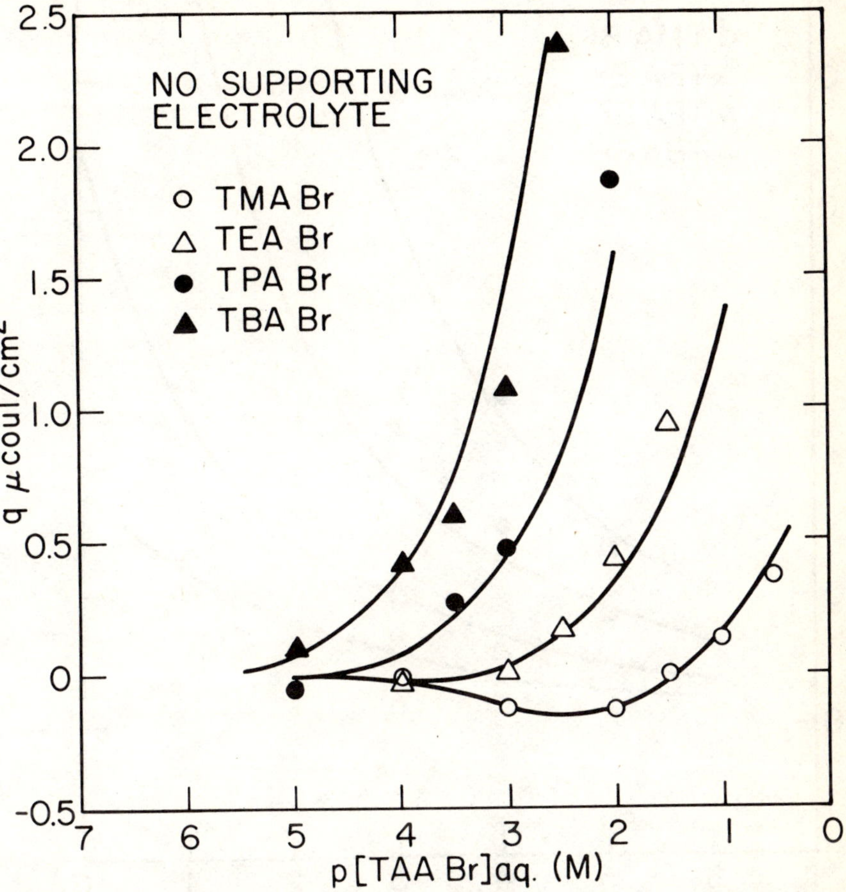

Figure 12. The quantity $(\sigma_o + \sigma_+ H_2O))$ calculated by GCVN theory is indicated by the lines for varying aqueous TAABr concentrations and constant (10^{-2} mol/l) concentration of TAATPB in nitrobenzene. No aqueous supporting electrolyte. Points are experimental; charge function values obtained from interfacial tension curves (Fig. 9).

difference is the fact that space charge at L/L interfaces is determined by relative salt solubilities in each phase. At metallic interfaces, charge can be controlled by

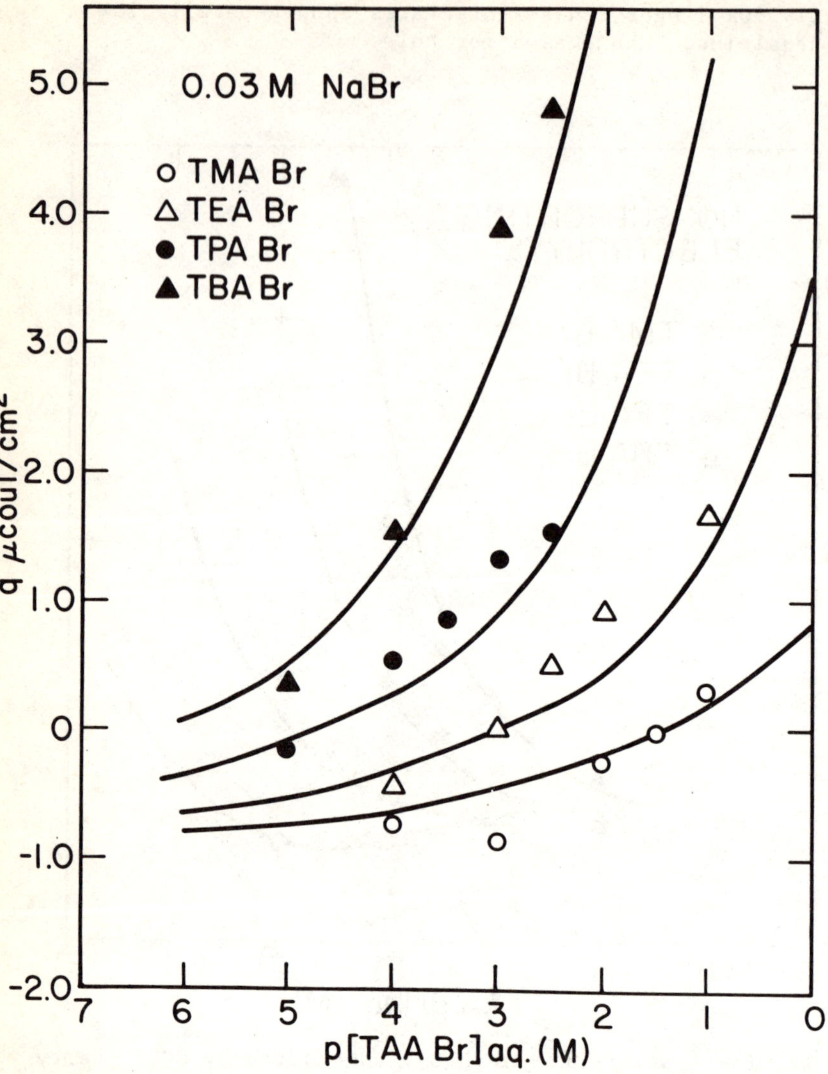

Figure 13. As Figure 12, but calculated lines are for 3×10^{-2} mol/l NaBr in the aqueous phase. Experimental points use this supporting electrolyte also.

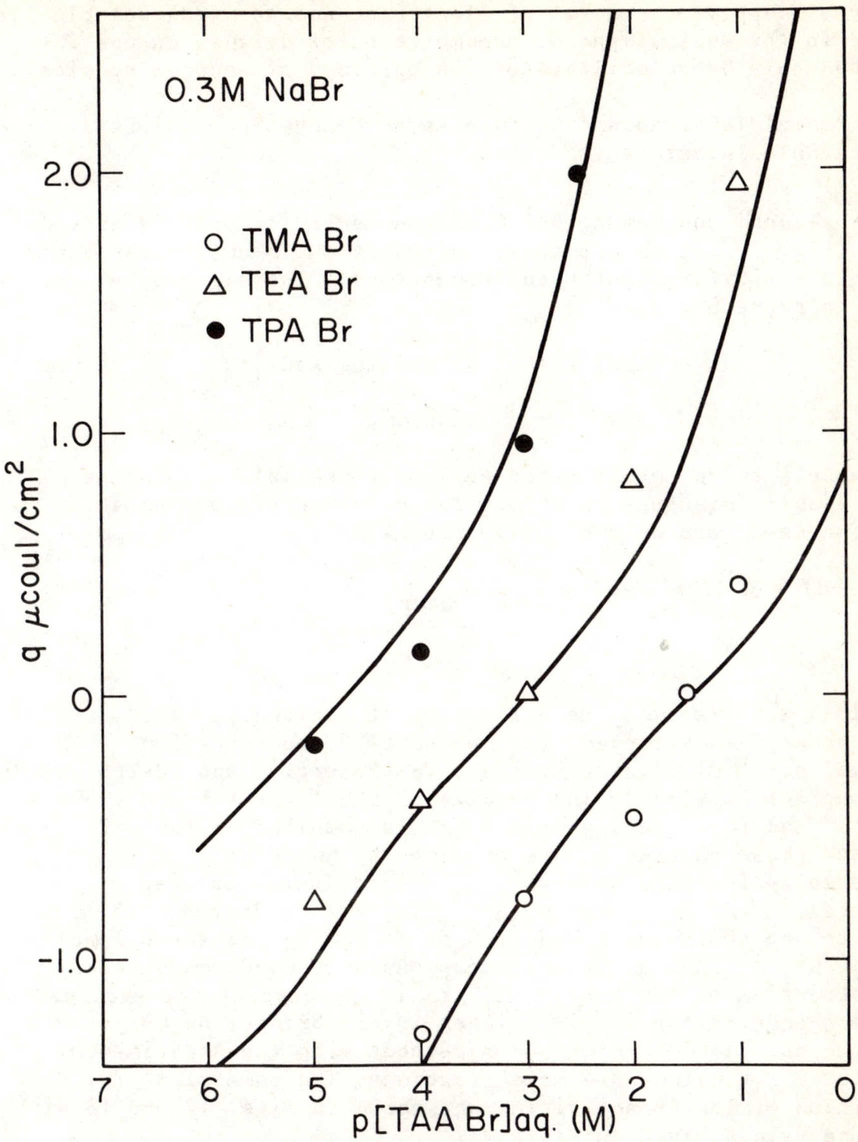

Figure 14. As Figure 12, but calculated lines are for 0.3 mol/l NaBr in the aqueous phase with corresponding experimental points.

external supply or removal of electrons, and for very soluble salts in the aqueous phase, accumulation of diffuse charge is limited only by solubility (or ion pairing) of aqueous species.

Interfacial Tensions and Diffuse Layer Charges at a Blocked, Polarizable Interface

Kakiuchi and Senda (54) used the ascending water electrode cell (Fig. 3) to measure the tensions of the nearly polarizable interface LiCl(aq)/TBATPB (nitrobenzene). The electrochemical cell is represented by

$$\mathrm{Ag}|\mathrm{AgCl}|\mathrm{TBACl}|\mathrm{TBATPB}|\mathrm{LiCl}|\mathrm{AgCl}|\mathrm{Ag}$$

(water) (nitrobenzene) (water)

A four electrode potentiostat applied potentials across the polarizable interface at 25°C. The electrocapillary equation has the same form as eqn. (24) above with the σ_+ term removed:

$$-d\gamma = q^w d\underline{\Delta}\phi(w\rightarrow o) + \Gamma_{Li^+(water)} d\mu_{LiCl} + \bar{\Gamma}_{TPB(org.)} d\mu_{TBATPB} \quad (31)$$

Tensions were measured as a function of applied voltage (electrocapillary curves) for two sets of concentration variations: LiCl(aq.) at constant TBATPB(org.), and TBATPB(org.) at constant LiCl(aq.) The results of the first set are shown in Fig. 15 and the location of the pzc is compiled in Table V. Aqueous phase charges q^w are obtained by numerical differentiation of γ vs log a_{LiCl}. Individual surfaces excesses $\Gamma_{Li^+(water)}$ and $\Gamma_{Cl^-(water)}$ were evaluated. The results are shown, as a function of q^w in Fig. 16 for 0.1 mol/l LiCl. Solid lines are values computed by Gouy-Chapman theory. A subtraction of the term $x_{LiCl}{}^w/x_w{}^w\Gamma_w$ to correct for excluded space charge in the ion-free inner layer, brings the Gouy-Chapman calculation into good agreement with the experimental points. The x values are mole fractions. The same plots for variation of TBATPB activities are shown in Figs. 17 and 18 with the pzc values given in Table VI.

In addition to these results, Esin-Markov plots and the potentials at the outer Helmholtz Planes in each phase were computed as a function of rational potential. There is no evidence for adsorption of ions at the Inner Helmholtz Plane. Again using the free energies of single ion transfer, an extra

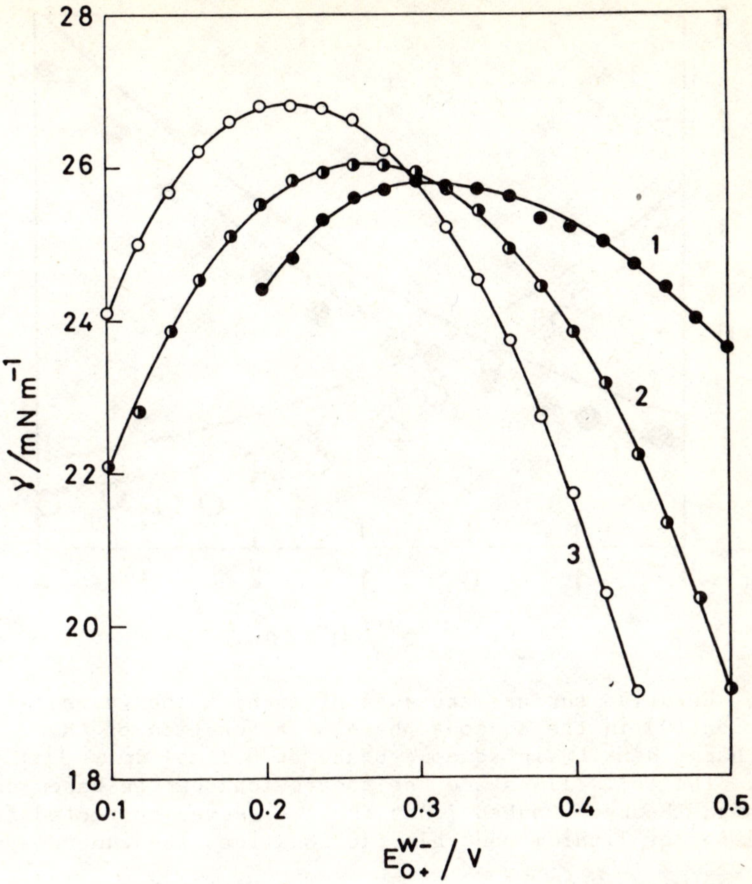

Figure 15. Electrocapillary curves for the interface between nitrobenzene solution of 0.1 mol dm^{-3} tetrabutylammonium tetraphenylborate and aqueous solution of (1): 0.01, (2): 0.10, and (3): 1.00 mol dm^{-3} lithium chloride at 25°C.

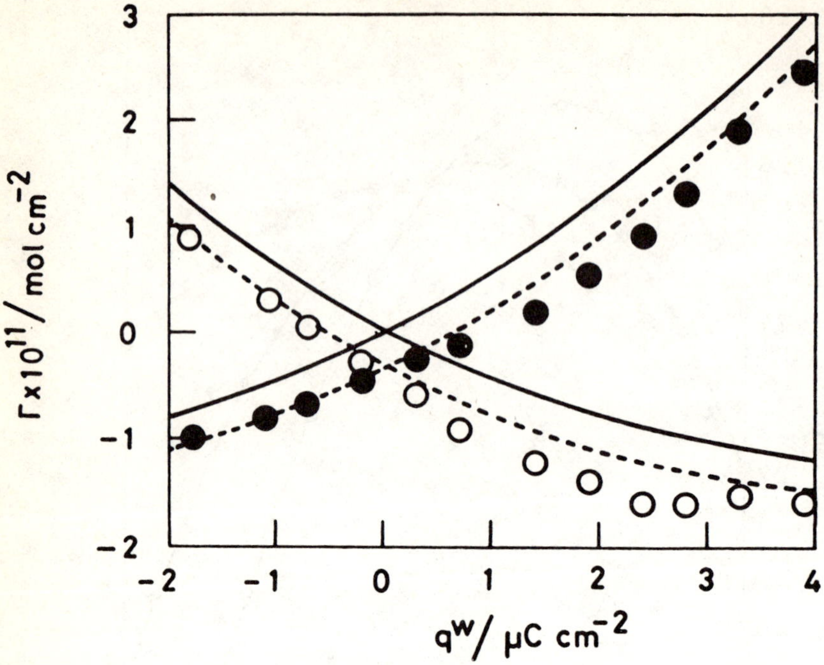

Figure 16. Relative surface excesses of lithium ion (●) and chloride ion (O) in the aqueous phase as a function of the surface charge density in aqueous phase at 0.1 mol dm^{-3} lithium chloride. The solid lines are the theoretical curves based on Gouy-Chapman theory. Dashed lines are the curves corrected for the exclusion of lithium and chloride ions from the inner layer.

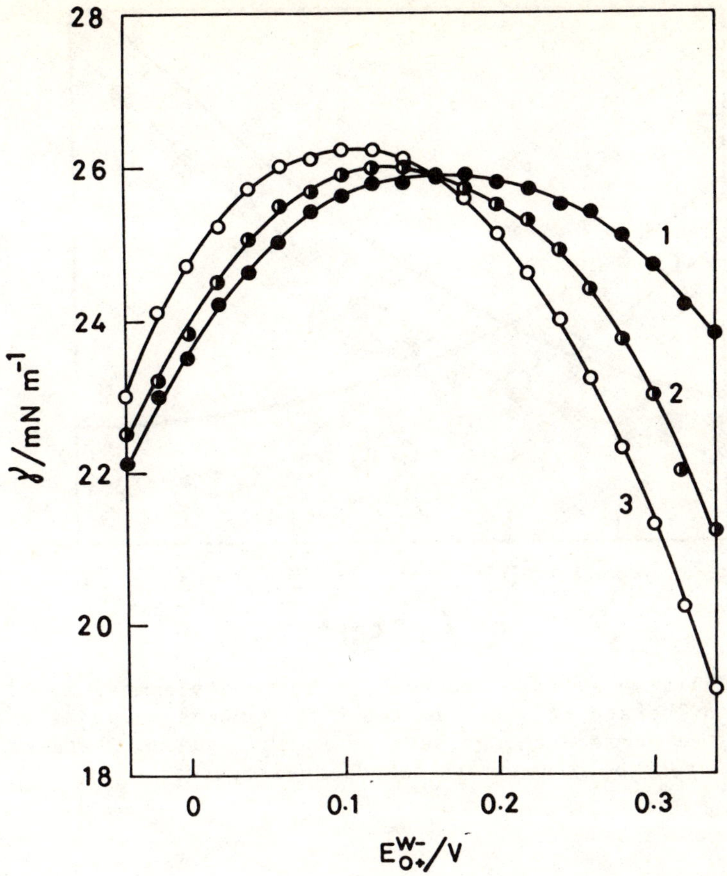

Figure 17. Electrocapillary curves for the interface between aqueous solution of 0.1 mol dm^{-3} lithium chloride and nitrobenzene solution of (1): 0.02, (2): .05, and (3): 0.17 mol dm^{-3} tetrabutylammonium tetraphenylborate at 25°C.

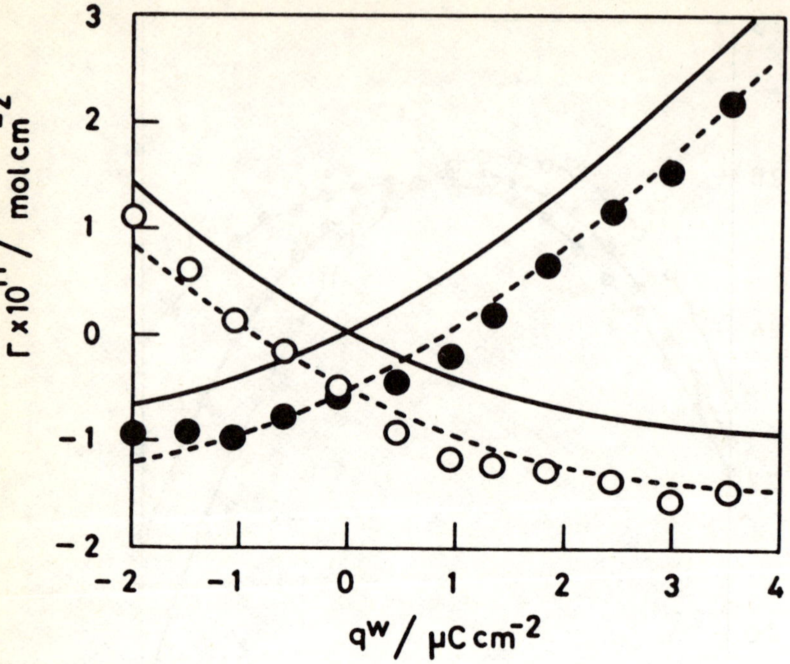

Figure 18. Relative surface excesses of tetrabutylammonium (O) and tetraphenylborate (●) ions in the nitrobenzene phase as a function of the surface charge density in the aqueous phase at 0.1 mol dm^{-3} tetrabutylammonium tetraphenylborate.

TABLE V. Coordinates of Electrocapillary Maximum for the Interface Between Nitrobenzene Solution of 0.1 mol dm^{-3} Tetrabutylammonium Tetraphenylborate and Aqueous Solution of 0.01 to 1.0 mol dm^{-3} Lithium Chloride at 25°C

c^{w}_{LiCl} / mol dm^{-3}	$a^{\pm,w}_{LiCl}$	γ_{pzc} / mN m^{-1}	E_{pzc} / V
0.01	0.009	25.8	0.309
0.02	0.017	25.7	0.289
0.05	0.041	25.8	0.278
0.10	0.079	26.0	0.265
0.20	0.151	26.2	0.245
0.50	0.379	26.5	0.227
1.00	0.776	26.8	0.217

TABLE VI. Coordinates of Electrocapillary Maximum for the Interface Between Aqueous Solution of 0.1 mol dm^{-3} Lithium Chloride and Nitrobenzene Solution of 0.01 to 0.17 mol dm^{-3} Tetrabutylammonium Tetraphenylborate at 25°C

c^{w}_{LiCl} / mol dm^{-3}	$a^{\pm,NB}_{TBATPB}$	γ_{pzc} / mN m^{-1}	E_{pzc} / V
0.01	0.008	25.9	0.167
0.02	0.014	25.9	0.153
0.03	0.021	26.0	0.142
0.05	0.033	26.0	0.131
0.07	0.044	26.0	0.129
0.10	0.061	26.1	0.122
0.17	0.099	26.3	0.108

thermodynamic assumption, it was possible to calculate the residual potential (χ-potential) between the two Inner Helmholtz Planes. These results are illustrated in Figs. 19 and 20. The residual compact layer potential difference is near zero for small charge values ± 2 μCcm^{-2}, but increases with positive q^{w}. The potential across the aqueous compact layer at the pzc is nearly independent of salt concentrations and has a value of about 20 mV, nearly 1/3 that

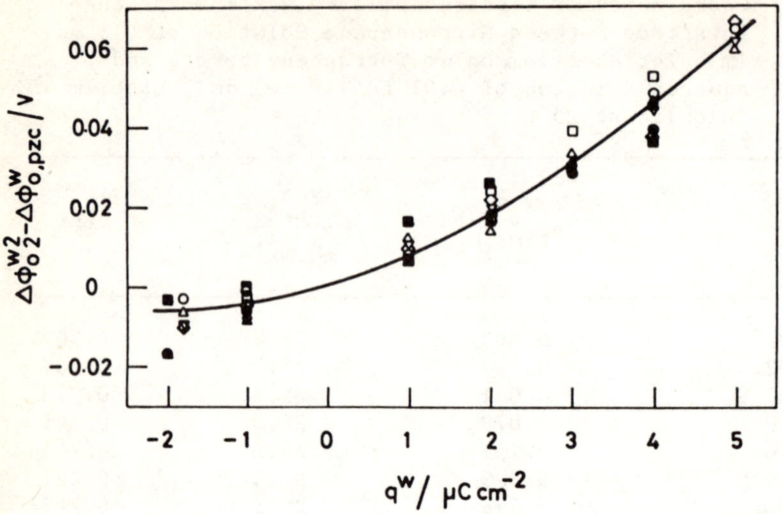

Figure 19. Change of the inner layer potential difference across the two outer Helmholtz planes $\Delta\phi_{o2}^{w2}$ with the change of the surface charge density in aqueous phase at various solution compositions. The concentration of lithium chloride is 0.05 (□), 0.1 (△), 0.2 (◇), 0.5 (▽) and 1.0 (O) mol dm^{-3} when the concentration of tetrabutylammonium tetraphenylborate is 0.1 mol dm^{-3} and the concentration of tetrabutylammonium tetraphenylborate is 0.05 (■), 0.1 (●), and 0.17 (▲) mol dm^{-3} when the concentration of lithium chloride is 0.1 mol dm^{-3}.

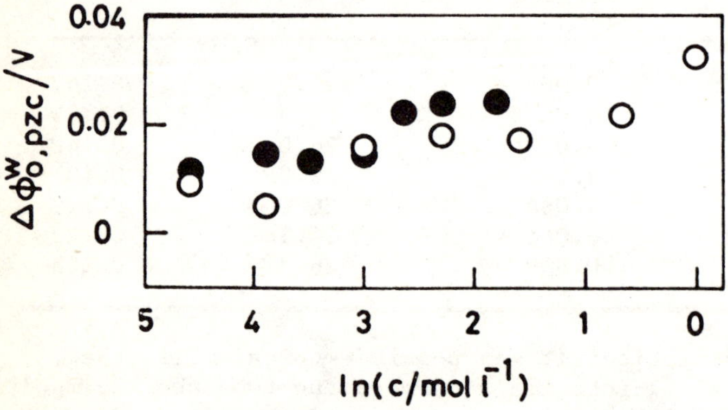

Figure 20. Potential drop across the two outer Helmholtz planes as a function of the concentrations of tetrabutylammonium tetraphenylborate (●) and of lithium chloride (O).

at an aqueous/mercury interface. Evidently the orientation of dipoles shows less alignment, or there are compensating water and nitrobenzene dipoles.

Kinetics of Ion Crossings, and Polarization Curves

The topic of ion crossing kinetics resembles superficially the similar topic in electron transfer theory. However, there are at least four important differences: (1) The L/L interface is doubly diffuse and two sets of Outer and Inner Helmholtz Planes appear in the model; (2) The moderate, if not small compact layer suggests that the straightforward dependence of rate constant at the OHP (or plane of ion transfer) on potential may not be significant and dominant as it is in electron transfer kinetics; 3) The Frumkin correction of reactant concentrations at the OHP (or plane of ion transfer) from bulk values may be a dominant consideration for ion transfers; and (4) the single salt interface is not adjustable in terms of concentrations, as pointed out above. One cannot legitimately identify standard states of unit activity simultaneously existing in both phases. Thus to define $\Delta\phi^{o}$ for unit activities of species M^+ in each phase requires use of two salts M^+X^-(aq.) and M^+Y^-(org.) with X^- ideally hydrophilic and Y^- ideally hydrophobic so that independent adjustment of $[M^+]$ and $[\overline{M}^+]$ can be achieved without recourse to equilibrium partitioning calculations. The quasi-equilibrium 4-state model is used for the derivation of potential dependent rate constants, fluxes, net current densities and exchange current densities.

To keep the arguments general, we will consider both potential-dependence of rate constants and concentrations as identifiable effects. The interface is then visualized in comparison with the electrolyte/metal interface with reference to the model in Figure 21.

If the current density for a kinetically controlled ion transfer is expressed as

$$I = \vec{i} - \overleftarrow{i} = zF(\vec{J} - \overleftarrow{J}) \tag{32}$$

where $\vec{i}$ and $\vec{J}$ are inward-going quantities (water to solvent) current and flux, two types of rate constant can appear:

$$I = zF[\vec{k}_t a(0) - \overleftarrow{k}_t \overline{a}(0)] \tag{33}$$

or

$$I = zF[\vec{k}a - \overleftarrow{k}\overline{a}] \tag{34}$$

where $\vec{k}_t$ and $\overleftarrow{k}_t$ are "true" rate constants at the "surface" positions (perhaps OHP) of ion transfer, within the space-charge region, and $\vec{k}$ and $\overleftarrow{k}$ are conventional formal rate

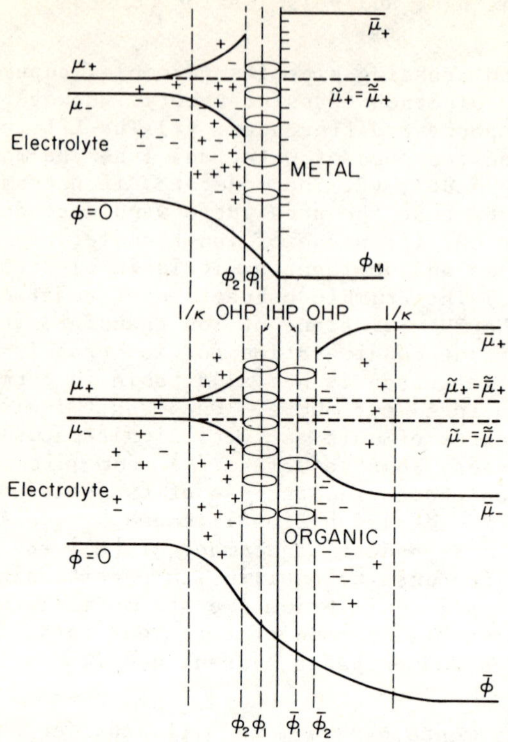

Figure 21. Comparative local structure for electrolyte/metal (upper) and immiscible liquid/liquid electrolytes (lower) equilibrium interfaces. (ϕ) Local electrostatic potential: (ϕ_2, ϕ_1) OHP and IHP potentials: (μ_+, μ_-) single ion chemical potentials. Tildes denote electrochemical potentials, and overbars organic phase parameters. Ellipses are possible aligned dipoles. $1/\kappa$ is the Debye screening length.

constants applying to bulk ion activities as modified by concentration polarization. These latter activities can be designated as surface activities that would exist if space-charge perturbations of activity profiles were not considered. For small currents and well-stirred solutions a and $\bar{a}$ can be considered bulk values. In this section z is $|z|$.

Considering an ion in electroneutral aqueous solution as state I, at the site of transfer as state II, at the final transfer state in the organic phase as III and at the electroneutral region of the hydrophobic liquid phase as IV, the standard energies for cations are

$$\tilde{\mu}_I^{\,o} = \mu^o + zF\phi; \; \tilde{\mu}_{II}^{\,o} = \mu^o + zF\phi_2 \tag{35}$$

$$\tilde{\bar{\mu}}_{III}{}^{o} = \bar{\mu}^{o} + zF\bar{\phi}_2; \quad \tilde{\bar{\mu}}_{IV}{}^{o} = \bar{\mu}^{o} + F\bar{\phi} \tag{36}$$

Assuming that the activation energy to reach the transition state is referred to the bulk state, and that the forward activation barrier involves a fraction α of the electrical part of the energy between states II and III, then the forward activation energy might be of the form

$$\Delta\vec{G}^{o} = G^{*} - \tilde{\mu}_{I} = G^{*} - \tilde{\mu}_{II}{}^{o} + (\tilde{\mu}_{II}{}^{o} - \tilde{\mu}_{I}{}^{o})$$
$$= \Delta\vec{G}(\text{chem.}) + \alpha(\tilde{\bar{\mu}}_{III}{}^{o} - \tilde{\mu}_{II}{}^{o}) + (\tilde{\mu}_{II}{}^{o} - \tilde{\mu}_{I}{}^{o}) \tag{37}$$

and

$$\Delta\overleftarrow{G}^{o} = \Delta\overleftarrow{G}(\text{chem.}) + (\alpha-1)(\tilde{\bar{\mu}}_{III}{}^{o} - \tilde{\mu}_{II}{}^{o}) + (\tilde{\bar{\mu}}_{III}{}^{o} - \tilde{\bar{\mu}}_{IV}{}^{o}) \tag{38}$$

The formal rate constants $\vec{k}$, $\overleftarrow{k}$, $\vec{k}_t$ and $\overleftarrow{k}_t$ are related to each other, and to the potential-independent chemical rate constants $\vec{k}_o$ and $\overleftarrow{k}_o$ by

$$\vec{k} = \vec{k}_o \exp[-\alpha zf(\bar{\phi}_2-\phi_2)]\exp[-zf(\phi_2-\phi)] \tag{39a}$$

$$= \vec{k}_t \exp[-zf(\phi_2-\phi)] \tag{39b}$$

$$\overleftarrow{k} = \overleftarrow{k}_o \exp[(1-\alpha)zf(\bar{\phi}_2-\phi_2)]\exp[-zf(\bar{\phi}_2-\bar{\phi})] \tag{40a}$$

$$= \overleftarrow{k}_t \exp[-zf(\bar{\phi}_2-\bar{\phi})] \tag{40b}$$

The energy parameters in eqns. (35) and (36) are illustrated as solid lines in Fig. 22. After writing current I in terms of the equilibrium rate constants, $\vec{k}^{eq}$ and $\overleftarrow{k}^{eq}$ for I = 0, and from the identity, we find

$$\Delta\phi_2 = \Delta\phi - (\bar{\phi}-\bar{\phi}_2) + (\phi-\phi_2) \tag{41}$$

$$I/zF = \vec{k}_o[\exp\{-\alpha zf\Delta\phi + (\alpha-1)zf(\phi_2-\phi) + \alpha zf(\bar{\phi}-\bar{\phi}_2)\}a] - \overleftarrow{k}_o[\exp\{(1-\alpha)zf\Delta\phi + (1-\alpha)zf(\phi-\phi_2) - \alpha zf(\bar{\phi}_2-\bar{\phi})\}\bar{a}] \tag{42}$$

which can be factored into the double-interface or membrane Frumkin correction, and the usual terms:

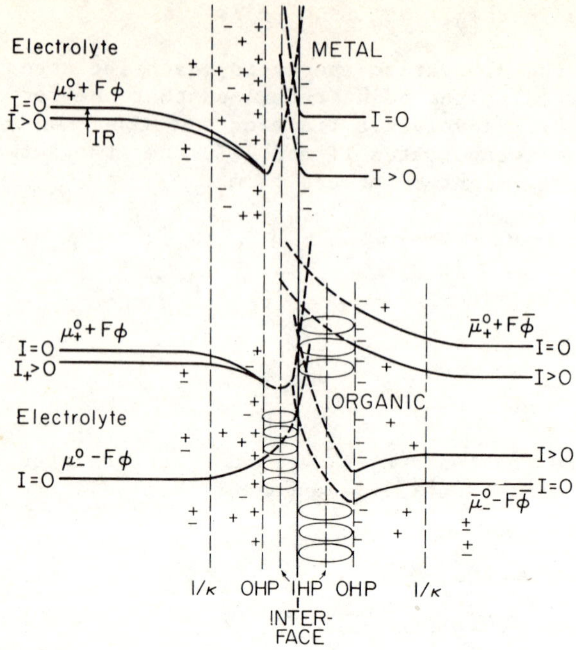

Figure 22. Energy barrier model for ions at equilibrium and positive current. (Upper) Electrolyte/metal interface; (lower) immiscible liquid/liquid electrolyte interface. Note that $\mu^o + zF\phi$ is the energy parameter for ions that is equivalent to the Fermi energy for electrons. Hypothetical barriers are sketched according to the ionic analog of the Erdey-Gruz-Volmer model. Bar parameters apply to the organic phase.

$$I/zF = [\exp\{(\alpha-1)zf(\phi_2-\phi) + \alpha zf(\bar{\phi}-\bar{\phi}_2)\}][\vec{k}_o\exp(-\alpha zf\underline{\Delta}\phi)a - \overleftarrow{k}_o\exp\{(1-\alpha)zf\underline{\Delta}\phi\}\bar{a}] \tag{43a}$$

$$FC = \exp\{(\alpha-1)zf(\phi_2-\phi) + \alpha zf(\bar{\phi}-\bar{\phi}_2)\} \tag{43b}$$

Noting that

$$\vec{k}_o \times \text{Frumkin correction} \times \gamma = k_f \tag{44a}$$

$$\overleftarrow{k}_o \times \text{Frumkin correction} \times \bar{\gamma} = k_b \tag{44b}$$

one has the familiar form:

$$I/zF = k_f \exp(-\alpha zf\underline{\Delta}\phi)c - k_b \exp\{(1-\alpha)zf\underline{\Delta}\phi\}\bar{c} \tag{45}$$

Except for the unusual Frumkin correction, these results could have been anticipated from ordinary potentiostatic kinetic theory.

When only one ion transfers from aqueous phase to organic phase there is no salt partition equilibrium involved. Activities a and $\bar{a}$ are independently adjustable. Both may be set at unit activity to establish $\underline{\Delta}\phi^o$, analogous to E^o, or any other $c = \bar{c}$ to establish $\underline{\Delta}\phi^{o'}$, analogous to $E^{o'}$. Rate constants at equilibrium and unit activities can be defined in more conventional forms:

$$k_o{}^o = \vec{k}_o \exp[-\alpha zf\underline{\Delta}\phi^o] = \overleftarrow{k}^o \exp[(1-\alpha)zf\underline{\Delta}\phi^o] \tag{46a}$$

$$k_o{}^{o'} = \vec{k}_o \exp[-\alpha zf\underline{\Delta}\phi^{o'}] = \overleftarrow{k}_o \exp[(1-\alpha)zf\underline{\Delta}\phi^{o'}] \tag{46b}$$

Then one can further simplify and refer to equilibrium potentials in eqn. (43):

$$I = i_t{}^o[\text{Frumkin correction}] \times [\exp\{-\alpha zf(\underline{\Delta}\phi-\underline{\Delta}\phi^{eq})\} - \exp\{(1-\alpha)zf(\underline{\Delta}\phi-\underline{\Delta}\phi^{eq})\}] \tag{47a}$$

$$I = i^o \times [\exp\{-\alpha zf(\underline{\Delta}\phi-\underline{\Delta}\phi^{eq})\} - \exp\{(1-\alpha)zf(\underline{\Delta}\phi-\underline{\Delta}\phi^{eq})\}] \tag{47b}$$

These forms can also be written in terms of the overpotential $\eta = \underline{\Delta}\phi - \underline{\Delta}\phi^{eq}$. The exchange currents $i_t{}^o$ and i^o are given by

$$i^o = zFk_o{}^o a^{1-\alpha}\bar{a}^{\alpha} \tag{48}$$

$$i_t{}^o = i^o(FC) \tag{49}$$

since $k_o{}^o = k_t{}^o(FC)$ (50).

The principal difference between ion transfer kinetics and electron transfer kinetics is the Frumkin Correction that contains the effect of the compact layer potential difference in the space between two OHPs. Thus ϕ_2 and $\bar{\phi}_2$ appear in the correction. For comparison with other treatments of ion transfer kinetics, Samec, et. al., use the symbols:

$$k_o^{o'} = k^o;\ k_o = k^{w\to o};\ k_t = k_t^{w\to o} \tag{51}$$

In the foregoing analysis, no partitioning of applied potential $\Delta\phi$ between diffuse and compact layers was assumed. Koryta has extended the analysis by using the Gouy-Chapman theory to relate $\phi_2-\phi$ and $\bar{\phi}_2-\bar{\phi}$ to the actual concentrations and dielectric constants of the phases. Excluding activity coefficients and restricting applied voltages to values large compared with RT/F,

$$I/zF = k_o^{o'}\frac{[\epsilon(w)c(w)]^{\pm\frac{1}{2}}}{[\epsilon(o)c(o)]} \exp\left(\tfrac{1}{2}-\alpha\right)\frac{F}{RT}\Delta\phi^o \times\{c_i(w)\exp\left[\beta\frac{F}{RT}(\Delta\phi-\Delta\phi^o)\right] - c_i(o)\exp\left[-(1-\beta)\frac{F}{RT}\times(\Delta\phi-\Delta\phi^o)\right]\} \tag{52}$$

where $\beta = \frac{1}{2}$ and the plus sign in the exponent of the first expression in square brackets applies for $\Delta\phi>0$ and minus sign for $\Delta\phi<0$. Thus, for k_o

$$k_o = k_o^{o'}\frac{[\epsilon(w)c(w)]^{\pm\frac{1}{2}}}{[\epsilon(o)c(o)} \exp\left[\left(\tfrac{1}{2}-\alpha\right)\frac{F}{RT}\Delta\phi^o\right] \tag{53}$$

and for an operational charge transfer coefficient $\beta = \frac{1}{2}$. It should be noted that this value does not follow from the "symmetry of energy barrier" for charge transfer, but merely from the formalism of the Gouy-Chapman theory.

There are no direct measurements of ion crossing rates using small voltage steps and the usual Tafel analyses. Instead, the bulk of data arise from large amplitude methods in which rates and mass transport are coupled. In addition, possible self-retardation by formation of surface films has not been taken into account. Nevertheless, there are trends in the rate constant data from Table VII. For the tetralkylammonium ions, the least oil soluble show the smallest forward rate constants for the w→o transfer. For the other ions, those with greater lipophilicity show greater rates into the organic phase. The value of the rates put transfers in the quasi-reversible range. It is possible, however, that all of the w→o rates are reversible and that surface film formation during the measurement actually slows the reactions to give the appearance of quasi-reversibility.

The question of the transfer coefficient α and its significance cannot be resolved. So far most of the data show $\alpha(w\to o)$ is less than $\frac{1}{2}$.

Impedance Characteristics of a Blocked L/L Interface and Determination of the Double Layer Capacitance.

TABLE VII. Electrochemical Kinetic Rate Constants for Water/Nitrobenzene Electrolyte Systems

Ion	α	$-\log k_s$	$-\log k_f^\circ$	$-\log k_b^\circ$	Direction	Method	Ref.
NMe_4^+	0.5	3.3			B aqueous	CV	37
	reversible				w→n	P	41
NEt_4^+	0.345	2.7	2.3	3.35	w→n	CP	24
			2.6	3.64	w→n	CP	28
$NProp_4^+$	0.272	2.63	1.91	4.57	w→n	CP	24
			1.94	4.69	w→n	CP	28
NBu_4^+	0.234	2.32	1.25	5.82	w→n	CP	24
			1.35	5.51	w→n	CP	28
$NPent_4^+$			0.96	6.86	w→n	CP	28
Na^+	.40		2.17		w→n	CP	21
Cs^+	.40	1.26			B aqueous	CV	39

TABLE VII. (continued)

Ion	α	$-\log k_s$	$-\log k_f^\circ$	$-\log k_b^\circ$	Direction	Method	Ref.
Br^-	.24	6.5	2.7	7.83	n→w	CP	19
	.29		3		n→w	CP	21
Pic^-	reversible				w→n	P, CV	106
					B aqueous	CV	
NEt_4^+	0.3	2.88			w→n	not specif.	107
$NProp_4^+$	0.3	2.76			w→n		107
NBu_4^+	0.3	2.60			w→n		107
$NPent_4^+$	0.3	2.73			w→n		107
$Choline^+$	0.3	1.92			w→n		107
$acetycholine^+$	0.3	1.80			w→n		107
$choline^+$	0.3	2.72			w→n		108
$acetylcholine^+$	0.3	1.85			w→n		108

TABLE VII. (continued)

k_s = calculated from k_f°, k_b°

P = polarography

CP = chronopotentiometry

CV = cyclic voltammetry

B = both

Only recently (66) have conventional impedance measurements been applied to blocked L/L systems. Previously, Gavach et al. (21) investigated impedances of the aqueous/nitrobenzene interface using systems in which one phase contained an ion reversibly transferable across the interface. A diffusion polarization or Warburg-like impedance was observed from 200 Hz to 20 kHz. Schuhmann and Seta (101) developed a theoretical model of L/L interfacial impedances allowing for ion adsorption at the interface prior to phase transfer.

More recent work of Samec et al. (44) dealt with AC impedance measurements of the capacitance of the aqueous/nitrobenzene interface. In that study impedance was measured by an AC polarographic technique and interfacial capacitance was evaluated as a function of potential difference between the two phases. The interface was polarized only in the potential range where transfer of the base electrolyte ions was barely detected by cyclic voltammetry.

The effect of AC current passing through the water/nitrobenzene interface was also investigated by Joos and Vanden Bogaert.(102) Their measurements however do not lead to impedance results since they studied a nonpolarizable interface using cetyltrimethylammonium bromide in both solvents. Because the distribution constant of this salt is almost unity, cetyltrimethylammonium cations and bromide anions can move freely from one phase to the other. In each case the interface will contribute little or no impedance.(103)

The polarizable (or nearly polarizable) interface is, on the basis of the earlier work, a good candidate for modeling in terms of a Randles circuit as in Figure 23.

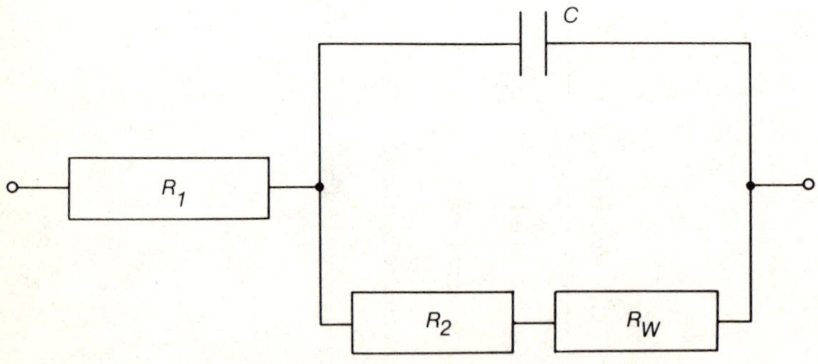

Figure 23. Randles equivalent circuit of the electrochemical interface. (R_1) ohmic electrolyte resistance; (R_2) transfer resistance; (C) double layer capacitance; (R_w) Warburg impedance.

In addition to electrolyte resistance two other constraints on the flow of electric current must be taken into account: the charge transfer resistance R_{ct} and the Warburg impedance R_w. For ideally reversible processes R_{ct} tends to zero. For a blocked interface R_{ct} assumes some finite value dependent upon the rate of ion transfer between the phases. Geometric impedances of sufficiently concentrated electrolytes will not interfere with measured impedances since time constants for these processes are on the order of 10^{-7} s. In Figure 23, R_1 stands for electrolyte resistance, R_2 is the transfer resistance, C is the total double layer (doubly diffuse) resistance, C is the total double layer (doubly diffuse) capacitance and R_w is the Warburg impedance. The Warburg impedance represents a kind of resistance to mass transfer. The classical semi-infinite Warburg impedance is a complex quantity equal to:

$$Z_w = \sigma w^{-\frac{1}{2}} + j\sigma w^{-\frac{1}{2}} \tag{54}$$

The admittance components therefore must be:

$$Y' = w^{\frac{1}{2}}/2\sigma \tag{55}$$

$$Y'' = w^{\frac{1}{2}}/2\sigma \tag{56}$$

where σ is the diffusion parameter with the dimension $\Omega\ s^{-\frac{1}{2}}$. For nearly blocked interfaces at low currents, Warburg or diffusion polarization impedances can be considered unimportant. The resistance to charge transfer across an interface is large and the circuit of Randles simplifies to a series RC network; R is the solution resistance and C the total diffuse layer capacitance. For this situation the imaginary admittance is given by:

$$Y'' = wC/[1+w^2C^2R^2] \tag{57}$$

When $w^2C^2R^2 >> 1$ this expression simplifies to:

$$Y'' = 1/wCR^2 \tag{58}$$

and when $w^2C^2R^2 << 1$

$$Y'' = wC \tag{59}$$

The real admittance is given by:

$$Y' = R/[R^2 + 1/w^2C^2] \tag{60}$$

and it may be similarly simplified.

The system chosen for study was water/nitrobenzene electrolytes containing 0.01 mol/l NaCl and 0.005 mol/l TPAsTPB, respectively. The steady state current dependence upon the applied DC potential is shown in Figure 24. The near zero current ($<1\mu A/cm^2$) from −375 mV to 400 mV indicates the interface remains nearly blocked within this potential range. The positive and negative potential limits correspond respectively to Na^+ and Cl^- transfer from water to nitrobenzene. $\Delta G(w\to o)$ values of 34.2 and 31.4 kJ/mol for these transfers are more favorable than the value of -35.9 kJ/mol for both $TPAs^+$ and TPB^-. Real and imaginary admittances were measured as a function of applied potential difference across the

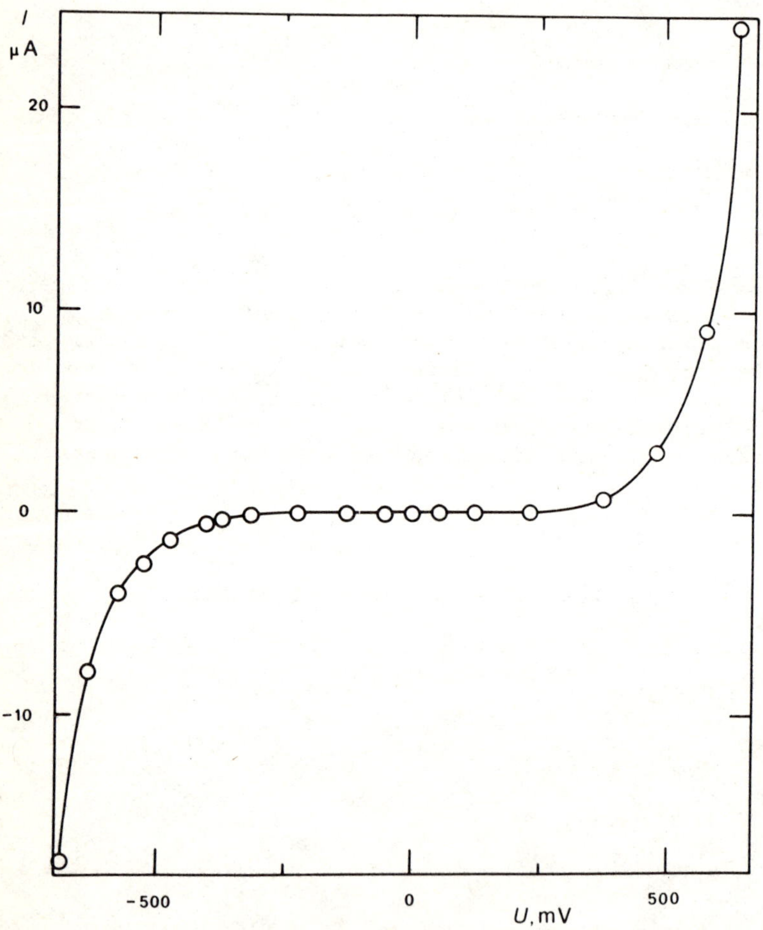

Figure 24. Steady state voltage-current characteristics of the interface. Electrolytes: 0.01 mol/l NaCl in water, 0.005 mol/l TPAsTPB in nitrobenzene; interface area 0.5 cm^2.

liquid/liquid interface. An example of an admittance vs DC potential plot, obtained at 1 Hz is shown in Figure 25.

This rather unusual looking response is consistent with a DC-dependent capacitance and a constant resistance. The real admittance approaches the conductivity value far from the pzc as predicted from eqn. (60), while the real admittance followed eqn. (59) near the pzc, but becomes attenuated according to eqn. (57) far from the pzc. Bulk IR drops were calculated and subtracted from the applied potential to determine interfacial potential drops. Because of low currents the corrections were <1 mV for applied potentials from -325 to 275 mV.

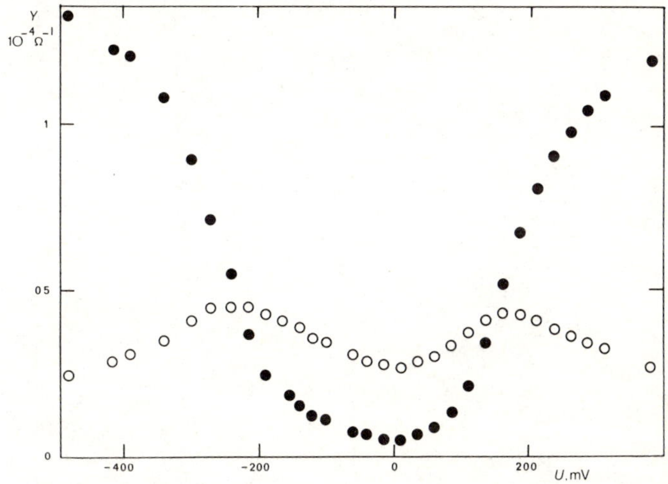

Figure 25. 1 Hz admittance values of the system vs. applied potential. (●) Real admittance; (O) imaginary admittance.

Impedance plane plots for 0.0, -220 and -400 mV are shown in Figure 26. At 0 V the plot approaches the characteristic vertical line of a series RC network. Lack of a vertical line may reflect a small but measureable concentration polarization. The high frequency solution resistance is given by the real axis intercept of the line. At -220 mV the impedance plane plot begins to show a Warburg-like behavior characterized by a line of greater than unity slope that merges into a vertical line corresponding to the double layer capacitance impedance. A greater than unity Warburg slope has previously been observed for unblocked interfaces and was attributed to an adsorption step prior to phase transfer.(101) Our system may show a similar adsorption dependence as the interface becomes unblocked by virtue of DC current flow at the larger applied potential differences. At -400 mV the impedance plane plot shows classical Warburg behavior characterized by a line of unity slope. The appearance of Warburg behavior over this frequency range is clearly a result of the decrease in cut off frequency corresponding to the increased double layer capacitance with potential. Investigation of the interfacial impedance at frequencies extending below 0.5 Hz should reveal a vertical line in impedance plane plots for higher interfacial potentials, provided the assumed series RC circuit is correct.

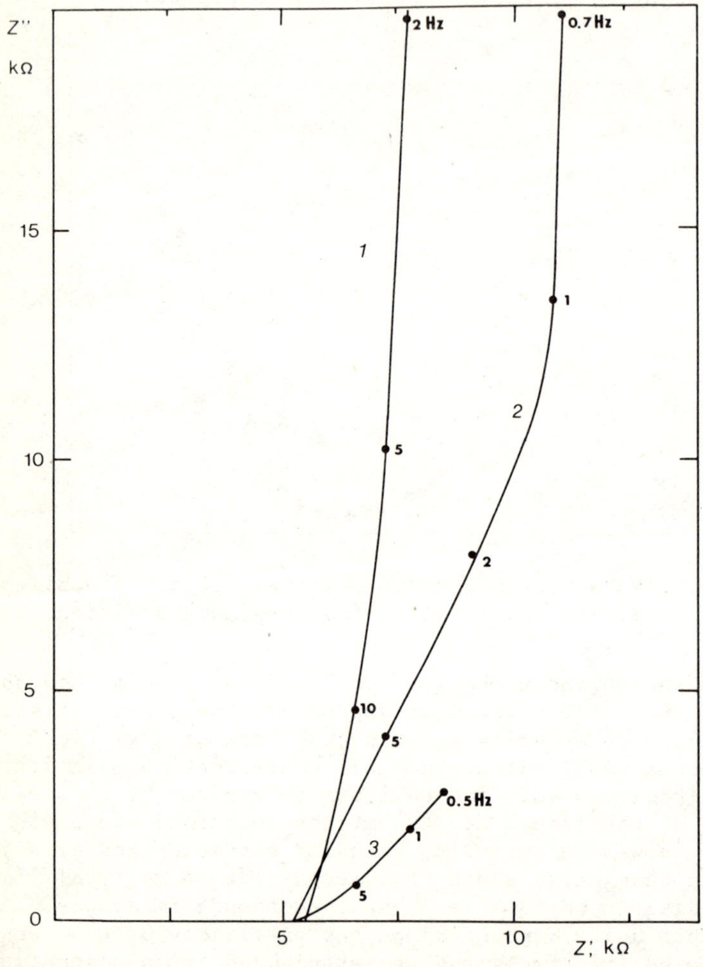

Figure 26. Complex plane impedance plot (Argand diagram) for experimental data. 1 - 0.0 mV of applied voltage, 2 - -220 mV, 3 - —400 mV. Bold figures refer to the measuring frequency.

Three general methods may be employed to obtain double layer capacitance values from the admittance data. Real and imaginary admittance values allow direct calculation of C. At the frequency which $Y_I = Y_R$ the relation $\tau = RC$ also allows calculation of C since R and w are known.

The solid line in Figure 27 shows the dependence of total double layer capacitance upon interfacial potential as calculated from Gouy-Chapman theory for the system studied. Capacitance values shown are those calculated from real admittance at 1.0 Hz, imaginary admittance at 10 Hz, use of

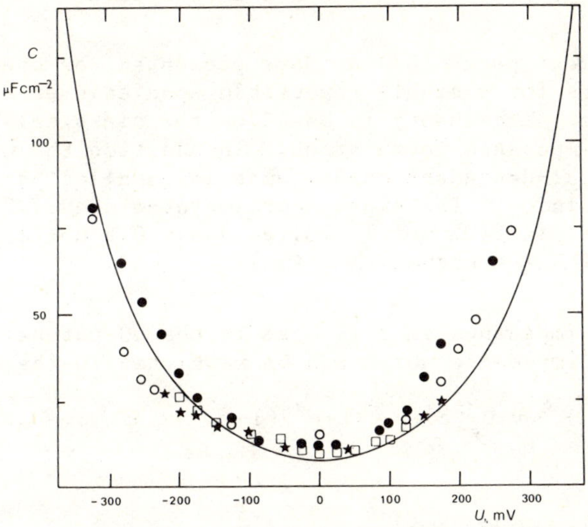

Figure 27. Specific capacitance of the interface obtained from the experimental data vs. applied interfacial potential. * - calculations from 0.5 Hz data, O - 10 Hz data, □ - $\tau = RC$ data, ● - 1 Hz data. Calculated (Gouy-Chapman) capacitance is indicated by solid line.

the relation $\tau = RC$ for frequencies at which $Y_I = Y_R$, and from 0.5 Hz imaginary admittance data. The asymmetry of the calculated capacitance values should be anticipated since opposite polarities will result in depletion layers of a different nature. Use of the relation $\tau = RC$ is limited to the potential range in which $1/\tau > 0.5$ Hz. Within this range the agreement with calculated capacitance is quite good. Use of the relations between real of imaginary admittance and capacitance is limited to frequencies at which the change in measured admittance is much larger than signal uncertainties. Also, these relations only strictly apply when the Warburg contribution to measured impedance is insignificant. This is true over the approximately 600 mV potential range that the interface remains blocked.

These data show the unexpected result that the Gouy-Chapman model is obeyed over an unusually wide DC potential range. Subsequently we studied another blocked interface 0.05 mol/l LiCl(aq.) and 0.1 mol/l TBATPB in nitrobenzene. In addition, in a recent paper (47) capacitance of a blocked L/L interface has been determined with results comparable with ours.

Impedance Characterics of an Unblocked L/L Interface

In a very recent paper (67) we have presented the theory of faradaic impedances for a single reversible species over a wide DC potential range. The theory is based on the classical reversible redox impedance formulation. In addition the Gouy-Chapman DC potential-dependent capacitance in parallel has been considered and included. The experimental system used 1.2 and 10×10^{-5} mol/l TMACl in 0.05 mol/l LiCl(aq.) and 0.3 mol/l THATPB in nitrobenzene (THA = tetrahexylamonium).

The faradaic impedance in this case is the DC potential-dependent Warburg impedance which can be rewritten in the form:

$$Z_f = (RT/\sqrt{2}n^2F^2Aw^{\frac{1}{2}}D_w^{\frac{1}{2}}c)(e^{-a}+2+e^{a})(1-j) \qquad (61)$$

where:

$$a = (nF/RT)(E_{dc}-\underline{\Delta}\phi°) \qquad (62)$$

The amplitude of faradaic current for a reversible system is:

$$I = n^2F^2Aw^{\frac{1}{2}}D_w^{\frac{1}{2}}c\underline{\Delta}E/4RT\ \cosh^2(a/2) \qquad (63)$$

Substituting:

$$H = (RT/\sqrt{2}n^2F^2Aw^{\frac{1}{2}}D_w^{\frac{1}{2}}c)(e^{-a}+2+e^{a}) \qquad (64)$$

the impedance appears in a more workable form:

$$Z_f = H(1-j) \qquad (65)$$

In real systems however, one must also take into account the bulk resistance of electrolytes and the L/L interfacial capacitance contributions to total system impedance.

We use the equivalent circuit for the interfacial impedance, which includes the capacitance in parallel with the Warburg impedance, as described by the circuit diagram in Figure 23. The total impedance is then:

$$Z = R_b + H/(1+2wCH+2w^2C^2H^2) - j(H+2wCH^2)/(1+2wCH+2w^2C^2H^2) \quad (66)$$

To interpret the effect of reversible interfacial ion transfers in this system, the imaginary impedance of this anticipated circuit has been calculated for many situations. The imaginary portion has been chosen because it strongly reflects interfacial capacitance and does not reflect the bulk resistance. Figure 28 shows calculated 1 Hz imaginary

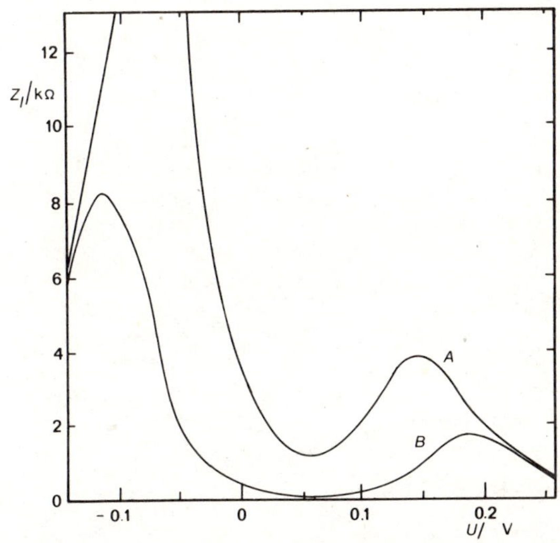

Figure 28. 1 Hz imaginary impedances calculated as a function of potential (relative to pzc) for 0.1 mol l^{-1} base electrolyte concentrations in each phase. (A) 10^{-4} reversible aqueous ion; (B) 10^{-5} mol l^{-1} reversible aqueous ion.

impedance values (Z_I) as a function of interfacial potential for tetramethylammonium ion transport across the water/nitrobenzene interface with base electrolyte concentrations of 0.1 mol l^{-1}. Aqueous TMA^+ concentrations of 10^{-5} and 10^{-4} mol l^{-1} were used in these calculations. The pzc was, for this calculation selected to be 60 mV more negative than the standard potential of transfer of the transported ion. The values of diffusion coefficients in water and nitrobenzene were assumed to be 10^{-5}

$cm^2\ s^{-1}$ and $0.33 \times 10^{-5}\ cm^2\ s^{-1}$. To account for the varying capacitance C, the Gouy-Chapman theory was used.

The potential dependence of imaginary impedances at several frequencies are shown in Figure 29 for a system with 0.05 mol/l^{-1} LiCl and 0.03 $mol\ l^{-1}$ THATPB and an aqueous concentration of $10^{-4}\ mol\ l^{-1}$ TMACl. The larger potential window of a more blocked base electrolyte supresses background currents and allows better resolution of the Z_I minimum corresponding to TMA^+ transfer. Z_I data yield a pzc capacitance of 20.3 $\mu F/cm^2$ for this base electrolyte system in reasonable agreement with the Gouy-Chapman value of 17.5 $\mu F/cm^2$.

Imaginary impedances calculated as described above are indicated by dashed lines in Figure 29. Experimental and theoretical impedances are in good agreement within the

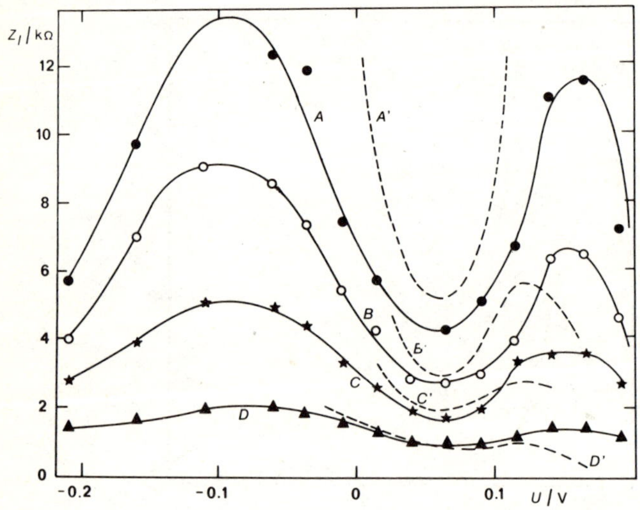

Figure 29. Imaginary impedance as a function of applied interfacial potential at (▲) 10 Hz; (*) 2.5 Hz; (O) 1.0 Hz; (●) 0.4 Hz. Dashed lines indicate Z_I calculated for the above frequencies. Aqueous electrolytes; $5 \times 10^{-2}\ mol\ l^{-1}$ LiCl and $10^{-4}\ mol\ l^{-1}$ TMACl; nitrobenzene electrolyte: $3 \times 10^{-2}\ mol\ l^{-1}$ THATPB.

potential region in which TMA^+ dominates impedance behavior. Divergence of experimental and calculated curves away from the pzc is still anticipated since calculations do not account for the small but significant base electrolyte transfer and inappropriate use of ideal static Gouy-Chapman capacitances.

In Figure 30 are shown imaginary impedances as a function of potential for the THATPB electrolyte system and a more dilute aqueous concentration of 2 x 10^{-5} TMACl. Dashed lines indicate calculated imaginary impedances. Agreement between calculated and experimental impedances is again very good near the standard potential for TMA^+ transfer.

The theoretical calculations predict only a shoulder corresponding to TMA^+ transfer, while experimental results show well-resolved Z_I peaks comparable with those observed at higher TMA^+ concentrations. There is a probable cause for this surprising result. Over the potential range where a shoulder is anticipated, a DC TMA^+ current of about 3 μA/cm^2 flows from the aqueous phase to the nitrobenzene phase. As DC currents flow, films or adsorbed layers are frequently formed at the interface.(68) These films become more pronounced and occasionally become visible as DC currents approach diffusion

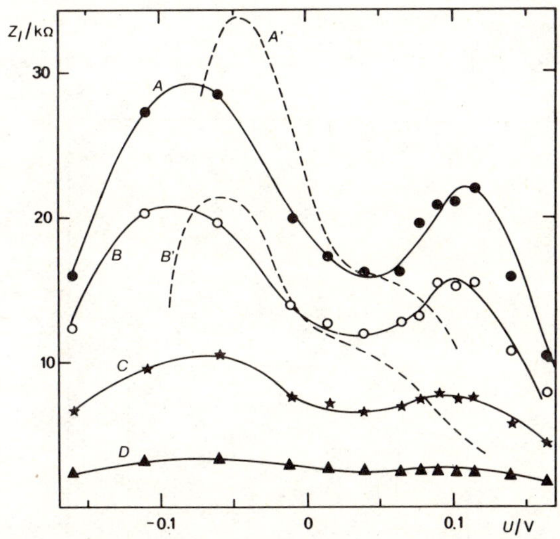

Figure 30. Imaginary impedance as a function of applied interfacial potential at (/\) 10 Hz; (*) 2.5 Hz; (O) 1 Hz; (●) 0.625 Hz. Dashed lines indicate calculated impedances at 1.0 Hz and 0.625 Ha. Aqueous electrolytes: 5x10^{-2} mol l^{-1} LiCl and 2x10^{-1} TMACl; nitrobenzene electrolyte: 3x10^{-2} mol l^{-1} THATPB.

control, and may act progressively to reduce interfacial capacitance. Decreased interfacial capacitance leads to an increased Z_I for a given frequency. A finite compact layer capacitance accompanying current flow would result in decreased interfacial capacitance. In addition, an enhanced effective surface concentration of TMA^+ during film formation may also act to decrease Z_I through increased AC TMA^+ currents. A peak could appear if an initial Z_I increase due to decreased capacitance is diminished at potentials where Warburg impedance becomes dominant. In Figure 29 experimental imaginary impedances also increase more sharply than those calculated values at potentials positive of the Z_I minimum. This results supports the above hypothesis.

Impedance theory for liquid membranes (104) includes, among other factors, high frequency geometric charging, intermediate frequency interfacial ion-crossing-kinetics, and at low frequencies, the Warburg or faradaic impedance. If the admittance data are correctly interpreted, two semicircles and the Warburg 45° line are expected in the impedance plane plot ($-Z_I$ vs Z_R). When the time constants for the individual processes are widely separated, well-resolved semicircles are expected. However, for this system the kinetics are quasi-reversible so the intermediate frequency arc should be merged into the high frequency side of the Warburg line.

Indeed, as shown in Figure 31, wide frequency range

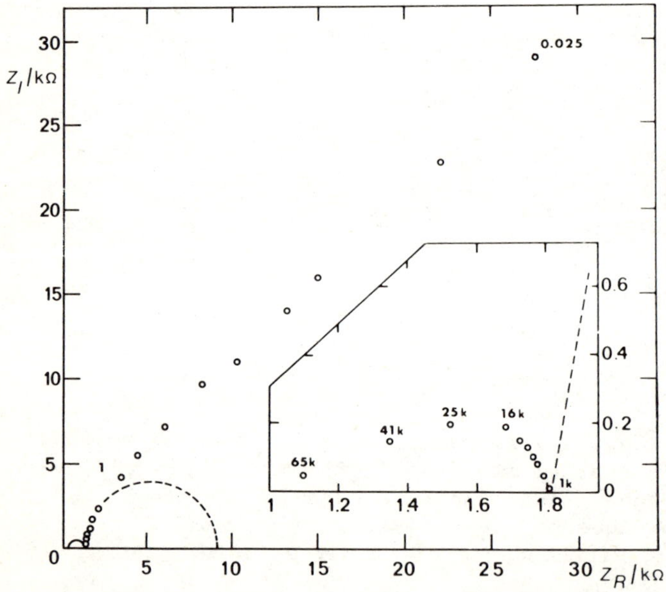

Figure 31. Impedance plane plot from 0.025 to 65000 Hz at −200 mV applied interfacial potential for a 0.01 mol l^{-1} LiCl/0.01 mol l^{-1} TBATPB electrolyte system. The dashed semicircle indicates a possible adsorption or kinetic impedance. The inset shows the high frequency geometric nitrobenzene impedance.

impedance data confirm this hypothesis. In comparison with blocked interface data, Figure 26, reversible ion transfer at an otherwise polarizable nitrobenzene/water interface has been shown to influence interfacial impedance by introduction of a faradaic mass transport impedance. Interfacial impedances have been reported for frequencies as low as 0.025 Hz, somewhat lower than previously studied.(105) The dual diffuse layer interfacial model is supported by reasonable agreement of pzc capacitances with those predicted by Gouy-Chapman theory. However, capacitances deviating from static Gouy-Chapman values have been observed at the limits of the blocked electrode "window", where base supporting electrolytes begin to carry current. The hypothesis is made that adsorption processes and increased surface concentrations of ions, associated with film formation during faradaic current passage, influence observed capacitances and impedances. We have confirmed the existence of an intermediate frequency semicircle in the impedance plane. This semicircle, which we have not previously investigated, may involve adsorption and an activation overpotential from a surface barrier.

Summary of Galvanostatic (Reversible and Irreversible) Theory For One- and Two-Ion (Same or Opposite Sign) Transfer

In 1902 Nernst and Riesenfeld (1) treated the problem of single ion transport under constant current across an immiscible L/L interface. They considered the reversibly partitioned case (KI_3) at water/phenol in the absence of supporting electrolyte. Their solutions for concentration-distance and concentration-time profiles predate the better known work of Sand on the closely related topic of solid electrodes. Their initial transport equation is

$$I(t_+ - \bar{t}_+)/F = \left[-D\frac{\partial c}{\partial x} + \bar{D}\frac{\partial \bar{c}}{\partial x}\right]_{x=0} \qquad (67)$$

in which transference numbers (t_+ and $\bar{t}_+$ for positive aqueous and organic phase ions), salt diffusion coefficients and concentrations (D, $\bar{D}$, c and $\bar{c}$ aqueous and organic respectively) appear. This equation does not apply in the presence of suporting electrolytes, but other appropriate equations can be found.

Transport of ions across immiscible liquid-electrolyte interfaces differs in several significant ways from electrolyte-metal interface processes. Diffusion-migration via Nernst-Planck equation must be considered on both sides of an interface.(59) Interfacial transport of ions of opposite sign must be accounted for with consideration of salt partition equilibria.

Partition equilibria lead to nonlinear boundary conditions that preclude use of linear transform methods for the general transport equation solution. Linearization conditions can be found. These factors cause a somewhat wider range of behavior, even for reversible systems, than is observed for electrolyte-metal examples. Coupling of ionic motion in partition-controlled cases causes a great difference between transport of a single salt component in the absence or presence of a supporting, nonpartitioning, inert electrolyte in one or both phases. Absence of inert salt can render a single salt system with a well-defined transition time to be completely ohmic with no transition time at all. This difference is more than a prolonging of transition times by migration. There is a fundamental change in the shapes of concentration-distance profiles caused by inert salts for systems that involve migration and partitioning.

For reversible and irreversible single ion transport in the presence of excess, inert, nonpartitioning electrolytes, depletion concentration polarization on the side of the exiting ion determines the transition time. Since ion fluxes are continuous across the interface, concentration profile slopes have the same sign in each phase. Accumulation of the transporting ion occurs on the entering side. For a reversibly partitioned salt in supporting electrolyte, ions of both signs carry current across the interface. For all the cases computed, as a consequence of the limited accessible potential range, depletion concentration polarization occurs on both sides of an interface in a coordinated way

A further unusual effect in the absence of supporting electrolyte is electrosorption, identified by Blank.(106) For a single reversible salt, mass transport with partitioning produces accumulation or depletion of salt in the diffusion layer, depending on current direction and relative ion transference numbers in the two phases.

The procedure (59,61,64) for the solution of the linear problems uses semi-infinite boundary conditions and Laplace transformation methods for the arrangement:

Aqueous solution	Organic solution
Concentrationn c_i	Concentration $\bar{c}_i$
Diffusion coefficient D_i	Diffusion coefficient $\bar{D}_i$
Increasing negative x	Increasing x

$$x = 0$$

Fick's second law is obeyed in all cases with proper definition of salt diffusion coefficients. The latter eliminates the field term in the Nernst-Planck equation. Positive currents are positive charges moving in the positive x-direction. Concentration in the organic phase, and the corresponding transport parameters have super bars. The basic Nernst-Planck equation is

$$J_i = -D_i\frac{\partial c_i}{\partial x} - \frac{z_iD_iF}{RT}\frac{\partial \phi}{\partial x}c_i \tag{69}$$

In supporting electrolytes the second term can be omitted, and equations for both ions combined with electroneutrality of fluxes (equality of partial currents) gives

$$J_i = -D\frac{\partial c_i}{\partial x} + z_it_i\frac{I}{F} \tag{70}$$

with $D = 2D_+D_-/(D_++D_-)$, the "salt" diffusion coefficient. The transference numbers are bulk values: $t_i = D_i/(D_i+D_j)$ for aqueous ions and $\bar{t}_i = \bar{D}_i/(\bar{D}_i+\bar{D}_j)$ for organic ions. Boundary conditions for all ion transfers are continuity of fluxes at x = 0, and for salt (two ions) transfers

$$K^2 = \left[\frac{\bar{c}_+\bar{c}_-}{c_+c_-}\right] \text{at } x = 0 \tag{71}$$

is obeyed by surface concentrations. In addition, $c_i = c_i^o$ at $x = -\infty$ and $\bar{c}_i = \bar{c}_i^o$ at $x = \infty$. It is this boundary condition that leads to non-linear equations. Linearization is achieved by high loading of the transferring ion in the receiving phase so that the concentration in that phase during current flow, remains sensibly constant.

Methods for solution are straightforward. Only the use of boundary conditions differs from case to case. From Fick's second law, for unsupported cases

$$c_i = c_i^o/s + A_i\exp[(s/D_i)^{\frac{1}{2}}x] \tag{72}$$

$$\bar{c}_i = \bar{c}_i^o/s + B_i\exp[-(s/\bar{D}_i)^{\frac{1}{2}}x] \tag{73}$$

For the unsupported conditions, replace D_i by D, and $\bar{D}_i$ by $\bar{D}$. For two-ion transport cases there will be two A unknowns and two B unknowns to be established from other boundary conditions. Here "s" is the transform variable.

The following cases have been studied (59) by analytical and by numerical methods.

Case I. Single ion of charge ±1 reversibly transferred from aqueous to organic phases (or the reverse) in supported electrolyte.

Case II. Single ion of charge ±1 reversibly transferred from aqueous to organic phases in the absence of supporting electrolyte.

Case III. Two ions of charge +1 and −1, respectively, reversibly transferred from aqueous-organic phases in the presence of supporting electrolyte.

Case IV. Two ions of charge +1 and −1, respectively, transferred from aqueous-organic phases in the absence of supporting electrolyte.

Case V. Single ion of charge ±1 reversibly transferred from aqueous to organic (or the reverse) in aqueous supporting electrolyte, but no organic-phase supporting electrolyte.

Case VI. Two ions of charge +1 and −1, respectively reversible transferred from aqueous-organic phases with aqueous supporting electrolyte, but no organic-phase supporting electrolyte.

Digital simulations were used to evaluate numerically the nonlinear problems, Cases III and VI.

Concentration-Distance-Time Profiles.

For the simplest classical case of one cation moving from aqueous to organic phase (left to right), typical depletion and accumulation regions are seen in Figure 32.

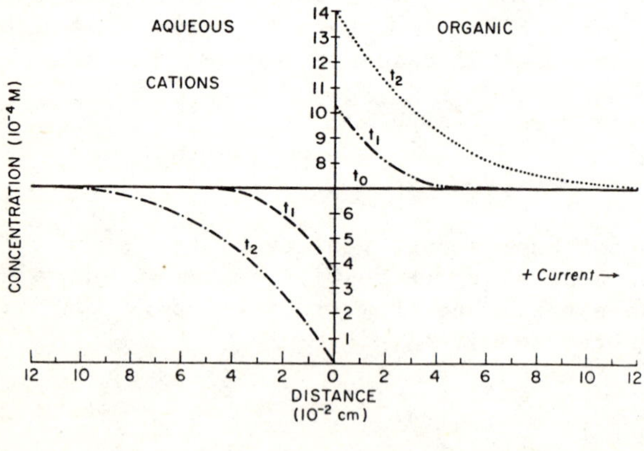

Figure 32. Concentration-distance-time profiles for one univalent cation moving reversibly from aqueous to organic with inert, blocked supporting electrolyte. K = 1, I = 1x 10^{-4} A/cm^2.

Ion	Charge	D_{aq}	D_{org}	c_{aq}	c_{org}
Rev. cation	1	0.3	0.75	1x10^{-3}	1x10^{-3}
Inert	1	0.3	0.3	1	1
Inert	−1	0.3	0.3	1.001	1.001

$t_o = 0$; $t_1 = 30$; $t_2 = 48.2 = \tau$.

These results are essentially the same as those described by Delahay.(107) When the supporting electrolyte is removed as in Case II, the results are similar for the single ion case. Transition times are significantly different.

In Figure 33, the transport of two ions of opposite

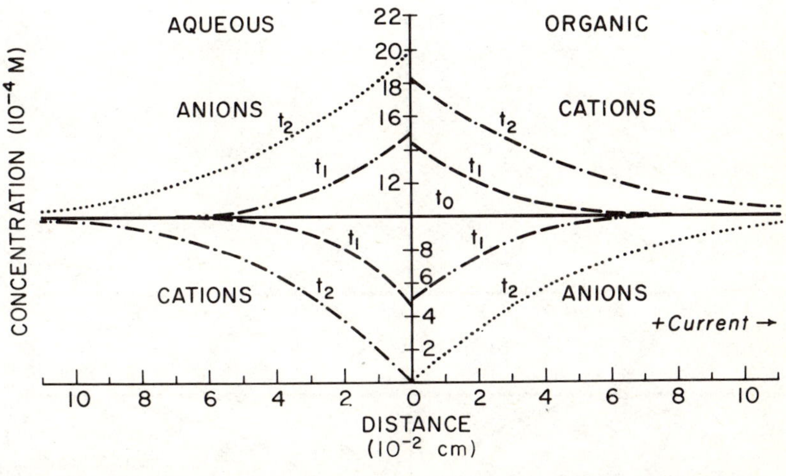

Figure 33. Concentration-distance-time profiles for two univalent, reversible ions moving: cations aqueous to organic; anions moving organic to aqueous with supporting, inert, blocked electrolyte. K = 1, I = 1x10^{-4} A/cm^2.

Ion	Charge	D_{aq}	D_{org}	c_{aq}	c_{org}
Rev. cation	1	0.3	0.3	1x10^{-3}	1x10^{-3}
Rev. anion	−1	0.3	0.3	1x10^{-3}	1x10^{-3}

Inert	1	0.3	0.3	1	1
Inert	1	0.3	0.3	1	1

$t_o = 0$; $t_1 = 30$; $t_2 = 78.1 = \tau$.

charge in supporting electrolyte (Case III) is illustrated. For positive current, depletion of aqueous cations and organic anions occur simultaneously. The transition time occurs when both surface concentrations reach zero. The concentration profiles for Case III are similar to Cases I and II, in the sense that cation depletion and accumulation occur in aqueous and organic phases, respectively while simultaneous anion depletion and accumulation happen in the organic and aqueous phases, respectively.

Conversely, in Case IV, the most interesting results for a single salt without supporting electrolyte, is shown in Figure 34. The response depends on all diffusion

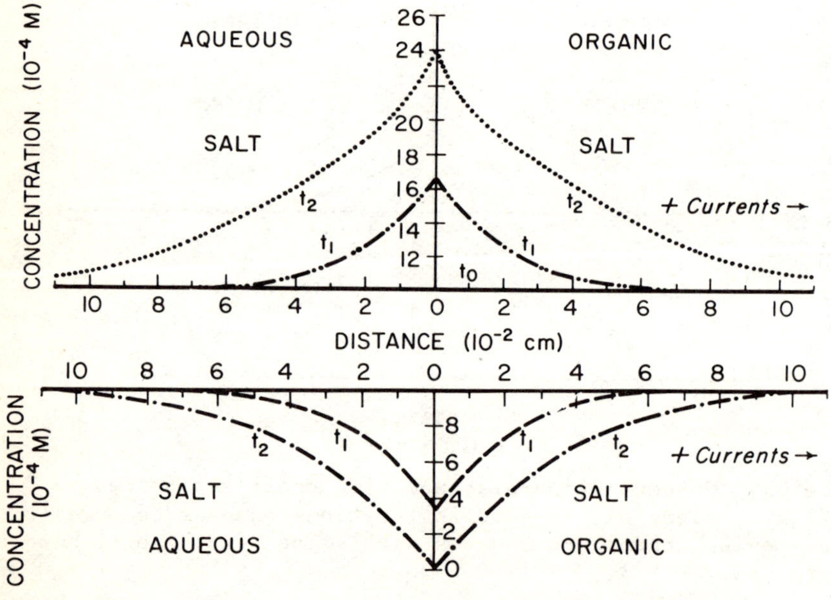

Figure 34. Concentration-distance-time profiles for two, univalent, reversible ions moving as in Figure 33. No supporting electrolyte. K = 1, I = 1×10^{-4} A/cm^2.

Ion	Charge	D_{aq}	D_{org}	c_{aq}	c_{org}
(a, top) Electrosorption					
Rev. cation	1	0.3	0.75	$1x10^{-3}$	$1x10^{-3}$
Rev. anion	−1	0.075	0.4	$1x10^{-3}$	$1x10^{-3}$
$t_o = 0$; $t_1 = 50$; $t_2 = 192 = \tau$.					
(b, bottom) Electrodesorption					
Rev. cation	1	0.075	0.3	$1x10^{-3}$	$1x10^{-3}$
Rev. anion	−1	0.4	0.075	$1x10^{-3}$	$1x10^{-3}$
$t_o = 0$; $t_1 = 50$; $t_2 = 94.5$.					

coefficients, current, concentrations, and partition coefficients. For example, here the current is positive and K = 1; consequently surface concentrations are equal initially, and profiles show simultaneous depletion or accumulation in both phases. Depletion occurs (for positive current) when $t_+ < t_+$.

Large Amplitude Perturbation Methods.

The theory of mass transport-limited reversible one- and two-ion transfers can be supplemented by a corresponding theory for kinetic-mass transport (irreversible) coupled processes.(61-64) Methods used to test the theory include chronoamperometry, chronopotentiometry and cyclic voltammetry. Some chronoamperometric results are given by Senda et. al. (52), while Koryta et. al. have used cyclic voltammetry extensively. In this section only constant current results will be presented.

The principal results of the theory using digital simulations are hypothetical chronopotentiograms and log analytical plots: $\bar{\phi}-\phi$ vs time, and $\bar{\phi}-\phi$ vs log $[1-(t/\tau)^2]$, respectively. The latter plot is strictly valid for total irreversibility. However, high loading of the transferring ion in one phase linearizes the transport problem and removes the term for time-dependent concentration in the entering phase. Consequently, the log analytical plot above applies to both reversible and irreversible cases.

The characteristic behavior from distinguishable two-wave kinetically-limited cases to merged undistinguishable, single-wave behavior has been illustrated. When both ions are slow ($10^{-11} < k_f < 10^{-2}$ cm/sec), but one of the ions has a forward rate constant four orders of magnitude or more larger than the other, two distinct waves are predicted. As k_f for each ion approaches a common value, a single totally irreversible wave is found. Merging of two waves into an unusual single-transition time chronopotentiogram is the intermediate result. Merging of two irreversible waves into a single, reversible wave can also occur when the rate constants are allowed to increase.

Log analytical plots of separated waves give expected irreversible slopes, $RT/\alpha F$. Sub-Nernstian curvature at times near the transition times is characteristic of interference by a second transporting ion of the same or opposite charge.

Digital simultations of ion transfers at low (or no) supporting electrolytes give linear analytical log plots with spurious slopes, as a result of the time-dependent IR drops generated between the pair of sensor electrodes.(60) Since the latter cannot be placed within the concentration polarization regions on either side of the interface, IR drop contributions that depend on $\ln[1-(t/\tau)^{\frac{1}{2}}]$ are included in the measured signal. While these effects are concentration-dependent, they are theoretically independent of applied current.

Comparison of Various Cation Crossings.

Representative chronopotentiograms for the transport of several TAACl's are shown in Figure 35.

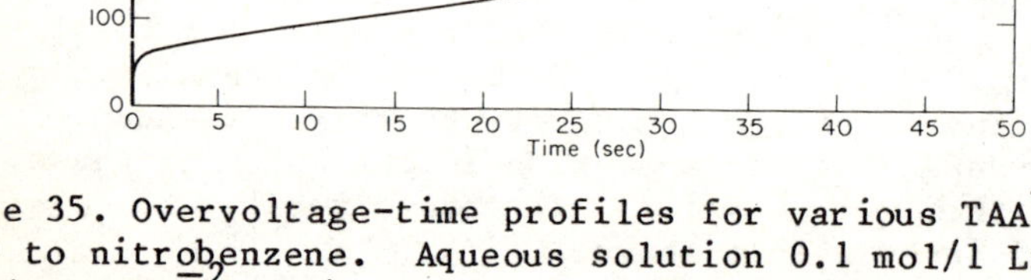

Figure 35. Overvoltage-time profiles for various TAA^+ ions from water to nitrobenzene. Aqueous solution 0.1 mol/l LiCl, organic solution $2x10^{-2}$ mol/l TAATPB; current density 62.5 μA/cm^2. Curve a, $1x10^{-3}$ mol/l TBACl; b, $1x10^{-3}$ mol/l tetrapropylammonium chloride; c, $1x10^{-3}$ mol/l tetraethylammonium chloride; and d, $1x10^{-3}$ mol/l tetramethylammonium chloride.

As expected, as the TAA^+ becomes less oil soluble, a more negative potential must initially be applied to pass current. In the case of tetramethylammonium ion (TMA^+), the potential difference between the transport of TMA^+ and that of TPB^- is too small to provide an accurate transition time.

Distortion of chronopotentiogram shape is greater when TAA^+ ions larger than tetrabutyl are studied. Figure 36 shows the results for the transport of tetrapentylammonium ions. The shape of this transient clearly shows adsorption of the cation at the interface.

There is very little freedom in choice of acceptable aqueous supporting electrolytes for general ion transport studies. The tendency for support cations to carry current, in addition to the cation being studied, is related to the free energy of transfer for the co-cation. The most hydrophilic co-cations are required to obtain the best negative-going potential limit. The limited potential range is illustrated by comparing the chronopotentiograms of TBA^+ transport from water to nitrobenzene using various aqueous alkali chloride solutions, Figure 37. The ultimate negative potential limit is determined by TPB^- transport. The TBATPB electrolyte gives a range of about 375 mV for observation of

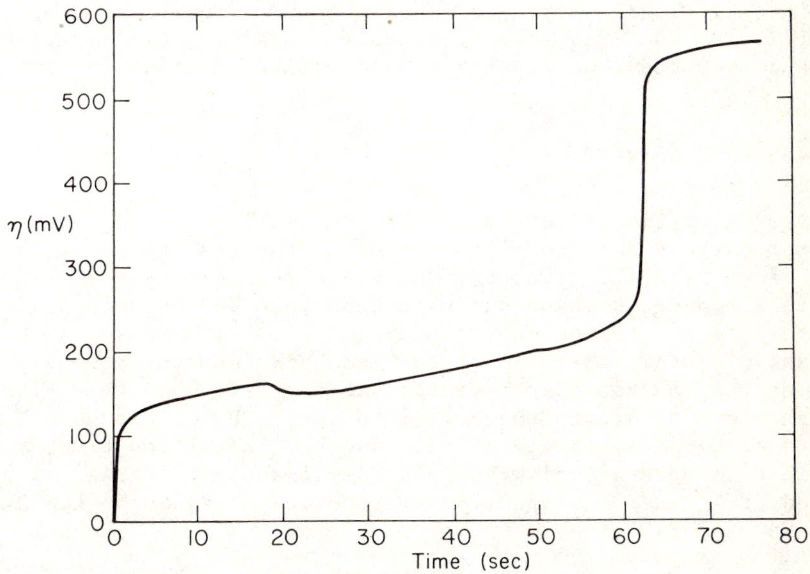

Figure 36. Overvoltage-time profiles for tetrapentylammonium$^+$ from water to nitrobenzene. Aqueous solution 0.1 M NaF and 2.5×10^{-4} mol/l tetrapentylammonium fluoride; organic solution

$2x10^{-2}$ mol/l tetrapentylammonium tetraphenylborate; current density 78 μA/cm^2.

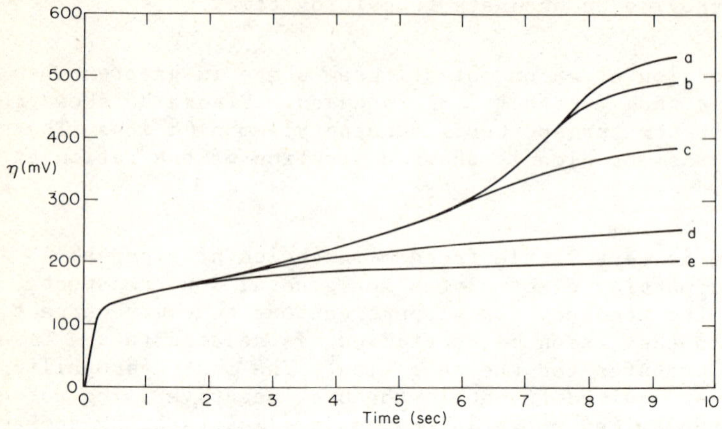

Figure 37. Overvoltage-time curves for aqueous solution of TBACl $8x10^{-4}$ mol/l ; nitrobenzene solution of TBATPB $2x10^{-2}$ mol/l; current density 59 μA/cm^2. Curve a, 0.1 mol/l LiCl; b, 0.1 mol/l NaCl; c, 0.1 mol/l KCl; d, 0.1 mol/l RbCl; and e, 0.1 mol/l CsCl.

the TBA chronopotentiogram. This accessible range, required for sharp breaks with easily determined τ values, is attainable with LiCl, and NaCl is also acceptable. Clearly, Cs^+, Rb^+, and K^+ are unacceptable as supporting electrolyte cations in many transport studies.

Aqueous Anion Effects.

There are two effects of anions on observed chronopotentiometric transition times of the TAA^+ cation. These arise from relative anion hydrophobicities. First, the inert salt in the aqueous phase can partition into the organic phase. This results in transportable anions, e.g., I^- from NaI, ClO_4^- from $NaClO_4$, etc., that can be carried back into the aqueous phase as TAA^+ moves under positive current. Second, the salt TAAX can partition into the organic phase. The effect on chronopotentiograms is one of altering the transition time, but not creation of a second wave. If only transport of TAA^+ is desired, the aqueous supporting electrolyte needs to be LiF, LiCl, NaF, or NaCl.

Typically, partitioning causes transport by both TAA^+ and X^-. The result may be a longer or a shorter transition time depending on D_+ vs D_-. When all of the TAA^+ is extracted into the organic phase, all current is carried by X^-. If excess X^- is in the organic phase, as in the case of TAAI in NaI, the transition time will be increased over that of TAA^+ alone. For supporting electrolytes containing oil-soluble anions such as I^- and ClO_4^-, the dominant factor is the solubility of TAAX in nitrobenzene. As shown in Figure 38

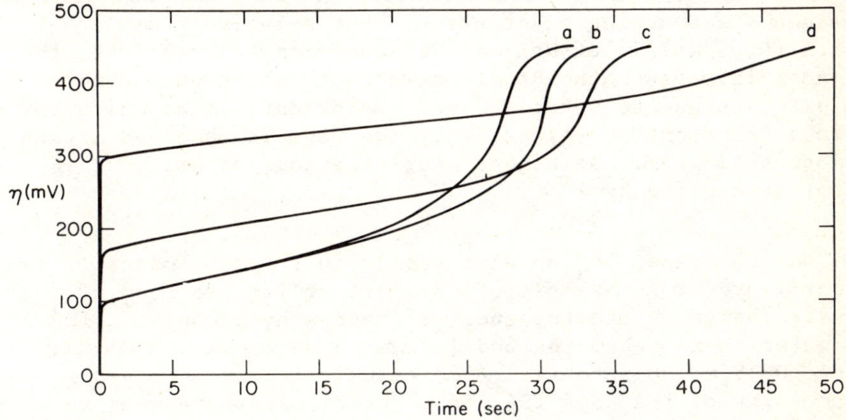

Figure 38. Overvoltage-time curves for aqueous solutions of TBACl $1x10^{-3}$ mol/l ; organic solutions of TBATPB $2x10^{-2}$ mol/l; current density 62.5 μA/cm^2. Curve a, 0.1 mol/l NaCl, b, 0.1 mol/l NaBr; c, 0.1 mol/l NaI; and d, 0.1 mol/l $NaClO_4$.

when NaI or $NaClO_4$ is used as the supporting electrolyte, all of the TBA^+ has partitioned into the nitrobenzene as TBAX. When current is passed, only I^- is transported from the nitrobenzene to water (or ClO_4^- if $NaClO_4$ is used). The change in the initial potential required to pass current (TBA^+ vs I^- vs ClO_4^-) provides a key to the transporting ion and reflects the relative solubility of the transporting ion in the two phases.

A complicated intermediate case arises when a slightly oil-soluble anion is used as the supporting electrolyte. An example is Br^- that partitions (as TAABr) into the nitrobenzene and increases the net concentration of transporting ions. The resulting increase in the transition time is shown in Figure 38. Since the concentration of supporting electrolyte is two to three orders of magnitude larger than the concentration of the ion to be studied, the salt partition coefficient of the supporting electrolyte does not need to be very large to yield prolonged transition times.

Examples of Two-Ion Transport

When the forward rate constants of two ions differ by one order or magnitude of less, only one wave is predicted. This behavior is seen in the co-transport of TBA^+ and I^-. Simultaneous transport of two oppositely charged ions can be achieved through the use of mixed suporting electrolytes. For example, Figure 39 shows the chronopotentiograms for 9×10^{-4} mol l^{-1} solution of TBACl with combinations of NaCl and NaI used as the aqueous supporting electrolyte. For relatively small amounts (0.04 mol/l) of NaI in the supporting electrolyte, the transport involves both TBA^+ from water to nitrobenzene and I^- from nitrobenzene to water. Since the potentials at which the two ions transport are close, only one wave is observed in the chronopotentiogram. At higher concentrations of NaI, all current is carried by I^-.

Two-ion transport can also result in the appearance of two separate waves. An example of this effect for two ions of opposite charge is the transport of tetraethylammonium$^+$ (TEA^+) from water to nitrobenzene and I^- from nitrobenzene to water, Figure 40. Similar potential-time transients are obtained for the crossing of TBA^+ and TEA^+ from water to nitrobenzene shown in Figure 41. In supported electrolyte solutions, simultaneous crossing of two oppositely charged ions is virtually indistinguishable from the transport of two ions of the same charge.

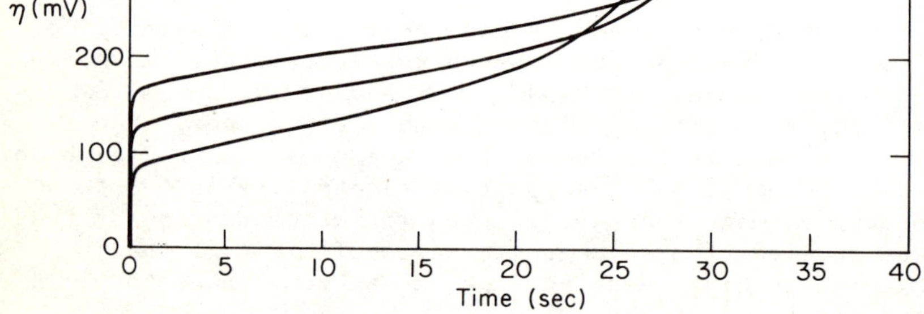

Figure 39. Simultaneous two-ion transport; aqueous solution 9×10^{-4} mol/l TBACl; organic solution 2×10^{-2} mol/l TBATPB; current density 59 $\mu A/cm^2$. Curve a, 0.1 mol/l NaCl; b, 0.1 mol/l NaCl + 1.4×10^{-2} mol/l NaI; and c, 0.1 mol/l NaCl + 0.1 mol/l NaI.

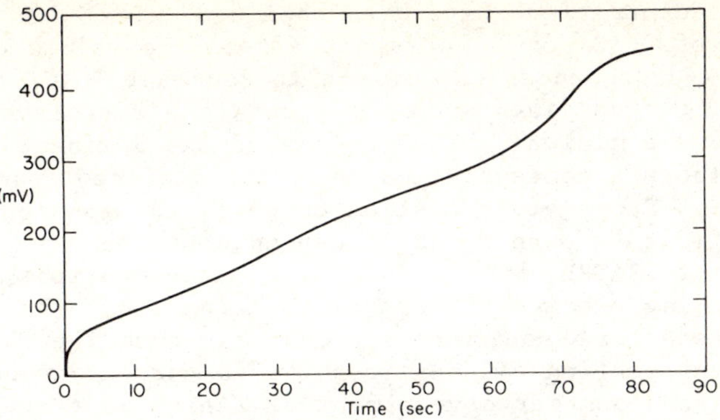

Figure 40. Two-step ion transport. Aqueous solution of 2.8×10^{-4} mol/l TEACl and 0.1 mol/l LiCl; organic solution of tetraheptylammonium iodide 6.6×10^{-4} mol/l and 7.0×10^{-3} mol/l tetraheptylammonium tetraphenylborate; current density 59 $\mu A/cm^2$. First process I^-(org) → I^-(aq). Second process TEA^+(aq) → TEA^+(org).

Possible Induction of Irreversibility by Surface Film Formation.

For both TBA^+ and I^-, α was found to vary with applied currents over the current range 1.5 to 25 μA. This strong effect is seen in a typical analytical plot for TBA^+ in Figure 42. The corresponding iodide responses are shown in Figure 43.

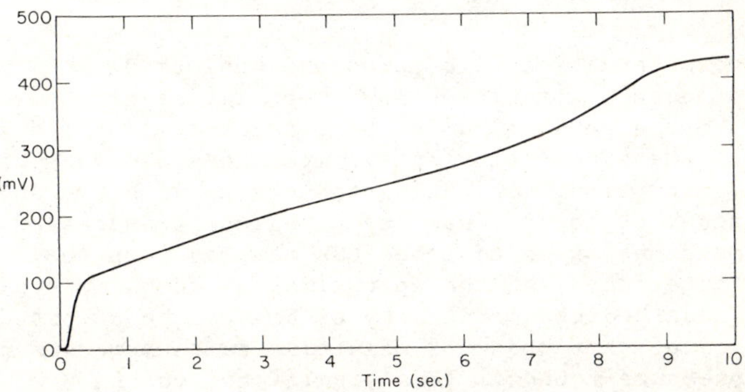

Figure 41. Two-step ion transport. Aqueous solution of 4.8×10^{-4} mol/l TBACl, 2.6×10^{-4} mol/l TEACl, and 0.1 mol/l LiCl; organic solution of 2×10^{-2} mol/l TBATPB; current density 125 $\mu A/cm^2$.

The accumulated slopes (α values) for many currents at several concentrations of TBACl and for I^- at constant supporting electrolyte concentrations have been obtained. The concentration of TBACl or I^- does not affect the values of α. Likewise the α-dependences on current at constant TBACl or I^- concentration, but varying aqueous supporting electrolyte concentrations are minimal. The pronounced and dominant effect of tetraphenylborate concentration on α for different currents, has been shown. For a given applied current, the apparent transfer coefficients were found to depend upon the concentration of TBATPB, but not upon the concentrations of aqueous supporting electrolyte or transporting ion. The I^- data are less scattered and more reproducible than the TBA^+ data. An interpretation of these results is current-induced film formation and barrier development by TAATPB salt at the interface.

Properties of the L/L Interface in the Presence of Proteins.

The influence of adsorbed proteins on electrochemical properties of the nitrobenzene-aqueous interface has been investigated using cyclic voltammetric and impedance techniques. Upon addition of 4 µg/ml ovalbumin cyclic voltammetric ΔE_p values for Cs^+ transfer shift from nearly temperature-independent reversible values to more temperature-dependent irreversible values. These effects are especially pronounced at lower temperatures and suggested formation of a difficultly-permeable surface film. The cyclic voltammetry of a wide range of concentrations of bovine serum albumin in base L/L electrolytes was also studied.

Impedance characteristics of blocked and unblocked nitrobenzene/aqueous interfaces have been investigated as a function of bovine serum albumin (BSA) concentration and applied DC potential. Measured interfacial impedances decrease upon initial BSA adsorption, suggesting the charge density, and thus the capacitance, of the diffuse layer region is increased. From measured dimensions of an adsorbed BSA monolayer on the heptane/air interface, and the approximately known molecular charge, the adsorbed charge density of BSA can be computed. This value is the same order of magnitude as that measured by the bubble pressure method. Thus significant charge accumulation by BSA adsorption is possible. Upon further BSA addition (up to 150 ppm) increasing interfacial impedances indicates formation of a substantial adsorbed film, which decreases total interfacial capacitance. At still higher BSA concentrations faradaic base electrolyte leakage becomes significant under otherwise blocked conditions. An interfacial phase which facilitates ion transfer is apparently formed. Adsorption effects on interfacial impedance have been modeled by addition of a BSA concentration-dependent capacitor to a Randles equivalent circuit of the interface.

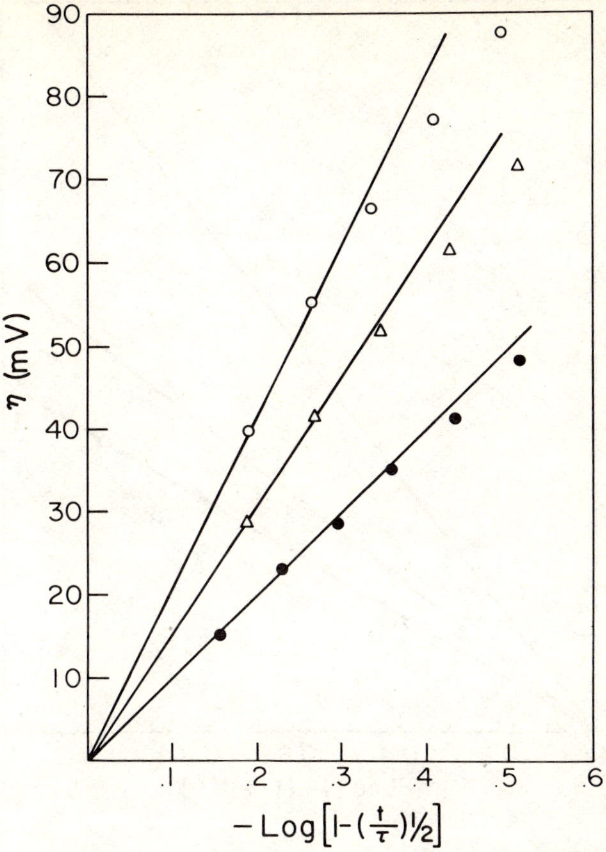

Figure 42. Log analysis plots of TBA^+ transport from water to nitrobenzene. Aqueous solution: 0.1 mol/l NaCl and 9.6×10^{-5} mol/l TBACl. Nitrobenzene solution: 0.01 mo/1/ TBATPB. O = 3.5 μA, /\ = 8.0 μA, O = 12 μA.

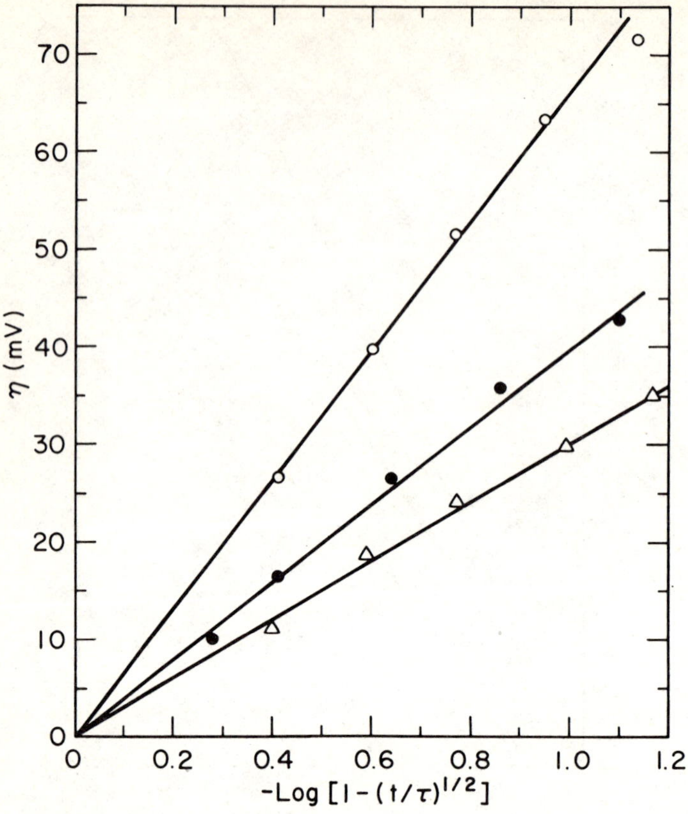

Figure 43. Log analysis plots of I^- transport from nitrobenzene to water. Aqueous solution: 0.1 mol/l NaI. Nitrobenzene solution: 0.01 mol/l TBATPB and 5×10^{-4} mol/l TBAI. $\triangle$ = 3.5 μA, O = 6.4 μA, O = 19.0 μA.

REFERENCES

1. Nernst, W. and Riesenfeld, E.H., 1902, Ann. Phys., 8, pp. 600-608.
2. Guastalla, J, (1956-7), Mem. Serv. Chim. etat. (Paris) 41, pp. 317-321; (CA 53, 9776); (1956) J. Chim. Phys., 53, p. 470.
3. Guastalla, J., 1969, C.R. Acad. Sci. Ser. C., 269, pp. 1360-1363.
4. Guastalla, J., 1970, Nature, 227, pp. 485-486.
5. Guastalla, J., 1971, Experientia, Suppl., 18, pp. 333-342.
6. Guastalla, J., 1973, C.R. Acad. Sci. Ser. C, 276, pp. 17-19.
7. Boguslavskii, L.I., Frumkin, A.N.and Gugeshashvili, M.I., 1974, Bioelectrochem. Bioenerg., 1, pp. 506-514.
8. Manvelyan, M.A., Neugodova, G.L. and Boguslavskii, L.I., 1976, Elektrokhimiya, 12, pp. 309-313 (Eng. Trans. 310-313).
9. Frumkin, A.N., Boguslavskii, L.I. and Manvelyan, M.A., 1976, ibid, 12, pp. 767-769 (Eng. Trans. 723-725).
10. Boguslavskii, L.I., Frumkin, A.N. and Gugeshashvili, M. I., 1976, ibid, 12, pp. 856-860 (Eng. Trans. pp. 799-802).
11. Manvelyan, M.A., Neugodova, G.L. and Boguslavskii, 1976, ibid., 12, pp. 1250-1254 (Eng. Trans. 1145-1149).
12. Manvelyan, M.A., Boguslavskii, L.I., 1976, ibid, 12, pp. 1693-1696 (Eng. Trans. 1541-1544).
13. Manvelyan, M.A. and Boguslavskii, L.I., 1976, ibid, 12, pp. 1805-1808 (Eng. Trans. 1636-1639).
14. Boguslavskii, L.I., Frumkin, A.N. and Manvelyan, M.A., 1977, Dokl. Akad. Nauk SSSR, 233, pp. 144-147 (Eng. Trans. pp. 242-246).
15. Gavach, C.,Mlodnicka, T. and Guastalla, J., 1968, C.R. Acad. Sci. Ser. C., 266, pp. 1196-1199.
16. Gavach, C., 1969, ibid, 269, pp. 1356-1359.
17. Gavach, C. and d'Epenoux, B., 1971, ibid, 272, pp. 872-875.
18. Gavach, C. and Henry, F., 1972, ibid, 274, pp. 1545-1548.
19. Gavach, C., Henry, F., and Sandeaux, R., 1974, ibid, 278, pp. 491-494.
20. Gavach, C, 1973, J. Chim. Phys. Physicochim. Biol., 70, pp. 1478-1482.
21. Gavach, C., Seta, P. and Henry, F., 1974, Bioelectrochem. Bioenerg., 1, pp. 329-342.
22. Gavach, C. and Henry, F., 1974, J. Electroanal. Chem. Interfacial Electrochem., 54, pp. 361-370.
23. Gavach, C. and d'Epenoux, B., 1974, ibid, 55, pp. 59-67.
24. Gavach, C. and d'Epenoux, B. and Henry, F., 1975, ibid, 64, pp. 107-115.
25. d'Epenoux, B. and Gavach, C., 1976, J. Colloid Interfac. Sci., 56, pp. 138-145.

26. Gavach, C., Seta, P., and d'Epenoux, B., 1977, J. Electroanal. Chem., 83, pp. 225-235.
27. Gros, M., Gromb, S. and Gavach, C., 1978, ibid, 89, pp. 29-36.
28. d'Epenoux, B., Seta, P., Amblard, G. and Gavach, C., 1979, ibid, 99, 77-84.
29. Koryta, J., Vanysek, P. and Brezina, M., 1976, ibid, 67, pp. 263-266; 1977, ibid, 75, pp. 211-228.
30. Koryta, J., Vanysek, P., Khalil, M.W., Marecek, V., Samec, Z. and Brezina, M., in "Ion Selective Electrodes," E. Pungor and I. Buzas, Editors, pp. 441-444, Akademiai Kiado, Budapest, (1978).
31. Koryta, J., 1979, Electrochim. Acta, 24, pp. 293-303.
32. Koryta, J., Brezina, M., Hofmanova, A., Homolka, D., Hung, L.Q., Khalil, W., Marecek, V., Samec, S., Sen, S.K., Vanysek, P. and Weber, J., 1980, J. Electroanal. Chem., 116, pp. 61-68.
33. Homolka, D., Hung, L.Q., Hofmanova, A., Khalil, M.W., Koryta, J., Marecek, V., Samec, Z., Sen, S.K., Vanysek, P., Weber, J., Brezina, M., Janda, M. and Stibor, I., 1980, Anal. Chem., 52, pp. 1606-1610.
34. Koryta, J., 1983 Ion-Selective Electrode Revs., 5, pp. 131-164 based on a Symposium, "Ion-Selective Electrodes", Oct. 13-15, 1980, E. Pungor, Editor, pp. 53-72, Akademiai Kiado, Budapest (1981).
35. Koryta, J. and Vanysek, P., in "Advances in Electrochemistry and Electrochemical Engineering", Vol. 12, H. Gerischer and C. Tobias, Editors, pp. 113-176, John Wiley & Sons, Inc., New York (1981).
36. Koryta, J., Hung, L.Q. and Hofmanova, A., 1982, Studia Biophysics, 90, pp. 25-29.
37. Samec, Z., Marecek, V., Koryta, J. and Khalil, M.W., 1977, J. Electroanal. Chem., 83, pp. 393-397.
38. Samec, Z., Marecek, V. and Weber, J., 1977, ibid, 96, pp. 245-247.
39. Samec, Z., Marecek, V. and Weber, J., 1979, ibid, 100, pp. 841-852.
40. Samec, Z., Marecek, V. and Weber, J. 1979, ibid, 103, pp. 11-18.
41. Samec, Z., 1979, ibid, 99 pp. 197-205 and pp. 385-389.
42. Samec, Z., 1979, ibid, 103, pp. 1-9.
43. Samec, Z., 1980, ibid, 111, pp. 211-216.
44. Samec, Z., Marecek, V., Weber, J. and Homolka, D., 1981, ibid, 126, pp. 115-119 and pp. 121-129.
45. Samec, Z., Homolka, D. and Marecek, V., 1982, ibid, 135, pp. 265-283.
46. Samec, Z., Homolka, D., Marecek, V. and Kavan, L., 1983, ibid, 145, pp. 213-218.
47. Marecek, V. and Samec, Z., 1983, ibid, 149, pp. 185-192.

48. Samec, Z., Marecek, V., and Homolka, D., 1983, ibid, 158, pp. 25-36.
49. Senda, M. and Kakutani, 1980, Hyomen, 18, pp. 535-542.
50. Senda, M., Kakutani, T. and Osakai, T., 1981, Denki Kagaku, 49, pp. 322-329.
51. Osakai, T., Kakutani, T., Nishiwaki, Y. and Senda, M., 1983, Bunseki Kagaku, 32, pp. E81-E84.
52. Kakutani, T., Osakai, T. and Senda, M., 1983. Bull. Chem. Soc. Jpn., 56, pp. 991-996.
53. Kakiuchi, T. and Senda, M., 1983, ibid, 56, pp. 1322-1326.
54. Kakiuchi, T. and Senda, M., 1983, ibid, 56, pp. 1753-1760.
55. Mathis, D.E. and Buck, R.P., 1979, J. Membrane Sci., 4, pp. 379-394.
56. Mathis, D.E., Stover, F.S. and R.P. Buck, 1979, ibid, 4, pp. 395-414.
57. Mathis, D.E., Freeman, R.M., Clark, S.T. and Buck, R.P., 1979, ibid, 5, pp. 103-116.
58. Buck, R.P., Stover, F.S. and Mathis, D.E., 1979, J. Elecanal. Chem., 100, pp. 63-70.
59. Melroy, O.R., Buck, R.P., Stover, F.S. and Hughes, H.C., 1981, J. Electroanal. Chem., 121, pp. 98-114.
60. Melroy, O.R. and Buck, R.P., 1982, ibid, 136, pp. 19-37. Errata in J. Electroanal. Chem., 1983, 147, p. 351.
61. Melroy, O.R. and Buck, R.P., 1983, ibid, 143, pp. 23-36.
62. Melroy, O.R., Bronner, W.E. and Buck, R.P., 1983, J. Electrochem. Soc., 130, pp. 373-380.
63. Melroy, O.R., Bronner, W.E. and Buck, R.P., 1982, Chem. Biomed. Environ. Instr., 12, pp. 77-91.
64. Melroy, O.R. and Buck, R.P., 1983. ibid., 151, pp. 1-9.
65. Reid, J.D., Vanysek, P. and Buck, R.P., 1983, J. Electroanal. Chem., 147, pp. 71-82.
66. Reid, J.D., Vanysek, P. and Buck, R.P., 1984, ibid, 161, pp. 1-15.
67. Reid, J.D., Vanysek, P. and Buck, R.P., 1984, ibid, in press.
68. Bronner, W.E. Melroy, O.R. and Buck, R.P., 1984, ibid, 162, pp. 263-274.
69. Koczorowski, Z., 1981, ibid, 127, pp. 11-16.
70. Koczorowski, Z., Kalinska, A. and Figaszewski, Z., 1982, ibid, pp. 303-307.
71. Figaszewski, Z., 1982, ibid, 139, pp. 309-315.
72. Koczorowski, Z. and Geblewicz, G. 1982, ibid, 139, pp. 177-191.
73. Figaszewski, Z., Koczorowski, Z. and Geblewicz, G., 1982, ibid, 139, pp. 317-322.
74. Koczorowski, Z. and Geblewicz, G., 1983, ibid, 152, pp. 55-66.

75. Koczorowski, Z. and Zagorska, I., 1983, ibid, 159, pp. 183-193.
76. Hundhammer, B., Solomon, T. and Alemayehu, B., 1982, ibid, 135, pp. 301-304.
77. Kihara, S., Yoshida Z. and Fujinaga, T., 1982, Bunseki Kagaku, 31, pp. E297-E300.
78. Kihara, S. and Yoshida, Z., 1982, ibid, 31, pp. E301-E304.
79. Yoshida, Z. and Freiser, H., 1984, ibid, in press.
80. Yoshida, Z. and Freiser, H., 1984, personal communication.
81. Koryta, J., 1983, Ion Selective Electrode Rev., 5, pp. 131-164.
82. Vanysek, P. and Buck, R.P., 1984, J. Electroanal. Chem., 163, pp. 1-15.
83. Parker, A.J., 1969, Chem. Rev., 69, pp. 1-32.
84. Popovych, O, 1970, Crit. Rev. Anal. Chem., 1, pp. 73-117.
85. Karpfen, F.M. and Randles, J.E.B., 1953, Trans. Faraday Soc., 49, pp. 823-831.
86. Hung, L.W., 1980, J. Electroanal. Chem., 115, pp. 159-174.
87. Danil de Namor, A.F., Hill, T. and Sigstad, E., 1983, J. Chem. Soc. Faraday Trans., pp. 2713-2722.
88. Rais, J., 1971, Coll.Czech. Chem. Comm., 36, pp. 3253-3262.
89. Gerin, M. and Fresco, J., 1978, Anal. Chim. Acta, 97, pp. 165-176.
90. Mathis, D.E. and Buck, R.P., 1976, Anal. Chem., 48, pp. 2033-2035.
91. Bronner, W.E. and Buck, R.P., 1984, unpublished results.
92. Vanysek, P., Thesis, Czechoslovakian Academy of Sciences, 1982.
93. Marecek, V. and Samec, Z., 1981, Analyt. Letters, 14, pp. 1241-1753.
94. Marecek, V., Samec, Z., 1983, Anal. Chim. Acta, 151, pp. 265-269.
95. Homolka, D., Marecek, V., Samec., Z., Ryba, O., Petranek, J., 1981, J. Electroanal. Chem. Interfacial Electrochem., 125, pp. 243-247.
96. Vanysek, P., Behrendt, M., 1981, ibid, 130, pp. 287-292.
97. Vanysek, P., 1981, ibid, 121, p. 149.
98. Hundhammer, B., Solomon, T., Alemu, H., 1983, ibid, 149, pp. 179-183.
99. Abraham, M.H. and Liszi, J., 1981, J. Inorg. Nucl. Chem., 43, pp. 143-152.
100. Verwey, E.J.W. and Niessen, K.F., 1939, Phil. Mag., 28, pp. 435-446.
101. Schuhmann, D. and Seta, P., 1979, J. Colloid Interfac. Sci., 69, pp. 448-459.

102. Joos, P. and Vanden Bogaert, R., 1976, ibid, 56, pp. 206-217.
103. Joos. P., 1978, J. Electroanal. Chem., 86, pp. 75-80.
104. Buck, R.P., 1982, Ion-Selective Electrode Revs., 4, pp. 3-74.
105. Berlouis, J.E.A., Chatham, J., Girault, H.H.J. and Schiffrin, D.G., 1983, The Electrochem. Soc. Spring Meeting, San Francisco, CA, May 8-13, 1983, Ext. Abs., Vol. 83-1, p. 935.
106. Homolka, D. and Marecek, V., 1980, J. Electroanal. Chem., 112, pp. 91-96.
107. Samec,Z., Marecek, V. and Homolka, D., Proc. 34th ISE Meeting, Erlangen, W. Germany, Sept. 18-23, 1983, pp. 916-917.
108. Geblewicz, G. and Koczorowski, Z., 1983, J. Electroanal. Chem., 158, pp. 37-46.

ELECTRICAL DOUBLE LAYERS AT INTERFACES BETWEEN COLLOIDAL MATERIALS AND IONIC SOLUTIONS

B.H. Bijsterbosch

Laboratory for Physical and Colloid Chemistry,
Agricultural University, De Dreijen 6, 6703 BC
Wageningen, The Netherlands.

ABSTRACT

The study of electrical double layers plays a central role in colloid chemistry, because their occurrence is usually responsible for the stability of colloidal systems.

For two types of reversible systems, AgI and solid oxides, it is described how the relation between surface charge density and composition of the solution is established. It appears that many qualitative similarities exist with results obtained for the polarizable Hg/electrolyte solution interface.

Current models for the interpretation of experimental results are described. Since the potential determining ions Ag^+ and I^- in principle can reside anywhere within the solid AgI, but the charge of the oxides is strictly located at their surfaces, the models for the two representatives are rather different.

1. INTRODUCTION

Colloidal systems are dispersions of very small particles in a continuous medium. Some well-known examples are clays, paints, inks, emulsions and mist, illustrating that solid/liquid, liquid/liquid as well as solid/gas interfaces can be involved. A typical characteristic of stable colloidal systems is that the constituting particles retain their individuality and do not accumulate into larger entities that subsequently sediment or cream. This stability is either ensured by steric effects or by the presence of electrical double layers, or by a combination of

A. F. Silva (ed.), Trends in Interfacial Electrochemistry, 187–204.

the two. It is thus logical that in colloid chemistry since long much attention has been paid to electrical double layers at (mainly solid/liquid) interfaces. This interest is yet enhanced because of the substantial role electrical effects play in many applications, which can be as diverse as electrophoresis, adhesion, flotation and photography.

Of course, when studying electrical double layers at solid/liquid interfaces, well-characterized materials and carefully selected conditions are required. As to the solids, the emphasis has been on silver iodide and various oxides, so that we will restrict ourselves to the main results obtained for these two types of solid in contact with aqueous solutions.

2. SILVER IODIDE

2.1. Suitability as a model colloid

The traditional interest of colloid chemists in AgI as a model colloid has various reasons. Prominent among these is that the particles in a AgI sol or suspension have a well-defined stoechiometric composition and can be prepared in a highly reproducible and pure way. Also the link to AgI electrodes, the behaviour of which is well-established, is obvious.

Silver iodide particles acquire their charge through adsorption of constituent ions (Ag^+ and I^-) from the solution in which the particles are dispersed. Whichever of the two ions is present in excess determines the sign of the surface charge. Along with this adsorption, a potential difference between the two phases develops, illustrating that we are dealing with a <u>reversible</u> interface, as opposed to the <u>ideally polarizable</u> mercury/electrolyte solution interface. In the latter case the potential difference between the two phases can be externally imposed for any composition of the (aqueous) solution, without actual charge transfer between the two phases taking place. For AgI, on the other hand, it is the composition of the solution, especially with regard to the potential determining Ag^+ and I^- ions, that governs the adsorbed amount and the potential difference.

2.2. Titration method

Just as for polarizable interfaces, for silver iodide it is possible to experimentally determine the relation between surface charge density, σ_o, and surface potential, ψ_o, or, for that matter, the capacity. This information is usually obtained from potentiometric titrations in the cell

reference electrode ‖	AgI precipitate in a solution of specified composition	AgI	Ag

Ag/AgI electrodes show Nernstian behaviour, implying that at

25°C, E_{cell} varies by an amount of 59 mV when the equilibrium Ag^+ or I^- concentration changes by a factor of ten. In passing it may be reminded that the concentrations of Ag^+ and I^- are not independent, because their product always has to equal the solubility product at the prevailing temperature and ionic strength. The relation between E_{cell} and pAg(= $-\log c_{Ag^+}$) or pI(= $-\log c_{I^-}$) being known, from variations in E_{cell} one infers to which extent a known amount of added potential determining ions gets distributed over the precipitate and the solution. In principle, one could do without the precipitate since the AgI electrodes could serve as the adsorbent. But the ensuing variations in concentration would be far too small to be accurately determined, so that the precipitate is included for analytical reasons. This approach does imply the assumption, however, that the precipitate/solution interface in all respects behaves identical to the interface between the electrode and the solution.

In order to express adsorbed amounts as surface charge densities, the specific surface area ($m^2.g^{-1}$) of the precipitate has to be known. For the evaluation of this parameter various methods are available (1). Unfortunately, the results thus obtained break up into two groups, between which the data differ by a factor of as much as 3. This discrepancy is still a matter of discussion, but in this paper we will adopt the value obtained from measurements of the negative adsorption of co-ions. It closely matches the data obtained when identifying the experimental value of the differential capacity (in Farads. g^{-1}) at the point of zero charge in 10^{-3} M KNO_3 with the value inferred from the Gouy-Chapman theory (in Farads.m^{-2}).

In order to calculate $\sigma_o = F(\Gamma_{Ag^+} - \Gamma_{I^-})$, where F is the Faraday and Γ denotes a surface excess in moles per unit area, the pAg corresponding with the point of zero charge (p.z.c.) is another quantity to be known. This parameter has also been assessed by various means (2).

2.3. Results for electrolyte solutions

An example of σ_o(pAg) curves referring to an electrolyte solution of variable ionic strength is presented in fig. 1. Assuming that variations in the Galvani potential difference $\Delta\phi$ between solid and liquid (as expressed by variations in the pAg) can exclusively be ascribed to variations in the Volta potential ψ_o at the solution side of the interface, these curves essentially express differential capacities. The correctness of the assumption $\Delta\Delta\phi = \Delta\psi_o$ is supported by the fact that capacities measured directly by a bridge technique (3) and those inferred from coulostatic pulse experiments (4) confirm the values inferred from fig. 1.

These capacities appear to show the same trends as those observed for the Hg/electrolyte solution interface. As compared to the latter system, it is a disadvantage, however, that the available charge- and potential-range is much more limited.

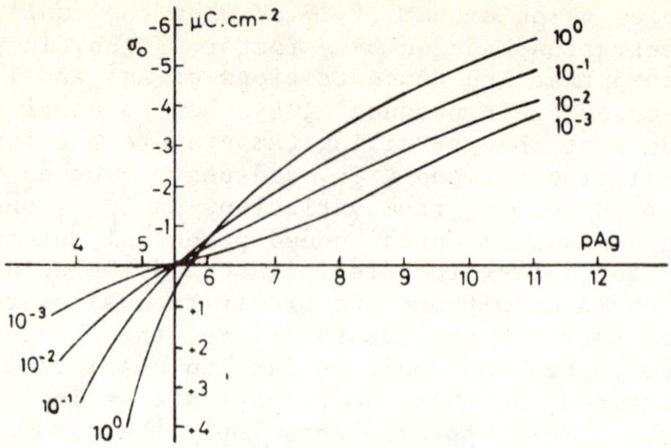

Fig. 1 Surface charge density of silver iodide as a function of pAg for some KNO_3 concentrations ($mol.l^{-1}$); T = 25°C.

Despite this limitation, a number of interesting aspects can be noticed, of which especially the more pronounced lyotropic effects are worth mentioning. Whereas for mercury in 10^{-1} M solutions the difference in capacity between Li^+ and Rb^+ at ψ_o = - 300 mV only amounts to about 3%, for AgI it is as high as 30%. This pronounced lyotropic effect has been ascribed to the specific water structure on the AgI surface (2).

The surface charge density σ_o and the charge density at the solution side of the interface are of identical magnitude, though of opposite sign. Since this condition is identical to that for the Hg/solution interface, Lijklema (5) has analysed his results in terms of the familiar Gouy-Chapman-Stern-Grahame model of electrical double layers. He showed that for potentials at which not yet specific adsorption occurs, the capacity of the Stern layer, C_s, is almost independent of the ionic strength and that variations in the overall capacity C are due to variations in the diffuse layer capacity C_d. Since $C^{-1} = C_s^{-1} + C_d^{-1}$ and since C_d strongly increases with ionic strength, in effect C_s is obtained by identifying it with the experimental C for a 10^{-1} M electrolyte solution.

Lijklema's approach has been criticized by Honig (6), who noted that silver iodide exhibits a fair amount of electrical conductance. Consequently, the "adsorbed" potential determining ions may actually create a space charge distribution inside the solid (in terms of Ag^+ interstitials and vacancies), thus denying the idea of a simple surface charge. Of course, a space charge inside the solid also implies that part of $\Delta\Delta\phi$ has to be ascribed to this phase. Actually, by introducing a double diffuse double layer (one in the solid and one in the liquid)

Honig can describe the data of fig. 1 rather well. In his model both space charge densities are dependent on the pAg, whereas a special feature is yet that at the interface a constant (i.e. pAg-independent) surface charge density of $-1.2\ \mu C.cm^{-2}$ resides. This latter quantity is ascribed to the difference in free energy of formation of interstitials and vacancies in the bulk of the solid and at the surface, resp.. Obviously, Honig can refrain from a Stern layer, which has the serious disadvantage, however, that the observed lyotropic effects can not be explained.

2.4. Uncharged low molecular organic compounds

Not only with respect to plain electrolyte solutions have qualitative similarities between electrical double layers on AgI and Hg been observed, also for solutions containing organic substances comparable trends emerge. The σ_o (pAg) curves in fig. 2 (7), pertaining to various concentrations of n-propanol in 10^{-1} M KNO_3 solutions, e.g., are very much alike to those obtained for Hg (8). Three important trends are noteworthy:

- alcohol addition causes the capacitances to decrease over the entire potential range
- the p.z.c. shifts to lower pAg values
- all σ_o(pAg) curves pass through a common intersection point.

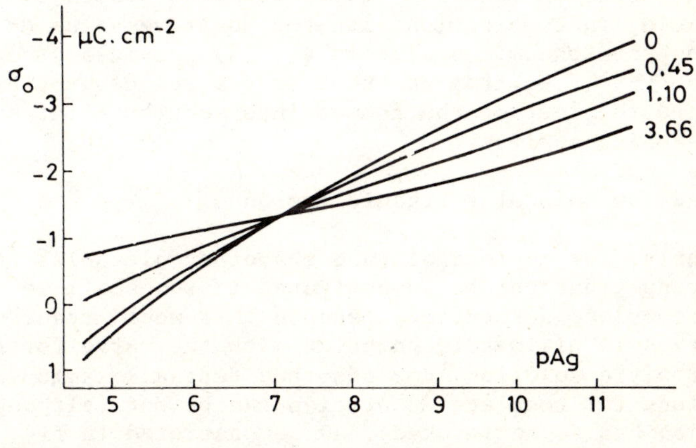

Fig. 2 Surface charge density of silver iodide as a function of pAg for some n-propanol concentrations ($mol.l^{-1}$) in aqueous 10^{-1} M KNO_3 solutions; T = 20°C.

The decrease in the capacitance is related to interfacial displacement of water by alcohol molecules, and the concomitant decrease in the dielectric constant, resp. increase in the thickness, of the Stern layer. Moreover, any specifically adsorbed ions being removed in this exchange process contribute to this trend.

The p.z.c.-shift to lower pAg is a general phenomenon for alcohols, and for other organic compounds too, but its magnitude depends rather strongly on the geometry of the adsorbed molecules. For AgI (7) as well as for Hg (8) it has been concluded that the adsorbed straight chain alcohols are oriented perpendicular to the interface, with the hydrocarbon tails touching the adsorbent and the hydroxyl groups remaining in the solution. From the available data it has also been inferred that the water molecules at the p.z.c. are oriented with the negative side of their dipole towards the AgI (7).

The common intersection point, to be denoted as pAg_m, actually represents a maximum in the adsorption, as can be proved as follows. Analysis of the Gibbs equation for the interfacial free enthalpy (2) shows that the surface excess of the organic compound at any arbitrary value of the pAg, $\Gamma_A^{(s)}$, is related to its value at the common intersection point, $\Gamma_{A,m}^{(s)}$, according to

$$\Gamma_A^{(s)} - \Gamma_{A,m}^{(s)} = -\frac{1}{F}\int_{pAg_m}^{pAg} (\delta\sigma_o/\delta\mu_A)_{\mu_s}\, dpAg \qquad (1)$$

μ denotes the chemical potential of a component, whereas the subscripts A and s refer to the organic substance and the electrolyte, resp.. From fig. 2 it can easily be deduced that to the right of the common intersection point increasing the alcohol concentration causes σ_o to become more positive. This positive ratio, in combination with the positive value of dpAg in that range and the minus sign in eq. (1) provides a negative sign for the LHS of eq. (1) and thus proves the desorption. For the region to the left of the common intersection point a similar reasoning applies.

2.5. Charged low molecular organic compounds

Apparently, for n-propanol (and the other alcohols) the potential range that can be investigated is too small to accomplish complete desorption, because this would require the σ_o(pAg) curves to ultimately coincide with the curve for the plain electrolyte solution. For adsorbed tetraalkylammonium ions the conditions for complete desorption can be met (although some extrapolation has to be invoked), as demonstrated in fig. 3 (9). Within experimental error, the maximum adsorbed amounts that are inferred from fig. 3 by applying eq. (1) appeared to be identical to the amounts that have been obtained in actual adsorption experiments. Not only does this observation prove the correctness of the thermodynamic analysis underlying eq. (1), it also demonstrates one of the advantages of using AgI over Hg, since the latter system, through its deficiency to provide a sufficient amount of interfacial area, does not enable direct adsorption measurements.

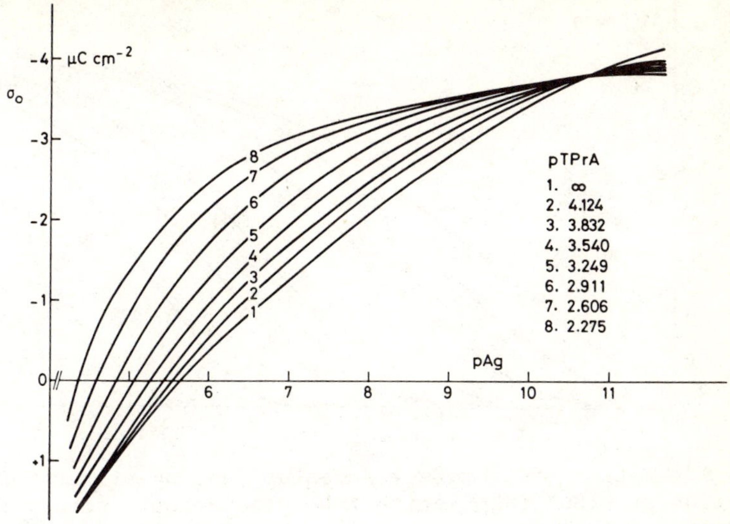

Fig. 3 Surface charge density of silver iodide as a function of pAg for some tetrapropylammonium nitrate concentrations in aqueous 10^{-1} M KNO_3 solutions; T = 20°C.

De Keizer (9) also analysed whether in the adsorption of tetraalkylammonium ions onto AgI the potential (i.e. the pAg) or the charge is the dominant factor. In the former case, known as potential congruency, the adsorption isotherm equation can be written as

$$F(pAg)\, c_A = f\, (\Gamma_A) \tag{2}$$

whereas in the latter case (charge congruency)

$$G(\sigma_o)\, c_A = g\, (\Gamma_A) \tag{3}$$

f,g,F and G are functions of the given parameters.

Eq. (3) implies that if charge congruency occurs plots of Γ_A vs. pA (= $-\log c_A$) at different values of σ_o must coincide when shifting them along the pA-axis. Potential congruency, on the other hand, exists when Γ_A vs. pA plots at different pAg have to be shifted in order to make them coincide. Figures 4 and 5 show the relevant plots for adsorption of tetraalkylammonium ions from 10^{-1} M KNO_3, demonstrating that the isotherms are neither fully charge- nor potential-congruent, but that a tendency towards the first dependency exists. This would mean that the contribution of the dipole of the adsorbing species to the free energy of adsorption is more important than that of their charge.

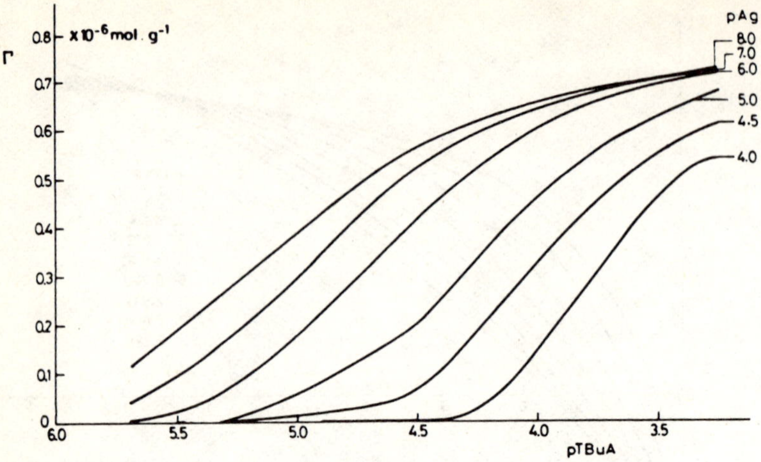

Fig. 4 Adsorption of tetrabutylammonium ions on silver iodide as a function of pTBuA (= - log(tetrabutylammonium concentration)) at some pAg values in aqueous 10^{-1} M KNO_3 solutions.

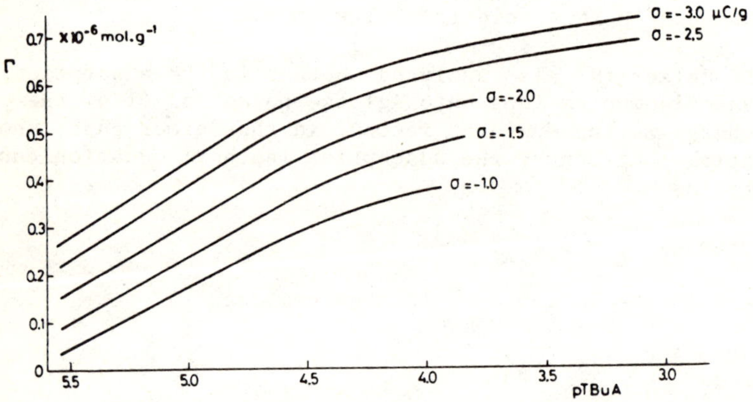

Fig. 5 Adsorption of tetrabutylammonium ions on silver iodide as a function of pTBuA at some surface charge densities in aqueous 10^{-1} M KNO_3 solutions.

A definite advantage of AgI over Hg as a model system is the possibility to obtain additional information form measurements on colloidal stability and electrokinetic phenomena. As to the stability, we notice that it is usually characterized by the coagulation concentration c_c, being that particular electrolyte concentration at which every collision between two dispersed particles leads to sticking together, so that the sol gets completely destabilized. According to the DLVO-theory for colloidal stability (10,11) c_c is related to the Gouy-Chapman potential ψ_d according to

$$c_c = (C/A^2)\ [\tanh\ (z_i e\psi_d/4\ kT)]^4 \qquad (4)$$

C is a constant, among other things depending on the geometry of the particles, A the Hamaker constant which is a measure of the Van der Waals attraction between two particles in a continuous medium, z_i the valency of the counterions, e the elementary charge, k the Boltzmann constant and T the absolute temperature. Thus, a increase in c_c indicates an enhanced value of ψ_d.

With this information it is interesting to consider how, for a negatively charged AgI sol, the coagulation concentrations of the alkali nitrates depend upon the concentration of n-butanol in solution, fig. 6 (12). First of all, it is noted that in the absence of alcohol, just as in the capacities, the lyotropic series clearly shows up. The capacity in the presence of specifically adsorbed Rb^+ causes a larger potentialdrop across the Stern layer and thus a lower value of ψ_d. This conclusion fits in with the stability data of fig. 6 since a lower value of ψ_d corresponds with a lower value of c_c. Upon adding the alcohol, the c_c values gradually increase and their differences diminish. Apparently, the specifically adsorbed counterions are displaced by the adsorbing alcohol molecules, resulting in a increase in ψ_d. This process proceeds until all counterions have been displaced. From that alcohol concentration on, c_c, and thus ψ_d, decreases due to the decrease in σ_o upon alcohol addition (cf. fig. 2). It is very striking that beyond the maximum the lyotropic sequence is reversed. This aspect can be explained by considering the radius of the hydrated ions in solution. Since a

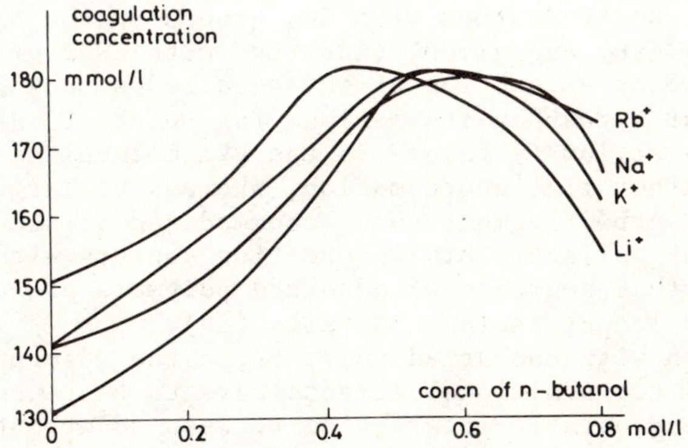

Fig. 6 Coagulation concentrations of a number of monovalent counterions as a function of the n-butanol concentration in aqueous solutions. AgI sol; pI = 6.5; T = 20°C.

hydrated Li^+ ion is larger than a hydrated Rb^+ ion, its distance of closest approach to the surface is larger, causing the potentialdrop in the Stern layer to be larger and thus ψ_d and c_c to be smaller.

2.6. Charged and uncharged high-molecular organic compounds

Qualitatively, σ_o(pAg) curves referring to solutions of polymers or polyelectrolytes are very similar to those pertaining to low-molecular compounds. This similarity is exemplified in fig. 7 for uncharged poly(vinyl alcohol) containing some acetate groups (13) and in fig. 8 for positively charged poly-L-lysine (14). Definite quantitative differences occur, as becomes obvious, e.g., when comparing the σ_o values at the adsorption maxima in figures 2 and 7. In fig. 8 this maximum is even beyond reach, an aspect to be discussed below.

A quantitative comparison between σ_o(pAg) curves for various organic compounds enables conclusions with regard to the conformation of the adsorbed polymers or polyelectrolytes to be drawn. The indifferent electrolyte concentration should be 10^{-1} M for this purpose, since at this concentration virtually only the capacity of the Stern layer is measured. It is this layer close to the surface where adsorbed organic molecules exert their effect. In analysing their PVA data, Koopal and Lyklema (13) obtained the degree of occupancy at the interface, θ, by employing the equation

$$\sigma_o(\theta) = (1 - \theta)\sigma_o(\theta = 0) + \theta\sigma_o(\theta = 1)$$

at constant pAg. $\sigma_o(\theta = 1)$, the value of the surface charge density at complete coverage, can not be obtained experimentally, since complete coverage with polymers can not be achieved. Given the molecular structure of PVA, it was assumed that $\sigma_o(\theta = 1)$ equals the arithmetic mean of the corresponding values for n-butanol (adsorbed with CH_3 groups on the AgI surface (7)) and ethylene glycol (adsorbed with CHOH groups (15)). Plotting θ as a function of the adsorbed amount Γ_p demonstrated that θ reaches its maximum far before Γ_p does so. Apparently, only at low Γ_p values do the PVA molecules get adsorbed in a rather flat conformation, whereas at larger Γ_p the additionnally adsorbed segments are accommodated at some distance from the surface. This information concurs with the current concept that segments of adsorbed polymers are distributed over loops, trains and tails (16).

In comparison with uncharged polymers, in the adsorption of polyelectrolytes two additional effects have to be taken into account: the electrostatic interaction between adsorbent and adsorbate and the mutual repulsion of charged segments in the polymer chain. In an extensive theoretical analysis, Van der Schee (14) has recently shown that as a result of these electrostatic contributions, polyelectrolytes get adsorbed in a much more extended (flat) conformation than uncharged polymers.

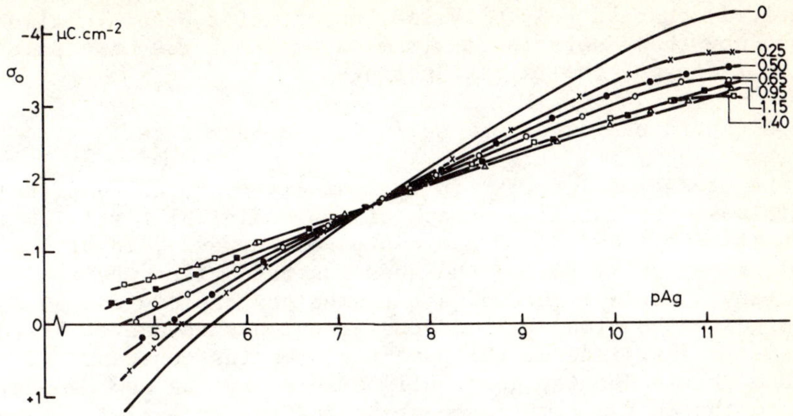

Fig. 7 Surface charge density of AgI as a function of pAg for some amounts of adsorbed poly(vinylalcohol) (in $mg.m^{-2}$). Aqueous 10^{-1} M KNO_3 solutions; T = 20°C.

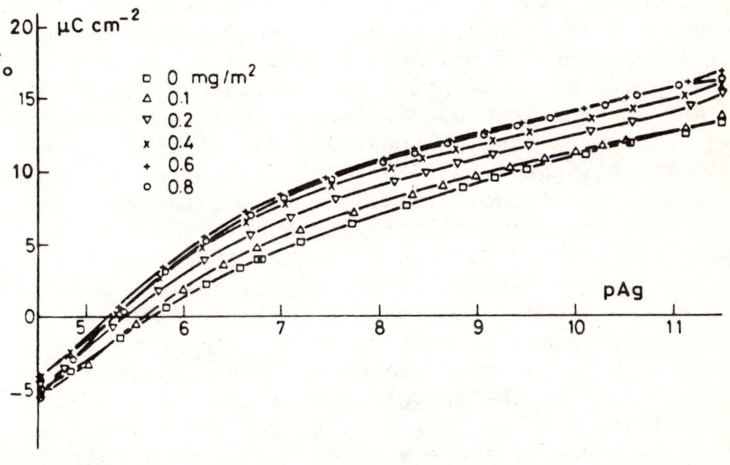

Fig. 8 Surface charge density of AgI as a function of pAg for some amounts of adsorbed poly-L-lysine. Aqueous 10^{-1} M KNO_3 solutions; pH = 3; T = 20°C.

Since adsorption isotherms of poly-L-lysine appear to be virtually independent of the molecular weight, the absence of loops and tails is confirmed. This flat conformation is caused by the strong attraction between negatively charged AgI and positively charged poly-L-lysine, which, of course, is also the reason why the adsorption maximum (common intersection point) in fig. 8 is beyond the titration range.

2.7. Electrodynamics of the AgI/solution interface

The preceding information exclusively concerns equilibrium conditions. The mutual approach of two colloidal particles in a coagulating system is a dynamic process, however, involving double layer overlap, and the question has arisen whether electrical double layers adjust themselves rapidly enough as compared to the time-scale of the Brownian encounters. If this process is fast indeed, the potential remains constant and the surface charge density gets fully adapted to the new conditions. If not, the charge remains constant and the potential changes. Rephrased, the question is whether coagulation takes place under conditions of constant potential or of constant charge, the corresponding interaction free energies being quantitatively different. In order to obtain insight in the timescale of double layer adjustment, coulostatic experiments with AgI electrodes have been performed (4,17,18). In these experiments a charge pulse of short duration (<1 μs) is applied to two identical electrodes and the transient potential difference recorded. This information is transformed into the frequency dependence of the interfacial impedance, Z(ω). Analysis of Z(ω) has demonstrated the existence of two consecutive relaxation processes (17), one on the order of microseconds related to the exchange current at the Ag/AgI interface, and one on the order of milliseconds, related to the AgI/solution interface. The latter is of the same order of magnitude as the relaxation time of the Brownian encounter. Assuming again all processes at the interface between dispersed particles and continuous medium to be identical to those at the electrode/solution interface, Lyklema and Van Leeuwen (19) concluded these results to imply that in AgI sols interaction takes place at constant total charge rather than at constant potential, but that the charge density along the surface varies during the coagulation process.

The impedance Z(ω) and the admittance Y(ω) can be interpreted in terms of the equivalent Randles circuit depicted in fig. 9 (4). From the relevant equations it appears that the imaginary component of the admittance, Y", is related to the capacity C, the charge transfer resistance θ, the Warburg coefficient σ and the angular frequence ω according to

$$Y''/\omega = \frac{\sigma\,\omega^{-\frac{1}{2}}}{\omega(\Theta + \sigma\omega^{-\frac{1}{2}})^2 + \sigma^2} + C \qquad (5)$$

which reduces to

$$Y''/\omega = \omega^{-\frac{1}{2}}/2\sigma + C \qquad (6)$$

if θ is negligible.

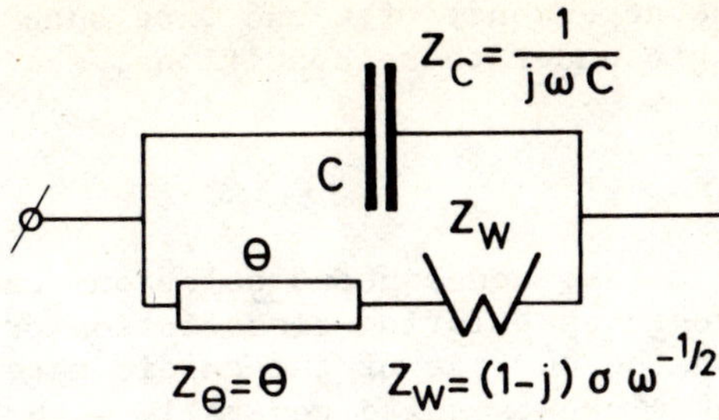

Fig. 9 Randles equivalent circuit for a AgI electrode/aqueous solution interface.

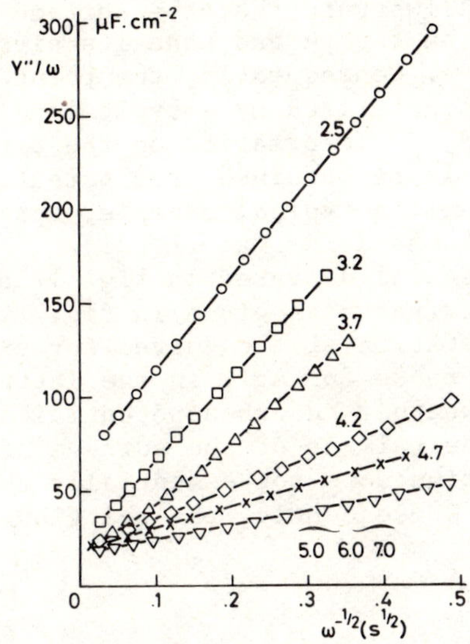

Fig.10 Y''/ω as a function of $\omega^{-\frac{1}{2}}$ for a number of pI values in aqueous 10^{-1} M KNO_3 solutions.

In fig. 10 experimentally obtained plots of Y''/ω vs. $\omega^{-\frac{1}{2}}$ are presented for various pI values in solutions of 10^{-1} M KNO_3 (4). As in all cases perfectly straight lines are obtained, Θ is negligible, indeed. The C values thus inferred are in reasonable agreement with those obtained by the titration and bridge techniques.

It appears that eq. (6) is also valid when low- and high-molecular organic substances are adsorbed on the AgI electrodes.

For a discussion of the dependency of C and σ on such variables as pAg and ionic strength under these conditions, the reader is referred to (20).

3. OXIDES

Interfaces between oxides and aqueous solutions can acquire electrical charges through association-dissociation processes of amphoteric OH groups on the surface of the oxidic materials. Indicating the relevant elements quite generally with the symbol M, the following reactions can be depicted

$$MOH_2^+ \rightleftarrows MOH + H^+ \quad (7)$$

$$MOH \rightleftarrows MO^- + H^+ \quad (8)$$

These two equilibria illustrate that the surface charge density will be dependent on the pH and that its sign can be either negative or positive. Consequently, the point of zero charge of any oxide is characterized by a typical pH value of the adjoining solution, pH_{pzc}. Information on the dependence between σ_o and pH can easily be obtained from potentiometric titration, just like for AgI. A typical example, applying to a non-porous surface, is presented in fig. 11 (21).

The σ_o (pH) dependencies illustrated in fig. 11 are quite similar to their AgI counterparts as given in fig. 1. Nevertheless, the interpretation of the curves for oxides is rather more involved than those for AgI. In the latter case, it can not be discriminated whether adsorbed potential determining ions constitute part of the bulk or of the surface of the AgI, which implies that adsorption does not essentially change the composition of the solid. Consequently, the relation

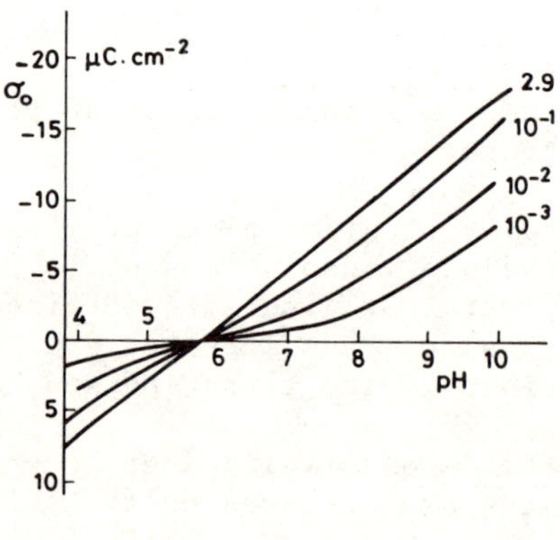

Fig. 11 Surface charge density of titaniumdioxide (rutile) as a function of pH for some KNO_3 concentrations ($mol.l^{-1}$); T = 25°C.

$$\psi_o = 2.3\ (RT/F)(pAg_{pzc} - pAg) \tag{9}$$

holds, and any variation in pAg can be translated into a variation in ψ_o, which enables the cited differential capacities to be calculated. The amphoteric hydroxyl groups only occur at the surfaces of the oxides, however, and any variation in their degree of dissociation of ionization changes the composition of this phase. Due to the corresponding variation in thermodynamic potential

$$\psi_o \neq 2.3\ (RT/F)(pH_{pzc} - pH) \tag{10}$$

and σ_o(pH) curves for oxides cannot be interpreted in terms of capacities.

For many other systems, such as clays, proteins, ionized monolayers, polystyrene latices and coals the origin of, and variation in, σ_o are rather comparable to those for oxides. For the interpretation of the σ_o (pH) curves, recently the so-called ionization-complexation model has received a fair amount of attention. This theory is too much involved to be presented here in any detail; comprehensive versions can be found, e.g., in two recent reviews (22,23). But the principles are as follows.

The dissociations (7) and (8) are specified by the dissociation constants K_{d1}^{int} and K_{d2}^{int} (int being an abbreviation of intrinsic) and the protonconcentration at the surface, $[H^+]_s$. $[H^+]_s$ is related to the bulk proton concentration $[H^+]$ according to

$$[H^+]_s = [H^+]\ \exp(-y_o) \tag{11}$$

where $y_o = F\psi_o/RT$.

Part of the negative surface charge that is due to MO^- groups can be neutralized by complexation with cations from the solution, say Na^+. Of course, the same holds for the positively charged sites and anions, say Cl^-. The complexation equilibria are formulated in terms of the constants K_{Na}^{int} and K_{Cl}^{int} and the concentrations $[Na^+]_\beta$ and $[Cl^-]_\beta$ in the inner Helmholtz layer. The latter are related to the bulk concentrations $[Na^+]$ and $[Cl^-]$ by

$$[Na^+]_\beta = [Na^+]\ \exp(-y_\beta) \tag{12}$$

$$[Cl^-]_\beta = [Cl^-]\ \exp(+y_\beta) \tag{13}$$

with $y_\beta = F\psi_\beta/RT$ and ψ_β is the potential in the inner Helmholtz layer.

ψ_o and ψ_β are correlated through

$$\psi_o - \psi_\beta = \sigma_o/K_1 \tag{14}$$

and ψ_β and ψ_d (the potential of the diffuse layer) through

$$\psi_\beta - \psi_d = - \sigma_d/K_2 \tag{15}$$

where K_1 and K_2 are the integral capacities of the inner and outer Helmholtz layer, resp..

The surface site density, N_s, is related to its various constituents and Avogadro's number N through

$$N_s = N(\Gamma_{MOH} + \Gamma_{MOH_2} + \Gamma_{MOH_2Cl} + \Gamma_{MO} + \Gamma_{MONa}) \tag{16}$$

whereas σ_o and σ_β are defined by

$$\sigma_o = F(\Gamma_{MOH_2} + \Gamma_{MOH_2Cl} - \Gamma_{MO} - \Gamma_{MONa}) \tag{17}$$

$$\sigma_\beta = F(\Gamma_{MONa} - \Gamma_{MOH_2Cl}) \tag{18}$$

Taking into account that electroneutrality requires

$$\sigma_o + \sigma_\beta = - \sigma_d \tag{19}$$

and that σ_d and ψ_d can be mutually converted using the Gouy-Chapman theory, finally a very complicated equation can be

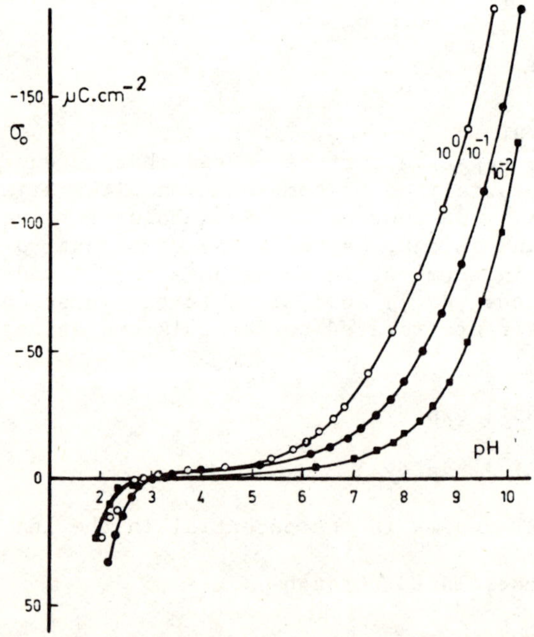

Fig. 12 Surface charge density of precipitated silica as a function of pH for some KCl concentrations (mol.l^{-1}); T = 20°C.

formulated that enables to calculate σ_o as a function of pH for a given electrolyte concentration and a given set of N_s, K_{d1}^{int}, K_{d2}^{int}, K_{Na}^{int}, K_{Cl}^{int}, K_1 and K_2. Most of these parameters are not known a priori or can not be established independently, however. So, in practice usually the other way round is chosen and by very elaborate graphical or computational methods it is inferred which set of parameters most adequately describes the experimental σ_o(pH) data. Due to the large number of required parameters, it is not unusual that more than one combination provides fair agreement between theory and experiment. Since the physical meaning of the various constants thus becomes rather vague, a serious weakness of the model emerges. On the other hand, in its favour pleads that in some cases ψ_d as ensuing from this type of analysis appears to be close to the ζ-potential as determined independently by electrophoresis.

For porous surfaces the dissociation-complexation model is not applicable, since there an additional effect shows up. Fig. 12 (24) illustrates that for precipitated silica at pH values above about 8, σ_o inceases very sharply upon increasing the pH and actually starts to exceed the corresponding site density at the surface. Lyklema (25) has developed a theory expressing how micro-porosity of the surface can account for this phenomenon.

REFERENCES

(1) Van den Hul, H.J. and Lyklema, J. 1968, J. Am. Chem. Soc. 90, pp. 3010-3015.

(2) Bijsterbosch, B.H. and Lyklema, J. 1978, Adv. Colloid Interface Sci. 9, pp. 147-251.

(3) Oomen, J.J.C., 1965, Thesis State University Utrecht; Engel, D.J.C. 1968, Thesis State University Utrecht; Pieper, J.H.A. and De Vooys, D.A. 1974, J. Electroanal. Chem. 53, pp. 243-252.

(4) Polder, R.B., Bijsterbosch, B.H. and Van Leeuwen, H.P. 1984, J. Electroanal. Chem., in press.

(5) Lijklema, J. 1961, Kolloid-Z. 175, pp. 129-134.

(6) Honig, E.P. 1969, Trans. Faraday Soc. 65, pp. 2248-2259.

(7) Bijsterbosch, B.H. and Lyklema, J. 1965, J. Colloid Interface Sci. 20, pp. 665-678.

(8) Blomgren, E., Bockris, J.O'M. and Jesch, C. 1961, J. Phys. Chem. 65, pp. 2000-2010.

(9) De Keizer, A. and Lyklema, J. 1980, J. Colloid Interface Sci. 75, pp. 171-184.

(10) Derjaguin, B.V. and Landau, L. 1941, Acta Physicochim. U.R.S.S. 14, pp. 633.

(11) Verwey, E.J.W. and Overbeek, J.Th.G. 1948, "Theory of Stability of Lyophobic Colloids", Elsevier, Amsterdam.

(12) Vincent, B., Bijsterbosch, B.H. and Lyklema, J. 1971, J. Colloid Interface Sci. 37, pp. 171-178.

(13) Koopal, L.K. and Lyklema, J. 1979, J. Electroanal. Chem. 100, pp. 895-912.

(14) Van der Schee, H.A. 1984, Thesis Agricultural University, Wageningen.
(15) De Wit, J.N. and Lyklema, J. 1973, J. Electroanal. Chem. 41, pp. 259-271.
(16) Fleer, G.J. and Lyklema, J. 1983, in "Adsorption from Solution at the Solid/Liquid Interface", G.D. Parfitt, C.H. Rochester, eds., pp. 153-220.
(17) Peverelli, K.J.P. and Van Leeuwen, H.P. 1979, J. Electroanal. Chem. 99, pp. 157-171; 1980, ibid. 110, pp. 119-135.
(18) Kvastek, K. and Horvat, V. 1981, J. Electroanal. Chem. 130, pp. 67-79; 1983, ibid. 147, pp. 83-96.
(19) Lyklema, J. and Van Leeuwen, H.P. 1982, Adv. Colloid Interface Sci. 16, pp. 127-137.
(20) Polder,R.B., Bijsterbosch, B.H. and Van Leeuwen, H.P. 1984, J. Electroanal. Chem., submitted.
(21) Yates, D.E. and Healy, T.W. 1980, J.C.S.Faraday I 76, pp. 9-18.
(22) Healy, T.W. and White, L.R. 1978, Adv. Colloid Interface Sci. 9, pp. 303-345.
(23) James, R.O. and Parks, G.A. 1982, in "Surface and Colloid Science", E. Matijević, ed., Plenum Publishing Company, 12, pp. 119-216.
(24) Tadros, Th.F. and Lyklema, J. 1968, J. Electroanal. Chem. 17, pp. 267-275.
(25) Lyklema, J. 1968, J. Electroanal. Chem. 18, pp. 341-348; 1971, Croatica Chemica Acta 43, pp. 249-260.

PHASE TRANSITIONS IN ADSORBED LAYERS

C. BUESS-HERMAN

Laboratoire de Chimie Analytique,
Faculté des Sciences, C.P. 160,
Université Libre de Bruxelles,
50, av. F.D. Roosevelt, 1050 Bruxelles.

ABSTRACT

When they are adsorbed onto ideally polarizable electrodes, organic surfactants may condense into compact layers, within defined limits of potential, bulk activity and temperature. Thermodynamic considerations based on the surface free energy excess are useful in mapping the borderline of these domains, characterized by sharp transitions involving immiscible films. These films are of gaseous, liquid as well as quasi-solid nature, according to the systems and the circumstances. Phase transformation may proceed by various types of mechanisms. The nucleation and growth mechanism is discussed more specifically, with due consideration to the selection of methods leading to reliable data. Examples are produced, in order to illustrate to what extent the theoretical models are consonant with the experimental findings.

I. INTRODUCTION

The study of phase transformations is a vital part of many fields of science, medicine and engineering. Involved in the wide range of every-day life activities, a number of phase changes (i.e. the boiling of water, the melting of solids, the condensation of vapors) are familiar to everyone. For the sake of rationalization, scientists have classified these phenomena in different categories, according to selected characteristics of the discontinuous changes which affect the properties of the system. The usual cooperative processes (solid $\rightleftarrows$ liquid, liquid $\rightleftarrows$ vapour, solid $\rightleftarrows$ vapour) which possess a latent heat of transition and present a discontinuous volumic modification are called first-order

A. F. Silva (ed.), Trends in Interfacial Electrochemistry, 205–253.

transitions. The absence of any latent heat and density variation, but the existence of a discontinuity in the heat capacity-temperature curve are the experimental facts which are indicative of a second-order transition. This process is typified by the transformation of some metals from ferro- to para-magnetic solids at their Curie points or, at very low temperature, the transition to states showing electric superconductivity. However the so-called Ehrenfest classification cannot account for the transformation observed in the helium system close to 2 K (liquid HeI $\leftrightarrows$ liquid HeII) which behaves neither as a first nor a second order transition. This process has been called a λ transition, in line with the lambda shape of the heat capacity-temperature plot.

For all kinds of transitions, the system tends to "hesitate" between order and disorder and is prone to exhibit thermodynamic fluctuations which reflect the search of a compromise between the simultaneous requirements for minimum energy and maximum entropy. The existence of ordered states of matter is easily explained by the fact that the systems are three-dimensional. When the dimensionality of the system is reduced, the probability of thermodynamic fluctuations increases significantly. However, ordered states can still have a finite range of existence in two dimensions while they are no longer meaningful in a hypothetical one-dimensional space.

In order to explain the universal character of the transition classes, theoreticians must take simultaneously into account the dimensionality d of the system, the order parameter or symmetry change n associated with the transition and the range of underlying forces. As far as short-range forces are only considered, theoretical predictions, based on the powerful method of the so called "Renormalization Group", lead to the conclusion that a limiting behaviour is reached when d approaches 2. The effects of d and n are quite spectacular in deciding whether phase transitions will be observed or not. For d = 3, phase transitions are always observed irrespective of the value of the order parameter whereas for d = 1 ordered states cannot exist, since the happening of a single fluctuation results in the total destruction of the continuity in order. In two dimensional systems, it is the value of n which determines the possibility of a phase transition : when the order parameter is equal or superior to 2, these systems are of particular interest because they are extremely close to the borderline which delimits the domain where sharp transition occurs.

This brief overview of some of the basic concepts which are involved in transitions (more details can be found in references 1 - 5) will help to understand why so many theoreticians and experimentalists are increasingly interested by the field of 2D systems. The most common examples of 2D systems are provided by the very thin monolayers which can be deposited on the surface of a solid (e.g. 6). However the need of a support for these films introduces substrate effects which may easily blur the "genuine" behaviour. Phase transitions are also frequently observed with monolayers spread at the air-liquid interface (7, 8). Although a large number of studies concerned with the adsorption properties of surfactants at electrodes has been carried out, it is only recently (9 - 12) that the formation and the behaviour of a compact monolayer has emerged as a new domain, in the scientific discipline which encompasses all 2D physical transformations.

II. ELECTROCHEMICAL SYSTEMS : MONOLAYERS ADSORBED ON ELECTRODES

The last few decades have seen an increasing interest in the study of the formation of a new phase in interfacial Electrochemistry. This is particularly the case for electrocrystallization, anodic passivation and underpotential deposition, all processes which are controlled by a charge transfer reaction across the interface. According to the experimental conditions (13, 14), this reaction may be restricted to a single monoatomic layer or may result in the build-up of a thicker deposit.

The formation and transformation of a monolayer made of organic molecules adsorbed at the electrode-solution interface constitutes an alternate way to study phase changes in Electrochemistry. Compared to their counterparts in electrocrystallization, these electrocapillary processes, which are exclusively related to tensioactive properties, present the advantage that the new phase is generally limited to the thickness of one molecule and that there is no charge transfer. As such, the electrode plays merely the role of a physical support for the two-dimensional film. It is of course essential that the electrode remains ideally polarizable throughout a substantial potential range in order to deal with a substrate which remains immune from electrochemical modifications. Quite generally, solid metal electrodes display very complex surface structures in view of the multiplicity of crystallographic orientations and the hardly avoidable presence of defects which hinder any attempt to characterize properly the local configuration. Under this respect, a liquid electrode such as mercury or gallium appears to be the most appropriate choice. In addition, the mercury electrode has been already extensively studied and provides a wealth of interfacial parameters which can be usefully employed in order to understand the properties of the adsorbed layer.

Adsorption from a solution is a more complex process than adsorption from a gas phase (15, 16). It amounts to the replacement of solvent molecules by surfactant molecules, a process which is simultaneously governed by solvent-electrode, solvent-adsorbate and adsorbate-electrode interactions. Water, which is the most common solvent, possesses a high dielectric constant and as such, tends to reject at its bulk periphery all molecules with low dielectric constant.

The thermodynamic study of adsorption of solution components on mercury is based on the classical Gibbs equation which relates, at constant temperature and pressure, the interfacial tension (γ) to the electrode potential (E) and to the bulk activity of the surfactant :

$$d\gamma = -\sigma dE - RT \sum_i \Gamma_i \, d \ln a_i \tag{1}$$

Values of the superficial excess Γ_i can be derived from a set of interfacial tensions, obtained by varying the bulk activity of the adsorbate. In the case of a neutral species :

$$\Gamma_i = - \frac{1}{RT} \left(\frac{\partial \gamma}{\partial \ln a_i}\right)_E \tag{2}$$

If a well-defined adsorbed layer exists, the value of Γ (Γ_{max}) is invariant within a given concentration and potential range. The area occupied by one adsorbed molecule is then readily calculated from :

$$S = \frac{1}{N_{Av} \Gamma_{max}} \tag{3}$$

Provided that the molecule is rigid, this value can be matched with the various cross-sections of the molecule, yielding useful clues about the effective orientation of the surfactant with respect to the surface.
In addition to interfacial tension, the charge density and the differential capacity of the electrode can be directly measured or derived from the γ measurements performed at constant activity :

$$\sigma = -\left(\frac{\partial \gamma}{\partial E}\right)_{a_i} \quad \text{and} \quad C = -\left(\frac{\partial^2 \gamma}{\partial E^2}\right)_{a_i} \tag{4}$$

The thickness of the adsorbed layer can be correlated to the differential capacity (15) while the shift of the pzc (extrapolated or not) which characterizes an invariant film is indicative of the perpendicular vector of the dipolar moment of the adsorbed molecules (16, 17).

Some additional data can also be obtained from the examination of the inhibition displayed by the adsorbed film towards faradaic reactions (18 - 20). The sensitivity to inhibition depends obviously on the detailed mechanism of the faradaic process (which may include both chemical and electrochemical steps) and thus on the nature of the depolarizer. By selecting appropriate systems, inhibition can often reveal some properties and in particular the compactness of the organic film.
Ellipsometry is one of the additional methods which is eminently suited for the determination of packing densities and layer thicknesses (15, 21).

The adsorption of neutral organic molecules at an electrified interface is generally characterized by a smooth variation of the superficial excess and of the electrocapillary parameters (i.e. γ, σ, C) when the potential is scanned. The adsorption behaviour is the easiest to investigate when the solubility of the surfactant is of the order of 10^{-2} - 10^{-3}M. The lower limit remains large enough to insure adsorption kinetics which are not significantly diffusion-controlled, at the usual time scale of the measurements (10s - 100μs).
On the other hand, too large solubilities present other problems related to non-ideal bulk behaviour. It should be mentioned that comparison between the superficial properties of various compounds belonging to a homologous family is much more straightforward if all concentrations are normalized with respect to their respective saturation values (22 - 24).
Figure 1 sketches the most common adsorption behaviour : the adsorption range in the C-E curve is indicated by two pseudo-capacitive peaks caused by the desorption processes which occur at high electrical fields. However, such curves are not systematically observed, in view of the frequent masking interference of faradaic processes involving the surfactant, the supporting electrolyte or the electrode itself.

The adsorption of dipolar molecules is generally more complicated. The adsorbed molecule reorients gradually with the potential, in order to adjust its dipole with the imposed electric field. Smooth variations in Γ and wide capacity peaks indicate the progressiveness of the superficial changes in the adsorption range (figure 2).
As shown by all available electrocapillary data, the superficial state may range from more or less bidimensional solutions to condensed layers, which may differ by the mode of packing and the orientation of the adsorbed molecules, which may be several.

The adsorption features can differ drastically from those depicted in figure 1 and 2, each time that the surfactant may be present at the interface under the form of two different and immiscible adsorbed films which prevail in adjacent potential ranges.

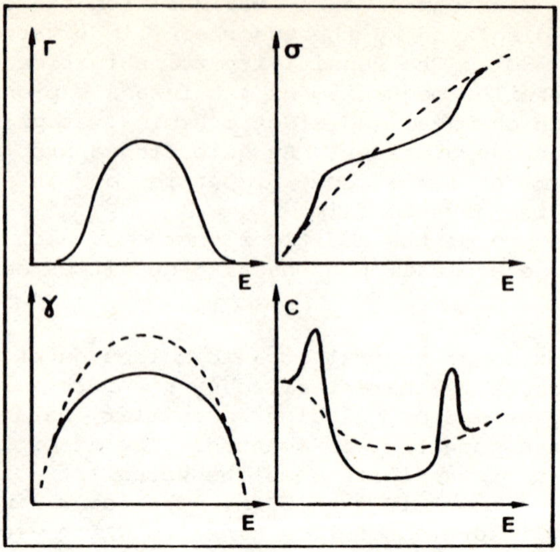

Figure 1 : Schematic representation of the usual "smooth" adsorption behaviour : superficial excess, charge density, superficial tension and differential capacity vs potential. The broken lines refer to the electrolyte alone.

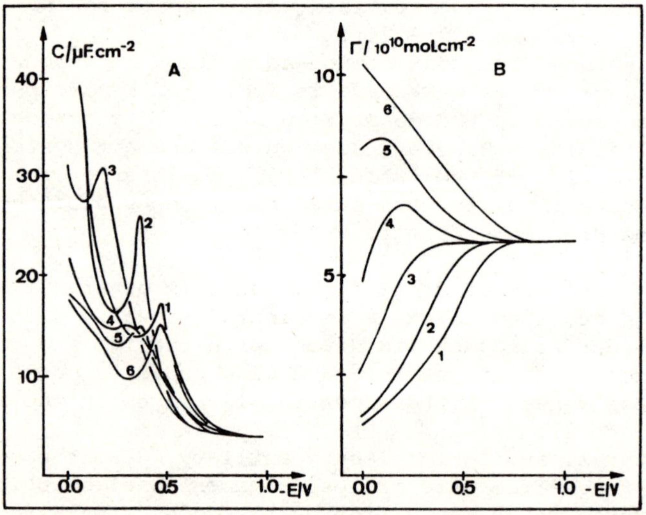

Figure 2 : The concentration effect of quinoline (referred to its saturation value c/c_S on the 117 Hz differential capacity (A) and superficial excess (B) : (1) 0.05; (2) 0.1; (3) 0.3; (4) 0.5; (5) 0.7; (6) 1.

Depending on the experimental conditions, it is not too unfrequent to observe at the mercury-aqueous solution interface (25 - 32) molecules which show a non-ideal behaviour characterized by the occurrence of a sharp phase transition which separates two immiscible superficial states. These structural changes result in sudden variations of the differential capacity, the charge density and the superficial excess.
Some typical phase transitions are sketched in figure 3, in which the superficial excess is plotted in function of the electrode potential. Up to four discontinuities have been detected for some systems such as adenosine (33) and coumarin (34). The availability of surfactants characterized by discontinuous changes in the interfacial properties provides a new framework for the investigation of the thermodynamics and kinetics of 2D transformation. However, although a monolayer at the electrode-solution interface can be considered as a 2D system, it is worth emphasizing that adjacent 3D phases (both the electrode and the solution) are present.

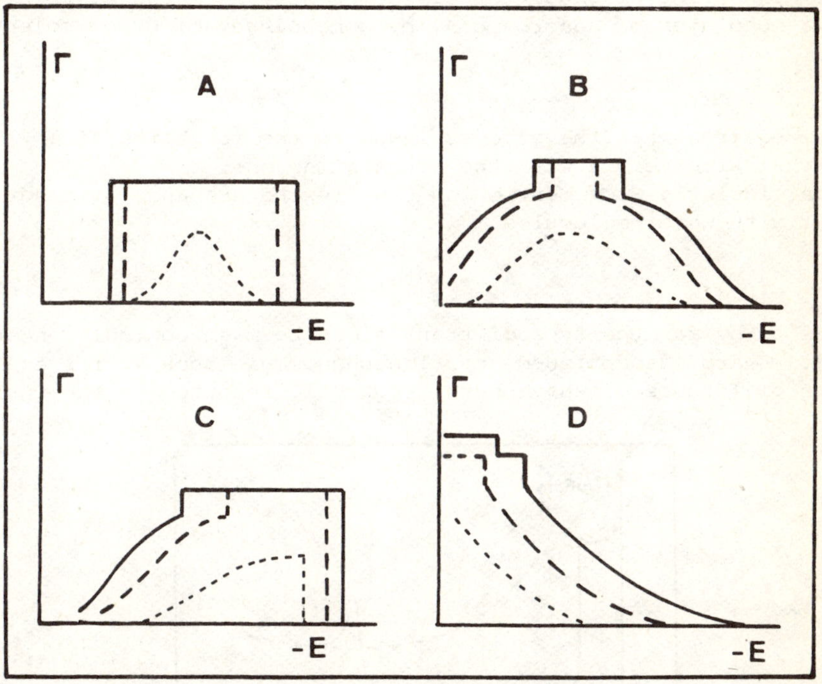

Figure 3 : Schematic superficial excess - potential curves associated to different phase transitions at various concentrations (increasing from----, ---to—). Experimental examples are for A : borneol, adamantanol; B : adenine, uracil, thymine; C : isoquinoline and some methyl derivates; D : thiourea.

These two phases play an important role by acting as heat sinks as well as composition buffers. Moreover, the electrical polarization of the electrode and the measurement of all local parameters are made possible by the favourable electric conductibility which prevails at both sides of the adsorbed layer.

III. THE MAIN CLASSES OF TRANSITIONS

The discontinuities which are observed in the $\Gamma = f(c)$ and $\Gamma = f(E)$ curves (and simultaneously in the C-E and σ-E plots) may correspond to a wide variety of interfacial modifications. At the present state of our knowledge, the various transitions arising from the local immiscibility of two superficial states prevailing at each side of a discontinuity belong to three different categories, according to the nature of the interfacial change undergone by the adsorbed layer.

The first class of transitions result from the *substitution* of a monolayer of adsorbate A_1 by a monolayer of other molecules A_2,

$$A_1 \rightleftarrows A_2$$

and requires that the process leads to the formation of a film which is immiscible with the preexisting one.
A particularly worth mentioning case is the exchange of a surfactant with water molecules.

$$A_1 \rightleftarrows \text{water}$$

This is by far the type of transition the most commonly encountered. Sharp adsorption-desorption phenomena (such as for example borneol (figure 4)) have been frequently reported in the current

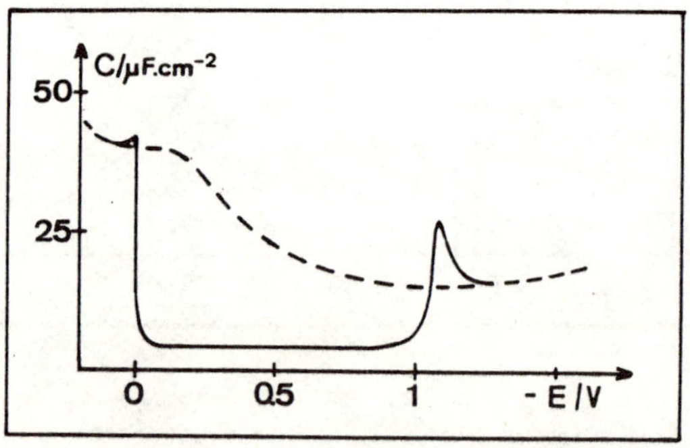

Figure 4 : Capacity plots of Na_2SO_4 1N alone (---), and after saturation with borneol (—) at 25°C.

literature (e.g. 30, 31).
Whereas wide adsorption peaks in the C-E curves are associated with the progressive formation of a dilute or "gas-like" film, sudden steps of the capacity (most of the time devoid of peaks) are indicative of the existence of a well-defined limiting superficial layer made of organic molecules.
The formation of condensed phases is consistently observed when phase transitions occur. These phases can be easily identified from the facts that (1) Γ and C reach a value which is invariant with respect to the potential and the concentration and (2) that σ is no longer concentration dependent. The condensed phase can be either of the "liquid-like" or "solid-like" type. This distinction can be sometimes deduced from the interfacial properties of the film (electrocapillary data, extent of inhibition) but is generally much more easily grounded on kinetic considerations related to the transformation from one state to the other, as it will be discussed later.

The second class of transitions involves the *rearrangement* of a film composed of only one adsorbate; A_1 and A_1' being two superficial states differing only by their ordering. Reorganization of the interfacial region may consist in the simple condensation of the molecules already present at the electrode, as well as in reorientation processes allowing dipolar molecules to adjust their dipole with the increasing electrical field and (or) favouring the establishment of more efficient lateral interactions between adsorbed molecules. Figure 5 is an example of a reorientation transition.

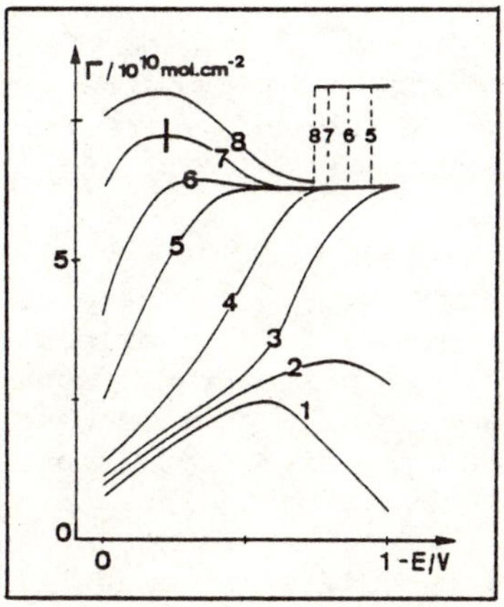

$$A_1 \rightleftarrows A_1'$$

Figure 5 : Reorientation transition of isoquinoline in 1N Na_2SO_4. 1-8 refer to isoQ concentrations (c/c_S) : 0.01, 0.02, 0.03, 0.1, 0.2, 0.3, 0.5, 0.7 and 1 respectively.

The third class of transitions embodies more complex processes such that *coadsorption* of two (or more than two) species is requested for the formation of at least one of the immiscible layers.

$$A_1 + A_2 \rightleftarrows A_1$$

and

$$A_1 + A_2 \rightleftarrows A_1'$$

Coadsorption leads here to the formation of a compact layer composed of the two adsorbates. A_2 may be another organic surfactant or an adsorbable ion from the supporting electrolyte. Joint adsorption involving anions such as NO_3^-, ClO_4^-, Cl^- is frequently observed (35 - 38) and may severely distort the "genuine" behaviour of the organic surfactant as shown in figure 6.

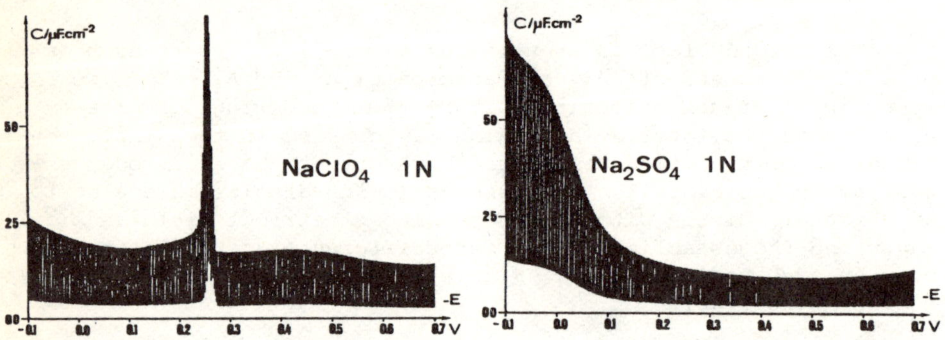

Figure 6 : Additions of perchlorate ions modify drastically the adsorption behaviour of 4-4'bipyridine, as reflected by a capacity plot showing a transition of class III (left curve).

The classification which is proposed here is a prerequisite step before attempting any meaningful comparison between the kinetic characteristics of all 2D phase transitions presently at hand. Its necessity is particularly obvious when a given experimental system presents several discontinuities. A good working example is afforded by 3-methyl-isoquinoline (3-MeisoQ) adsorbed on mercury (39) (figure 7). In the presence of an electrolyte such as Na_2SO_4, NaOH, NaF, two different transitions (A and B) are detected. At the most negative potential, transition A corresponds to the discontinuous replacement of 3-MeisoQ by water molecules (class I), whereas the transition B at about -0,2V is indicative of an inner rearrangement ($\beta \rightleftarrows \gamma$) of the organic film (class II). In the presence of a small concentration of tensioactive anions such as ClO_4^-, an additional transition (C) appears at a more positive potential. This transition separates a mixed layer (δ) of 3-MeisoQ and ClO_4^- from a film (β) where these anions are excluded (class III). When the concentration of ClO_4^- is increased, the range of prevalence of the mixed film δ increases.

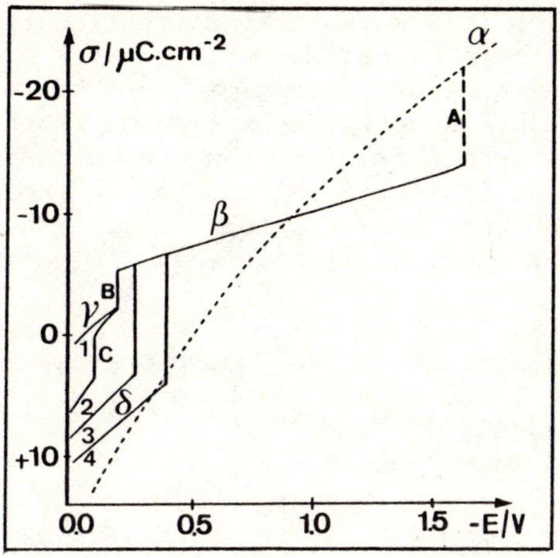

Figure 7 : A given system may present many phase transitions (A, B, C) : a case is 3-MeisoQ at 25°C in xN NaF + yN $NaClO_4$. (1) x = 0.9, y = 0; (2) x = 0.9, y = 10^{-4}; (3) x = 0.89, y = 0.01; (4) x = 0, y = 0.9. α, β, γ, δ refer to superficial layers made of water, 3-MeisoQ, 3-MeisoQ molecules reoriented and a mixture of 3-MeisoQ and ClO_4^- respectively.

Eventually, transition B is engulfed by the mixed layer δ which is the new competitor for the compact phase β. Although still of the class III, the transition is now a new one, since it involves a layer of different properties.

IV. THERMODYNAMIC STABILITY OF IMMISCIBLE 2D STATES

IV.1. Isotherms

The discontinuities along the potential scale which affect the capacity, charge and superficial excess values can be associated with similar steps observed when the *bulk activity* is varied at constant potential. Isotherms presenting sudden steps can be adequately described by the Frumkin model (40, 41) :

$$Bc = \frac{\theta}{1-\theta} \exp.(-2a\theta) \quad (5)$$

where B, c and θ are respectively the adsorption equilibrium constant, the bulk concentration and the degree of surface coverage. Depending on the sign of the interaction constant a, eqn.(5) may account for both attractive (a > 0) or repulsive (a < 0) interactions between adsorbed molecules. For a > 2 the isotherm no longer assumes the sigmoid shape characteristic of attractive interactions but degenerates into a vertical discontinuity. This reflects the occurrence of a transition process at a concentration for which the free energies of adsorption just before and after the step in θ are equal. However, this type of isotherm, widely used in classical adsorption studies, can account only for the description of a transition corresponding to the substitution of an organic compound by water ($A_1 \rightleftarrows$ water) despite the fact that this isotherm does not specifically consider the solvent as a competing surfactant.

When the organic surfactant can adsorb on the electrode in two distinct positions, the system can be described by a set of isotherms of the type (42, 43) :

$$B_1 c = \frac{\theta_1}{n_1(1-\theta_1-\theta_2)^{n_1}} \exp(-2n_1 a_{11}\theta_1 - 2n_1 a_{12}\theta_2) \qquad (6)$$

$$B_2 c = \frac{\theta_2}{n_2(1-\theta_1-\theta_2)^{n_2}} \exp(-2n_2 a_{22}\theta_2 - 2n_2 a_{12}\theta_1) \qquad (7)$$

where subscripts 1 and 2 correspond to each of the two positions. n_1 and n_2 stand for the number of solvent molecules or clusters displaced by the adsorbate in position 1 and 2, whereas a_{11}, a_{22} and a_{12} characterize the intermolecular interactions of molecules in orientation 1, orientation 2 and in the cross-wise configuration.
Similar equations apply also when two surface-active compounds (44) (1 and 2) adsorb at the electrode according to their respective concentrations c_1 and c_2. Instead of the classical two capacitor model, a three parallel capacitor network is required in order to simulate the inner layer when reorientation ($A_1 \rightleftarrows A_1'$) takes place. The basic equation for such a model is simply :

$$\sigma = \sigma_o(1-\theta_1-\theta_2) + C_1(E-E_{N1})\theta_1 + C_2(E-E_{N2})\theta_2 \qquad (8)$$

where σ_o is the electrode charge density in the absence of surfactant, C_1 and C_2 are the double layer capacities at θ_1 and θ_2 = 1 and E_{N1}, E_{N2} are the shift in the pzc when switching from $\theta_1 = \theta_2 = 0$ to $\theta_1 = 1$ and $\theta_2 = 1$.
Damaskin (43, 45) showed that such a model constitutes a good theoretical approach for the description of 2D condensation involving reorientation of the adsorbed molecules. Indeed, the shape of

the differential capacity curve predicted from this model matches very satisfactorily the curves obtained for example with uracil or adenine (as it is apparent from figure 8).

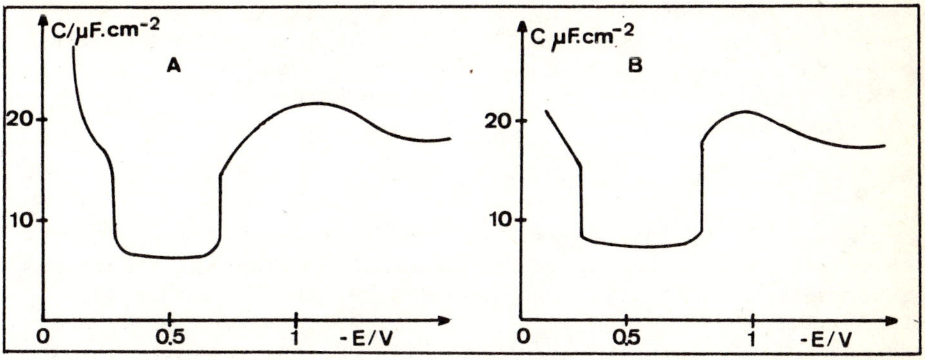

Figure 8 : Capacity pit resulting from the condensation of an organic film at 25°C (A) adenine 5 mM; (B) uracil (36mM) (both in 0.5 NaF).

The adsorption isotherms have been calculated by Damaskin in the simplified case where $a_{22} = a_{12} = 0$ and $a_{11} > 0$. Discontinuous isotherms can then arise even for values of a smaller than two provided that the value of the ratio B_2/B_1 is adequate.
A more general discussion including the case $a_{11} > 0$, $a_{22} > 0$, $a_{12} = 0$ has been since carried out by Kharkats (46). The system considered is such that besides a "rarefied 2D gas" at low adsorbate concentration and a condensed 2D liquid consisting of molecules oriented normally with respect to the electrode at high concentrations, a third configuration of molecule laying parallel to the surface and corresponding to a 2D liquid state is possible at intermediate concentrations. Depending on the adsorption characteristics of the different films, abrupt as well as gradual changes between the three states can be predicted as well as a behaviour such that one of the condensed states is no longer stable. The occurrence of double transitions may also be predicted ($G \rightarrow L_1$ followed by $L_1 \rightarrow L_2$) even in the case when only the attraction constant corresponding to interaction of molecules in the orientation parallel to the surface is not equal to zero.

The general conclusion which can be drawn from these different works is that the probability of 2D condensation depends on the ratio between the attraction parameters and the ratio of the adsorption energy in the two states. Moreover, it results from the difference in the dependence of B_1 and B_2 with the electrode potential that a suitable shift of the applied potential may cau-

se 2D phase transition, even when the interaction parameters remain invariant.
Discontinuous changes in the degree of coverage occuring when the *temperature* is varied at constant potential and concentration have also been considered by Kharkats (47, 48) on the basis of a generalized Frumkin isotherm. The calculations predict that the temperature dependence of the degree of coverage $\theta(T)$ can follow different trends. This may lead in some cases to the appearance of temperature induced sharp phase transitions.

IV.2. The equilibrium transition potential

Similarly to 3D systems, the domain of existence of 2D condensed states is subordinated to precise experimental conditions. At constant concentration and temperature, 2D "liquid or solid like" films are observed in a potential range delimited by sharp variations occuring in the superficial excess, capacity and charge density values at a characteristic potential E_θ. Rigorously speaking the equilibrium or reversible transition potential E_θ which separates two distinct superficial states must always correspond to the intersection of the electrocapillary curves representative of each of the two adsorbed films (a and b). The γ-E plot in figure 9 defining E_θ is very similar to the 3D diagram where the free energy of a substance in its solid and liquid state is plotted as a function of the temperature, the intersection corresponding then to the equilibrium melting temperature.

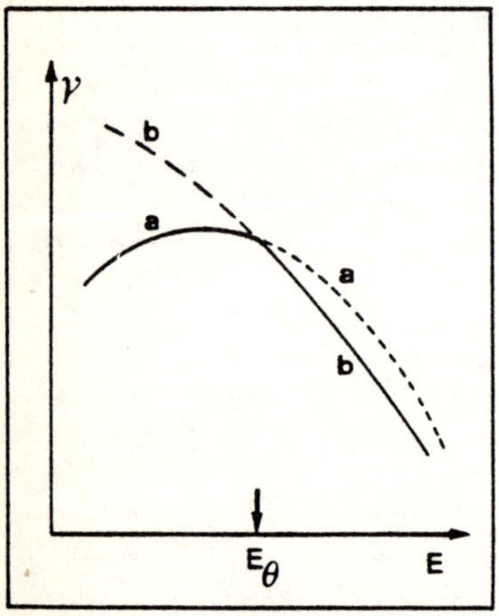

Figure 9 : The equilibrium transition potential E_θ is at the intersection of the γ-E curves related to the two superficial states a and b. Broken lines are metastable states.

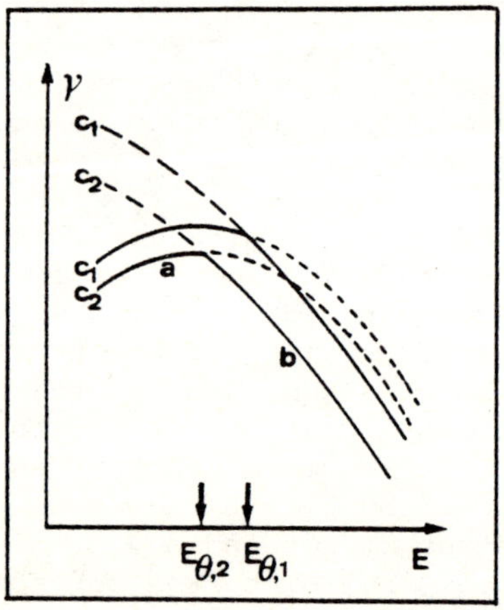

Figure 10 : Shift of the equilibrium transition potential with concentration ($c_2 > c_1$). The vertical shifts are determined by Γ_a and Γ_b. Here $\Gamma_b > \Gamma_a$.

Apart from its dependence on the nature of the surfactant, E_θ is a function of several other variables such as *its concentration, the temperature and the concentration and nature of the supporting electrolyte.*
For the sake of simplicity, we shall consider here the case of the reorientation of a surfactant which is capable of presenting at the interface two distinct superficial states a and b assumed to remain invariant with the concentration and potential in the range under investigation. As far as the small potential contribution of the diffuse part (ϕ_2) is neglected, each film can then simply be related to an electrocapillary parabola with its own set of parameters :

$$\gamma_a = \gamma_a^{max} - \frac{1}{2} C_a (E-E_a^{pzc})^2 \tag{9}$$

$$\gamma_b = \gamma_b^{max} - \frac{1}{2} C_b (E-E_b^{pzc})^2 \tag{10}$$

for a given bulk concentration and temperature.
According to their set of C, E^{pzc} and γ^{max} values, the two parabola may present two, one or no intersections. Consequently, in most cases, the existence of a condensed phase results in a capacity pit encompassed by two transitions. Although this case is often experimentally observed, more complex situations may nevertheless happen. One of the two transitions can escape detection if faradaic interferences limit the potential range where adsorption can be studied. With some systems, several transitions (2 or more) can be displayed (i.e. isoquinoline (49), 3-methyl-isoquinoline (50), coumarin (34)) as the result of the intersection of several electrocapillary curves characterizing various adsorption states in the potential range under examination.

A variation of the *bulk concentration of the surfactant* causes a modification of E_θ according to the difference $\Delta\Gamma = \Gamma_b - \Gamma_a$ in the superficial excesses of the two films which prevails on each side of the transition. A change in concentration results in a vertical relative shift of the two electrocapillary curves according to the difference between the two superficial excess values. When the transition involves the substitution of an organic surfactant by water (Γ_a or $\Gamma_b = 0$), a modification of the concentration affects only the electrocapillary curve of the organic surfactant which intersects with the electrolyte curve.
When a surfactant adsorbs under two different states a and b for which $\Gamma_b > \Gamma_a$, an increase in bulk concentration always extends the existence of the most populated film b (figure 10). More generally, by increasing the concentration of the surfactant, the most condensed film (with the highest superficial excess) will expand at the expense of the other and thus become thermodynamically stable in a larger potential range.

According to the Gibbs equation, the lowering of the interfacial tension is also a function of the *temperature*. The difference in the entropic terms characterizing each of the two films is responsible for the shift of E_θ. As can be expected, a decrease in temperature promotes the extension of phase b which is the most condensed of the two phases considered here. Typical examples of such a behaviour are the adsorption of thymine (51), adenosine (52), 5-iodocytosine (53), coumarin (34) and thiourea (figure 11 (54)).

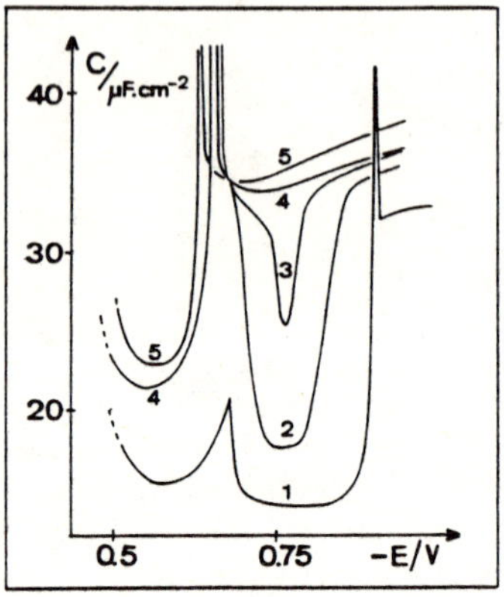

Figure 11 : A decrease in temperature favours the formation of condensed superficial states as shown by the capacity plots of saturated thiourea in 1N Na_2SO_4, at various temperatures : 1. 15°C; 2. 23°C; 3. 24.5°C; 4. 25°C; 5. 30°C.

The transition potential E_θ has also been found to depend significantly on the *concentration* and the *nature* of the supporting electrolyte. Salting out effects are responsible for detectable changes in the transition potential, as shown in Table 1, which refers to the desorption of pyridine. The discrepancy between the desorption potentials vanishes when the adsorption behaviour

Electrolyte	Desorption potentials	Electrolyte	Desorption potentials
Na_2SO_4 1N	-1.65 V/ECS	NaF 0.9N	-1.65 V/ECS
$NaNO_3$ 1N	-1.58 V/ECS	NaCl 1N	-1.60 V/ECS
$NaClO_4$ 1N	-1.56 V/ECS	NaBr 1N	-1.59 V/ECS

Table 1 : Anion effect on the desorption potential of 0.4N pyridine

in different electrolytes is compared for identical surfactant concentrations referred to their saturation value in each electrolyte, i.e. at roughly constant activity.
It is worth stressing again here that the *nature* of the electrolyte plays also a cardinal role in relation with the possible tensioactive properties of its ions.
The salting in - or - out effect and the change of the electrode charge density at constant potential are the two essential factors by which the adsorption of organic molecules is related to the concentration of a surface inactive base electrolyte. It is well known that the inner potential of the electrode $E - \phi_2$ is a function of the electrolyte concentration. This factor has been discussed for classical adsorption phenomena by Damaskin (55) on the basis of the Gouy-Chapman theory. When sharp phase transitions take place, the evaluation of the correction term ϕ_2 is not straightforward. Indeed, at the transition, the potential is associated with a finite span of different charge values (limited by the amplitude of the charge step) leading to distinct ϕ_2 values. This deviation can be easily minimized by using a high electrolyte concentration, for which ϕ_2 becomes negligible.

The equilibrium transition potential E_θ, whose physical meaning has been previously established, is a quantity which is difficult to measure with accuracy. In most experimental conditions, the steps arising in the differential capacity and charge curves deviates from the true equilibrium potential.
In similitude with many faradaic processes, the imposition of an "overvoltage" is required to allow the transformation (figure 12)

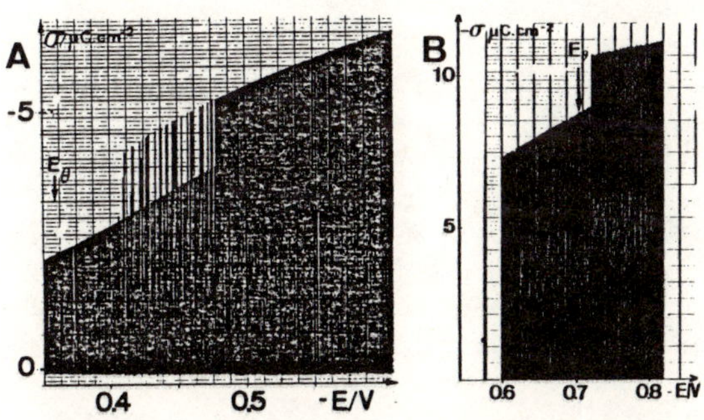

Figure 12 : Steps in charge measurements (here after 4s) do not usually coincide with the true equilibrium transition potential E_θ. (A) isoquinoline and (B) 6-methylisoquinoline saturated in 1N NaOH at 25°C.

to take place. Hysteresis effects (11, 12, 25) frequently present and deviations of the experimental transition potential with respect to equilibrium potential E_θ are not uncommon. As will be shown at a later stage, it is only by understanding properly the kinetics of the phase transition that E_θ can be correctly evaluated, with the help of experiments devised in a way such that both films are constrained to coexist effectively in a state of metastable equilibrium.

V. HOW TO TRIGGER AND FOLLOW A PHASE TRANSITION

V.1. The triggering of the transition

As indicated previously, a condensed 2D phase is confined within a defined potential-concentration-temperature range where it corresponds to the adsorption configuration of the lowest energy. Triggering a phase change requires a departure from the preexisting equilibrium of the system such that spontaneous formation of the new superficial phase is made possible. The difference in interfacial tension $\Delta\gamma^{\#}$ between the values relative to the final equilibrium γ_b and the metastable state γ_a, under controlled experimental conditions, must by necessity be negative in order to fulfill the thermodynamic condition required for the transformation. Since each of the two electrocapillary curves pertaining to the films a and b depend on the surfactant concentration, the temperature and the electrical variable, $\Delta\gamma^{\#}$ is necessarily also a function of these three variables. Accordingly, phase transitions can be triggered (in principle, at least) by an adequate modification of the concentration, the temperature or the electrical parameter. Kinetic studies are however only meaningful if the perturbation is much faster than the process that it triggers.

Condensation can be induced by a fast increase in concentration, a sudden decrease in temperature or by any suitable electrode polarization, the latter being by far the simplest experimentally. In view of the straightforward relation which links the potential to the free energy of the interface, the potentiostatic techniques are intrinsically much more worthwhile than their galvanostatic or coulostatic counterparts (56, 57) for which the potential shifts gradually during the relaxation process.

V.2. Rate measurements

Phase transitions are necessarily coupled to changes in the electrocapillary parameters which can be followed in function of the elapsed time, in order to study the kinetics of the process. Recording the decrease in interfacial tension triggered by a potential step is not a workable method because the measurement of γ takes a considerable time and also because all the relevant methods are essentially mechanical, in a way such that what is measured is purely local and is not related to the state of the whole electrode. On the other hand, the extent of coverage by the new phase is directly linked to the variation in the charge density or the differential capacity (9 - 11), while charging current transients are representative of the instantaneous rate of the invasion process. The variation of the faradaic inhibition displayed by the adsorbed layers has been recently demonstrated to be another reliable parameter (58) which can be related to the instantaneous coverage.

When the potential is stepped from one side of the transition to the other (figure 13), the resulting current and charge

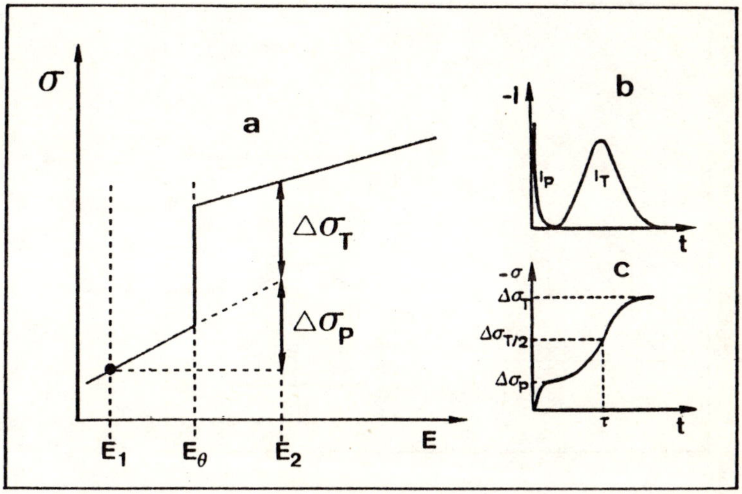

Figure 13 : Current (b) and charge (c) transients resulting from a single potential step experiment ($E_1 \rightarrow E_2$). τ is the time required for a half coverage of the electrode by the new phase.

transients present two distinct and successive parts. At short times, often determined in practice by the overall time constant of the equipment, a rapidly decaying current associated to the charge jump $\Delta\sigma_P$ is observed. This phenomenon is related to the fast incremental variation of the dielectric polarization, the dipolar orientation and (or) the superficial composition, all processes which do not alter significantly the structure of the interface. The second part of the transients (which can present different shapes according to the system) which involves the charge $\Delta\sigma_T$ is more specifically the result of the slow superficial rearrangement which constitutes the phase transition proper.

The first part, which will be called the "polarization" contri-

bution for the sake of brevity, is always present in charge and current transients.
Problems may sometimes happen when the experimental conditions are such that there is overlap between the initial step and the subsequent relaxation. Such a behaviour must be avoided, because the accurate value of the polarization component is of paramount importance, in relation with the evaluation of $\Delta\gamma^{\neq}$ which in turn is essential for the kinetic evaluation of the transients (see Section VII). It should be noted incidentally that charge or current transients become devoid of any kinetic content at the potential (E_2 in figure 14) for which $\sigma_a = \sigma_b$, so that the

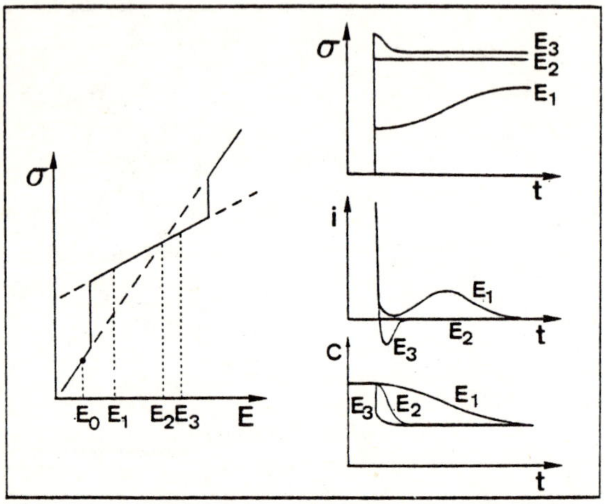

Figure 14 : Schematic charge, current or differential capacity transients resulting from a potential step from E_o to three typical values E_1, E_2, E_3. The kinetics at E_2 can only be followed on the capacity-time plot.

charges of the films before and after the rearrangement are the same (i.e. $\Delta\sigma_T = 0$, see figure 14). Such a situation is observed when the final potential coincide with the intersection of the σ-E curves representative of the two adsorbed states. When the potential step exceeds this particular value of the potential, $\Delta\sigma_P$ and $\Delta\sigma_T$ have opposite signs. This leads to a rate transient presenting a shape which is inverted with respect to the direction of the fast charging part.
The lack of information at the potential where $\sigma_a = \sigma_b$ can be circumvented by recording differential capacity transients. They remain of course meaningful only as far as the AC period is kept significantly shorter than the time needed for the completion of the process under investigation.

Differential capacity transients are restricted to adsorption kinetics, whereas charge and current transients are used in electrocrystallization studies. Phenomenologically, capacitive and faradaic transitions present three major differences :
1) The charge and current involved in adsorption processes ($\Delta\sigma_T$ $\sim$1-2μC.cm^{-2}) are considerably smaller than in faradaic processes ($\sim$500μC.cm^{-2}).
2) The additional amount of surfactant needed to build up a condensed layer is usually small compared to its bulk concentration. This condition can be easily fulfilled by appropriate selection of the concentration and the initial potential. For instance, when a condensed layer is confined within a potential region delimited by a rearrangement transition (class II) and a desorption transition (class I), the formation of the condensed film requires much less material when the initial potential is chosen in the adsorption domain of the surfactant.
3) The adsorption kinetics are considerably slower and less potential dependent than it is the case for faradaic processes, for which rates may increase by one order of magnitude per a few millivolts.

From this brief comparison with electrocrystallization, it follows that problems related to ohmic drop and mass transfer interferences are much less important in adsorption studies. Moreover, the wide potential range which is commonly available for the kinetic studies ($\sim$100-200 mV) can be used to implement more sophisticated potential programmes (such as double or triple potential steps) which might be extremely difficult to adapt to the study of faradaic processes.

VI. THE KINETICS OF ADSORPTION

VI.1. Brief overview

The processes through which the adsorption equilibrium is established have been widely discussed during the past three decades, mostly in relation with the frequency dependence of the capacitive peaks (59 - 61). In AC measurements, a small periodic signal is surimposed on a fixed potential and the superficial excess Γ varies periodically with time (except of course in regions where Γ is potential invariant). It may happen, in some instances, that the term $\partial\Gamma/\partial E$ does not attain its expected value because the adsorption process presents a slow step. In the majority of the cases, the process is then controlled by mass transfer, which has been investigated in a number of works (62 - 65). In the opposite case, the rate of adsorption is very low compared to mass transfer, so that the latter no longer influences the overall rate (66). Besides these two extreme situations, mixed control by adsorption and mass transfer has also been considered (60, 67).

The very first attempt to describe the kinetics of formation of a 2D condensed film was made by W. Lorenz (67) who introduced molecular association into the reaction scheme, written as :

$$\frac{1}{\upsilon} A_{\upsilon,ads} \rightleftarrows A_{ads} \rightleftarrows A_{sol}$$

The kinetic equations derived from this model which takes into account the simultaneous influence of the rates of mass transfer, intrinsic adsorption and association embody a considerable number of parameters which are not easily extractable from experiments.

VI.2. A preview of the main mechanisms controlling 2D phase transitions

This part and the following sections will be devoted to recent advances in the kinetic study of phase transitions occuring in adsorbed layers in the absence of mass transfer control.
As already indicated in section IV.2, the formation of a condensed phase does generally not take place at the equilibrium transition potential, but requires an overvoltage which depends largely on the experimental conditions. In 3D systems, it is well-known (e.g. 68 - 70) that phase changes such as the condensation of a vapour to its liquid or solid state, the freezing of a liquid, do not necessarily occur at the equilibrium value. In most cases, they require a significant activation energy. Supersaturation, which stays as a necessary condition, results from the fact that the transformation starts from a very small volume of the new phase, which possesses an excess energy with respect to the same number of particles belonging to a quasi-infinite sample of the new phase.

The kinetics of conversion of a given phase into another of higher ordering are usually explained by the classical nucleation theory, which rests on the concept of "critical nuclei" and on that of "equilibrium distribution of clusters". Nuclei of the critical size (or critical nuclei) can originate from a preexisting equilibrium population of clusters of sub-critical size, owing to favorable time-space fluctuations in density and orientation of the molecules. An appropriate perturbation of the system may shift the equilibrium distribution of clusters towards larger sizes, so that there is a concomittant increase of critical nuclei which, in turn, enhances the rate of nucleation J (number of nuclei per unit area and time). As soon as a cluster has reached its critical size, it has a one-half probability to expand by incorporating additional molecules and become then macroscopically detectable.

It has been recently discovered that condensation processes in 2D layers on electrodes are also controlled by a nucleation and growth mechanism. According to the relative magnitude of the rate of nucleation J and growth rate v, three basic cases can be distinguished. When the time needed to form a critical cluster (τ_N) is much shorter than the time required for a single nucleus to cover the whole surface of the electrode (τ_G), the transformation involves a large number of nuclei and is called the "polynucleation regime". On the other hand, when $\tau_N >> \tau_G$, one single nucleus can be fully grown without the formation of additional nuclei during the time needed for the complete transformation ("mononucleation regime"). The intermediate case which involves a few number of nuclei is called the "oligonucleation regime".

Since the nucleation and the growth rates vary very differently with the supersaturation of the system (imposed by the applied perturbation), each of the three above-mentioned regimes can be observed experimentally for a given overall composition (figure 15). At small supersaturations, the transformation is too

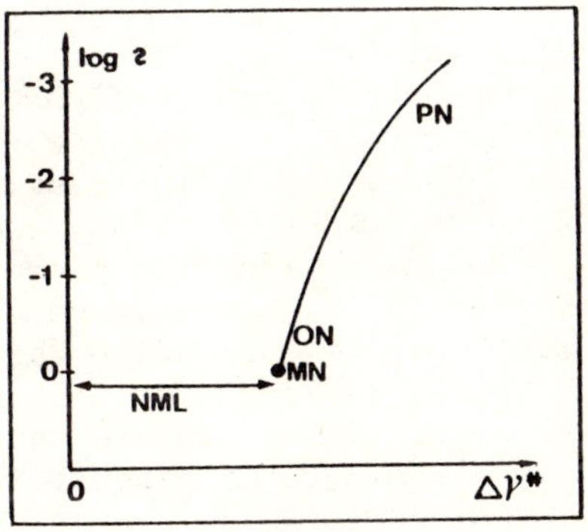

Figure 15 : Schematic log τ-$\Delta\gamma^{\#}$ plot. With increasing $\Delta\gamma^{\#}$, no-man's land (NML), mononucleation (MN), oligonucleation (ON) and polynucleation (PN) appear successively.

improbable to be detected within the observation time window (∿1ms - 10 or more s) and a no man's land region is observed where nothing happens. By increasing the supersaturation, the probability of formation of a nucleus increases and mononucleation transients can be observed from time to time.
Further increase of the supersaturation triggers the oligo- and the polynucleation processes in succession. The domain of prevalence of the different regimes is conspicuously affected by the possible presence of extraneous nucleating agents ("heterogeneous" nucleation) and by the type of potential programme (single or double step). According to the programme, the condensation pro-

cess may thus be controlled by the *progressive birth of nuclei* or by a *preexisting population of nuclei*.
Each of the two nucleation and growth mechanisms produces peaked i-t curves (or sigmoid σ-t curves). Their morphology which differs slightly will be discussed extensively in the following sections with particular emphasis on the kinetic parameters which can be extracted in each case.

In sharp contrast with the formation of a condensed layer, the reverse process (i.e. the melting or redissolution of the initial film) is much faster and shows current transients which decay exponentially. Similarly to 3D melting (68), the loss of ordering is linked with the *presence of defects* in the initial "solid-like" phase. This third mechanism which is also involved in 2D transitions will be reexamined in section X on the basis of experimental findings.

VII. THE NUCLEATION PROCESS

VII.1. The Classical Nucleation Theory (CNT)

It is widely recognized that the process of condensation is initiated by the formation and growth of "self-nucleated" or "homogeneous nucleated" critical clusters of the new phase resulting from local fluctuations in density and in the packing of the metastable film. By analogy with 3D phenomena and electrocrystallization, the mechanism of 2D phase transitions controlled by a nucleation and growth process can be quantitatively explained on the basis of the classical theory of nucleation which has been successively developed by Volmer (71), Becker-Döring (72), Farkas (73), Zeldovich (74) and Frenkel (75). In this theory, the size distribution of clusters is assumed to be essentially Boltzmannian :

$$n_N = n_1 \exp - \frac{\Delta G_F}{kT} \qquad (11)$$

where n_1 and n_N are respectively the number of non-clustered molecules and the number of clusters containing N molecules.

The rate-determining step of the mechanism is then assumed to be the incorporation of one additional molecule to the critical cluster, so that the steady-state nucleation rate J (the number of critical nuclei created per time and area units) can be expressed by :

$$J = Z\omega n_N^{*} \qquad (12)$$

where n_N^{*} is the equilibrium number of critical clusters, ω the frequency of incorporation of an additional molecule to the critical nucleus and Z the Zeldovich factor (74) which corrects for the departure from the Boltzmannian distribution. Coupling of equations 11 and 12 leads to a nucleation rate law of the form :

$$J = J_{\infty} \exp - \frac{\Delta G_F^*}{kT} \tag{13}$$

J_{∞} being a collision factor and ΔG_F^* the free energy of formation of the critical nucleus.

The relation between the free energy of formation of a 2D cluster in function of its size can be established by considering the respective weights of two antagonist energetic contributions, irrespective of the particular type of the phase change involved in the transition :

$$\Delta G_F = \varepsilon l + \Delta\gamma^{\#} A \tag{14}$$

The product $\Delta\gamma^{\#}A$ is the favourable term : A is the cluster area and $\Delta\gamma^{\#}$ the driving factor or "supersaturation" which is externally imparted to the system in order to trigger the transformation. The unfavourable energetic term is εl, where l is the length of the periphery and ε the line tension which has to be overcome by the system for the expansion of the nuclei. This last contribution takes into account the cohesion energy of each of the two adjacent phases and also the adhesion energy between them.
For circular clusters (figure 16), the free energy of formation of a critical cluster (for which $d\Delta G_F/dr = 0$) and its radius are given by :

$$\Delta G_F^* = - \frac{\pi\varepsilon^2}{\Delta\gamma^{\#}} \quad \text{and} \quad r^* = - \frac{\varepsilon}{\Delta\gamma^{\#}} \tag{15}$$

Similarly to the three-dimensional case and to electrocrystallization, the evaluation of ΔG_F^* rely on several assumptions (76) :

- Use is made of the Gibbs instead of the Helmholtz free energy.
- The clusters are considered as immobile. No contributions resulting from their possible translational or rotational modes are taken into account.
- The line tension is assumed to be constant and even at vanishingly small nuclei for which the number of hostile molecules adjacent to the periphery must become size dependent.
- The critical nucleus may contain a small number of molecules and it is far from obvious that $\Delta\gamma^{\#}$ retains at such a scale its macroscopic value.

According to the CNT, the fundamental expression for the rate

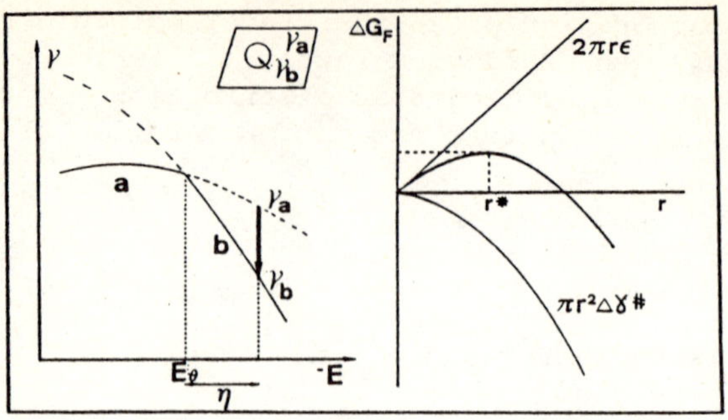

Figure 16 : Free energy of formation of a circular cluster resulting from two antagonist contributions. ε is the line tension to be overcome and $\Delta\gamma^{\#} = \gamma_b - \gamma_a$ is the surface driving energy.

of nucleation involved in the condensation of monolayers is simply :

$$J = J_\infty \exp \frac{\pi\varepsilon^2}{kT\Delta\gamma^{\#}} \qquad (16)$$

where J_∞ is the collision factor or the limiting rate of nucleation for infinite "supersaturation". Key parameters such as ε, r^* and ΔG_F^* can be obtained from the slope d ln J/d $(1/\Delta\gamma^{\#})$ and J_∞ from the intersection of the ln J versus $1/\Delta\gamma^{\#}$ straight line provided that the dependence of ε and J_∞ with $\Delta\gamma^{\#}$ can be held as negligible.

In view of the limitations already pointed out in relation with ΔG_F^* and other approximations in the overall kinetic scheme considered (hypothesis of micro-reversibility for the steps leading to the formation of the critical nucleus, ruling out of possible reaction paths involving nuclei of different sizes...), this simple model constitutes only a semi-quantitative treatment. It is nevertheless still widely used in 3D systems and in electrocrystallization, with the major aim of comparing experimental data and interpreting the dependence of J with the supersaturation. Most of the variance between this model and more elaborated theories developed for 3D processes affects only the collision factor, which may scatter by several orders of magnitude (77) according to the assumptions made. The present controversy about its evaluation indicates that the discrepancy is less than it was initially considered, and still lesser for 2D systems.

VII.2 The determination of the "supersaturation" $\Delta\gamma^{\#}$

As expected from eqn. 16, the rate of nucleation increases

when the "supersaturation" $|\Delta\gamma^{\#}|$ is made larger ($\Delta\gamma^{\#}$ being always a negative quantity). Opposite to the case of electrocrystallization, the overpotential $\eta(E-E_\theta)$ is no longer the primary driving factor when the nucleation process involves only electrocapillary forces. Table 2 gives the expressions of the supersaturation and free energy of formation of a critical cluster for several types of transformation (3D condensation and crystallization). electrocrystallization and 2D electrocapillary processes).

Process		« Supersaturation»	ΔG^{*}_{F}
3D condensation		RT ln p/p_o	$\dfrac{b\text{-}\gamma^3 V^2}{(RT \ln p/p_o)^2}$
3D crystallization		RT ln c/c_{sat}	$\dfrac{b\,\gamma^3 V^2}{(RT \ln c/c_{sat})^2}$
Electrocrystallization		$z\,e_o\,\eta$	$\dfrac{b\,S\,\varepsilon}{z\,e_o\,\eta}$
Adsorption	class I	$\Delta\gamma^{\#}=-RT\,\Gamma \ln c/c_o$	$-\dfrac{b\varepsilon^2}{\Delta\gamma^{\#}}$
	class II	$\Delta\gamma^{\#}=-RT\Delta\Gamma \ln c/c_o$	

Table 2 : "Supersaturation" and free energy of formation of a critical cluster. p_o, c_{sat} and c_o correspond respectively to the minimum pressure and concentration needed for the transformation. b is a shape factor : $b=\pi$ for a circle and $16\pi/3$ for a sphere. V and S are the volume and surface per molecule, e_o the charge of the electron and z the valency of the depositing ion.

At constant concentration and temperature, $\Delta\gamma^{\#}$ is however related to the imposed experimental variable η. Occasionally, a fairly good linearity may be found between $\Delta\gamma^{\#}$ and η, if the prospected domain is restricted to a narrow potential range close to E_θ. However, a more complex relation is the rule when the potential domain under investigation is more extended. The lack of a direct proportionality between η and $\Delta\gamma^{\#}$ is obvious when the condensed monolayer is bounded by two dilute films which prevail more cathodically and anodically respectively as illustrated by the adsorption of adenine (figure 17). In such a case, $\Delta\gamma^{\#}$ is necessarily equal to zero at the two equilibrium transition potentials E_θ but the curve $\Delta\gamma^{\#}-\eta$ must go through a maximum. The experimental fact that in such conditions $\Delta\gamma^{\#}$ and consequently J become smaller by increasing sufficiently the overpotential is a clear-cut demonstration of the true identity of the driving force.

The evaluation of $\Delta\gamma^{\#} = (\gamma_b-\gamma_a)$ which is required for the comparison and interpretation of transitions displayed by various systems raises some experimental difficulties.
Indeed, opposite to the final equilibrium interfacial tension γ_b, the absolute value of the interfacial tension of the metastable

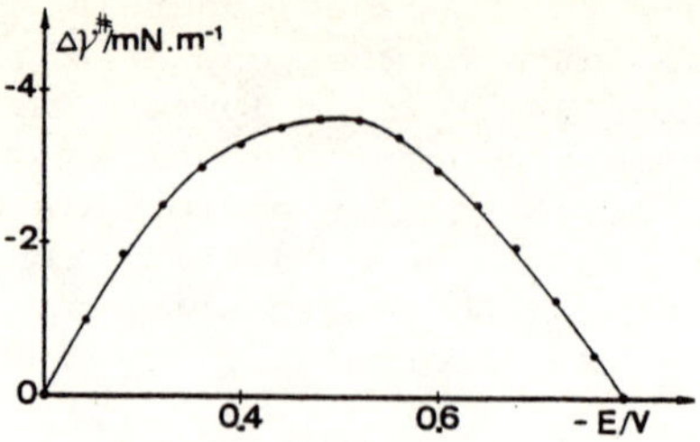

Figure 17 : The driving factor $\Delta\gamma^{\#}$ does not necessarily increase monotonously with the overpotential. This is illustrated by the adsorption behaviour of adenine (5.4 mM) for which $\Delta\gamma^{\#}$ values are plotted in function of the potential.

state γ_a cannot be measured properly by the classical electrocapillary methods.
In some instances (well defined initial film independent of E and of the concentration) the γ_a-E parabola can be reconstructed in the potential range where the film becomes metastable.
A more general way to evaluate $\Delta\gamma^{\#}$ can be based on the analysis of the complete charge transients. As described previously, the σ-transients are composed of two distinct parts $\Delta\sigma_P$ and $\Delta\sigma_T$. From transients obtained at different η, it is easy to reconstruct on the same plot the $\Delta\sigma_P$-η and $(\Delta\sigma_P + \Delta\sigma_T)$-η curves which are respectively representative of the charge of the metastable film and the final equilibrium state (figure 18). Since $\Delta\gamma^{\#}$ is equal to zero at the equilibrium transition potential E_θ, the driving factor can be thus calculated at any given overpotential $(E-E_\theta)$ by integrating from $E-E_\theta$ the charge densities pertaining to each of the two films a and b :

$$\Delta\gamma^{\#} = \int_{E_\theta}^{E} \sigma_a dE - \int_{E_\theta}^{E} \sigma_b dE \qquad (17)$$

The value of $\Delta\gamma^{\#}$ at any overpotential η_1 is given by the shaded area in figure 18.

VII.3. The accurate localization of E_θ

It has been already mentioned that the evaluation of the equilibrium transition potential E_θ requires a polarization programme such that each of the two antagonist films coexist effectively in metastable equilibrium in the absence of any net charging current. Such a condition can be fulfilled by resorting to multipotential step methods (figure 19). The first step from E_1 to E_2, required to trigger the transition a → b, is followed after a time τ (when $\Delta\sigma_T$ is half-consumed) by a second step from

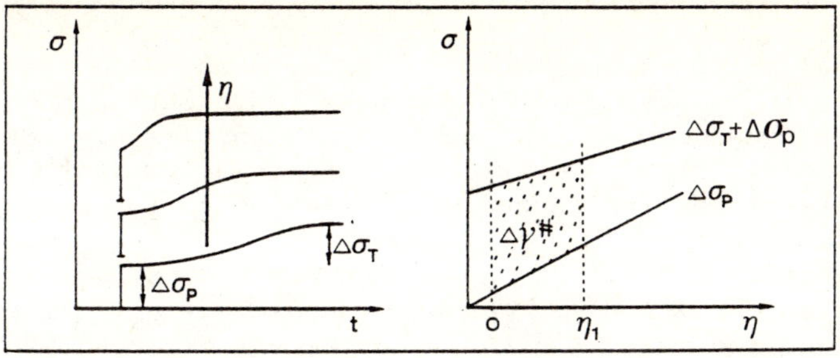

Figure 18 : A σ-η plot can be constructed from the $\Delta\sigma_p$ and $\Delta\sigma_R$ values obtained from a series of charge transients at increasing overpotentials. $\Delta\gamma^{\#}$ at η_1 is given by the dashed area.

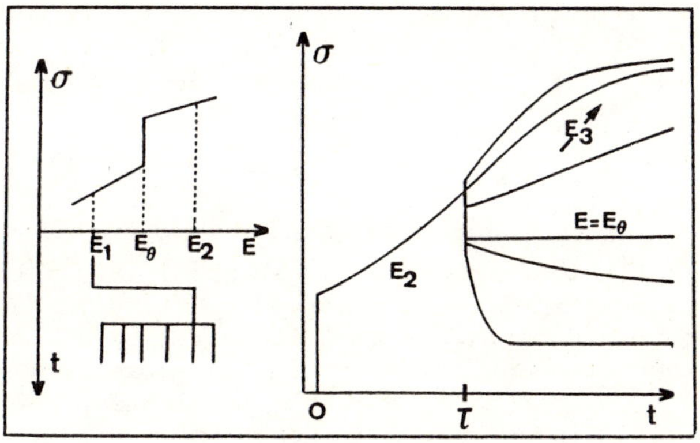

Figure 19 : Determination of E_θ by double potential step experiments. The first step from E_1 to E_2 triggers the transition. At half-coverage the potential is stepped to E_3. It is only when $E_3 = E_\theta$ that the charge density remains constant.

E_2 to E_3. The value of E_3 is varied at each successive drop in order to locate the E_θ value which is the potential at which the final charge (figure 19) or capacity remains invariant with time. In some instances, when the reverse process (S or L → G) is much faster than the condensation itself, single potential step experiments starting from the "solid-like" phase provide also an accurate value of E_θ (12).

VIII. THE FORMATION OF CONDENSED PHASES BY "INTRINSIC" NUCLEATION

The evaluation of the "self", "2D homogeneous" or "intrinsic" nucleation rate of a system undergoing a condensation process requires often demanding experimental conditions (highly purified solutions, well-defined electrodes...) in order to deal with processes which are not influenced by the action of foreign nucleants.
At low nucleation rates, the single nucleus needed for triggering the whole transformation may appear stochastically in time (the mononucleation regime). A direct measurement of J can be then based on the statistical analysis of the times elapsed before the birth of each nucleus in a succession of separate experiments under strictly comparable conditions. Such a regime is favoured by decreasing the area of the electrode; the probability of nucleation being directly proportional to the available area.
Increasing the electrode size or the supersaturation leads, on the contrary, to the progressive birth of a finite number of nuclei. Under such conditions (the oligo- and poly-nucleation regimes) the nucleation and growth processes take place simultaneously.

VIII.1. The mononucleation regime

Experiments can sometimes be devised in such a way that one single nucleus can trigger the full superficial transformation. The mononucleation regime can be observed provided that the growth rate of a nucleus is much larger than its birth rate. Such a nucleus appears then stochastically in time and space. The probability of birth or no-birth on an electrode of known area A and within a given time window is easily determined from a set of repetitive experiments. The statistical analysis of the data uses the Poisson distribution, which leads to the determination of the nucleation rate J, according to :

$$P_N = \frac{\mu^N (\exp - \mu)}{N!} \qquad \text{where} \quad \mu = J\,A\,t \qquad (18)$$

Since it is not always easy to observe mononuclear transients in the total absence of bi- or trinucleations which have a much lower but finite probability (figure 20), it is generally preferred to analyze the data related to the absence of nucleation (non-events). Plotting ln P_o or ln$(1-P_o)$ in function of the time gives a straight line whose slope is directly related to the rate of nucleation J (figure 21).
In some cases, distorted Poisson distributions (78) are found, in relation with the existence of a time-lag in the nucleation process. This time-lag can have several origins such as response limitations of the electronics, slow reequilibrium from the bulk

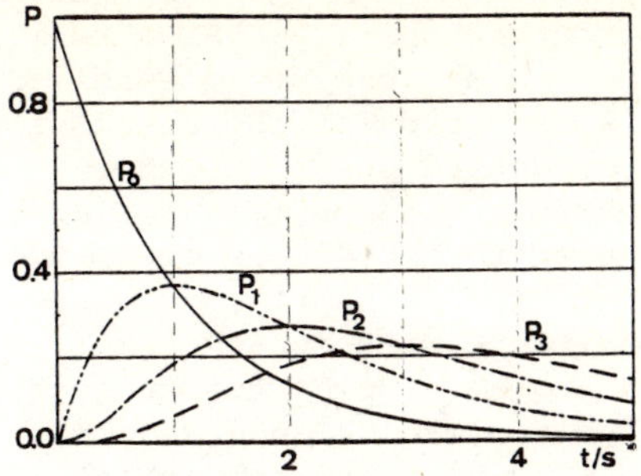

Figure 20

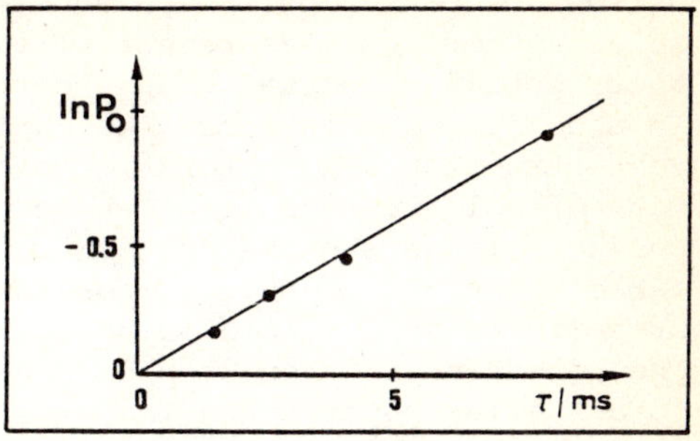

Figure 21

Figure 20 : Probabilities P_o, P_1, P_2, P_3 in function of time for the occurrence of 0, 1, 2 or 3 events respectively. Values calculated from a Poisson distribution where λ=1 event per second.

Figure 21 : Determination of J from the slope of the plot ln P_o vs t. This case (λ = 116 $N.s^{-1}$ and J = 9.8 10^3 $N.s^{-1}$ cm^{-2}) corresponds to the rearrangement transition of 6-methyl-isoquinoline at η = -0.185V. Values of ln P_o have been obtained from double potential step experiments : nucleations occuring during the first step (η = -0.185 V and time τ) are detected at a lower overpotential.

of the solution or sluggish restoration of the Boltzmann distribution of clusters.

A similar type of statistical analysis has already been performed in the field of electrocrystallization where mononucleations are usually seldom events (79, 80). The first case of observation of single events has been reported for the electrodeposition of Ag^+ on a perfect silver surface (81). In the case of layer by layer deposition, data can be collected easily by polarizing the electrode at a fixed overpotential and recording the current. Under such conditions, as soon as a first layer is completed, the initial substrate perpetuates itself and the probability of formation of a following nucleus initiating the formation of the second layer remains unchanged.
In adsorption processes such experiments have no equivalent, because completion of the first layer brings the full process to its definitive stop. Accordingly, the potential has to be shifted back to a value such that there is "dissolution" or "melting"

of the new phase, before being stepped again to the overpotential selected for the measurement of the birth probability. As a result, double potential step experiments are an absolute necessity. In addition to this programme, which is required for the successive destruction and restoration of the condensed phase, another mode of polarization is often very powerful. It consists in imposing a large supersaturation to the electrode by means of a suitable potential prepulse, whose duration (τ_P) is short enough to preserve an appreciable number of experiments with non-events. The potential is then stepped backwards to a value corresponding to a smaller supersaturation, inside the "no'man's land" region, where the nucleation probability is so slow that the only thing which may happen is the growth of those nuclei occasionnally formed during the prepulse. By varying τ_P, it is possible to act upon the probability of non-event and to compute the rate of nucleation.

The mononucleation regime provides an unique opportunity for evaluating simultaneously the absolute values of the rate of nucleation and growth : the distribution of the birth time is that of the start of the transients and gives J while the analysis of the shape of the transient can be correlated with the values of the growth rate. A possible complication is that mononuclear transients may present different shapes according to the position of the nucleus with respect to the geometry of the electrode (81, 82). The particular case considered here (that of a stationary drop) is the most easily tractable since the formation of a circular nucleus on a sphere involves sites which are all equivalent. The instantaneous coverage in then simply given by the area of a spherical cap with an angular aperture increasing linearly with time.
In terms of instantaneous charge :

$$\sigma = \frac{\Delta\sigma_T}{2}\left(1 - \cos\frac{vt}{r}\right) \qquad (19)$$

where $\Delta\sigma_T$, r and v are the charge required by the full process involving the whole sphere, the radius of the electrode and the radial growth rate respectively. The necessary proportionality which exists between the total duration of the transient and the radius of the electrode can be easily checked by performing experiments with various drop sizes. Experimental verification of eqn.(19) is the proof that the radial growth related to a given overpotential is effectively a constant, irrespective of the size of the expanding nucleus. In addition to these diagnosis criteria, the occurrence of mononucleation is indicated each time one observes a succession of transients of the same shape and duration but randomly distributed along the time coordinate (figure 22). This feature is specific for the growth of a circle on a sphere. For nuclei and (or) electrodes presenting other geometries, mononuclear transients may present rather contrasted shapes according to the location of the nucleus on the surface (81, 82).

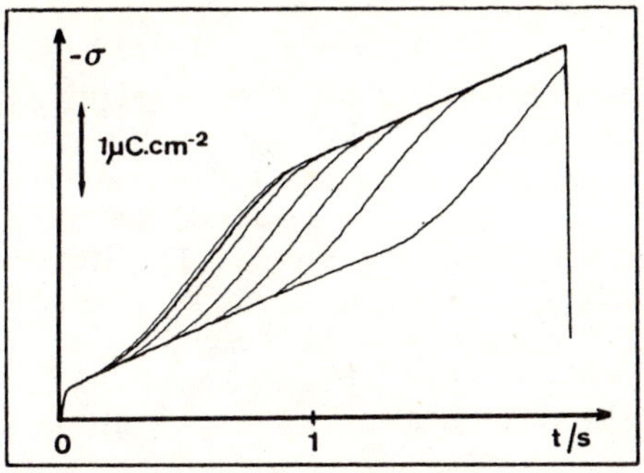

Figure 22 : Monomolecular transients observed in the presence of a solution of 1M NaOH saturated with isoquinoline. The overpotential is -20 mV.

Since there are no overlapping effects to be considered in the case of mononucleation, the rate of growth can be calculated straightforwardly. Figure 23 summarizes a few recent data obtained with different systems. It may be noted that the rate of incorporation of additional molecules into the fringe of the expan-

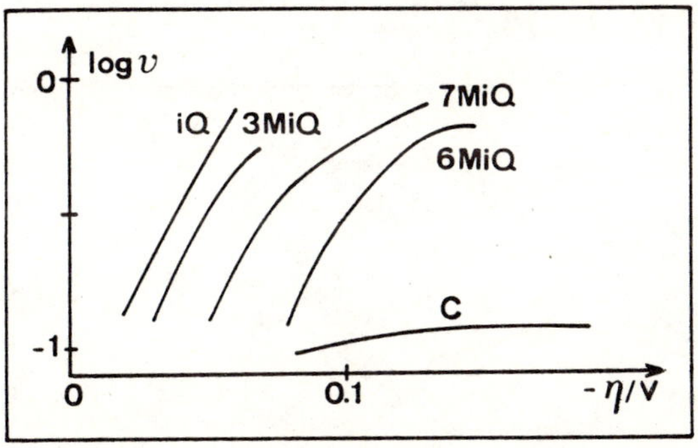

Figure 23 : The case of reorientation transitions : variation of the logarithm of the growth rate v with η for 1N NaOH solutions at 25°C saturated with isoquinoline (iQ) and various methyl isoquinolines (3 MiQ, 6 MiQ, 7 MiQ). Coumarin (C) was studied at 3°C in NaF 0.9N.

ding nucleus is typically of the order of a few mm per second at an overpotential of -50mV. Growth rates of a similar order of magnitude have been observed in the case of the electrodeposition of Ag^+ on silver, for which the growth rate constant K_v varies according to the crystallographic plane from 1.0 to 1.9 $cm.s^{-1}V^{-1}$ (83).

At the present state of our knowledge about growth processes (which is still rather fragmentary), it is not yet possible to elucidate what are the rate-determining mechanisms. A constant radial growth excludes of course the hypothesis of spiral expansion from the nucleus. Stochastic peripheral formation of protruding bimolecular double-kinks, followed by fast lateral propagation may be considered as a possible microscopic mechanism, if the reasonable assumption is made that the double-kink expectancy is a constant, irrespective of the length of the periphery.

VIII.2. The oligonucleation regime

A further small increase in the supersaturation of the system leads to the appearence of transients which are no longer mononuclear. The irreproducibility which is now observed (figure 24) obeys to stochastics involving a small number of nuclei and no longer only one. Transients of that type are difficult to analyze mathematically. However if the growth rate can be obtained at a chosen η by a lucky case of mononucleation (or if it can be extrapolated from mononuclear experiments performed at lower supersaturations), it becomes possible to evaluate the average number of nucleations provided that the time window is made sufficiently short to avoid any significant overlap effect. Figure 24 shows a series of transients of the "oligo" type from

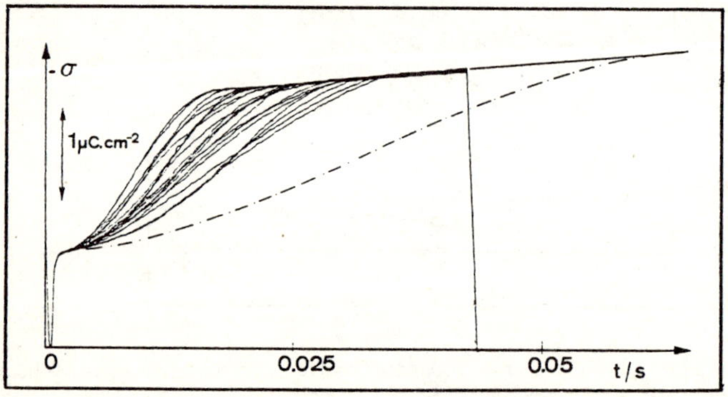

Figure 24 : Oligonuclear transients observed with the rearrangement transition of isoquinoline (sat) at η = -70 mV.

which J has been evaluated for a total coverage not exceeding 5%. This determination of J is only approximative since the stochastic expression for the transient becomes extremely cumbersome.

A case which is easily tractable is the process which involves two circular nuclei allowed to grow on a sphere. A few simulated transients are shown in figure 25. Only two angular parameters are needed for the full description of the transient : angle α defines the relative position of the two birth sites while angle β is representative of the radius reached by the first nucleus at the instant of the second birth. From experiments such that binucleations are still seldom events, so that trinucleations can be neglected, it is thus possible to evaluate the position of one nucleus with respect to the other and the time elapsed between their birth (84). This may help to assess whether or not the circumstances of formation of the second nucleus are truly random in time and space, despite the presence of the first expanding nucleus.

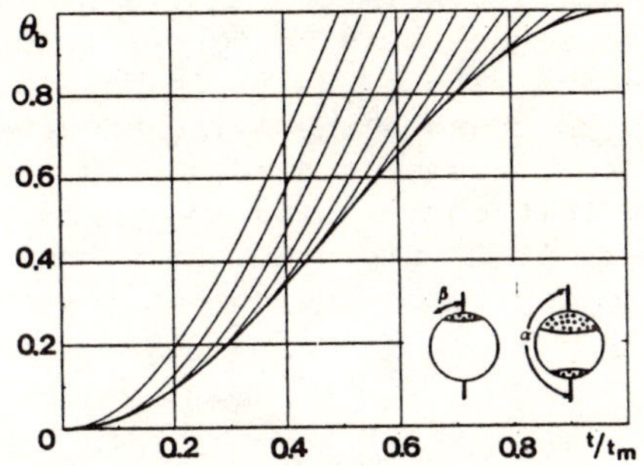

Figure 25 : Total coverage by the new phase θ_b from 2 nuclei in function of t normalized with respect to the time one nucleus needs to cover the surface (t_m). Angle α is constant (180°). Each curve from left to right corresponds to an increment in β of 15° (from 0 to 180°).

VIII.3. The polynucleation regime

At still larger supersaturations or electrode area, the nuclei created independently and stochastically in the time and space coordinates become so numerous that the transients triggered by a single potential step are no longer dependent on the shape and area of the electrode. Such a deterministic behaviour is described by the classical Avrami equation valid for "Progressive Polynucleation and Growth" (85 - 88). This equation takes into account the fact that the growth of the new phase is progressively hindered by the collisions endured by the expanding nuclei. This equation which can be easily adapted to electrocapillary processes takes then the following expressions :

$$i = \Delta\sigma_T \pi J \upsilon^2 t^2 \exp\left(-\frac{\pi}{3} J \upsilon^2 t^3\right) \qquad (20)$$

$$\sigma = \Delta\sigma_T \ (1 - \exp\ (- \frac{\pi}{3}\ Jv^2 t^3)) \qquad (21)$$

where $\Delta\sigma_T$ is the charge involved in the total film transformation. The half-life of the transient τ (for which $\sigma = \Delta\sigma_T/2$) is given by :

$$\log \tau = -0.060 - \frac{1}{3} \log Jv^2 \qquad (22)$$

The polynuclear regime (PN) can be easily differentiated from the mononuclear regime (MN) by two distinct criteria : (1) Opposite to MN, PN curves always start in synchronism with the potential step. (2) Moreover, they are independent of the size and shape of the electrode. The logarithm of the half-life of various PN transients are plotted (figure 26) in function of the imposed experimental parameter η, for a wide variety of surfactants. It appears clearly that η is not the actual driving force, since, in some cases, the half lifes increase with η, as it is for instance the case during the superficial changes involving adenine and adenosine. The $\Delta\gamma^{\#}$-η plot for adenosine and adenine given in figure 27 confirms convincingly that correlating log τ and $\Delta\gamma^{\#}$ is much more meaningful. However, it is important to stress that PN transients are not purely controlled by the nucleation rate but

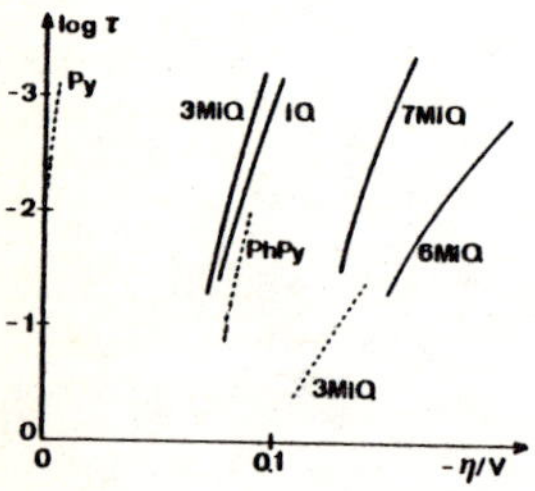

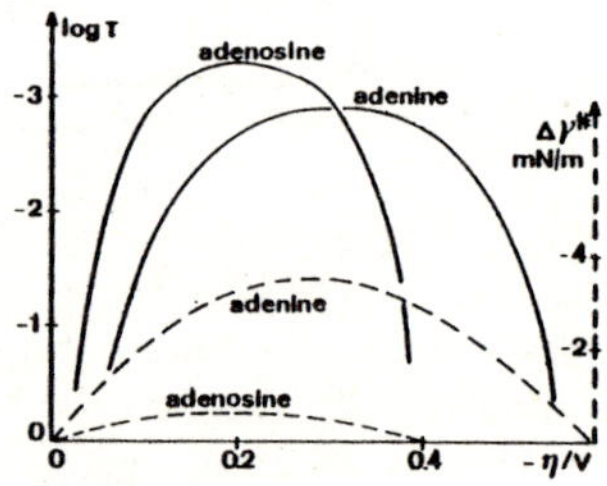

Figure 26 : Half-times relative to condensation controlled by a progressive polynucleation. Dotted lines refer to the adsorption transition of pyridine (Py), phenyl-pyridine (PhPy) and 3-methyl-isoquinoline (3 MiQ) while full lines are for the reorientation transition of isoquinoline and a few of its methyl derivates.

Figure 27 : The behaviour of adenine (5,4 mM) and adenosine (7 mM) shows conclusively that the kinetics of condensation are controlled by $\Delta\gamma^{\#}$ and not by η. Full lines refer to log τ-η while broken lines give the $\Delta\gamma^{\#}$-η curves.

that they also depend on the growth rate, since both processes occur simultaneously in time. The rates of nucleation and of growth appear always as a product Jv^2 in equations 20-22 and they cannot

be separated from each other. In fact, the evaluation of J at a given potential is only possible by eqn. (22) if the absolute growth rate is known at the same potential. It can be sometimes estimated by extrapolating the growth rates calculated from mononuclear events at smaller supersaturations. As will be shown at a later stage, appropriate double potential step experiments may be performed in order to determine the slope of the $\upsilon = f(\eta)$ curve at any value of η.
It may be added that, as it is apparent from the present set of experimental data, the log τ versus η or $\Delta\gamma^{\#}$ plots reflect mostly variations of the rate of nucleation, υ being generally much less potential-dependent.

One may wonder about the social behaviour of the expanding nuclei : do they grow independently from each other, till a collision will bring their borderlines to a stop, or is the merging a more complex process, involving coalescence into a new reorganized moeity ? A partial answer to this question may be provided by experiments which involve a three-potential step, the median potential being E_θ at which there is no longer any increase of the coverage. Staying at E_θ for various periods of time before resuming the nucleation and growth process provides useful clues about the behaviour of the nuclei during the rest period Collective merging, for instance, must result in a reduced number of active nuclei and accordingly in a slower final transient. The few data presently available indicates that there is no detectable alteration in the repartition of sizes and positions during the rest period.

IX. THE PROCESS OF CONDENSATION FROM PREEXISTING NUCLEI

In most experiments, even after extensive pretreatment of the electrode and the solution, unknown extraneous nucleants are often still present. They may trigger the transition and disturb considerably the evaluation of the intrinsic rate of nucleation of the system. Foreign nucleation is caused either by the existence of privileged sites at the electrode (imperfections of the surface of solid electrodes, 3-phase boundaries created by a vessel, a holder or a capillary) or by the presence of unsuspected traces of foreign surfactants in the solution. Transitions initiated by a single heterogeneous nucleus or by a large number of preexisting nuclei are two particular situations which will be more specifically discussed. Although they complicate the accurate evaluation of J, each of these two cases deserves close examination because it may lead to useful kinetic data not extractable otherwise.

IX.1. Mononucleation regime

The sum of all kinetic data related to phase transitions gathered up to now shows that, in contrast with the growth rate, the nucleation rate is an extremely steep function of the supersaturation. This fact restricts drastically the experimental conditions adequate for the detection of "homogeneous" mononucleations and limits at the same time the domain of supersaturation where υ can be determined. It is an experimental fact that, contrary to the "homogeneous" nucleations, the frequency of "heterogeneous" nucleations is fortunately much less dependent on the supersaturation. It is therefore not too unfrequent to observe "heterogeneous" mononucleations in a wide range of $\Delta\gamma^{\#}$ values. Transitions can thus happen in the so-called "no-man's land" (figure 15) where it would take years or centuries to record a single self-nucleated event. It results from this circumstance that "heterogeneous" mononucleations are much more than a simple annoying artefact, because the rate of growth becomes now accessible within a fairly extended overpotential domain. A suitable extrapolation of the growth rate curve gives then access to values pertaining to potentials where the measurements are no longer possible, in view of the prevalence of the polynucleation regime.

In the particular case of the mercury hanging drop, the formation of a single nucleus can be sometimes traced back to the unavoidable presence of a capillary in contact with the electrode and the solution. Surfactant molecules adsorbed along the lumen of the capillary are necessarily subjected to additional interfacial forces which must probably affect their local orientation and promote them as seeds. It is also probable that the boundary between the capillary and the metal delimits a region where smaller critical sizes are needed (89). Subnuclei favoured by the lumen of the capillary are not necessarily promoted to their critical size when the potential is stepped to a value such that the phase change is thermodynamically possible. This has been clearly shown in the case of the rearrangement transition involving isoquinoline (84). After a complete clean-up of the cell and several purifications steps for the mercury and the solution, a random time distribution of transients of identical shape is observed for a given η. After a few hours of experiments, a marked shortening of the delay times is noticed for the same η. Eventually, without any modification in their shape, all transients are synchronized with the potential step. Replacing the capillary by a fresh one restores immediately the initial random distribution. Such a behaviour demonstrates that the transients, although they may appear randomly in time, are not necessarily random in space. As such they are not representative of the intrinsic process of nucleation. Another evidence for the capillary effect is provided by experiments performed on drops of various sizes formed at the tip of the same capillary. The probability of occurrence of the transients must remain constant for "capillary induced" nucleations. If this is not the case, tensioactive molecules or aggregates present in the solution must be responsible for the formation of foreign nuclei at the electrode.

Experimental conditions such that mononuclear transients start in synchronism with the potential step may sometimes be deliberately searched in order to obtain the essential parameters needed to understand and describe the nucleation and growth process.

In some instances, the effective birth of a second nucleus able to grow is detected by a conspicuous distortion of the shape of the mononuclear transients. As already discussed previously, the transients which cumulate the growth of two circular nuclei on a sphere can be described mathematically with the help of two parameters. As a result, the effective randomness of the time and space coordinates can be easily checked by analyzing the shape of a collection of binuclear transients. It is only rather recently that the spatial distribution of single nucleations has been submitted to statistical evaluation in the case of the deposition of Ag^+ on silver (90, 91).

The so-called "equatorial configuration" is much more easily obtained when conditions are such that the mononucleation is synchronized with the potential step. The principle is as follows. During the first E step, a single nucleus is allowed to grow untill it invades exactly one-half of the electrode. At this "equatorial" state, one hemisphere of the drop is covered by the new phase, while the other remains in its initial state. At this precise moment, the potential is stepped to the value E_2 which is precisely the equilibrium potential. By keeping the electrode at E_θ during increasing times and stepping afterwards to a potential E_3 at which growth proceeds again, it is possible to assess whether or not there has been gradual desaggregation of the two initial hemispheres, keeping in mind that each of the two superficial films must keep its 50% occupancy.

Experiments performed with isoquinoline (figure 28) show that transients observed at E_3 remain identical irrespective of the time spent at E_θ. This fact indicates that the further growth proceeds from the equatorial frontier and not from islands of the new phase scattered all over the drop, a situation which would have completely modified the shape and duration of the final transient.

IX.2. Polynucleation regime

Growth from a large number of preexisting nuclei (N_N), without formation of additional nuclei, is described by the Avrami equation for "Instantaneous Polynucleation and Growth" (85 - 88) which can be rewritten as :

$$i = \Delta\sigma_T \; 2\pi \; N_N \; \upsilon^2 t \; \exp\,(-\, N_N \upsilon^2 t^2) \qquad (23)$$

$$\sigma = \Delta\sigma_T \; [1 - \exp\,(-\pi N_N \; \upsilon^2 t^2)] \qquad (24)$$

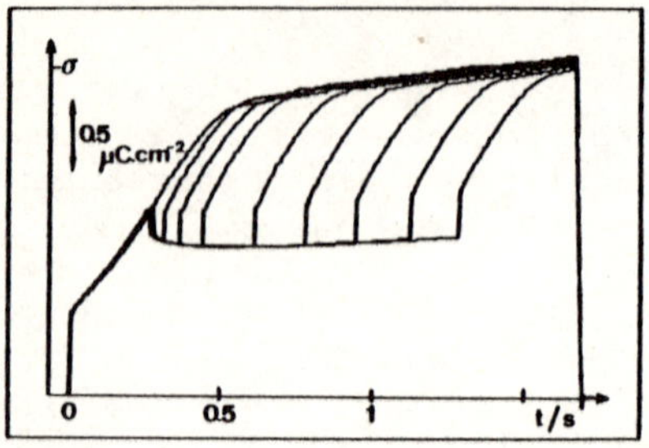

Figure 28 : Absence of a significant desaggregation of the equatorial configuration of isoquinoline (sat in 1N NaOH). A potential step triggers a mononuclear transient at η = -43 mV, followed at τ by variable times spent at E_θ before stepping back to η = -43 mV.

As it is already the case for progressive polynucleation and growth, the two parameters N_N and υ related to the present situation cannot be determined separately by the mathematical analysis of the transients.

When a large number of foreign nucleants preexist at the electrode, a single potential step triggers necessarily an instantaneous polynucleation (IPN) transient from which only the product $N_N\upsilon^2$ can be determined. By increasing the amplitude of the potential step, maintained however in the region of small values of J, the half-life of the transients is given by :

$$\log \tau = -0.328 - \frac{1}{2} \log N_N \; \upsilon^2 \qquad (25)$$

which expresses the dependence of υ with the overpotential.

As most experiments on mercury are, fortunately, carried out with reasonably clean solutions and cells, IPN transients are rarely triggered by a single potential step. The double potential step method can however be used in order to create artificially, during the prepulse period, a large number of homogeneous nuclei. The number of nuclei created is well-defined only if the nucleation reaches quasi-instantly its steady-state rate. If this is the case, the number of nuclei formed must be proportional to the duration of the prepulse (τ_p). Since J is a very steep function of the potential, very few additional nuclei are brought to life after the prepulse provided that the potential is stepped to a value such that the rate of nucleation becomes relatively insi-

gnificant. Under such conditions, the growth of the preexisting nuclei contributes exclusively to the final process.
By repeating such experiments with constant prepulses (E_1 and τ_p) shifting of the final potential (figure 29a) results in variations of the half-life which reflects the dependence of the rate of growth with the overpotential E_2 (or its corresponding supersaturation.

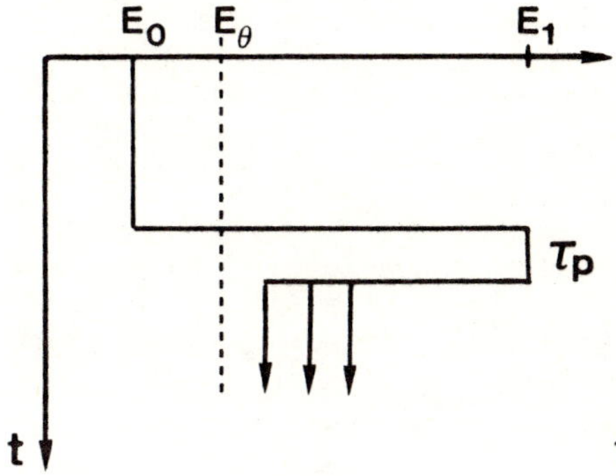

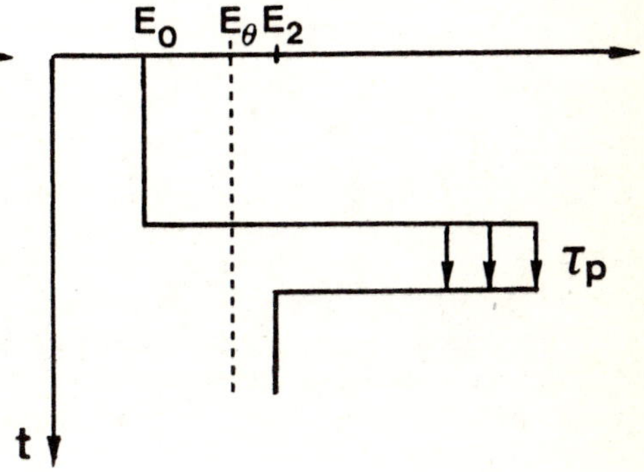

Figure 29a Figure 29b

Figure 29a : Potential step programme for the evaluation of the effect of the supersaturation on the growth rate. The first pulse ($E_o \rightarrow E_1$) is aimed at creating a large number of nuclei (large $\Delta\gamma^{\#}$ does not necessarily mean large η) during a time τ_p. The growth is then followed at various smaller $\Delta\gamma^{\#}$ for which no additional nuclei are formed.

Figure 29b : A similar potential step programme allows the evaluation of the dependence of J with $\Delta\gamma^{\#}$. For this purpose, the potential is stepped to various values during the first pulse τ_p and the growth of the nuclei formed is observed at a constant smaller supersaturation.

The dependence of the nucleation rate with the supersaturation can be derived from experiments such that the prepulse potential E_1 is the only variable (figure 29b) all other parameters being kept constant. According to the classical nucleation theory, key parameters such as the line tension ε, the radius r^{*} and the free energy of formation ΔG_F^{*} of the critical cluster can be easily calculated from the slope of the ln J vs $1/\Delta\gamma^{\#}$ plot. Table 3 gives values of the line tension for different surfactants and classes of transition. Correlation between ε and the complexity of the molecules is only meaningful for transitions involving similar molecular rearrangements as it seems to be the case for the

	isoquinoline	3-methyl isoquinoline	6-methyl isoquinoline	7-methyl isoquinoline	adenosine
$d\ln J/d(1/\Delta\gamma^{\#})$ 10^2 (J/m^2)	-4.8	-12.4	-8.3	-7.0	-0.16
$\varepsilon \times 10^{12}$ (J/m)	7.9	13	10	9.6	1.4

Table 3 : Line tensions derived from the variation of J with $\Delta\gamma^{\#}$.

different methyl-isoquinolines. It is incidentally worth noting that the line tensions involved in these electrocapillary processes and in electrocrystallization ($\varepsilon \simeq 2.10^{-11}$ J/m (83)) present quite similar values. The radius of a critical nucleus for a given $\Delta\gamma^{\#}$ can also be calculated. Typically, in the case of isoquinolines (transitions of class II), at a supersaturation of -1.5 mN/m, the radius is of the order of 6-8 nm and corresponds to the agglomeration of several hundred molecules.

X. PHASE TRANSFORMATION INITIATED BY DEFECTS IN THE SUPERFICIAL LAYER

Since a transition separates two distinct superficial states, potential step experiments can be performed in each of the two directions accordingly as the aim is to trigger condensation or the reverse process (melting or dissolution) (12, 92). In contrast with condensation, the reverse process is not automatically characterized by peaked i-t or sigmoïd σ-t curves. In most cases indeed, the current transients are of the "falling" type with very little or no indication of a peaked component. Accordingly, charge transients display a shape which varies monotonously with time, as shown in figure 30.

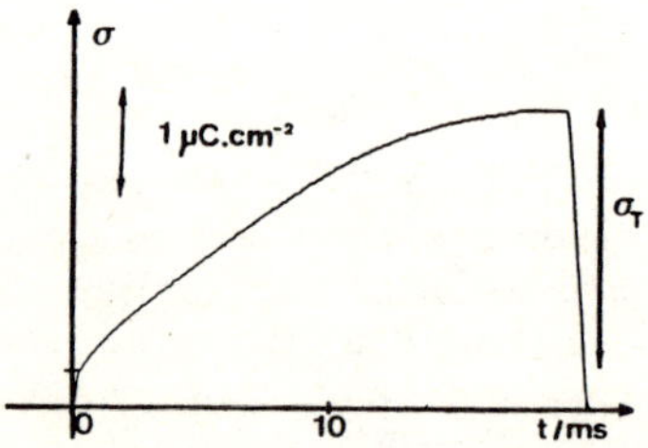

Figure 30 : Dissolution of the solid-like film by a single potential step experiment. (3 methyl isoquinoline (sat) at an overpotential of -10 mV).

The half-life time decreases very rapidly with the overpotential. It can be concluded from this observation that the melting or dissolution process requires a much less activation energy than is needed for condensation. This fact is consonant with the behaviour of 3D systems for which the melting of the solid phase (such as ice for example) is much easier than its formation, the latter being subordinated to the necessity of creating a population of critical clusters representative of the new phase. In the case of 2D melting, there is no significant increase of the rate if the system is submitted to a large-amplitude prepulse, which would be expected to act as a multinucleation step.
Another typical trait is that significant differences in the half-life of the transients are related to the circumstances of formation of the condensed phase. What looks as a kind of "pretransition" state (75) is often observed in the C-E curve, under the form of a monotonous increase of the capacity towards the transition potential. This phenomenon seems to be related to some partial relaxation of the condensed film (generally of the "solid" type) and is characterized by a half-life (of the order of 1 or 2 ms) which varies very little with the potential amplitude. As long as the half-life of the full transition is larger than a few milliseconds, this relaxation proceeds before the transition proper, while at shorter times the two processes overlap.
By resorting to more sophisticated potential-time programmes, it can furthermore be concluded that the age and modalities of formation of the condensed film are structural factors which affect in a detectable way the half-life of the decaying transient.

By correlating the ensemble of all pertaining kinetic data presently available, with the nature of the phases involved in the transition process, one reaches the conclusion that there is no longer formation and expansion of circular nuclei but that the phase change now proceeds from imperfections present in a "solid-like" film. These imperfections are probably boundary regions which have been formed by collision between the growing nuclei during the edification of the film (figure 31). It goes without

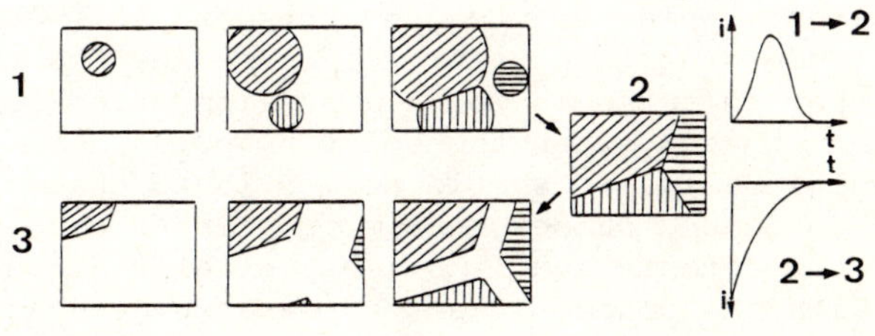

Figure 31 : Schematic illustration of the successive formation and dissolution of a solid-like film (shaded area).

saying that the geometry of these line defects, as well as their ability to act as seeds is critically dependent on the circumstances of formation of the "solid-like" film.

It results from these considerations that the absence of falling transients indicates that the initial film does not present the characteristics of an imperfect "2D solid" and is thus a condensed "liquid-like" film. This seems to be the case for pyridine which presents peaked i-t curves during its desorption, as reported by Armstrong (31). On the other hand, the desorption of more complex molecular films (such as phenyl-pyridines and substituted isoquinolines) gives falling i-t curves which plead for an initial film with the properties of a "quasi-solid". In contrast with pyridine, phenyl-pyridine and isoquinoline can condense much easily together owing to stronger lateral interactions between the molecules. It remains however that much more exploratory work with other surfactants should help in confirming more firmly the correlation which has been found between the shape of the transients and the structure of the initial film.

XI. CONCLUDING COMMENTS

Two-dimensional condensation processes at electrode surfaces have received little attention up to now. We are presently witnessing fast progresses which encompass many problems : elucidation of the nature of the adsorbed films, assessment of the thermodynamic factors which control their stability, selection of the most adequate methods for investigating kinetic aspects, understanding of the various mechanisms which are involved in the process of formation and destructuration, proper identification of the driving force and simultaneously screening research of an increasing number of model experimental systems.

Despite this intensive activity however, the field is very far from being totally exhausted. Among other necessary developments, it remains to expand as far as possible the experimental framework, by analyzing more closely determining parameters such as surfaces, solvents, electrolytes and surfactants. The few available data concerning temperature effects (see Chapter by F. Silva) indicate that there seems to be a proliferation of unexpected new surfactant phases, when the solvent approaches its freezing point. Additionally, it should be mentioned that a fairly large number of condensed phases result from intense coadsorption between neutral surfactants and ions (only briefly alluded to in this chapter), with the concommittant appearance of sharp phase transitions. Very little is still known about these mixed phases, and on the circumstances of their formation. All these processes and many others deserve to be studied systematically, in order to assess more firmly what are the general trends or rules

which relate the molecular nature of the phase change to its mechanism and rate parameters.
Further developments of the classical theory of nucleation, as applied to the case considered here, are also highly desirable, if it is sought to establish meaningful correlations between molecular dynamics and experimental parameters like the line tension, the critical radius and the rate of growth.

A considerable amount of work is clearly still needed, and in many directions. The lecturer's hope is that she has contributed to stimulate cooperation and efforts in the line of these exciting perspectives. It is clear that the investigation of surfactant transition processes has presently emerged as a full and lively discipline, in that far-reaching domain of physical chemistry which embraces all aspects of phase transformations.

ACKNOWLEDGEMENTS

The author is particularly indebted to Prof. L. Gierst for many fruitful discussions and wishes to thank Dr. G. Quarin, Dr. F. Thomas (on leave from James Cook University, Australia), C. Franck and M. Gonze for their active collaboration in the experimental work.

REFERENCES

1. Prigogine, I., Defay, R., *"Chemical Thermodynamics"*, Longmans, Green and Co., LTD., London, 1954.
2. Domany, E., in *"Interfacial Aspects of Phase Transformations"*, ed. Mutaftschiev, B., D. Reidel, Dordrecht, 1982, p.119.
3. Kosterlitz, J.M., Thouless, D.J., 1973, J. Phys. C 6, p.1181.
4. Kosterlitz, J.M., in *"Phase Transitions in Surface Films"*, ed. Dash, J.G., Ruwalds, J., Plenum Press, 1980, p.193.
5. Katz, J.L., in *"Interfacial Aspects of Phase Transformations"*, ed. Mutaftschiev, B., D. Reidel, Dordrecht, 1982, p.261.
6. Bienfait, M., in *"Phase Transitions in Surface Films"*, ed. Dash, J.G., Ruwalds, J., Plenum Press, 1980, p.29.
7. Gaines, G.L., *"Insoluble Monolayers at Liquid-Gas Interfaces"*, Interscience Publishers, 1966, p.177.
8. Nagle, J.F., 1980, Ann. Rev. Phys. Chem. 31, p.157.
9. Retter, U., 1978, J. Electroanal. Chem. 87, p.181.
10. Quarin, G., Buess-Herman, Cl., Gierst, L., 1981, J. Electroanal. Chem. 123, p.35.
11. Sridharan, R., de Levie, R., 1982, J. Phys. Chem. 86, p.4489.

12. Buess-Herman, Cl., Quarin, G., Gierst, L., 1983, J. Electro-anal. Chem. 148, p.97.

13. Proceedings of the Symposium on *"Electrocrystallization"*, ed. Weil, R., Barradas, R.G., The Electrochemical Society, 1981.

14. Vetter, K.J., *"Electrochemical Kinetics"*, Academic Press, New-York, 1967.

15. Damaskin, B.B., Petrii, D.A., Batrakov, V.V., *"Adsorption of Organic Compounds on Electrodes"*, Plenum Press, 1971.

16. Trasatti, S., 1974, J. Electroanal. Chem. 53, p.335.

17. Conway, B.E., Mathieson, J.G., Dhar, H.P., 1974, J. Phys. Chem. 78, p.1226.

18. Chevalet, J., Rouelle, F., Gierst, L., Lambert, J.P., 1972, J. Electroanal. Chem. 39, p.201.

19. Meurée, N., Gierst, L., 1971, Coll. Czech. Chem. Commun. 36, p.389.

20. Lipkowski, J., Galus, Z., 1975, J. Electroanal. Chem. 61, p.11.

21. Parsons, R., 1977, Sci. Prog. (Oxford) 64, p.253.

22. Gierst, L., in *"Transactions of the Symposium on Electrode Processes"*, ed. Yeager, E., J. Wiley, New York, 1961, p.294.

23. Loutfy, R.O., 1973, J. Electroanal. Chem. 41, p.27.

24. Gileadi, E., 1966, J. Electroanal. Chem. 11, 137.

25. Retter, U., Jehring, H., Vetterl, V., 1974, J. Electroanal. Chem. 57, p.391.

26. Retter, U., J. Electroanal.'Chem., 1980, 106, p.371;
1982, 136, p.167.

27. Krznaric, D., Valenta, P., Nurnberg, H.W., Branica, M., 1978, J. Electroanal. Chem. 93, p.41.

28. Brabec, V., Kim, M.H., Christian, S.D., Dryhurst, G., 1979, 100, p.111.

29. Dyatkina, S.L., Damaskin, B.B., Fedorovich, N.V., Stenina, E.V., Yusupova, V.A., 1973, Elektrokhimiya 9, p.1283.

30. Dyatkina, S.L., Damaskin, B.B., Stenina, E.V., Fedorovich, N.V. 1974, Elektrokhimiya 10, p.318.

31. Gierst, L., Pecasse, C., in *"Polarography"*, 1964, ed. Hills, G.J., Mac Millan, London, p.305.

32. Damaskin, B.B., Dyatkina, S.L., Petrochenko, S.I., 1969, Elektrokhimiya 5, p.935.

33. Vetterl, V., 1968, J. Electroanal. Chem. 19, p.169.

34. Cl. Buess-Herman, F. Thomas, in preparation.

35. Buess-Herman, Cl., Gierst, L., Vanlaethem-Meurée, N., 1981, J. Electroanal. Chem. 123, p.1.

36. Vanlaethem-Meurée, N., Lambert, J.P., Gierst, L., 1972, Chem. Ing. Technol. 4, p.219.

37. Dyatkina, S.L., Damaskin, B.B., 1972, Elektrokhimiya 8, p.1123.

38. Trasatti, S., 1970, J. Electroanal. Chem. 28, p.257.

39. Buess-Herman, Cl., 1978, Doctorate Thesis, Université Libre de Bruxelles, Belgium.

40. Frumkin, A.N., 1925, Z. Phys. Chem. 116, p.466.

41. Damaskin, B.B., Dyatkina, S.L., 1966, Elektrokhimiya 2, p.981.

42. Damaskin, B.B., 1969, J. Electroanal. Chem. 21, p.149.

43. Damaskin, B.B., 1977, Elektrokhimiya 13, p.816.

44. Damaskin, B.B., Frumkin, A.N., Borovaya, N.A., 1972, Elektrokhimiya 8, p.807.

45. Damaskin, B.B., Akhmetov, N.K., 1979, Elektrokhimiya 15, p.1691.

46. Kharkats, Yu. I., 1980, J. Electroanal. Chem. 115, p.75.

47. Gurevich, Yu.Ya., Kharkats, Yu.I., 1978, J. Electroanal. Chem. 86, p.245.

48. Gurevich, Yu.Ya., Kharkats, Yu.I., 1978, Elektrokhimiya 14, p.821.

49. Buess-Herman, Cl., Vanlaethem-Meurée, N., Quarin, G., Gierst, L., 1981, J. Electroanal. Chem. 123, p.21.

50. Buess-Herman, Cl., Quarin, G., Gierst, L., 1983, J. Electroanal. Chem. 148, p.79.

51. Sridharan, R., 1983, Doctorate Thesis, Georgetown University, Washington D.C.

52. Retter, U., Lohse, H., 1982, J. Electroanal. Chem. 134, p.243.

53. Retter, U., 1984, J. Electroanal. Chem. 165, p.221.

54. Hills, G., Silva, F., 1982, J. Electroanal. Chem. 137, p.387.

55. Survila, A.A., Damaskin, B.B., 1967, Elektrokhimiya 3, p.1138.

56. Hills, G.J., Peter, L.N., Scharifker, B.R., Da Silva Pereira, M.I., 1981, J. Electroanal. Chem. 124, p.247.

57. Jenard, A., Hurwitz, H.D., 1976, J. Electroanal. Chem. 70, p.27.

58. Pospíšil, L., Müller, E. Dörfler, H.D., 1984, Electrochimica Acta 29, p.773.

59. Frumkin, A.N., Melik-Gaikazyan, V.I., 1951, Dokl. Akad. Nauk. S.S.S.R. 77, p.855.

60. Lorenz, W., Möckel, F., 1956, Z. Elektrochem. 60, p.507, 939.

61. Armstrong, R.D., 1969, J. Electroanal. Chem. 20, 168.

62. Delahay, P., Trachtenberg, I., 1957, J. Am. Chem. Soc. 79, p.2355.

63. Delahay, P., Fike, C.T., 1958, J. Am. Chem. Soc. 80, p.2628.

64. Holub, K., Němec, L., 1968, J. Electroanal. Chem. 18, p.209

65. Guidelli, R., Moncelli, M., 1978, J. Electroanal. Chem. 89, p.261.

66. Delahay, P., Mohilner, D.M., 1962, J. Am. Chem. Soc. 84, p.4247.

67. Lorenz, W., 1958, Z. Elektrochem. 62, p.192.

68. Woodruff, D.P., *"The Solid-Liquid Interface"*, Cambridge University Press, 1973.

69. Zettlemoyer, A.C., *"Nucleation"*, Marcel Dekker, New York, 1969.

70. Franks, F., *"Water - a comprehensive Treatise"*, Plenum Press, New York, vol. 7, 1982.

71. Volmer, M., Weber, A., 1926, Z. Phys. Chem. 119, p.277.

72. Becker, R., Döring, W., 1935, Ann. Phys. 24, p.719.

73. Farkas, L., 1927, Z. Phys. Chem. A125, p.236.

74. Zeldovich, J.B., 1943, Acta Physicochim. (U.R.S.S.), 18, p.1.

75. Frenkel, J., 1939, J. Chem. Phys. 7, p.200, 538.

76. Abraham, F.F., *"Homogeneous Nucleation Theory"*, Academic Press, New York, 1974.

77. Reiss, H., 1977, Advances in Colloid and Interface Science 7, p.1.

78. Sridharan, R., de Levie, R., 1984, J. Electroanal. Chem. 169, p.59.

79. Bostanov, V., Obretenov, W., Staikov, G. Budevski, E., 1983, J. Electroanal. Chem. 146, p.303.

80. Gunawardena, G., Hills, G., Scharifker, B., 1981, J. Electroanal. Chem. 130, p.99.

81. Budevski, E., Bostanov, V., Vitanov, T., Stoynov Z., Kotzeva, A., Kaischew, R., 1966, Electrochimica Acta 11, p.1697.

82. Smith, A., Fletcher, S., 1980, Electrochimica Acta 25, p.583, p.889.

83. Budevski, E., Bostanov, V., Staikov, G., 1980, Ann. Rev. Mater. Sci. 10, p.85.

84. Gierst, L., Franck, C., Quarin, G., Buess-Herman, Cl., 1981, J. Electroanal. Chem. 129, p.353.

85. Canac, F., 1933, C.R. Acad. Sci. Paris 196, p.51.

86. Avrami, M., J. Chem. Phys. 7 (1939), p.1103; 8 (1940), p.212; 9 (1941), p.177.

87. Evans, U.R., 1945, Trans. Faraday Soc. 41, p.365.

88. Fleischmann, H., Thirsk, H.R. in *"Adv. in Electrochemistry and Electrochemical Engineering"*, ed. Delahay, P., Interscience, New York, 1963, vol. 3.

89. Kern, R., in *"Interfacial Aspects of Phase Transformations"*, ed. Mutaftschiev, B., D. Reidel, Dordrecht, 1982, p.287.

90. Barradas, R.G., Boag, P., Porter, J.P., Vander Noot, T., Budevski, E., Bostanov, V., Obretenov, W., Kossev, T., 1983, J. Electroanal. Chem. 154, p.1.

91. Obretenov, W., Bostanov, V., Budevski, E., 1984, J. Electroanal. Chem. 170, p.51.

92. Pospíšil, L., Kůta, J., Müller, E., Dörfler, H., 1980, J. Electroanal. Chem. 106, p.359.

SURFACE SCIENCE MODEL STUDIES OF THE ELECTROCHEMICAL INTERFACE

E. M. Stuve*, K. Bange**, and J. K. Sass

Fritz-Haber-Institut der Max-Planck-Gesellschaft,
Faradayweg 4-6, D-1000 Berlin 33, West Germany

Adsorption experiments in ultra-high vacuum (UHV) involving electrolyte solution components (solvents and counter ions) on metal surfaces are described. Interactions between H_2O and coadsorbed electropositive (Cs) and electronegative (Br, O) species on Ag(110) and Cu(110) were studied with thermal desorption spectroscopy (TDS), measurements of the work function change ($\Delta\phi$), and low energy electron diffraction (LEED). New results are presented for Cs adsorption and Cs and H_2O coadsorption on Ag(110). A short review of H_2O and O coadsorption on Ag(110) and Cu(110) and H_2O and Br coadsorption on Cu(110) is also given. The binding energy of H_2O was enhanced by coadsorption of either Cs, Br, or O as evidenced by high temperature H_2O TDS states not present for H_2O adsorption on the clean surface. All of the coadsorbates studied produced changes in the initial dipole moment for H_2O adsorption. The work function increased (positive initial dipole moment) for H_2O adsorption on Cs covered Ag(110) and a model of hydrated Cs, i.e., H_2O bound directly to adsorbed Cs, with the hydrogen atoms of H_2O near the surface is proposed. Low coverages of the electronegative coadsorbates, Br and O, caused the initial dipole moment for H_2O to be more negative; this result may be due to either preferential orientation of the H_2O dipoles or to enhanced charge transfer from H_2O to the surface. The relevance of these results to electrochemistry is discussed.

* Present address: Department of Chemical Engineering, BF-10, University of Washington, Seattle, WA 98195, USA

**Present address: Siemens AG, WRS ESR B, Postfach 120, D-1000 Berlin 13, Federal Republic of Germany

A. F. Silva (ed.), Trends in Interfacial Electrochemistry, 255–280.

1. INTRODUCTION

The study of electrochemical processes focusses on the interfacial region of the electrode and the electrolyte with particular interest in (i) the structure of the electrode, (ii) the structure of the electrolyte, (iii) the interactions between the electrode and ionic species (specific adsorption), (iv) the interactions between the solvent and ionic species (specific and non-specific adsorption), and (v) the details of the spatial variation of the electrostatic potential in the interfacial region. The interfacial region has been traditionally described by various forms of double layer models (1) which were designed to reproduce differential capacity measurements of liquid mercury electrodes (2). As these models are based upon measurements of macroscopic quantities, they contain information regarding only the relationship between electrode potential and the excess charge in the double layer. To obtain a better understanding of electrochemical processes, it is necessary to study the microscopic details listed above to determine whether the double layer model is appropriate and, if so, how it can be extended to describe the molecular structure of the electrode/electrolyte interface.

Substantial progress has been made in fundamental studies of the liquid-solid interface in recent years. These studies can be classified into two groups: in-situ spectroscopic studies in which the liquid-solid interface is directly observed under actual electrochemical conditions, and ex-situ studies in which the electrode with some or all of its double layer is removed from the electrochemical cell and transferred to an ultra-high (UHV) vacuum chamber for analysis (3,4). The in-situ studies, of course, provide more direct information, but the ex-situ studies allow the use of the more highly developed UHV spectroscopic techniques, although the extent to which the system is altered by the transfer process must always be considered.

Our research group has taken a slightly different approach to fundamental electrochemical studies by attempting to simulate the electrochemical double layer in an UHV environment (5). This has the advantage of providing powerful surface analysis techniques to the researcher, but without the uncertainties that arise in the transfer process of ex-situ studies. Although the information obtained is not as direct as the in-situ studies, the double layer simulation method allows preparation of well defined adsorption systems including some that may not occur in a real electrochemical environment which is, for example, limited by the range of potentials over which the electrolyte is stable. Double layer simulation experiments provide knowledge of the intrinsic interactions between the electrode surface and the electrolyte, and thus present the most basic information about this interface. The ability to study systems not obtainable in electrochemical cells certainly will assist in the development of truly robust double layer models, since a good model should predict why some systems may be obtained both in-situ and in UHV whereas others can only be synthesized in the latter environment.

In the next section we will briefly review some of the important concepts of the double layer and discuss the relevance of UHV double layer simulation experiments in more detail. The primary techniques used in these studies are work function change measurements ($\Delta\phi$) which provide information about the composite, i.e. substrate plus adsorbate, dipole layer at the surface and thermal desorption spectroscopy (TDS) which provides information about the bond strength, coverage, and stability of the adsorbed species. The electrolyte is composed of a solvent, usually H_2O, and current carrying cations and anions. A primary step in understanding the behavior of a H_2O-ion-metal system is knowledge of the H_2O-metal and ion-metal interactions. We consider H_2O adsorption on Ag(110) and Cu(110) surfaces in section 3. We have chosen Cs for a cationic species and Br and O for anionic species, and in section 4 we discuss Cs and O adsorption on Ag(110) and Br and O adsorption on Cu(110). Finally, in section 5 we consider the coadsorption of H_2O with the species studied in section 4. The adsorption energy of H_2O is enhanced by all of the ion model species: Cs, Br, and O; but the type of interactions between H_2O and Cs are fundamentally different compared to those between H_2O and either Br or O, and in at least one case, namely oxygen, a reaction is observed to occur with H_2O.

2. SIMULATION OF THE DOUBLE LAYERS

Our interests in this section are to model the electrode/electrolyte interface, and we will restrict the present discussion to metal electrodes and aqueous electrolytes. Fig. 1 shows a plot of the electronic charge density vs. distance for the jellium model of a metal/vacuum interface (6). The jellium edge is located at z = 0 and the plane of the outermost metal nuclei is located one-half of an interplanar spacing to the left of the jellium edge (z = -1/2 d). The metal electrons tail out into the vacuum a distance of approximately 2 Å and can interact with an adsorbate or electrolyte as the case may be. If the adsorbate is a charged species, such as an alkali, the electron tail contains the screening, or image, charge (6). The response of the electron tail as the metal is charged (equivalent to an applied electric field normal to the surface) is such that the tail moves by several tenths of an angstrom out into the vacuum as the metal becomes negatively charged and closer to the metal atoms as the metal becomes positively charged (7). The position of the center of gravity of the image charge is less sensitive to movement of the

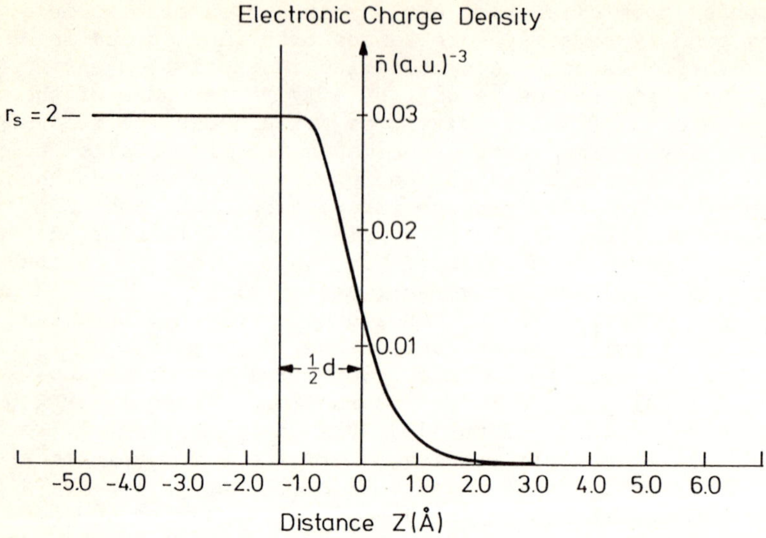

Figure 1: Plot of the electronic charge density as a function of distance perpendicular to the surface for a $r_s = 2$ jellium substrate. From ref. (6).

electron tail, however (8). Thus, adsorption of an electropositive species will cause the electron tail to move out into the vacuum (9), and conversely, adsorption of electronegative species will cause the tail to move towards the surface. According to the effective medium theory (10), an adsorbate atom is adsorbed at a certain distance from the surface determined by the optimum electronic charge density for the bonding interaction between the surface and the adsorbate. Thus, an adsorbed species can be thought of as "resting" at a certain density on the electron tail, and hence, will move with the electron tail. We can therefore expect that, in principle, the properties of an adsorbed species are altered by application of an electric field or coadsorption of some other species which modifies the electrostatic potential at the surface, since both actions may cause a change in the equilibrium distance between the surface and the adsorbed species.

The electrolyte consists of, say, a water solvent into which a salt has been dissolved to provide the current carrying ions. In the bulk of the solution the ions are solvated by water molecules such that the oxygen end complexes to cations and the hydrogen end to anions (Fig. 2). Recent neutron scattering measurements of electrolyte solutions (11,12) confirm this and provide detailed information about the structure of solvated ions. Models of the cation-D_2O and anion-D_2O solvation structure are shown in Fig. 3. For approximately 1 Molal $NiCl_2$ solutions the Ni^{2+}-O distance is approximately 2.1 Å, and the Ni^{2+}-O-D angle is $42^\circ \pm 8^\circ$.

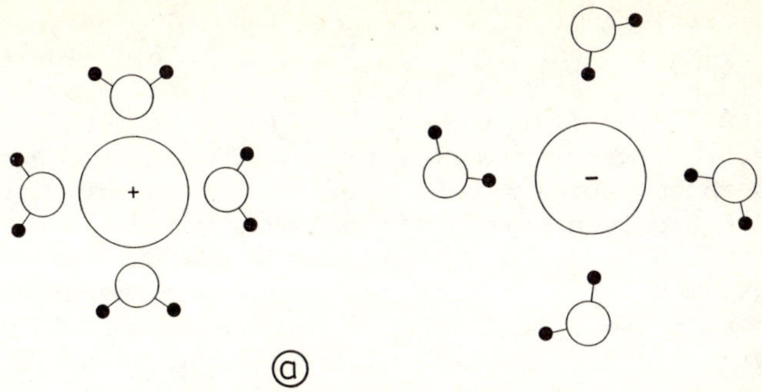

Figure 2: Illustration of hydration of cations (a) and anions (b) in aqueous solution according to (11).

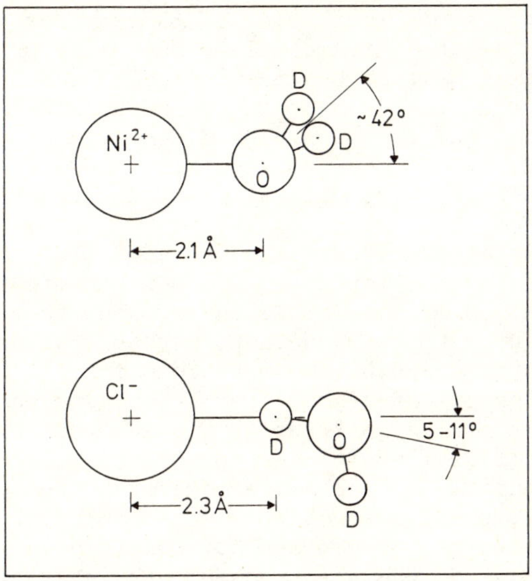

Figure 3: Structure of hydrated Ni^{2+} and Cl^{-} in $NiCl_2$ solutions according to (11).

The number of coordinating H_2O molecules per Ni^{2+} ion varies from 6 to 7 depending on the solution concentration. For the case of anion hydration the Cl^{-}-D distance is 2.3 Å, the Cl^{-}-D-O angle is 5°-11°, and the coordination number is approximately 6. Other cation and anion species have also been studied (11,12) at various concentrations. The general results are that the cation-O-D angle ranges from about 35° to 50° except at low concentrations, the anion-D-O angle is normally less than 20°, and the coordination number for both cations and anions is approximately 6 with some exceptions. The energetics of ion hydration have been reviewed by Märk and Castleman (13) and Kebarle (14). The absolute single-ion hydration enthalpy ΔH_h^o (in Kcal mole^{-1}) for alkali and halide ions ranges from 132 for Li^+ to 72 for Cs^+ and from 113 for F^- to 64 for I^- (see ref. (14) for more details). Furthermore, the energy gained for each H_2O molecule complexed to an ion becomes approximately equal to the energy required to remove a H_2O molecule from the hydrogen bonded liquid water phase once the number of H_2O molecules of hydration reaches five or six; attachment of additional H_2O molecules therefore ceases to be exothermic. We will discuss shortly how this structural and energetic information of the bulk electrolyte may become important in UHV double layer simulation experiments.

When the electrode and electrolyte are brought into contact a potential drop appears across the interface. This potential drop $\Delta\phi$ is determined by the extent to which the metal surface electrons tail out into the solution and any preferential orientation of the solvent dipolar molecules at the surface. It is often convenient to use the parallel plate capacitor model to approximate the potential drop across the interface (15). This model incorporates two charged plates: one in the metal and the other several angstroms into the solution. The dielectric is composed of the electron tail, the dipolar solvent molecules, and any counter ions in this region (specifically or non-specifically adsorbed).

The parallel plate capacitor, although admittedly crude, illustrates the primary connection between electrochemical experiments and UHV simulation of the double layer. When the electrode and electrolyte are brought into contact, a potential drop develops across the interface with contributions from the electron tail of the metal, the orientation of permanent dipoles at the interface, the amount of charge transfer for adsorbed species, and any changes in the dielectric constant of the solvent in this region. If this potential drop could be measured, some details about the dielectric, that is to say the metal-solution interface, could be inferred. This is, of course, not possible in an electrochemical cell. Instead, the electrode/electrolyte interface may be polarized by a variable external voltage source (*via* the counter and/or reference electrode) such that the potential drop at the interface can be *changed* by a known amount. The charge that flows into or out of the interface as the potential drop is changed (cyclic voltammetry) can then be used in attempts to infer structural details of the metal-solution interface (1).

In an UHV environment the potential drop across a metal-vacuum interface is the significant contribution to the work function and is readily measured *via* the Kelvin probe method, for example. The solution side of the interface can be simulated by adsorbing H_2O and various counter ions of interest. Within the framework of the parallel plate capacitor model, then, double layer simulation experiments in UHV entail measurement of the potential drop across the interface as the composition of the interface is changed. Because the experiments are carried out in UHV, the structure of the adsorbed layers, which simulate the dielectric medium of the capacitor model, can be relatively easily studied; this constitutes the main appeal of these experiments. We note that the differential capacity of the interface C_{diff}, given by

$$C_{diff}^{-1} = d(\Delta\phi)/d\sigma \quad (1)$$

where σ is the excess charge at the interface, can be measured by both double layer simulation and electrochemical experiments so that these two methods can, in principle, be directly related. We shall see, however, that σ is often difficult to determine in UHV experiments. Nevertheless, the combination of microscopic surface analysis techniques and macroscopic measurements (e.g., work function) provides an important link between microscopic details of the double layer and the construction of theoretical models.

Control of the potential drop at the electrode/electrolyte interface is essential for studying electrochemical processes. This is easily achieved with electrochemical cells, but is far more difficult in UHV studies since the potential drop $\Delta\phi$ depends on the nature of the metal-adsorbate-vacuum interface. Control of $\Delta\phi$ can be effected, however, by adsorption of electropositive or electronegative species (9). Alkali adsorption can decrease $\Delta\phi$ by as much as 3 eV, for example, and produces a positively charged adsorbate with a negative screening charge on the metal very similar to a cathodic change in electrode potential. Likewise, adsorption of electronegative species produces a negatively charged adsorbate, a positive screening charge, and is similar to an anodic change in electrode potential. Thus, by coadsorbing H_2O and electropositive or electronegative species, one can not only study the interactions between a specifically adsorbed species (anionic or cationic) and H_2O, but also the behavior of adsorbed H_2O with variable interfacial potential drop. (Separation of these two effects may not always be possible, however.)

The experiments that we report in this paper all illustrate the concept of H_2O coadsorbed with electropositive or electronegative species. Our goals are to define the properties of the individual adsorbates and then the interactions and possible reactions that occur in the coadsorption systems. The experimental methods used were Kelvin probe measurement of $\Delta\phi$, TDS, and low energy electron diffraction (LEED). We will present TDS experiments that show that reorientation of H_2O molecules in the presence of a coadsorbate is usually accompanied by an increased heat of desorption of H_2O. This is a fortunate result because the TDS spectra can then be used to determine the number of H_2O molecules per electropositive or electronegative species; we define this as the surface hydration number. Finally, in some cases, LEED studies provide information about the periodicity of the adsorbed species relative to the substrate periodicity, from which details about lateral interactions in the adsorbed layer can be deduced.

3. ADSORPTION OF H_2O ON CLEAN METAL SURFACES

Thermal desorption spectra of H_2O on clean Cu(110) and Ag(110) are shown in Fig. 4. The H_2O exposure was about 2 L on Ag(110) and 3.5 L on Cu(110) (1 L H_2O = 4.8×10^{14} molecules cm^{-2}). The spectrum for Cu(110) shows two thermal desorption peaks at 170 and 175 K. These are respectively attributed to desorption from bulk (ice) layers of H_2O and desorption of H_2O that is directly bound to the surface (16). Only one TDS peak at 170 K, representative of desorption from both bulk layer and directly adsorbed H_2O, was observed for Ag(110) (17). These findings illustrate a fundamental difference in H_2O adsorption between Cu(110) and Ag(110), namely, the binding energy for H_2O adsorbed directly on Cu(110) is stronger than for bulk layer sites, but no detectable difference exists between the two sites on Ag(110).

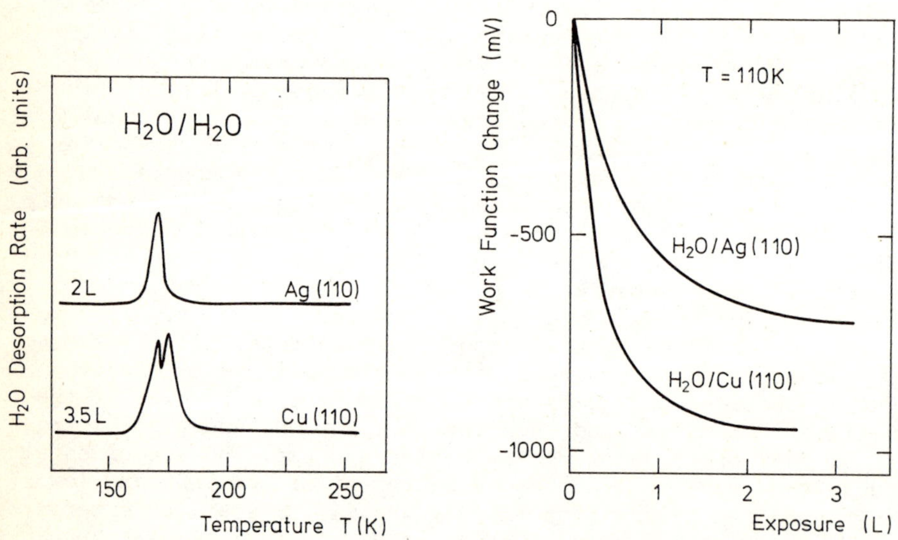

Figure 4: Thermal desorption spectra for H_2O from Ag(110) and Cu(110). The adsorption temperature was 120 K.

Figure 5: Work function change $\Delta\phi$ as a function of exposure for H_2O adsorption on Ag(110) at 80 K and Cu(110) at 110 K.

Fig. 5 shows the Kelvin probe measurements of the change in work function $\Delta\phi$ for H_2O adsorption on Cu(110) and Ag(110). Both curves show a monotonic decrease in $\Delta\phi$ with increasing water exposure. Following a H_2O exposure of 3 L, the change in the work function was -940 mV for Cu(110) and -680 mV for Ag(110). If the sticking coefficient is assumed to be unity for H_2O, then the initial dipole moment μ_o can be determined from the slope of the $\Delta\phi$ vs. exposure curve at 0 L H_2O. For Cu(110) μ_o is 0.9 D (D = Debye), but only 0.6 D for Ag(110).

The TDS, $\Delta\phi$, and LEED (see below) measurements allow some observations about the differences in H_2O adsorption between Cu(110) and Ag(110). Recall that the adsorption energy for H_2O on Cu(110) is larger than for either H_2O on H_2O (heat of sublimation) or H_2O on Ag(110). We can postulate, then, that H_2O is more strongly oriented on Cu(110) than on Ag(110). The meaning of this becomes clear by performing a thought experiment in which Ag or Cu is adsorbed on a single crystal of ice. According to the TDS data of Fig. 4, one would expect that the structure of the ice surface layer would not be perturbed by the adsorption of a layer of (110) oriented Ag, but that some restructuring of the topmost ice layer would occur for a layer of (110) oriented Cu. It is worth mentioning that the size of the Ag(110) unit cell provides a better fit than Cu(110) for an unperturbed ice layer (18). LEED observations provide additional evidence that H_2O is more strongly oriented on Cu(110). A c(2x2)-H_2O pattern was observed for Cu(110) (16), but no LEED patterns were detected for H_2O on Ag(110) (18) indicating that, in the latter case, no tendency for long-range order between the Ag(110) substrate and the H_2O adsorbate exists.

The $\Delta\phi$ measurements can also be interpreted in terms of more effective orientation of H_2O on Cu(110), albeit subject to certain assumptions. The change in the work function is sensitive to the net orientation of the H_2O permanent dipole, the amount of charge transfer from H_2O to the metal surface, and inter-molecular interactions. All of these effects cause the work function to decrease, assuming that H_2O binds to the surface <u>via</u> the oxygen end. Inter-molecular interactions can be neglected for consideration of the initial dipole moment since we are interested in the limit of zero coverage. That μ_o is less than the permanent H_2O dipole of 1.8 D indicates that the plane of the molecule is tilted with respect to the surface normal. (We will call this the angle of tilt for H_2O.) This is in agreement with electron stimulated desorption ion angular distribution (ESDIAD) measurements (18,19) and cluster calculations (20). If, for the sake of illustration, we consider only the dipole orientation effect on μ_o, then it is clear that the larger μ_o is, the smaller the angle of tilt for H_2O. Theoretical studies (20) have shown, however, that charge transfer from H_2O to the metal can be significant so that, in reality, use of μ_o as an indicator of the H_2O tilt angle is not straightforward. In the present case we can expect that charge transfer between H_2O and Cu(110) and H_2O and Ag(110) should at least be qualitatively similar, and we suggest that the difference in μ_o for these two surfaces does indeed represent a difference in the angle of tilt. On Cu(110), therefore, the angle of tilt for H_2O is most probably less than on Ag(110). In a recent review article, Thiel and Madey (21) discuss how the structure of adsorbed H_2O on several metals can be determined from TDS, ESDIAD, and LEED measurements.

The maximum change of the work function $\Delta\phi_{max}$ may also contain information about the first one or two layers of adsorbed H_2O. In Fig. 5 it can be seen that $\Delta\phi$ was nearly constant for exposures greater than approximately 2 L, which corresponds to a coverage of approximately one H_2O molecule per topmost metal atom. Thus, additional layers of H_2O have little effect on $\Delta\phi$, from which we conclude that there is no net orientation of the permanent dipole in the bulk ice layers. The $\Delta\phi$ that is measured at lower coverages must represent, in some way, the net orientation of the dipoles in the first layer of H_2O. A larger, more negative $\Delta\phi_{max}$ would be produced by a net orientation of dipoles more perpendicular to the surface. The problem with this interpretation is again that the possibility of differences in charge transfer of the two surfaces must be considered, and in addition, screening effects may play a role.

In summary of this section, we have seen that the TDS and LEED measurements provide good evidence that H_2O is more strongly oriented on Cu(110) than on Ag(110). We have shown how the $\Delta\phi$ results can be interpreted in support of this claim, but lack of understanding of the charge transfer and inter-molecular interactions for adsorbed H_2O make the $\Delta\phi$ measurements by themselves inconclusive. We include the discussion of the $\Delta\phi$ results to point out the importance of the charge transfer and possibly also the screening effects, in addition to the dipole orientation effect, and to stimulate interest in more theoretical and experimental studies of the $\Delta\phi$ of adsorbed water.

4. ADSORPTION OF ELECTROPOSITIVE AND ELECTRONEGATIVE SPECIES

In the first part of this section we present results of a study of adsorption of an electropositive element, cesium, on Ag(110). At the end we will summarize earlier studies of electronegative elements, bromine and oxygen, on Cu(110) and Ag(110). Studies of electropositive adsorbates, of which the alkali metals are the best examples, have been and continue to be the subject of much experimental and theoretical interest in surface science. It is not our intention here to review the literature; the interested reader should consult ref. (22) and references therein for studies of alkali adsorption.

4.1 Cs Adsorption on Ag(110)

Thermal desorption spectra for Cs following adsorption onto Ag(110) at 130 K are shown in Fig. 6. The Cs coverages were determined

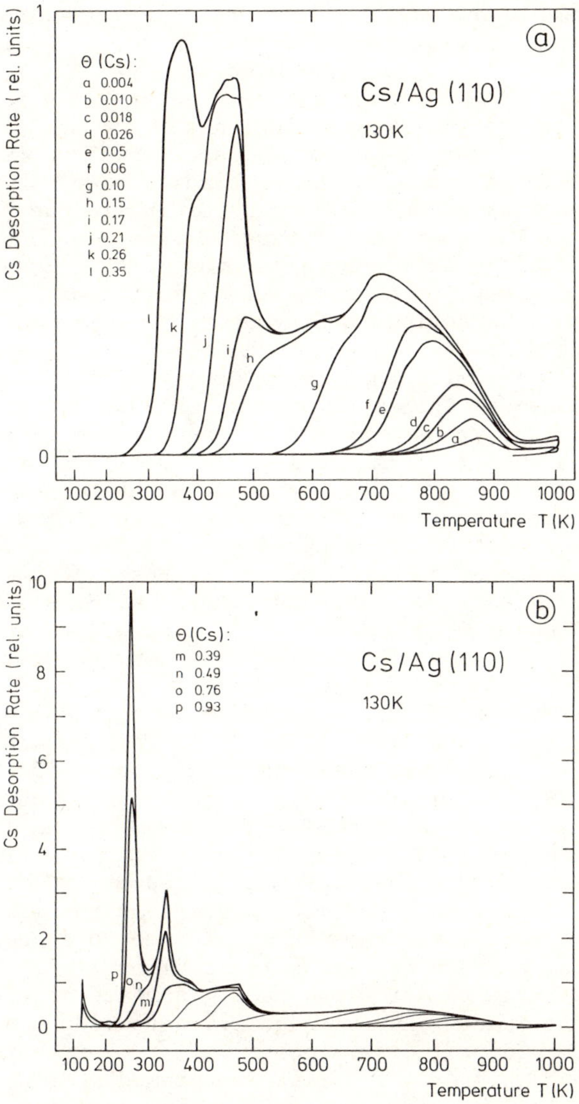

Figure 6: Cesium thermal desorption spectra from Ag(110) following Cs adsorption at 130 K: (a) low coverage and (b) high coverage of Cs. The coverages have been corrected for non-linear heating rates.

from the area under the TDS curves with the mass spectrometer signal calibrated as follows. The metallic radius of Cs is 2.69 Å from which the saturation coverage of Cs on Ag(110) can be estimated to be θ(Cs) = 0.48 Cs atoms per topmost Ag atom, assuming that Cs is adsorbed in a close-packed hexagonal layer. The desorption of Cs multilayers can be easily identified in the TDS spectra since this state must be zero order in Cs coverage. The coverage at which the multilayer TDS peak begins to be populated was therefore defined to be θ(Cs) = 0.48.

Cs desorption occurred in a state at about 875 K for θ(Cs) less than 0.01. If the pre-exponential factor for Cs desorption is assumed to be 10^{13} sec^{-1}, and if the spectrum for θ(Cs) = 0.004 is assumed to be first order, then the binding energy of Cs on Ag(110) can be estimated to be 2.4 eV according to the peak temperature of 875 K. The peak temperature of this low coverage Cs desorption state shifts to lower temperature with increasing coverage indicating repulsive interactions in the Cs adlayer. Indeed, repulsive interactions are evident for θ(Cs) as low as 0.01. The average distance between Cs adatoms at this coverage is approximately 35 Å assuming that Cs is evenly distributed about the surface. The Cs-Cs repulsive interactions are therefore truly long-range. For θ(Cs) greater than 0.04 the minimum temperature for Cs desorption shifted dramatically to lower temperature, also indicating strong repulsive interactions. As the coverage was increased still further, four TDS peaks were eventually resolved. The broad high temperature state has a peak temperature of 710 K and was populated for θ(Cs) less than 0.17. Peaks at 460 K and 340 K were observed as θ(Cs) exceeded 0.2 and 0.35, respectively. Finally, above θ(Cs) = 0.48 a peak at 275 K developed and is attributed to the desorption of multilayers of Cs, as previously mentioned.

Fig. 7 shows the work function change as a function of Cs coverage for adsorption on Ag(110) at 300 K. This curve is in excellent agreement with that measured by photoemission by Hayden, et al. (23), and exhibits the typical strong decrease in Δϕ to a minimum value of -2400 mV for θ(Cs) = 0.27 and increase to a constant value of -2150 mV for a nearly complete first layer of Cs. Since these measurements were performed at 300 K, the maximum coverage of Cs was θ(Cs) ∿ 0.4 according to Fig. 6. The decrease in Δϕ at low Cs coverage is attributed to charge transfer from Cs to the surface resulting in an almost fully ionized adsorbate. This charge transfer produces a surface dipole between the positively charged Cs and the negative screening charge in the metal, and the initial dipole moment of 8.3 D is a measure of the strength of this dipole. If it is assumed that Cs is fully ionized at low coverage, then the differential capacity of the surface in the limit of θ(Cs) = 0 can be estimated to be 5 μF cm^{-2} according to eqn. (1). Depolarization effects in the Cs adlayer, caused by dipole-dipole interactions, are evidenced by the in-

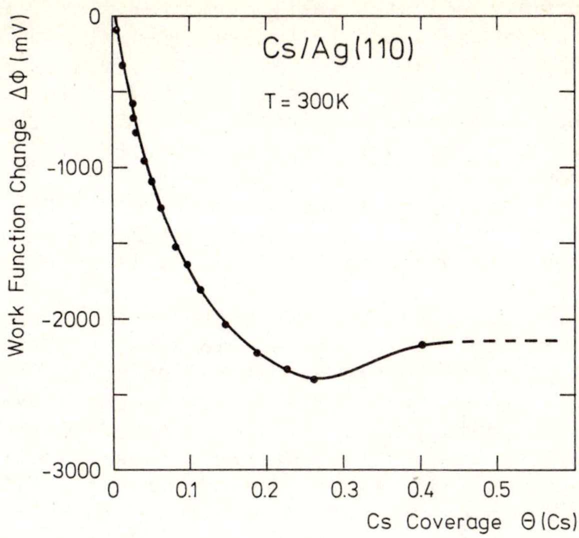

Figure 7: Work function change $\Delta\phi$ as a function of Cs coverage for adsorption onto Ag(110) at 300 K.

crease in the slope of the $\Delta\phi$ vs. θ(Cs) curve (i.e., the slope is less negative) as θ(Cs) was increased to 0.27. The increase in $\Delta\phi$ at greater Cs coverages is due to the more metallic nature of the Cs layer and the final value of -2150 mV is close to the difference in work function between clean Ag(110) and metallic Cs.

Hayden et al. (23), have previously studied the LEED characteristics of Cs adsorption on Ag(110). A p(1x2) pattern, attributed to a Cs-induced reconstruction of the Ag(110) surface, was observed following Cs adsorption at 130 K and annealing to 300 K for θ(Cs) between 0.05 and 0.25. For θ(Cs) greater than 0.25 but less than 0.4, a (1x3) pattern was observed and was also assigned to a reconstruction. The structures of the reconstructed surfaces are not as yet known, but preliminary results from our laboratory, which we will present in a later publication (24), support the missing row model for the (1x2) reconstruction.

4.2 Br Adsorption on Cu(110)

The details of Br adsorption on the (110), (111), and (100) faces of Cu have been previously presented (25,26). We will review the results for Br on Cu(110) here in order to complete the comparison of electronegative and electropositive adsorbates. The work function change upon adsorption of Br on Cu(110) at 110 K is shown by the dashed line in Fig. 8. The increase in $\Delta\phi$ with increasing Br exposure indicates that charge was transferred out of the surface and towards the Br adatom. The saturation value $\Delta\phi_{max}$ for Br adsorption was +1000 mV. Because Br_2 was dosed from an electrochemical cell mounted within the UHV chamber, the actual units of Br_2 exposure could not be determined. LEED results, discussed below, indicate that the saturation Br coverage was $\theta(Br) = 2/3$, from which the Br coverage in Fig. 8 can be estimated if $\Delta\phi$ is assumed to be linearly related to $\theta(Br)$. With this assumption, μ_o for Br is +0.4 D and, if Br is further assumed to be fully ionic, the differential capacity is approximately 110 $\mu F\ cm^{-2}$.

A well defined sequence of LEED patterns was observed with increasing Br coverage and is depicted in Fig. 9. At low coverage two extra spots appeared and were centered about the 1/2, 1/2 position of the reciprocal unit cell. These two spots were split along the (110) direction, that is, along the direction of the troughs of the (110) surface. We interpret these results as a c(2x2) Br structure with some compression along the (110) direction. As the Br coverage was increased the two spots moved to occupy the 1/3, 1/2 and 2/3, 1/2 positions. Additional spots also

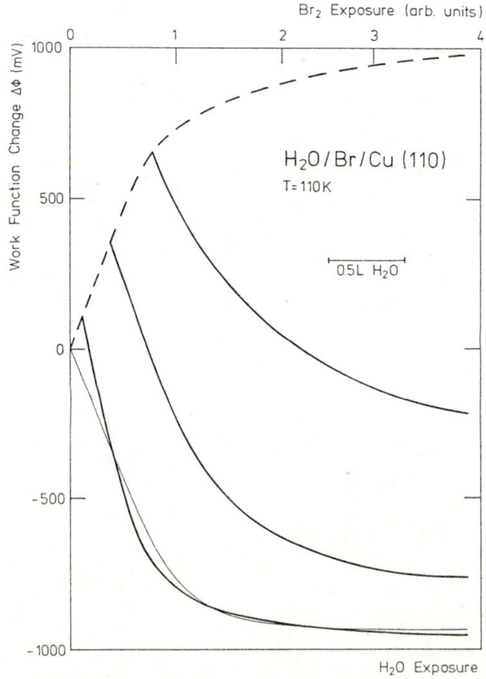

Figure 8: Work function change $\Delta\phi$ for H_2O adsorption on Br covered Cu(110). The dashed line represents $\Delta\phi$ for Br on clean Cu(110).

appeared such that a (3x2) pattern was formed. The real space interpretation of this pattern, shown in Fig. 9, corresponds to a coverage of $\theta(Br) = 2/3$. For purposes of comparison we note that Br on Ag(110) forms a p(2x1) pattern at submonolayer coverages and a c(4x2) pattern at saturation coverage (18).

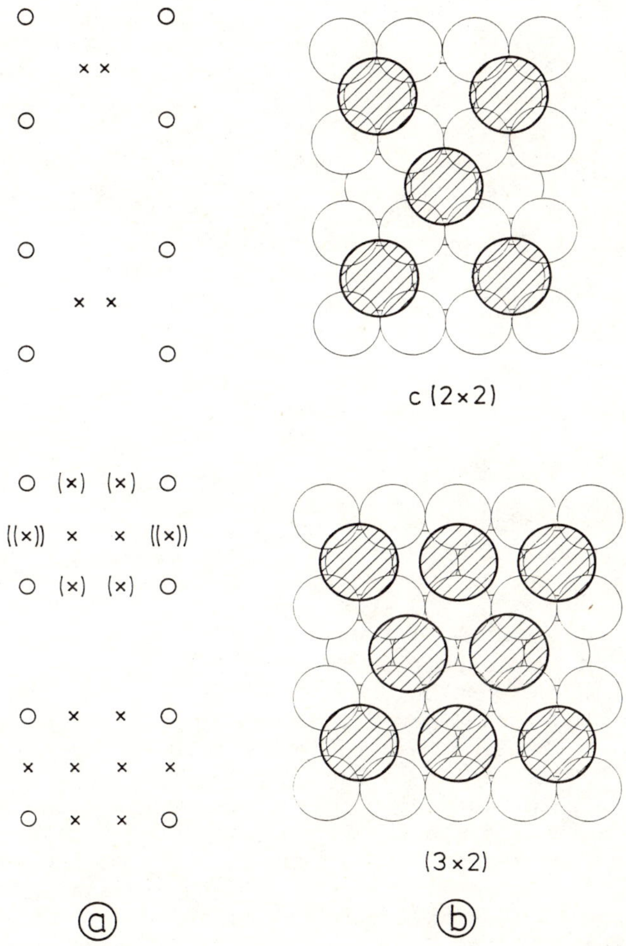

Figure 9: (a) Evolution of LEED patterns with increasing Br coverage on Cu(110). The adsorbate-induced spots are indicated by crosses ((x) weak; ((x)) very weak). (b) Adsorption model for Br/Cu(110): increasing Br coverage caused transformation of the c(2x2) structure into the (3x2) structure.

4.3 Oxygen Adsorption on Cu(110) and Ag(110)

Like bromine, oxygen is also an electronegative adsorbate, but exhibits a strong tendency to react with other coadsorbates. Oxygen adsorption on both the Cu(110) and Ag(110) surfaces has been the subject of tremendous interest in surface science (16, 18,27,28). Molecular oxygen is stable on Ag(110) at 100 K and probably also on Cu(110), but dissociation occurs on both surfaces upon heating to 300 K. Dissociative oxygen adsorption at 300 K causes an increase in $\Delta\phi$ on both surfaces with values of $\Delta\phi_{max}$ of +370 mV for Cu(110) (16) and +850 mV on Ag(110) (28). The adsorption kinetics are extremely different for the two surfaces as indicated by the initial sticking coefficients of approximately 0.6 on Cu(110) (16) and 10^{-3} on Ag(110) (28). On Cu(110) a (2x1) LEED pattern appears at an oxygen coverage corresponding to $\Delta\phi_{max}$ and has been attributed to a reconstruction in which all (100) atom rows adjacent to oxygen-containing rows are missing (29,30). On Ag(110) a series of (nx1) LEED patterns with n between 7 and 2 are seen as for $\theta(O)$ ranging from 1/7 to 1/2 (28). No O-induced reconstruction of Ag(110) has as yet been reported, although recent data suggest that a local O-induced reconstruction should at least be considered (18).

5. COADSORPTION OF H_2O AND ELECTROPOSITIVE/NEGATIVE SPECIES

5.1 H_2O and Cs on Ag(110)

The influence of coadsorbed Cs on H_2O adsorption on Ag(110) is readily seen in the D_2O TDS spectra and the $\Delta\phi$ vs. $\theta(H_2O)$ curves shown in Figs. 10 and 11, respectively. For these experiments the clean Ag(110) surface at 300 K was exposed to Cs, cooled to 130 K, and then exposed to water. The work function was monitored after Cs adsorption, H_2O adsorption, and after heating to 300 K for the TDS experiments. The work function of the Cs covered surface before H_2O adsorption and after H_2O desorption did not change indicating that no chemical reaction between Cs and H_2O occurred. Cs stabilized D_2O as seen by the higher temperature TDS states between 180 and 225 K depending on the Cs coverage. The $\Delta\phi$ curves in Fig. 11 show that $\Delta\phi$ actually increased for H_2O adsorption on the Cs covered surface. For $\theta(Cs)$ less than 0.1 $\Delta\phi$ increased and then decreased through a single maximum with increasing $\theta(H_2O)$. The H_2O coverage at the maximum in the $\Delta\phi$ vs. $\theta(H_2O)$ curve also increased with increasing Cs coverage for $\theta(Cs)$ less than 0.1. For $\theta(Cs)$ between 0.1 and 0.25 $\Delta\phi$ increased monotonically with increasing $\theta(H_2O)$ and eventually reached a plateau.

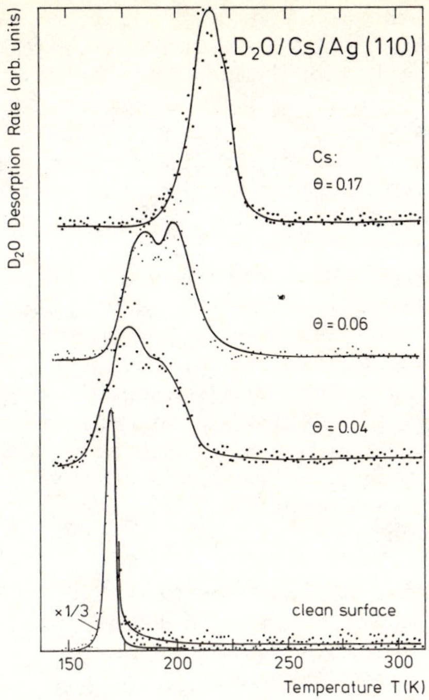

Figure 10: Thermal desorption spectra for D_2O as a function of Cs coverage θ(Cs) following exposure of the Cs covered Ag(110) surface at 130 K to approximately 0.2 L of D_2O.

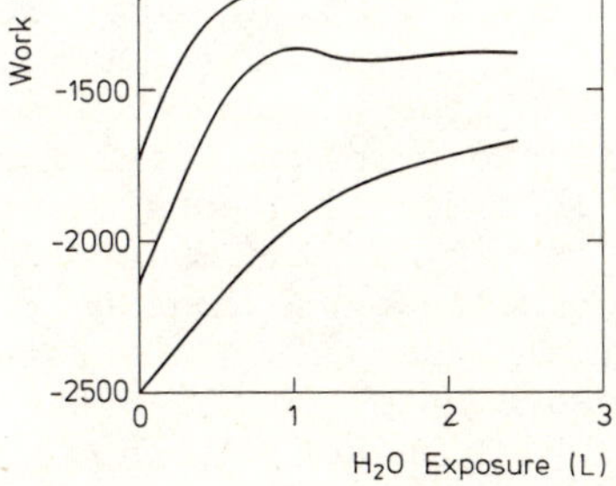

Figure 11: Work function change Δϕ for H_2O adsorption as a function of Cs coverage on Ag(110) at 130 K.

Examination of the TDS and $\Delta\phi$ results provides some interesting information about the structure of H_2O in the presence of Cs on Ag(110). The higher temperature desorption states clearly show that H_2O is <u>stabilized</u> by coadsorbed Cs. At first glance, this finding appears to conflict with the observations of Nørskov, et al. (31), in their theoretical study of poisoning and promotion of adsorption rates by electropositive and electronegative species. In that work adsorbates (N_2, CO, H_2, etc.) were classified as either charge donating (to the surface) or charge withdrawing. Consideration of the electrostatic effects between coadsorbed species lead to the observation that the adsorption of charge donating (withdrawing) species should be promoted by electronegative (electropositive) coadsorbates. Conversely, adsorption of charge donating (withdrawing) species should be poisoned by electropositive (electronegative) coadsorbates. According to cluster calculations (20), water is a charge donor with charge going from the lone pair electrons of the oxygen atom toward the surface to form the chemisorption bond. Thus, H_2O should be <u>destabilized</u> by coadsorbed Cs, since the latter is an electropositive element.

The above argument presupposes that H_2O is bound to the surface <u>via</u> the oxygen atom, but the increase in $\Delta\phi$ upon H_2O adsorption (Fig. 11) indicates that this may not be so. In section 3 we considered the different reasons for the change in work function with water adsorption and stated that all three; dipole orientation, charge transfer, and inter-molecular interactions; cause the work function to decrease. Because there is also coadsorbed Cs in these experiments, the effect of dielectric screening of the Cs atoms by H_2O must also be considered. Depolarization effects between Cs atoms cause the dipole moment per adsorbed Cs atom to decrease as the Cs coverage increases, for coverages less than the minimum in the $\Delta\phi$ vs. θ(Cs) curve. The slope of the $\Delta\phi$ vs. θ(Cs) curve therefore becomes less negative in this coverage range. If the Cs atoms were screened from each other by some polarizable coadsorbate, then Cs-Cs depolarization would be reduced and $\Delta\phi$ would further decrease upon adsorption of the shielding coadsorbate, neglecting the effect of the coadsorbate on $\Delta\phi$ itself (32). The only remaining explanation of the increase in $\Delta\phi$ is that the H_2O molecule is reoriented by coadsorbed Cs such that the permanent dipole moment points <u>towards</u> the surface, as shown in the model for hydration of adsorbed Cs in Fig. 12. This model is consistent with the cation-water structure shown in Fig. 3. The main H_2O chemisorption bond is the interaction between the lone pair electrons of the oxygen with the positively charged Cs adsorbate. Doering et al. (33), have also proposed a model in which the orientation of H_2O is strongly perturbed by Na on Ru(001), but the hydrogen atoms of the H_2O molecule were located away from the surface. Location

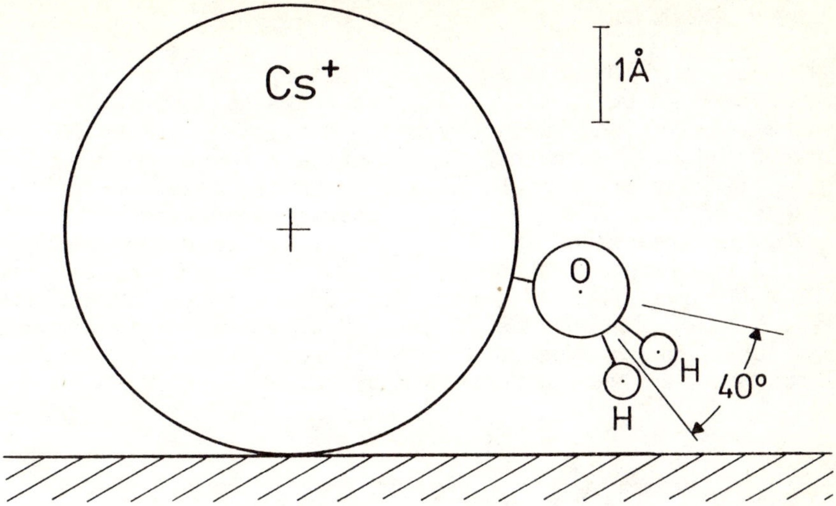

Figure 12: Model for Cs hydration on Ag(110). The individual Ag atoms are not shown due to uncertainties in the structure of adsorbed Cs on this surface.

of the hydrogen atoms near the surface, required by the necessary dipole reorientation, is consistent with calculations which show that a positive electrostatic electron potential (i.e., a potential which attracts positive charge) exists at the next nearest neighbor site to an alkali adsorbate at distances of 0 to 1 Å above a jellium surface (31,34). H_2O, configured as shown in Fig. 12, does not come under the classification of charge donating or withdrawing since it is technically bound directly to the Cs atom and not to the Ag(110) surface, and thus does not violate the aforementioned poisoning and promotion relationships between the direction of charge transfer for a particular adsorbate and electropositive or electronegative coadsorbates.

Within the model for Cs hydration (Fig. 12), the $\Delta\phi$ vs. $\Theta(H_2O)$ curve can be interpreted as follows. The maximum in the $\Delta\phi$ vs. $\theta(H_2O)$ curve at low coverages indicates that H_2O was preferentially complexed to the Cs atoms. When all of the Cs atoms were fully hydrated, the work function again decreased due to H_2O adsorption at sites on the Ag(110) surface that were unperturbed by Cs. ((We are currently performing experiments to determine the number of H_2O molecules that can complex to a Cs atom). For θ(Cs) greater than 0.1 there were apparently not enough unperturbed Ag(110) adsorption sites for H_2O and only a plateau in the $\Delta\phi$ vs. $\theta(H_2O)$ curve was seen.

5.2 H_2O and Br on Cu(110)

Fig. 13 shows the TDS spectra for H_2O desorption as a function of Br precoverage on Cu(110) (25). The peaks at 165 and 175 K are the multilayer and first layer H_2O desorption states observed on clean Cu(110), and the higher temperature peaks are due to desorption of H_2O that is stabilized by coadsorbed Br. One or two of these stabilized H_2O peaks were observed depending on the Br coverage. The area under the TDS peaks at temperatures greater than 175 K represents the number of H_2O molecules stabilized by Br. For θ(Br) less than 1/3, the ratio of H_2O to Br is approximately 1:1 (25). The absence of the 175 K peak as θ(Br) exceeded 1/3 was due to the disappearance of H_2O adsorption sites on the metal surface that were unperturbed by Br.

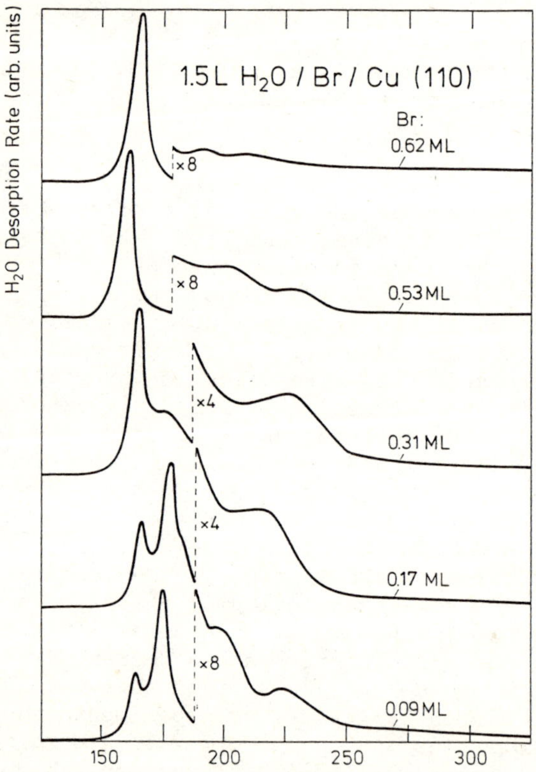

Figure 13: Thermal desorption spectra for H_2O as a function of Br coverage following exposure of the Br covered Cu(110) surface at 110 K to 1.5 L of H_2O.

Only minor changes in the behavior of the work function for water adsorption as a function of bromine coverage were observed, but of particular interest is the change in μ_o for H_2O (Fig. 8). For small amounts of Br μ_o decreased (i.e., became more negative) relative to the clean surface and then increased at larger Br exposures. As discussed above, this can be attributed to orientation of the permanent dipole more perpendicular to the surface compared H_2O on clean Cu(110) or to increased charge transfer to the surface.

As in the case of Cs on Ag(110), some knowledge of the structure of H_2O on Br covered Cu(110) may help distinguish between the dipole orientation and charge transfer contributions to the change in work function. Regardless of which of the four LEED patterns shown in Fig. 11 were present for Br-covered Cu(110), adsorption of H_2O produced a sharp (3x2) pattern. The 1:1 ratio for H_2O:Br in the stabilized H_2O TDS peaks of Fig. 13 indicates that H_2O and Br formed mixed (3x2) islands as illustrated in Fig. 14 (25). This model of coadsorbed H_2O and Br has H_2O bonded directly to the surface so that hydration of Br, in the sense of direct bonding between H_2O and Br, does not occur in this system.

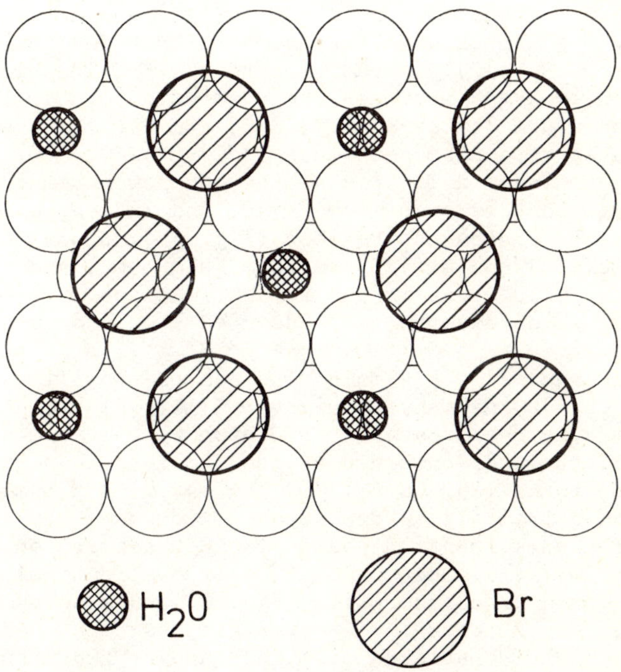

Figure 14: Arrangement of Br and H_2O in a composite surface layer, consistent with the observed (3x2) LEED pattern and the 1:1 coverage ratio determined by TDS.

Nevertheless, the chemisorption bond for H_2O is stronger in the mixed H_2O and Br islands than on sites unperturbed by Br. Stabilization of H_2O by coadsorbed Br can be interpreted in terms of electrostatic interactions between the two species (31,34) or in terms of increased charge transfer to the surface in a manner similar to the chemical picture of surface acid-base interactions described by Barteau and Madix (35). The electrostatic effect involves alignment of the H_2O dipoles along the field created between the negatively charged Br and the screening charge on the metal. The charge transfer effect involves through-surface interactions between the Lewis base (electron acceptor) Br and the Lewis acid (electron donor) H_2O. We cannot distinguish between these two explanations, and we note that both effects may be operative in this case. Similarly, we cannot distinguish between the dipole orientation and charge transfer effects to explain the change in μ_o for H_2O due to coadsorbed Br, although preliminary theoretical evidence seems to favor the former.

5.3 H_2O and O on Cu(110) and Ag(110): Surface Reactions

The H_2O and O system is an example of H_2O adsorption with a reactive electronegative coadsorbate. Many similarities exist between H_2O and O coadsorption on the Cu(110) and Ag(110) surfaces (16,17,36). Fig. 15 shows TDS spectra for H_2O desorption from oxygen covered Cu(110). In addition to the H_2O TDS states for clean Cu(110), oxygen promotes the adsorption of H_2O into three higher temperature states at 200, 235, and 290 K; three O-induced H_2O TDS states were also seen on Ag(110) at 200, 240, and 320 K (17). Of the O-induced TDS peaks, the two lower temperature peaks represent H_2O that is stabilized by O, and indeed, similarities with the H_2O-Br system can be seen (Fig. 13). H_2O also reacts with O on Cu(110) and Ag(110) to form OH groups, and the highest temperature H_2O TDS peak is due to the reaction of two OH groups to form H_2O, which desorbs, and adsorbed atomic oxygen, which is left behind.

A large number of H_2O molecules can be stabilized by atomic oxygen. On Cu(110) as many as eight H_2O molecules (including the H_2O molecule that reacts) per O atom were stabilized at low O coverages, but this number decreased to less than one for θ(O) greater than 0.25. On Ag(110) the ratio of H_2O molecules to O atoms varied from six to one as the O coverage was increased to 0.5 (18). The work function change for H_2O adsorption on O covered Cu(110) as a function of θ(O) is qualitatively similar to that for H_2O and Br coadsorption (Fig. 8) and also shows a decrease in μ_o for H_2O for low O coverages, as discussed in the previous section. LEED patterns of p(2x1) and c(2x2) structure were observed for O and H_2O on Cu(110) and mixed adsorption of H_2O and O, similar to H_2O and Br, was also proposed (16). The number of

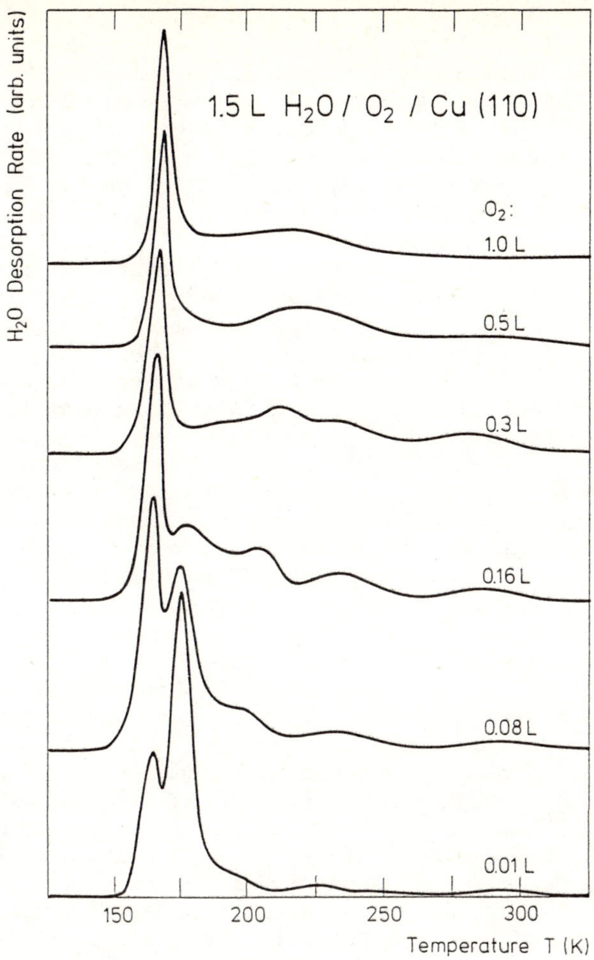

Figure 15: Thermal desorption spectra for H_2O as a function of O coverage following exposure of the O covered Cu(110) surface to 2L of H_2O.

similarities between the O and H_2O and the Br and H_2O coadsorption systems suggest a common mechanism for stabilization of H_2O by O and Br. The chemical reaction between O and H_2O apparently does not significantly alter the stabilization mechanism between H_2O and electronegative adsorbates. More research into the mechanism of H_2O stabilization by electronegative species is required before we can fully understand these interactions.

4. CONCLUSIONS

We have discussed the concept of double layer simulation experiments in UHV and how $\Delta\phi$, TDS, and LEED measurements can be used to probe the interactions among coadsorbed solvent and solute species. Our primary interest was the characterization of interactions between H_2O and electropositive (Cs) and electronegative (Br, O) species on Ag(110) and Cu(110) surfaces. H_2O adsorbs on the clean Cu(110) surface with a slightly higher binding energy than in the bulk ice layers, but on Ag(110), no significant difference in the binding energies between first layer and multilayer H_2O was found. On Cu(110) preferential orientation of H_2O with its permanent dipole situated more perpendicular to the surface normal was inferred from LEED and $\Delta\phi$ vs. $\Theta(H_2O)$ measurements, but uncertainties in the charge transfer and inter-molecular interactions of adsorbed H_2O make this observation tentative.

Substantial adsorbate-adsorbate and adsorbate-substrate interactions occur in the Cs, Br, and O single adsorbate systems. The binding energy of Cs on Ag(110) was determined to be 2.4 eV and strong repulsive interactions were observed among Cs atoms over distances as large as 35 Å. The differential capacity of Cs on Ag(110) in the limit of zero coverage was estimated to be 5 $\mu F\ cm^{-2}$ assuming that adsorbed Cs is fully ionic at low coverage. Bromine adsorbs into a c(2x2) structure on Cu(110) with the 1/2, 1/2 spot split along the (110) direction. A (3x2) structure forms at saturation Br coverage and a real space structure corresponding to $\theta(Br) = 2/3$ was proposed.

Water is more strongly adsorbed when coadsorbed with either electropositive or electronegative species as evidenced by the higher temperature H_2O TDS states that occur for coadsorbed H_2O and Cs, Br, or O. A large sensitivity of μ_o for H_2O to the coverage of the coadsorbate was observed in all cases studied. For $\theta(Cs)$ less than 0.4, μ_o was positive and a model for H_2O directly bound to Cs, that is, hydrated Cs, with the hydrogen atoms located near the surface was proposed. The opposite occurs with coadsorbed H_2O and either Br or O on Cu(110); μ_o becomes more negative for small coverages of either coadsorbate. LEED measurements of H_2O and Br and H_2O and O on Cu(110) indicate that mixed adsorption occurs between H_2O and electronegative species. The change in μ_o for H_2O with Br or O coverage can be explained by either preferential orientation of the H_2O dipoles or by enhanced, through-surface charge transfer involving H_2O and the coadsorbate. Many similarities exist between the H_2O and Br and H_2O and O coadsorption systems on Cu(110) suggesting a common mechanism for the interactions between H_2O and electronegative species on metal electrodes.

ACKNOWLEDGEMENTS

One of us (JKS) would like to thank Dr. Fernando de Silva for a rewarding and pleasant stay in Portugal. It is a pleasure to acknowledge significant contributions to the work described in this paper by R. Döhl-Oelze. We would also like to thank the participants of the Third Electrochemical Study Group (Berlin, 1984) for stimulating discussions and E. Piltz for expert technical assistance in the preparation of this manuscript. This work was supported by the Deutsche Forschungsgemeinschaft through Sonderforschungsbereich 6. One of us (EMS) gratefully acknowledges a fellowship from the Alexander von Humboldt-Stiftung.

REFERENCES

(1) See, for example: Parsons, R. 1975, J. Electroanal. Chem 59, pp. 229-240.
(2) Delahay, P. "Double Layer and Electrode Kinetics" (Interscience, New York, 1965).
(3) Duke, C.B. 1980, Surface Sci. 101, pp. 624-632.
(4) Yeager, E. 1980, Surface Sci. 101, pp. 1-22.
(5) Sass, J. K. 1983, Vacuum 33, pp. 741-751.
(6) Lang, N. D. and Kohn, W. 1973, Phys. Rev. B 7, pp. 3541-3550.
(7) This effect was first discussed by: Price, D. and Halley, J. W. 1983, J. Electroanal. Chem. 150, pp. 347-353.
(8) Lang, N. D. 1983, in "Theory of the Inhomogeneous Electron Gas," S. Lundqvist and N. H. March (Eds.), (Plenum, New York) pp. 309-389.
(9) Lang, N. D. 1971, Phys. Rev. B 4, pp. 4234-4245.
(10) Nørskov, J. 1982, Phys. Rev. B 26, pp. 2875-2885.
(11) Neilson, G. W. 1983, Physica 120B, pp. 325-334.
(12) Enderby, J. E. and Neilson, G. W. 1981, Rep. Prog. Phys. 44, pp. 593-653.
(13) Märk, T. D. and Castleman, A. W., in "Advances in Atomic and Molecular Physics", D. R. Bates and B. Bederson, Eds., Vol. 20 (1984).
(14) Kebarle, P., in "Modern Aspects of Electrochemistry," Vol. 9, B. E. Conway and J. O'M. Bockris, Eds. (Plenum, New York, 1974).
(15) Bockris, J. O'M. and Reddy, A. K. N., "Modern Electrochemistry" (Plenum, New York, 1970).
(16) Bange, K., Grider, D. E., Madey, T. E., and Sass, J. K. 1984, Surface Sci. 136, pp. 38-64.
(17) Stuve, E. M., Madix, R. J., and Sexton, B. A. 1981, Surface Sci. 111, pp. 11-25.
(18) Bange, K., Madey, T. E., and Sass, J. K., to be published.

(19) Doering, D. L. and Madey, T. E. 1982, Surface Sci. 123, pp. 305-337.
(20) Müller, J. E. and Harris, J., to be published; Ribarsky, M. W., Luedtke, W. D., and Landman, U., to be published.
(21) Thiel, P. A. and Madey, T. E., to be published.
(22) Bonzel, H. P. 1984, J. Vac. Sci. Technol. A 2, pp. 866-872.
(23) Hayden, B. E., Prince, K. C., Davie, P. J., Paolucci, G., and Bradshaw, A. M. 1983, Sol. State Comm. 48, pp. 325-328.
(24) Stuve, E. M., Döhl-Oelze, R., and Sass, J. K., to be published.
(25) Bange, K., Döhl, R., Grider, D. E., and Sass, J. K. 1983, Vacuum 33, pp. 757-761.
(26) Richardson, N. V. and Sass, J. K. 1981, Surface Sci. 103, pp. 496-509.
(27) Lin, J.-H. and Garrison, B. J. 1984, J. Chem. Phys. 80, pp. 2904-2913.
(28) Engelhardt, H. A. and Menzel, D. 1976, Surface Sci. 57, pp. 591-618.
(29) Bronckers, R. P. N. and de Wit, A. C. J. 1981, Surface Sci. 112, pp. 133-152.
(30) Lapujoulade, J. le Cruer, Y., Lejay, M., and Maurel, E. 1982, Surface Sci. 118, pp. 103-120.
(31) Nørskov, J. K, Holloway, S., and Lang, N. D. 1984, Surface Sci. 137, pp. 65-78.
(32) Luftman, H. S. and White, J. M. 1984, Surface Sci. 139, pp. 369-378.
(33) Doering, D. L., Semancik, S., and Madey, T. E. 1983, Surface Sci. 133, pp. 49-70.
(34) Lang, N. D., Holloway, S., and Nørskov, J. K., to be published.
(35) Barteau, M. A. and Madix, R. J., in "The Chemical Physics of Solid Surfaces and Heterogeneous Catalysis," Vo. 4., D. A. King and D. P. Woodruff, Eds. (Elsevier, Amsterdam, 1982).
(36) Bange, K., Madey, T. E., and Sass, J. K., Surface Sci., Proc. ECOSS-6, in press.

SOLVENT EFFECTS UPON ELECTROCHEMICAL KINETICS: INFLUENCES OF INTERFACIAL SOLVATION AND SOLVENT RELAXATION DYNAMICS

Michael J. Weaver
Department of Chemistry, Purdue University
West Lafayette, Indiana 47907, U.S.A.

Some likely influences of the solvent medium upon the kinetics of simple electrochemical reactions are described, and illustrated with recent results from the author's laboratory.

Introduction

The solvent has long been known to exert profound influences on the kinetics of electron-transfer reactions at electrode surfaces as well as in homogeneous solution. Nevertheless, our understanding of such solvent effects has remained demonstrably inadequate. This is undoubtedly due in large part to the frequency manifold origins of the effects observed upon variation of the solvent medium, arising in part from alterations in the chemical nature of the reactant as well as from longer-range reactant-solvent interactions.

A valuable class of model systems is provided by one-electron couples involving substitutionally inert inorganic and organometallic complexes. Important virtues of these systems include the opportunity to hold constant the free-energy barrier associated with metal-ligand vibrations ("inner-shell" barrier) as the solvent is altered, along with the ability to vary the nature of the reactant-solvent interactions by means of alterations in the ligand structure as well as the solvent.

The present conference paper is intended as a summary of some recent research in our laboratory aimed at gaining a fundamental understanding of solvent effects in electrochemical kinetics. Besides examining the effects of altering the bulk solvent, experiments are also described where the influence of

A. F. Silva (ed.), Trends in Interfacial Electrochemistry, 281–299.

the *interfacial* solvent structure in aqueous media is probed through variations in the hydrophility of the metal surface. Particularly since some of this recent experimental activity has been associated closely with concurrent theoretical development, it is useful at the outset to provide a brief summary of the underlying kinetic formulations.

I. Fundamental Rate Expressions

The observed rate constant for outer-sphere electron transfer, k_{ob}, can be related to the corresponding activation free energy for the elementary electron-transfer step, ΔG^*, by (1)

$$k_{ob} = K_p \nu_n \kappa_{el} \Gamma_n \exp(-\Delta G^*/RT) \qquad \text{Eq. 1}$$

where K_p (cm) is the effective equilibrium constant for forming the "precursor state" in the double layer from the bulk reactant, ν_n (sec^{-1}) is the nuclear frequency factor, κ_{el} is the electron-tunneling probability in the transition state, and Γ_n is the nuclear-tunneling factor. (The last term is close to unity for most reactions.) This relation is based on an "encounter preequilibrium" formulation of the preexponential factor, rather than the more conventional collisional model. The fundamental correctness of the former for electrochemical processes has been emphasised (1). It envisages electron transfer taking place via unimolecular activation of reactant in sufficiently close proximity to the electrode surface so that the electron-tunneling probability, κ_{el}, is suitably large. Since κ_{el} decreases sharply as the reactant-electrode distance increases, the inevitable integral of reaction sites will contain dominant contributions from those involving close approach of the reactant to the metal surface (2). For so-called "nonadiabatic pathways", $\kappa_{el} << 1$ even at the plane of closest approach, whereas for "adiabatic pathways" $\kappa_{el} \sim 1$ under these conditions. By analogy with homogeneous electron-transfer processes, heterogeneous outer-sphere pathways can be regarded as those where the reactant does not penetrate the inner layer ("coordination layer") of solvent molecules immediately adjacent to the metal surface. However, as discussed below the kinetics of such reactions may nonetheless be influenced significantly by the interfacial solvent environment.

The nature of solvent medium can yield important influences on several of these terms. These will now be considered in turn.

(i) *Precursor stability constant, K_p.* This quantity can be expressed for electrochemical reactions as

$$K_p = K_o \exp(-w_p/RT) \qquad \text{Eq. 2}$$

where w_p is the work of transporting the reactant from the bulk solution to the predominant reaction site, and K_o (cm) is a statistical term. The latter is numerically equal to the effective "reaction zone thickness", within which the reactant needs to be located in order to contribute importantly to the overall reaction rate (2). For nonadiabatic or weakly adiabatic reactions, $K_o \sim 0.5$ to 2 Å (2).

For *bona fide* outer-sphere reactions, it is convenient to express Eq. 1 instead in terms of the conventional "double-layer corrected" rate constant, k_{corr}:

$$k_{corr} = K_o \nu_n \kappa_{el} \Gamma_n \exp(-\Delta G^*_{corr}/RT) \qquad \text{Eq. 3}$$

where ΔG^*_{corr} is a work-corrected free energy of activation. Both k_{corr} and ΔG^*_{corr} are the quantities that would be observed in the absence of coulombic double-layer effects; k_{corr} is related to k_{ob} on the basis of the simple Frumkin model by (3)

$$\ln k_{corr} = \ln k_{ob} + (Z_r - \alpha_{corr})\phi_r F/RT \qquad \text{Eq. 4}$$

where Z_r is the charge number of the oxidized form of the redox couple, α_{corr} is the work-corrected cathodic transfer coefficient, and ϕ_r is the average potential at the reaction site with respect to the bulk solution. Equation 4 can be related to Eqs. 1 - 3 if it is assumed that $w_p = ZF\phi_r$ and $\Delta G^* = \Delta G^*_{corr} + \alpha_{corr} F\phi_r$.

Besides the niceties of discreteness-of-charge effects, additional contributions to w_p might be envisaged if the solvating environment in the vicinity of the reaction site in the double layer differs from that in the bulk solution. Such a "specific" work term contribution will occur if the reactant is partly desolvated in the transition state. This is considered briefly in Section II. This circumstance may apply even for outer-sphere reactions at interfaces where the local solvent structure is strongly perturbed by the metal surface (Section III).

(ii) *Electron Tunneling Probability*, κ_{el}. Although the probability of electron tunneling for outer-sphere systems is anticipated to be relatively insensitive to the intervening medium (4), the nature of the solvent may affect κ_{el} through its influence upon the reactant-electrode distance, x. This is because the reactant in the outer-sphere transition state is expected to be separated from the metal surface by a layer of solvent molecules. Therefore the size of the solvent molecules, together with their ability to solvate the reacting species, should influence x and hence κ_{el}.

(iii) *Nuclear Frequency Factor,* ν_n. This quantity denotes the effective frequency with which the reacting system surmounts the free-energy barrier. The magnitude of ν_n will be determined both by the characteristic inner-shell (reactant vibration) and outer-shell (solvent polarization) frequencies, ν_{is} and ν_{os} respectively, appropriately weighted according to the corresponding inner- and outer-shell components of ΔG^*_{corr}; ΔG^*_{is}, ΔG^*_{os}, respectively. When $\Delta G^*_{is} \to 0$, $\nu_n \approx \nu_{os}$, whereas when $\Delta G^*_{is} \sim \Delta G^*_{os}$, $\nu_n \approx \nu_{is}$ (1). The solvent dynamics should therefore influence the preexponential factor for reactions where the inner-shell barrier is small. This question is considered in Section III.

(iv) *Free-energy barrier,* ΔG^*. The influence of the surrounding solvent, as well as other environmental influences upon ΔG^*_{corr}, is conventionally divided into so-called "intrinsic" and "thermodynamic" (or "extrinsic") contributions (5,6). The latter is identified with the solvent influence upon the standard (or "formal") potential E_f of the redox couple, and hence upon the thermodynamic driving force at a given electrode potential E, whereas the latter is associated with the solvent contribution to the free-energy barrier when the driving force is zero, i.e., when $E = E_f$. For electrochemical reactions involving multi-electron transfer and/or coupled chemical steps, these contributions cannot be separated entirely since E_f for the required rate-determining electron-transfer step (as opposed to the overall reaction) will be unknown.

We shall consider here the influence of the solvent upon the intrinsic barrier, $\Delta G^*(int)$. The work-corrected standard rate constant, k^s_{corr}, (i.e., that measured at E_f) can be related to $\Delta G^*(int)$ by [c.f. Eq. 3]:

$$k^s_{corr} = K_o \nu_n \kappa_{el} \Gamma_n \exp[-\Delta G^*(int)/RT] \qquad \text{Eq. 5a}$$

where $$\Delta G^*(int) = \Delta G^*_{os}(int) + \Delta G^*_{is}(int) \qquad \text{Eq. 5b}$$

The inner-shell contribution to $\Delta G^*(int)$, $\Delta G^*_{is}(int)$, will remain essentially constant in different solvents providing that the coordination sphere remains unchanged. The solvent contribution to $\Delta G^*(int)$, $\Delta G^*_{os}(int)$, for electrochemical reactions can be estimated on the basis of the conventional dielectric continuum treatment from (7)

$$\Delta G^*_{os}(int) = \frac{e^2}{8}\left(\frac{1}{a} - \frac{1}{R}\right)\left(\frac{1}{\varepsilon_{op}} - \frac{1}{\varepsilon_s}\right) \qquad \text{Eq. 6}$$

where e is the electronic charge, a is the reactant radius, R is twice the reactant-electrode distance, and ε_{op} and ε_s are the optical and static dielectric constants, respectively (7). While the quantitative, or even qualitative, validity of the dielectric continuum treatment has been questioned (8), it has proved relatively successful in predicting the energetics of photo-induced electron transfer in homogeneous intramolecular systems, for which the barrier height can be extracted directly from spectral data (9). The approximate success of Eq. 6 is probably associated with the dominance of the ε_{op} term, since typically $\varepsilon_{op} << \varepsilon_s$. Thus although the "static" component of the barrier (the ε_s term) is liable to be seriously in error for many systems as a consequence of dielectric saturation, etc., the "optical" component (associated with ε_{op}) is anticipated to be relatively insensitive to the local solvent structure. The errors in the "static" component in Eq. 6 have been estimated to be no greater than 1 - 1.5 kcal mol^{-1} even for systems displaying strong specific ligand-solvent interactions (10). These and related considerations suggest that Eq. 6 tends to underestimate ΔG^*_{os}(int) (10); on the other hand, deviations in the opposite direction are suggested from a molecular dynamics solvent model (11).

In the following sections, some recent experimental data gathered in our laboratory that provide illustrative examples of these solvent effects will be presented. For brevity, experimental and related details are omitted; these can be found in the articles cited.

II. Interfacial Solvation Effects - Variation of Solvent

Several sets of measurements will be described that provide evidence for the influence of interfacial solvation to the electrode kinetics of simple inorganic redox couples. The first involve the comparison between the solvent dependence of standard rate constants, k^s_{corr}, for $Ru(NH_3)_6^{3+/2+}$ and $Co(en)_3^{3+/2+}$ (en = ethylenediamine) at mercury electrodes with the corresponding values, k^s_{calc}, calculated from theoretical models on the basis of Eq. 5. Details of the latter calculations are given in refs. 12 and 13. Values of ΔG^*_{is}(int) were obtained from vibrational and bond length data (13), the solvent-dependent ΔG^*_{os}(int) values were obtained from Eq. 6 with a = 3.5 Å, R = 14 Å in water, and R = 17 Å in nonaqueous media (12). The value of ν_n is equated with ν_{is} (*vide infra*), 1 x 10^{13} sec^{-1} for both couples (13); K_o is taken as 0.6 Å, Γ_n equals 1.05 and 3.0 for $Ru(NH_3)_6^{3+/2+}$ and $Co(en)_3^{3+/2+}$ in various solvents, taken from refs. 6, 12, 14, and 15, are listed in Table I. Whereas small increases in k^s_{calc} are predicted upon substitution of aqueous by nonaqueous (particularly aprotic) media, large (up to 10^3 fold) decreases

TABLE I Solvent Dependence of Standard Rate Constants for $Ru(NH_3)_6^{3+/2+}$ and $Co(en)_3^{3+/2+}$ (en = ethylenediamine) at Mercury Electrodes

Redox Couple	Solvent[a]	A_{corr}[b] cm sec^{-1}	k^s_{corr}[c] cm sec^{-1}	k^s_{calc}[d] cm sec^{-1}	DN[e]
$Ru(NH_3)_6^{3+/2+}$	H_2O	$5x10^3$	2.0	2.5	∿18
	PC		∿3	5	15.1
	DMF	25	0.25	10	26.6
	DMSO	$2.5x10^2$	$5x10^{-2}$	15	29.8
$Co(en)_3^{3+/2+}$	H_2O	0.7	$2.5x10^{-2}$	$3x10^{-4}$	∿18
	F	$8x10^5$	$1.5x10^{-3}$	$7x10^{-4}$	∿24
	NMF	$2x10^6$	$6x10^{-4}$	$7x10^{-4}$	27
	PC		$3x10^{-4}$	$7x10^{-4}$	15.1
	AN	$2x10^5$	$1.5x10^{-3}$	$5x10^{-4}$	14.1
	DMF	$2.5x10^6$	$1x10^{-5}$	$1.2x10^{-3}$	26.6
	DMSO	$1x10^7$	$1x10^{-5}$	$2x10^{-3}$	29.8

[a] PC = propylene carbonate, DMF = N,N-dimethylformamide, DMSO = dimethylsulfoxide, F = formamide, NMF = N-methylformamide, AN = acetonitrile

[b] Double-layer corrected preexponential factor, obtained from intercept of Arrhenius plot of $\ln k^s_{corr}$ versus (1/T) (see text).

[c] Double layer-corrected standard rate constant, obtained from observed rate constant (measured using either a.c. or normal pulse polarography), by using Eq. 4.

[d] Calculated standard rate constant, obtained as described in text.

[e] Solvent "donor number", from ref. 16.

in k^s_{corr} are obtained instead. This marked solvent dependence of k^s_{corr} for $Co(en)_3^{3+/2+}$ has previously been ascribed tentatively to increases in $\Delta G^*_{os}(int)$ associated with specific ligand-solvent interactions, perhaps with accompanying decreases in κ_{el} (6). This explanation is consistent with the observation that k^s_{ob} decreases with increasing ligand-solvent interactions as monitored by the solvent donor number (Table I) (6,7).

However, several pieces of additional information suggest that the substantial solvent dependences of k^s_{corr} for both $Co(en)_3^{3+/2+}$ and $Ru(NH)_6^{3+/2+}$ are due primarily to variations in K_o associated with differences in reactant solvation in the bulk and interfacial environments. In particular, we have found that sufficiently high two-dimensional concentrations (ca 1-5 x 10^{-11} mol cm^{-2}) of $Co(en)_3^{3+}$, $Co(NH_3)_6^{3+}$ and certain other ammine complexes can be induced via electrostatic attraction at the silver-aqueous interface containing adsorbed chloride or bromide anions so to enable their reduction kinetics to be monitored directly in the precursor state at the outer Helmholtz plane by means of rapid linear sweep voltammetry (18). (Indeed, such diffuse-layer cations also exhibit detectable surface-enhanced Raman spectra (19).) These measurements enable *unimolecular* rate constants for the elementary electron-transfer step, k_{et}, to be evaluated. (This is directly analogous to the measurement of k_{et} for outer-sphere cation-anion pairs in homogeneous solution (20).) After a correction for electrostatic double-layer effects, comparison of the resulting k_{et} values with k_{corr} for the same reaction obtained at the mercury-aqueous interface at the same electrode potential enables K_o for the latter to be estimated since (1)

$$k_{corr} = K_o k_{et} \qquad \text{Eq. 7}$$

The resulting estimates of K_o for $Co(en)_3^{3+/2+}$ and $Co(NH_3)_6^{3+}$, ca 200 and 10 Å, respectively, are markedly larger than the value of 0.6 Å, anticipated for weakly adiabatic processes (1,13), that was assumed when deriving the k_{calc} values in Table I. A similarly large estimate of K_o, ca 5 Å, has been obtained for reduction of Cr(III) ammines at the mercury-aqueous interface by using a related procedure whereby k_{corr} for the outer-sphere reaction is compared with k_{et} for a structurally related inner-sphere reaction (21). The same procedure yields a substantially smaller reaction zone thickness, ca 0.1 - 0.3 Å, for Cr(III) aquo reductions (21).

These results suggest that unexpectedly high concentrations of ethylenediamine and ammine reactants are present at mercury-aqueous interfaces, most likely associated with differences in the outer-shell solvation of the reactant in the bulk and interfacial environments, and possibly to partial reactant desolvation at the electrode surface. This is consistent with the data in Table I in that such surface environmental effects are less likely in nonaqueous solvents having high donor numbers than in aqueous media. Therefore "normal" outer-sphere pathways, associated with smaller values of K_o and hence k^s_{corr}, are expected in the former solvents. The general observation that $k^s_{corr} < k^s_{calc}$ in aprotic media (Table I) may be due to several factors, including the occurence of nonadiabatic pathways (i.e., $\kappa_{el} < 1$) as well as to outer-shell barriers that are larger than estimated using the dielectric continuum model.

Of interest in this regard are the experimental frequency factors, A_{corr} (cm sec^{-1}), also listed in Table I. These were obtained from the temperature dependence of k^s_{corr} by using

$$\ln A_{corr} = \ln k_{corr} - (1/T)[d\ln k_{corr}/d(1/T)] \qquad \text{Eq. 8}$$

To a first approximation A_{corr} can be identified with the combined preexponential factor in Eq. 5, $K_o \nu_n \kappa_{el} \Gamma_n$, provided that these terms are temperature independent. For the present systems, it is expected that $A_{corr} \simeq 10^4$-10^5 cm sec^{-1}. The abnormally small value of A_{corr}, 0.7 cm sec^{-1}, for $Co(en)_3^{3+/2+}$ in water (Table I) is further evidence of the deviation of this reaction from a "normal" outer-sphere pathway. The much larger A_{corr} values (10^5-10^7 cm sec^{-1}) seen for $Co(en)_3^{3+/2+}$ in nonaqueous media (Table I) are closer to the theoretical expectations and indicate that the correspondingly smaller values of k^s_{corr} are unlikely to be due to decreases in κ_{el}.

III. Interfacial Solvation Effects - Alteration of Electrode Material

As noted above, in principle an interesting way of varying the interfacial solvent environment while maintaining other factors, including bulk solvation, constant is to alter the chemical nature of the metal surface (22). Given the complex hydrogen-bonded nature of liquid water, marked changes in its structure are anticipated even over several molecular layers near surfaces known to interact specifically with water (i.e., "hydrophilic surfaces") (23). It is therefore of interest to examine outer-sphere rate data at surfaces of differing hydrophilicity. One difficulty with interpreting rate constants as a function of the electrode material is the uncertainties in the electrostatic double-layer corrections, especially at polycrystalline solid surfaces and in the presence of extensive specific adsorption of the supporting electrolyte. The results of some recent experiments designed to minimize these difficulties (24) will now be briefly described.

Table II contains representative electrochemical rate data for the irreversible reduction of four Cr(III) complexes at five metal surfaces in contact with aqueous solution. These reactions are all irreversible; a common potential of -1000 mV vs. saturated calomel electrode (s.c.e.) was chosen to facilitate intercomparison of the data. This potential is sufficiently negative so to essentially eliminate specific anion adsorption with the perchlorate or hexafluorophosphate electrolytes employed. The five metal surfaces; liquid mercury and gallium, lead, and underpotential deposited (upd) monolayers of lead and thallium at silver, all provide relatively well-defined electrodes yet with significant anticipated differences in their hydrophilicity (25).

The observed rate constants for each reaction in Table II vary substantially as the electrode material is altered. Part of these variations are undoubtedly due to electrostatic double-layer effects since the metals yielding smaller k_{ob} values also exhibit the most negative potentials of zero charge (p.z.c.) where ϕ_r will be the least negative [Eq. 4]. However, the corresponding k_{corr} values listed in Table II, obtained by applying Eq. 4 (24) also exhibit similar dependencies on the metal substrate as for the aquo reactants. This correction yields roughly substrate-independent values of k_{corr} for $Cr(NH_3)_6^{3+}$ reduction. Support for the validity of these double-layer corrections is provided by the similar substrate dependencies of k_{corr} for all three aquo complexes, even though the corrections are smaller for the fluoro and especially the sulfato complex on account of their smaller net cationic charge (26). The relative magnitudes of k_{corr} for each complex are also virtually independent of the electrode potential since the work-corrected transfer coefficients, α_{corr}, are also almost independent of the metal surface. Comparable results have also been observed for several other reactions involving aquo and ammine complexes (24).

Significantly, the observed dependence of k_{corr} for the aquo complexes on the metal surface, Hg > upd Pb $\sim$ upd Tl $\gtrsim$ Pb > Ga is consistent with the anticipated differences in their hydrophilities (24). The disparate behavior of mercury and gallium is particularly interesting since these both provide well-defined liquid surfaces; the former has only a small, and the latter a large, tendency to bind to the oxygen atom of water (25). The most likely, albeit not the sole, possibility is that K_0 and/or κ_{el} for the aquo complex reductions are decreased substantially (up to ca 10^3-fold) upon substituting more hydrophilic surfaces for mercury. This observed sensitivity of the aquo complexes to the surface environment, and the relative lack of such an

TABLE II Rate Constants for the Electroreduction of Cr(III) Complexes at -1000 mV vs s.c.e. at Various Metal Surfaces at 25°C

Reactant	Surface	k_{ob} [a] cm sec^{-1}	α_{ob} [b]	k_{corr} [c] cm sec^{-1}	α_{corr} [d]
$Cr(OH_2)_6^{3+}$	Hg	$5x10^{-2}$	0.61	$3x10^{-3}$	0.50
	Ga	$8x10^{-6}$	0.58	$3x10^{-6}$	0.50
	Pb	$3x10^{-5}$	0.61	$1.5x10^{-5}$	0.55
	upd Pb	$3x10^{-3}$	0.55	$1.5x10^{-3}$	0.52
	upd Tl	$3x10^{-4}$	0.50	$3x10^{-4}$	0.50
$Cr(OH_2)_5F^{2+}$	Hg	$2.5x10^{-4}$	0.58	$2.5x10^{-5}$	0.54
	Pb	$2x10^{-6}$	0.55	$1x10^{-6}$	0.52
	upd Pb	$\sim 2x10^{-5}$	0.55	$\sim 1x10^{-5}$	$\sim$0.55
	upd Tl	$1.5x10^{-6}$	0.65	$1.0x10^{-6}$	0.6
$Cr(OH_2)_5OSO_3^{+}$	Hg	$3.5x10^{-3}$	0.54	$2x10^{-3}$	0.52
	Ga	$\sim 7x10^{-6}$	$\sim$0.55	$\sim 5x10^{-6}$	$\simeq$0.55
	Pb	$3x10^{-5}$	0.5	$2.5x10^{-5}$	$\simeq$0.5
	upd Pb	$5x10^{-5}$	0.5	$3x10^{-5}$	$\simeq$0.5
$Cr(NH_3)_6^{3+}$	Hg	$2x10^{-2}$	0.84	$4x10^{-4}$	0.75
	Ga	$\sim 2x10^{-4}$	0.8	$1.5x10^{-4}$	$\simeq$0.7
	Pb	$6x10^{-6}$	0.76	$1.5x10^{-6}$	0.7
	upd Pb	$2.5x10^{-4}$	0.78	$7x10^{-5}$	0.70
	upd Tl	$8x10^{-5}$	0.70	$7x10^{-5}$	0.65

[a] Observed rate constant for one-electron electroreduction of complex at electrode potential E = -1000 mV at metal surface listed; electrolyte was 0.5 M $NaClO_4$ + 3 mM $HClO_4$, except for $Cr(NH_3)_6^{3+}$ reduction which was measured in 40 mM $La(ClO_4)_3$ + 3 mM $HClO_4$.

[b] Observed transfer coefficient, determined from $\alpha_{ob} = -(RT/F)(d\ln k_{ob}/dE)$.

[c] Rate constant at -1000 mV corrected for electrostatic work terms, determined from listed value of k_{ob} using Eq. 4; required values of the diffuse-layer potential ϕ_r determined as noted in ref. 24.

[d] Transfer coefficient corrected for electrostatic work terms(24).

effect for $Cr(NH_3)_6^{3+}$ reduction, may well be associated with the especially strong hydrogen bonding between aquo ligands and surrounding water molecules (27). Approach of such aquo reactants to hydrophilic metal surfaces may necessitate disturbance of this secondary solvation since the electrode will tend to orient water molecules so to impede solvation of nearby reactant cations.

The likelihood that the hydrophilic surfaces induce substantial perturbations upon the interfacial reactant solvation is supported by the decreases in the activation enthalpies, ΔH^*, that accompanying decreases in k_{corr} (24). Moreover, the observed rate constants and activation parameters for $Cr(OH_2)_6^{3+}$ reduction at mercury electrodes are substantially closer to the numerical predictions obtained from the above rate formalisms than those observed at the hydrophilic surfaces (12,24). This further indicates that the mercury surface provides an relatively "nonperturbing" environment for aquo reactants. The distinctly different behavior exhibited by reactions involving otherwise-similar aquo and ammine (and ethylenediamine) reactants is broadly speaking consistent with the distinctly weaker hydration of the latter. The former can usefully be regarded as "surface-structure sensitive" processes, associated with strong aquo ligand-solvent interactions. The weaker solvent interactions experienced by the ammine reactants lead to substantial deviations from "normal" outer-sphere behavior at the weakly solvated mercury surface, which nonetheless appear to be relatively insensitive to the surface hydrophilicity.

Another series of systems that illustrate the ligand-dependent influences of interfacial solvation upon electrochemical kinetics at the mercury-aqueous interface is provided by examinations of the electroreduction kinetics of $Co^{III}(NH_3)_5L^{2+}$ complexes, where L is a carboxylate ligand containing a variety of organic substituents that nonetheless lack a surface binding group (28). Large (up to 10^4-fold) variations in k_{corr} at a given electrode potential are observed as L is varied, even though similar reactivities are observed for the homogeneous reduction of all these complexes by the outer-sphere reductant $Ru(NH_3)_6^{2+}$ (28). Generally, k_{corr} increases as the "hydrophobic" nature of L is increased, especially when L contains one or more aromatic rings, suggesting that the cross-sectional reactant concentration at the interface, and hence K_o, increases as the strength of reactant-solvent interactions decreases. On the other hand, closely similar values of k_{corr} are observed for all these reactions at the mercury-dimethylsulfoxide interface. This again demonstrates the predominance of "normal" outer-sphere pathways, at least for cationic reactants, in this more strongly solvating medium.

Regardless of the physical details, it seems clear that the nature of the interfacial environment can yield significant and even substantial influences upon the reaction energetics of ostensibly outer-sphere electrode reactions in aqueous solution. Generally speaking, large deviations from the reactivities expected for normal outer-sphere pathways occur when the reactant-solvent interactions are relatively weak, even for reactants not expected to replace the inner-layer solvent molecules. This circumstance may encompass the large majority of electrode reactions involving cationic inorganic reactants in aqueous solution, as well as for anionic species which often proceed via adsorbed transition states under these conditions.

IV. Solvent Relaxation Dynamics

As is conventional, we have assumed so far that the frequency factor of the electron-transfer step is independent of the surrounding solvent. On the basis of the encounter pre-equilibrium formulation [Eqs. 1 or 5], the frequency for adiabatic pathways ($\kappa_{el} \sim 1$) is identified with ν_n. On the basis of the transition-state treatment (TST), ν_n can be expressed as a weighted mean of the characteristic inner-shell and outer-shell (solvent) frequencies, ν_{is} and ν_{os}, as (1,29)

$$\nu_n = \left(\frac{\nu_{os}^2 \Delta G^*_{os} + \nu_{is}^2 \Delta G^*_{is}}{\Delta G^*_{os} + \Delta G^*_{is}} \right)^{1/2} \qquad \text{Eq. 9}$$

Since it is commonly presumed that $\nu_{os} << \nu_{is}$, even when $\Delta G^*_{is} < \Delta G^*_{os}$ the overall frequency factor is anticipated to commonly be dominated by ν_{is} rather than by ν_{os}. Recent theoretical treatments, however, suggest instead that the dynamics of solvent reorganization may play an important role in the kinetics of outer-sphere electron-transfer reactions, at least when the inner-shell barrier is small (30,31). We now briefly summarize the resulting relationships, and utilize them to examine the solvent-dependent electrochemical exchange kinetics for some metallocene redox couples.

It has been pointed out that the effective value of ν_{os} can often be determined by the so-called longitudinal (or "constant charge") solvent relaxation time, τ_L (30,31). This quantity is related to the experimental Debye relaxation time, τ_D, obtained from dielectric loss measurements using

$$\tau_L = (\varepsilon_\infty / \varepsilon_s) \tau_D \qquad \text{Eq. 10}$$

where ε_∞ is the high-frequency solvent dielectric constant. For solvents where $\tau_L \gtrsim 10^{-12}$ sec^{-1}, ν_{os} is dominated by the solvent relaxation frequency, ν^L_{os}, such that for electron-exchange reactions (i.e., when the free-energy driving force is zero) (30,31):

$$\nu^L_{os} = \tau_L^{-1}\left(\frac{\Delta G^*_{os}}{4\pi k_B T}\right)^{1/2} \qquad \text{Eq. 11}$$

where k_B is the Boltzmann constant. Since τ_L typically falls in the range ca 10^{-11} to 2×10^{-13} sec for common electrochemical solvents (32), when $\Delta G^*_{os} \sim 5$ kcal mol^{-1} ν^L_{os} will lie in the range ca 10^{11} to 3×10^{12} sec^{-1}. Since ν_{is} for systems involving metal-ligand vibrations will typically be close to 1×10^{13} sec^{-1}, generally $\nu^L_{os} \lesssim \nu_{is}$. It is important to note, however, that in contrast to Eq. 9 which refers to the rate of barrier *crossing*, Eq. 11 describes deviations from TST where ν_{os} is controlled by the rate of *approaching* the barrier top associated with slow solvent relaxation. Consequently, ν^L_{os} will tend to dominate ν_n when $\nu^L_{os} << \nu_{is}$; i.e. the *slowest* activation mode controls the effective frequency factor. This is the opposite result to that anticipated from TST (33). For rapidly relaxing solvents, when $\nu^L_{os} \lesssim 10^{-12}$ sec^{-1}, the effective outer-shell frequency can be controlled partly by solvent rotation, i.e., "solvent inertial" effects (30). Approximate expressions are available with which to estimate this frequency, ν^R_{os}, which corresponds to the TST limit (30).

Similarly to the TST limit, the relative contributions of ν^L_{os} and ν_{is} to ν_n will be weighted according to the relative magnitudes of ΔG^*_{os} and ΔG^*_{is} to the overall barrier height, although Eq. 9 is no longer appropriate. The circumstance $\nu^L_{os} \approx \nu_n$ is anticipated to hold for exchange reactions when approximately (34)

$$(\Delta G^*_{is}/\Delta G^*_{int})^{1/2}\ \nu_{is}\ \exp(-\Delta G^*_{is}/k_B T) \gtrsim \tau_L^{-1} \qquad \text{Eq. 12}$$

For the typical values $\Delta G^*_{int} = 5$ kcal mol^{-1}, $\nu_{is} = 1 \times 10^{13}$ sec^{-1}, $\tau_L \lesssim 5 \times 10^{-12}$ sec, this inequality will hold when $\Delta G^*_{is} \lesssim 1$ kcal mol^{-1}. This last condition is expected to be the case for a variety of redox couples where the bond distortions required for electron transfer are relatively small (e.g., aromatic molecule-anion, transition-metal couples containing aromatic ligands). We therefore expect that for such "rapid" electron-exchange systems the effective frequency factor will be determined by solvent reorientation.

Since τ_L and hence ν^L_{os} can vary greatly (ca 50 fold) with the solvent size and intermolecular structure, this circumstance can provide a large and even dominant influence on the solvent dependence of k_{corr}. We have evaluated standard electrochemical rate constants as a function of solvent at mercury electrodes for several cationic metallocene couples having the general form $M(Cp)_2^{+/o}$ or $M(Cp')_2^{+/o}$, where M = Fe, Mn, Co, Cp = cyclopentadienyl, and Cp' = pentamethylcyclopentadienyl (32,35). These couples have several important virtues for such fundamental examinations of solvent effects. In particular, only small changes ($\lesssim$ 0.04 Å) in the metal-ring bond distances accompany electron transfer (36); together with corresponding Raman vibrational data these yield small yet varying inner-shell barriers, ΔG^*_{is}(int) $\sim$ 0.025 to 0.25 kcal mol^{-1}. Nevertheless, the relatively small size of the complexes (a = 3.5 - 4 Å) yields sufficiently large ΔG^*_{os} (int) values so that the k^s_{ob} values are conveniently measureable (ca 0.1 - 2 cm sec^{-1}) using a.c. polarography. In addition, the formal potentials E_f can be varied between 0 and -1.5 V vs s.c.e. by altering the metal and substituting methyl groups on the cyclopentadienyl rings. The magnitude of the electrostatic double-layer corrections appears to be small under most conditions, so that $k^s_{ob} \approx k^s_{corr}$ (35).

Table III contains values of k^s_{ob} for two representative metallocene couples, $Fe(Cp')_2^{+/o}$ and $Co(Cp)_2^{+/o}$, together with data for the closely related dibenzenechromium couple, $Cr(C_6H_6)_2^{+/o}$ in eight nonaqueous solvents; acetonitrile (ACN), acetone, methylene chloride (CH_2Cl_2), formamide, N-methylformamide (NMF), N,N-dimethylformamide (DMF), dimethylsulfoxide (DMSO) and benzonitrile (C_6H_5CN). These solvents were chosen not only for their suitability for electrochemical measurements but also because of the wide variations between the values of τ_L, and hence ν^L_{os} obtained from Eqs. 10 and 11. These ν^L_{os} values vary from 4.5 x 10^{12} sec^{-1} in acetonitrile to 1.3 x 10^{11} sec^{-1} in benzonitrile.

To the left of the experimental k^s_{ob} values are listed two sets of calculated rate constants, k_{calc}. The first set, labelled "Eq. 9", were obtained using Eq. 5 by calculating ν_n from Eq. 9, assuming that $\nu_{is} >> \nu_{os}$, and ΔG^*_{is}(int) = 0.2 kcal mol^{-1} (a typical value for the reactants considered here (32,35)). (Other calculational details are given in the footnotes to Table III and in refs. 32 and 35.) Since the frequency factor using this approach will be solvent independent, the resulting solvent dependence of k^s_{calc} arises entirely from the predicted variations in the outer-shell barrier, ΔG^*_{os}(int), obtained from Eq. 6. Comparison between the values of k^s_{ob} and k^s_{calc} (Eq. 9) shows that whereas rough agreement is seen in some solvents, the latter entirely fail to account for the observed solvent dependence of k^s_{ob}, the ca 13-fold *increase* in k^s_{calc} from acetonitrile to benzonitrile contrasting the observed ca 3-5 fold *decreases* in k^s_{ob}.

TABLE III Comparison between Calculated and Experimental Rate Constants ($cm\ sec^{-1}$) for Electrochemical Exchange at Mercury-Nonaqueous Interfaces at 23°C

Solvent	Calculated Rate Constants, k^s_{calc} (Eq. 9)[a]	(Eq. 11)[b]	Observed Rate Constants, k^s_{ob} [c] $Fe(Cp')_2^{+/o}$	$Co(Cp)_2^{+/o}$	$Cr(C_6H_6)_2^{+/o}$
ACN	0.30	1.6		1.0	1.8
Acetone	0.55	2.0	2.2	0.95	1.2
CH_2Cl_2	4.0	10	0.45	0.25	0.45
Formamide	0.9	0.38		0.19	0.17
NMF	0.7	0.20	0.45	0.32	0.40
DMF	1.0	0.60	0.8	0.38	0.55
DMSO	2.5	0.7	0.25	0.15	0.11
C_6H_5CN	4.2	0.65	0.45	0.32	0.3

[a] Standard rate constant ($cm\ sec^{-1}$) calculated from Eq. 5, with $\Delta G^*_{is}(int) = 0$, $\Delta G^*_{os}(int)$ from Eq. 6 with $a = 3.8$ Å, $R = \infty$, $\Gamma_n = 1$, $K_o\kappa_{el} = 0.6$ Å, ν_n obtained from Eq. 9, with $\Delta G^*_{is}(int) = 0.2$ kcal mol^{-1}, $\nu_{is} = 6 \times 10^{12}\ sec^{-1}$, and $\nu_{os} << \nu_{is}$ (see text).

[b] As footnote a, but with ν_n set equal to ν^L_{os} obtained from Eq. 11 using literature dielectric loss data (see refs. 32,35).

[c] Observed rate constant ($cm\ sec^{-1}$) obtained for stated redox couple in solvent indicated, containing 0.1 M TBAP. Cp = cyclopentadienyl; Cp' = pentamethylcyclopentadienyl.

The second set of k^s_{calc} values in Table III, labelled "Eq. 11" were obtained using the same procedure as the first set, but with ν_n estimated instead by equating it with ν^L_{os} obtained from Eq. 11. [This condition $\nu_n \approx \nu^L_{os}$ is predicted for the present systems from Eq. 12.] In contrast to k^s_{calc}(Eq 9), these latter calculated values not only approximate the k^s_{ob} values in most solvents, but also correctly mimic the observed solvent dependence of k^s_{ob} (the sole exception is the data in methylene chloride). Such reasonable agreement between experiment and the theoretical predictions also extends to the electrochemical activation parameters (35).

It is interesting to note that, broadly speaking, as ε_{op} increases [and hence $\Delta G^*_{os}(int)$ decreases] in a series of solvents (such as those in Table III), τ_L^{-1} and hence ν^L_{os} tend to decrease. Consequently, the solvent dependence of the free-energy barrier and relaxation dynamics associated with outer-shell reorganization tend to offset each other. This circumstance can account both for the otherwise-perplexing solvent independence of exchange kinetics for some systems (37) as well as solvent-dependent rate ratios that are opposite from the expectations of the conventional theoretical treatment. (For example, values of k^s_{ob} in acetonitrile are often observed to be larger than in DMF, despite the greater values of $\Delta G^*_{os}(\text{int})$ in the former solvent.) Most importantly, the present evidence suggests that the dielectric continuum model can provide a reasonable representation of the reaction energetics, at least for reactants (such as metallocenes) that interact nonspecifically with the surrounding solvent, providing that the role of the solvent in the preexponential as well as the exponential factor (free-energy barrier) is taken into account. Nevertheless, according to Eq. 12 the former should be limited to reactants having relatively small inner-shell barriers. If feasible, it would be most interesting to examine solvent-dependent exchange kinetics for a series of structurally related redox couples having inner-shell barriers that could be varied over the range, say, from zero to 2 kcal mol^{-1} where the circumstance $\nu_n \approx \nu_{os}$ should give way to $\nu_n \approx \nu_{is}$.

V. Concluding Remarks

The foregoing provides several lines of evidence that indicate that the chemical nature of the solvent can yield important influences on electron-transfer kinetics even for outer-sphere electrochemical reactions where the inner-shell barrier is held constant. Several other types of experiments, for example those involving mixed solvents with preferential solvation of the surface and/or the reacting species, also lead to the same general conclusions. The theoretical formalisms utilized here, especially the dielectric continuum solvent model, should be used with caution in view of the simplifying assumptions involved. Nonetheless, the observed large sensitivities of the electrochemical rate data to the solvent medium can be satisfyingly rationalized, at least semiquantitatively, in terms of these models.

As noted at the outset, our fundamental understanding of solvent effects in electrochemical and other electron-transfer reactions remains rudimentary. However, the renaissance of theoretical activity in condensed-phase reaction dynamics (38) together with further critical experimental work should provide fresh insight into the role of the solvating environment for electron-transfer reactions in the near future.

Acknowledgments

Dr. Joseph Hupp contributed importantly to several facets of the present work. Prof. J. T. Hynes provided some valuable insights into solvent dynamics. This work is supported in part by the Office of Naval Research and the Air Force Office of Scientific Research. The author acknowledges a fellowship from the Alfred P. Sloan Foundation.

References

1. J. T. Hupp, M. J. Weaver, J. Electroanal. Chem. 152, 1 (1983).

2. J. T. Hupp, M. J. Weaver, J. Phys. Chem. 88, 1463 (1984).

3. P. Delahay, "Double Layer and Electrode Kinetics", Interscience, New York, 1965, Chapter 9.

4. For example, J. Logan, M. D. Newton, J. Chem. Phys. 78(2), 4086 (1983).

5. R. A. Marcus, J. Phys. Chem. 72, 891 (1968).

6. S. Sahami, M. J. Weaver, J. Electroanal. Chem. 124, 35 (1981).

7. R. A. Marcus, J. Chem. Phys. 43, 679 (1965).

8. For example, J. O'M. Bockris, S. V. M. Khan, "Quantum Electrochemistry", Plenum Press, New York, 1979.

9. C. Creutz, Prog. Inorg. Chem. 30, 1 (1983); T. J. Meyer, ibid 30, 389 (1983).

10. J. T. Hupp, M. J. Weaver, submitted for publication; J. T. Hupp, M. J. Weaver, J. Phys. Chem. 88, 1860 (1984).

11. D. F. Calef, P. G. Wolynes, J. Chem. Phys. 78, 470 (1983).

12. J. T. Hupp, H. Y. Liu, J. K. Farmer, T. Gennett, M. J. Weaver, J. Electroanal. Chem., in press.

13. J. T. Hupp, M. J. Weaver, submitted.

14. J. K. Farmer, J. T. Hupp, T. Gennett, M. J. Weaver, in preparation.

15. T. Gennett, M. J. Weaver, Anal. Chem., in press.

16. V. Gutmann, "The Donor-Acceptor Approach to Molecular Interactions", Plenum Press, New York, 1978, Chapter 2.

17. S. Sahami, M. J. Weaver, J. Electroanal. Chem. 122, 171 (1981).

18. M. A. Tadayyoni, M. J. Weaver, to be published.

19. M. A. Tadayyoni, S. Farquharson, M. J. Weaver, J. Chem. Phys. 80, 1363 (1984).

20. D. Gaswick, A. Haim, J. Am. Chem. Soc. 93, 7347 (1971).

21. J. T. Hupp, M. J. Weaver, J. Phys. Chem. 88, 1463 (1984).

22. S. Trassati, in "Advances in Electrochemistry and Electrochemical Engineering", H. Gerischer, C. W. Tobias (eds), Wiley, New York, Vol. X, 1977, p. 279.

23. W. Drost-Hansen, Ind. Eng. Chem. 61 (11), 10 (1969).

24. H. Y. Liu, J. T. Hupp, M. J. Weaver, J. Electroanal. Chem., in press.

25. S. Trassati, in "Modern Aspects of Electrochemistry", Vol. 13, B. E. Conway, J. O'M. Bockris (eds), Plenum, New York, 1979, p. 81; S. Trassati, Electrochim. Acta 28, 1083 (1983).

26. M. J. Weaver, F. C. Anson, Inorg. Chem. 15, 1871 (1976); J. Phys. Chem. 80, 1861 (1976).

27. E. L. Yee, R. S. Cave, K. L. Guyer, P. D. Tyma, M. J. Weaver, J. Am. Chem. Soc. 101, 1131 (1979); M. J. Weaver, S. M. Nettles, Inorg. Chem. 19, 1641 (1980).

28. V. Srinivansan, S. W. Barr, M. J. Weaver, Inorg. Chem. 21, 3154 (1982); T. T-T. Li, M. J. Weaver, in preparation.

29. N. Sutin, Prog. Inorg. Chem. 30, 441 (1983).

30. D. F. Calef, P. G. Wolynes, J. Phys. Chem. 87, 3387 (1983).

31. L. D. Zusman, Chem. Phys. 49, 295 (1980); I. V. Alexandrov, Chem. Phys. 51, 449 (1980).

32. M. J. Weaver, T. Gennett, Chem. Phys. Lett., submitted.

33. B. L. Tembe, H. L. Friedman, M. D. Newton, J. Chem. Phys. 76, 1490 (1982).

34. M. Ya Ovchinnikova, Russ. Theor. Exp. Chem. 17, 507 (1981).

35. T. Gennett, M. J. Weaver, submitted.

36. A. Haaland, Acc. Chem. Res. 12, 415 (1979).

37. E. S. Yang, M-S. Chan, A. C. Wahl, J. Phys. Chem. 84, 3094 (1980).

38. For example, see D. G. Truhlar, W. L. Hase, J. T. Hynes, J. Phys. Chem. 87, 2664 (1983).

UV-VISIBLE REFLECTANCE SPECTROSCOPY IN THE STUDY OF THE METAL-ELECTROLYTE INTERFACE

D. M. Kolb

Fritz-Haber-Institut der Max-Planck-Gesellschaft,
Faradayweg 4-6, D-1000 Berlin 33, West Germany

ABSTRACT

A survey is given of the various techniques in reflectance spectroscopy which can be used to study in-situ the optical and electronic properties of the electrochemical interface. After a short introduction examining what we can learn from reflectance measurements in general and how they are performed in spectroelectrochemistry, a number of specific examples will be discussed. Electroreflectance is used to investigate the optical properties of bare metal surfaces and their change with electrode potential. It is demonstrated how this technique can be used to detect surface states at the metal-electrolyte interface. By differential reflectance spectroscopy we determine reflectance changes due to surface reactions or adsorbate formation. This yields information on the optical properties of adlayers and under favourable circumstances can give some clues about the geometric structure of adsorbates on single crystal substrates. Finally, the use of surface plasmon excitation for the study of electrode surfaces is briefly mentioned.

1. INTRODUCTION

In electrochemistry we are dealing with chemical reactions which are connected with an electron transfer. These reactions occur at the electrode-electrolyte interface where usually the character of the electric conductivity is changed. The electrode is typically an electronic conductor, while the electrolyte is an ionic one. The electrochemical reaction which facilitates the electron trans-

A. F. Silva (ed.), Trends in Interfacial Electrochemistry, 301–330.

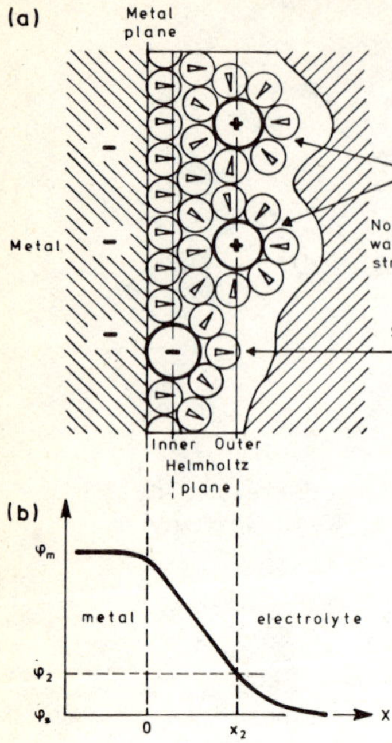

Fig. 1
(a) Model of the electrochemical double layer. After ref. (1).
(b) Potential distribution across the double layer.

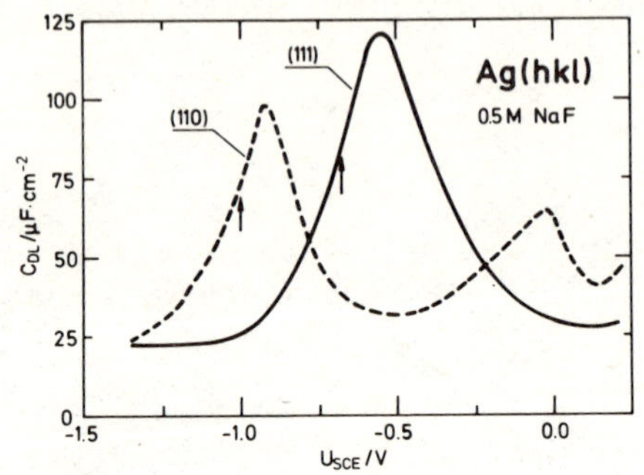

Fig. 2
Differential double layer capacity for Ag(111) and Ag(110) in 0.5 M NaF as a function of potential. Arrows indicate the pzc's.

fer through the interface is like a switch between both conductivities. In the absence of such a reaction the electrode-electrolyte interface is blocking; it behaves like a capacitor of molecular dimensions. A textbook model of the electrochemical interface is reproduced in Fig. 1 (1-6). The electrode surface is considered as one plate of a condensor, the other plate being built by solvated ions of the electrolyte at the so-called outer Helmholtz plane (OHP). The plate separation is given by the distance of closest approach for the solvated ions, which is typically of the order of 0.3 nm. For the case of non-specific adsorption (4), the electrostatic potential is assumed to drop linearly between the two plates, which define the region of the compact double layer. For dilute electrolytes, the chemical potential of the ionic species in solution counteracts the electric forces, leading to an extension of the potential into the electrolyte (diffuse part of the Helmholtz layer with potential φ_2 at the OHP; see Fig. 1b).

The double layer capacity usually varies with electrode potential, especially when single crystal surfaces are employed. As an example, the double layer capacity C_{DL} for Ag(111) and

Ag(110) in contact with 0.5 M NaF as a function of potential is reproduced in Fig. 2 (7). The interface has its largest capacity around the potential of zero charge (pzc) and drops markedly with charging in either direction. This behaviour was rationalized by a potential-dependent dielectric function of the interfacial water, reflecting various stages of association and orientation of the water molecules (8). However, the actual structure of the interfacial water is widely unknown. Gas phase studies on water adsorption have shown that the interaction of water with noble metal single crystal surfaces is weak, the adsorption energies being of the order of half an eV (about 10 Kcal/mole) for metals like Cu, Ag and Au (9). It means for these metals that the water-water bond in bulk water is about the same in strength as the metal-water bond. Stronger adsorption bonds for water have only been observed with adsorbate-covered (pre-dosed) surfaces (10,11). Hence the structure of the interfacial water should mainly be controlled by the hydrogen bonding between water molecules rather than by the direct interaction between water and the metal surface. In addition, the solvation shell of the ions and its distortion by the electrostatic forces at the metal surface will greatly influence the structure of the interfacial water.

Classical electrochemical experiments are based on the measurements of electrode potentials and currents. This has led to a thermodynamic description of the interface of truely remarkable details (2,12). However, the lack of structural information, geometric and electronic or vibronic, has hampered the development of a microscopic model for the electrochemical interface. Here, in-situ spectroscopic techniques such as reflectance spectroscopy were always believed to have the capability of yielding new information which is more specific on an atomic or molecular level, and hence are ideally suited to complement the information from classical electrochemical measurements with their innate ability to yield exact thermodynamic data. In the following a short survey is given of the various techniques used in the author's laboraratory.

2. SOME BASICS OF PHYSICAL OPTICS

The response of a solid to an incoming electromagnetic wave is usually characterized by one frequency-dependent quantity, the complex refractive index $\hat{n}(\omega) = n(\omega) - ik(\omega)$. Describing the electromagnetic plane wave by its time-dependent value of the electric vector,

$$\vec{E} = \vec{E}_o \exp[i(\omega t - \hat{n}\vec{k}\vec{r})] \tag{1a}$$

where $\vec{E}_o$ is the wave amplitude and $\vec{k}$ the wave vector ($|\vec{k}|=2\pi/\lambda$), it becomes evident that real and imaginary part of the complex refractive index represent, respectively, refraction and attenuation of the wave by the solid. For an electromagnetic wave propagating in z-direction, eq. (1a) can be rewritten as:

$$\vec{E} = \vec{E}_o \exp(-2\pi kz/\lambda)\exp[i(\omega t - 2\pi nz/\lambda)]. \quad (1b)$$

The first exp term represents the exponential attenuation of the wave amplitude with penetration into the solid, while the second term describes the time- and phase-dependence of the wave.

The two quantities, n and k, are the so-called refractive index and the extinction coefficient, and are commonly referred to as the optical constants. Since the intensity of the light is directly proportional to the mean-square electric field strength, $\langle\vec{E}^2\rangle$, Beer's law ($I = I_o \exp(-\alpha z)$) yields then the following expression for the absorption coefficient α:

$$\alpha = 4\pi k/\lambda. \quad (2)$$

Maxwell's relation, $\hat{\varepsilon} = \hat{n}^2$, leads us from the optical constants (n,k) to yet more fundamental parameters, the complex dielectric function, $\hat{\varepsilon}$. Real and imaginary part of $\hat{\varepsilon} = \varepsilon' - i\varepsilon''$ can be related to the optical constants:

$$\varepsilon' = n^2 - k^2 \quad (3a)$$

and

$$\varepsilon'' = 2nk. \quad (3b)$$

We should mention that ε' and ε'' are not independent variables, but are interconnected by the Kramers-Kronig relation (causality relation) (13).

Any optical excitation (absorption) process can in principle be described by a classical Lorentz oscillator or a sum of oscillators (14,15) yielding:

$$\hat{\varepsilon} = 1 + \frac{(4\pi Ne^2/m)}{(\omega_o^2 - \omega^2) + i\gamma\omega} \quad (4)$$

where ω_o is the resonance frequency of the oscillator, γ describes the damping and the other symbols have their usual meaning. The overall shape of real and imaginary part of $\hat{\varepsilon}$ of a single oscillator is sketched in Fig. 3. The physical origin for such an oscillator can be sought in free electrons, bound electrons (interband transitions), lattice vibrations (phonons) or others. In the UV-visible range, free and bound electrons are the main contributors to the overall optical response of a solid. Hence, the dielectric function in that photon energy range can be correlated with elementary electronic excitation processes and consequently with the electronic band structure of the solid.

The free electrons in the metal are treated as a collective of zero-frequency oscillators (ω_o=0), which leads to the well-known Drude relation (14,15):

$$\hat{\varepsilon}_f = 1 - \frac{\omega_p^2}{\omega^2 - i\omega/\tau} \qquad (5)$$

with $\omega_p = \sqrt{4\pi ne^2/m}$ being the plasma frequency and τ a collision time. For metals like Ag or Au, the free-electron contribution to the dielectric constant in the visible and near UV range (where $\tau^{-1} << \omega$) can be approximated by: $\varepsilon_f' = 1 - \omega_p^2/\omega^2$ and $\varepsilon_f'' \approx 0$.

The reflectivity of a solid is determined by its optical constants, the state of polarization and the angle of incidence of the light. For the simplest case of normal incidence, the reflectivity R for a solid-vacuum interface is given by:

$$R = \frac{(n-1)^2 + k^2}{(n+1)^2 + k^2} \qquad (6)$$

or, for a solid adjacent to a dielectric medium with refractive index n_o (e.g., glass, water):

$$R = \frac{(n-n_o)^2 + k^2}{(n+n_o)^2 + k^2} . \qquad (7)$$

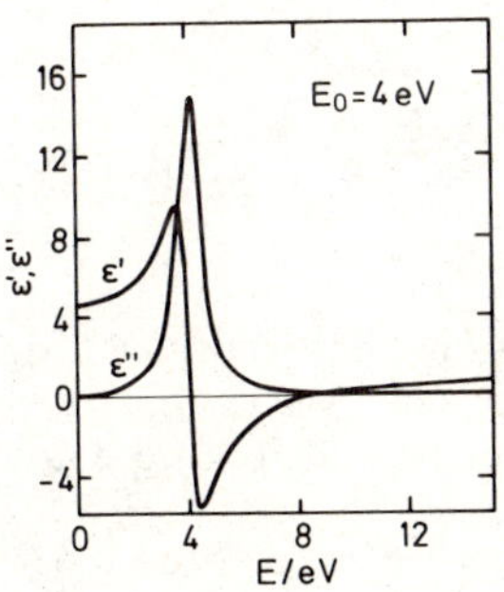

Fig. 3
Frequency dependence of ε' and ε" for a Lorentz oscillator with resonance at E_0=4 eV. After ref. (14).

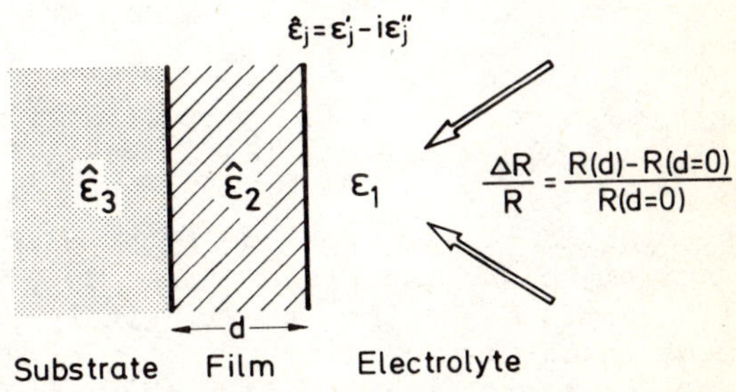

Fig. 4
Three-phase model with sharp boundaries for the electrode-electrolyte interface. After ref. (16,17).

In spectroelectrochemistry, however, we are usually dealing with three phases (bulk electrolyte, interfacial region, bulk electrode) rather than with two (16). We define the interfacial region as having optical properties different from those of the two semi-infinite bulk phases. This can be, e.g., the Helmholtz layer proper, an adsorbate or film deposit on the electrode surface or even the electrode surface region itself. Such a 3-phase system is sketched in Fig. 4, together with the phase notation (indices 1,2 and 3 of dielectric constants refer in the following always to electrolyte, film and substrate phase, respectively). As will be shown in the next section, it is often most convenient to determine normalized reflectance changes $\Delta R/R$. In the case of adsorption studies, $\Delta R/R$ represents the reflectance change for a bare surface due to the formation of an adlayer:

$$\Delta R/R = R_{123}/R_{13}-1 = [R(d)-R(0)]/R(0) \tag{8}$$

With the help of Fresnel's equations the exact expressions for $\Delta R/R$ can be obtained by calculating the reflection coefficients for the 3-phase system (R_{123}) and for the 2-phase system (R_{13}) (16). Such calculations, although quite cumbersome when done by hand, are easily performed by computers, However, for very thin adsorbate layers ($d \ll \lambda$), a linear approximation in d/λ leads to rather simple expressions in $\Delta R/R$ (16,17). These equations are very valuable since they allow a more direct insight into the correlations between optical constants and measured quantities than the exact treatment. For $d \ll \lambda$, it is:

$$\left(\frac{\Delta R}{R}\right)_{\perp} = \frac{8\pi d\sqrt{\varepsilon_1}\cos\varphi_1}{\lambda}\,\underset{\sim}{\mathrm{Im}}\left(\frac{\hat{\varepsilon}_2-\hat{\varepsilon}_3}{\varepsilon_1-\hat{\varepsilon}_3}\right) \tag{9}$$

$$\left(\frac{\Delta R}{R}\right)_{||} = \frac{8\pi d\sqrt{\varepsilon_1}\cos\varphi_1}{\lambda}\,\underset{\sim}{\mathrm{Im}}\left\{\left(\frac{\hat{\varepsilon}_2-\hat{\varepsilon}_3}{\varepsilon_1-\hat{\varepsilon}_3}\right)\left[\frac{1-(\varepsilon_1/\hat{\varepsilon}_2\hat{\varepsilon}_3)(\hat{\varepsilon}_2+\hat{\varepsilon}_3)\sin^2\varphi_1}{1-(1/\hat{\varepsilon}_3)(\varepsilon_1+\hat{\varepsilon}_3)\sin^2\varphi_1}\right]\right\}. \tag{10}$$

The symbols $\perp$, $||$ refer to light polarized perpendicular (s) or parallel (p) to the plane of incidence, φ_1 is the angle of incidence. $\underset{\sim}{\mathrm{Im}}$ means that the imaginary part of the following has to be taken.

For normal incidence ($\varphi_1=0$), eqs. (9) and (10) reduce to:

$$\left(\frac{\Delta R}{R}\right)_{\varphi_1=0} = 8\pi\sqrt{\varepsilon_1}\,\frac{d}{\lambda}\left[\frac{(\varepsilon_1-\varepsilon_2')\varepsilon_3''-(\varepsilon_1-\varepsilon_3')\varepsilon_2''}{(\varepsilon_1-\varepsilon_3')^2+\varepsilon_3''^2}\right]. \tag{11}$$

Eq.(11) turns out to be very useful in obtaining numbers for $\Delta R/R$. E.g., it is easily shown that reflectance spectroscopy is not very sensitive in detecting dielectric (non-absorbing) films

($\varepsilon_2''=0$) while absorbing films are more readily discernable in $\Delta R/R$. We demonstrate this for two different adlayers on an Ag substrate ($\varepsilon_3' = -24$, $\varepsilon_3'' = 1$ for $\lambda = 800$ nm) in contact with an electrolyte for which $\varepsilon_1 = 1.77$. The first example is a Pb monolayer formed at underpotentials. For convenience, we assume bulk optical properties for the lead monolayer ($\varepsilon_2' = -15$, $\varepsilon_2'' = 20$, d = 0.4 nm) which is certainly not correct but serves the purpose. Eq.(11) yields a reflectance decrease of about 1% for the Pb monolayer which is in good agreement with experimental findings (18). The second example deals with the Helmholtz layer as a purely non-absorbing film in the visible-near-UV range. We assume that the optical properties of the interfacial water change from those of bulk water to, say, $\varepsilon_2' = 1.87$ and $\varepsilon_2'' = 0$ by stepping the potential of the Ag electrode from the pzc to some extreme value. This yields a $\Delta R/R$-value for d=0.4 nm of -2.5×10^{-6}! In the latter case the effect is about 3 to 4 orders of magnitude smaller, which demonstrates the difficulties encountered with the study of the Helmholtz layer in the wavelength region of optical transparency.

3. EXPERIMENTAL

It is our aim to determine to a high accuracy minute reflectance changes which are caused by defined electrochemical manipulation. The measuring of normalized reflectance changes ($\Delta R/R$) is to be preferred over the extremely difficult task of determining in-situ absolute reflectance values, the latter being almost impossible because of unknown reflectance losses at cell windows and light scattering and absorption in the electrolyte. (For absolute in-situ reflectance measurements it would be desirable to have the photo-detector inside the electrochemical cell and as close to the electrode as possible. Such a technique is in principle feasible by employing a second, photosensitive electrode in the same electrochemical cell and using the photo current as a probe for light intensity (19)). The high sensitivity required in the study of electrode surfaces calls for modulation techniques and the use of phase-sensitive detection (lock-in amplifiers). Two different ways of electrochemical modulation spectroscopy are readily envisaged: The potential modulation and the coverage modulation.

Potential modulation in the double-layer charging region is used in electroreflectance studies, where the reflectance change induced by a potential variation (in the absence of any Faradaic reaction) is measured. The optical set-up is quite simple since a single-beam technique as shown in Fig. 5a suffices. Monochromatic and (preferably) linearly polarized light is reflected from the electrode surface and focussed onto a photomultiplier tube (PMT). The modulated PMT signal ($\sim\Delta R$) is amplified by a lock-in amplifier and subsequently divided by the DC signal ($\sim R$) of the PMT to yield the desired $\Delta R/R$. Signal-to-noise ratios of 10^{-6} in

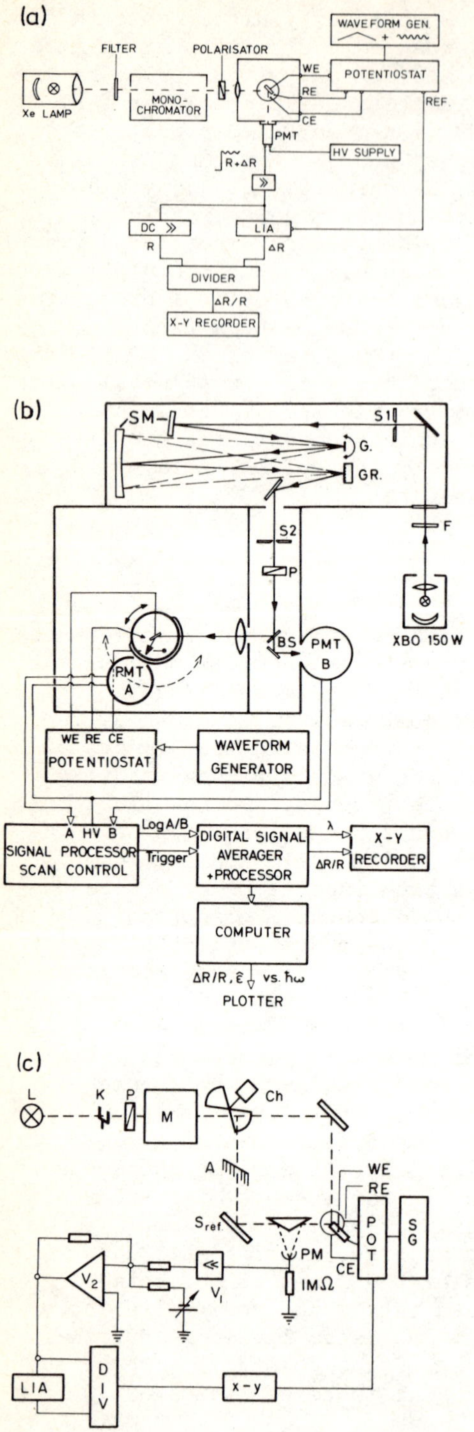

Fig. 5
Schematic diagrams of three different arrangements for in-situ reflectance studies. (a) Single-beam set-up for electroreflectance spectroscopy. (b) Rapid-scan spectrometer. S1, S2: entrance and exit slits. SM1, SM2: spherical mirrors. G: vibrating galvanometer mirror. GR: grating. P: polarizing prism. BS: beam splitter. After ref. (23). (c) Dual-beam set-up for adsorbate studies. L: lamp. Ch: chopper. A: variable light attenuator. S_{ref} : reference mirror. PM: photo multiplier tube. DIV: divider. LIA: lock-in amplifier. After ref. (25).

$\Delta R/R$ are routinely achieved with stabilized light sources (150 W Xe arc lamp for 220 nm < λ < 800 nm or 100 W tungsten lamp for 500 nm < λ < 1400 nm) and high-quality lock-in amplifiers.

While potential modulation is primarily used for the study of bare surfaces, coverage modulation is employed for investigating adsorbates. The coverage is, of course, also changed by a potential variation; however, the corresponding reflectance change is thought to be due to the change in coverage rather than potential. In most cases one would like to achieve an "on-off" condition by square-wave potential modulation, for which the resulting $\Delta R/R$ is caused by adlayer formation. The above described set-up (Fig. 5a), however, will only work if the adsorption-desorption reaction is fast enough to allow frequencies of at least several Hertz for the coverage modulation. Such an adsorption reaction is hydrogen on Pt (20,21).

In many cases, the adsorption or film formation reaction will be too slow to be modulated at a frequency suitable for a lock-in amplifier. Besides, when single crystal electrodes are employed, surface restructuring or even loss of single crystallinity may be endangered when the surface reaction under study is performed many times. An example is oxygen adsorption on Pt(111) electrodes, which is known to restructure the surface into steps with a few potential cycles only (22). For the study of adsorbates, the arrangement of Fig. 5b has proven to be quite successful (23,24). It is a rapid-scan spectrometer, which allows a single wavelength scan to be recorded within a few milliseconds. This is done by virtue of a small vibrating galvanometer mirror which reflects the light beam via a large spherical mirror onto the grating. Unlike in conventional monochromators, the grating is not rotated, but the angle of incidence for the incoming beam varies continuously with the vibrating mirror motion, yielding the desired wavelength scan at the exit slit. The fast scanning allows the use of signal averaging for noise reduction. By sampling about 1000 spectra at each potential (for bare and adsorbate-covered surface), which is done in a few seconds and which is usually enough for signal detection in the 10^{-3} range, the spectrum of $\Delta R/R$ can be obtained by one single potential step.

Fig. 5c shows another dual-beam set-up for recording reflectance changes due to a surface reaction as a function of potential, for fixed wavelength. When the electrode surface is bare, both light beams are made equal in intensity with the help of an optical attenuator (rotating comb for variable attenuation). When by a potential excursion an adsorbate is formed at the electrode, causing its reflectivity to change, the difference ΔR between bare and adsorbate-covered surface shows up directly as AC signal because the reference beam still represents the bare-surface condition.

The wavelength range for in-situ studies is limited to 220 nm < λ < 1400 nm, mainly by the transmission properties of the aqueous electrolyte. The high-energy limit is usually determined by the onset of absorption in the electrolyte and in the polarizing prism (Glan-Thompson), and by the rapidly decaying light intensity of the Xe arc lamp. The low-energy limit of 1400 nm is caused solely by the onset of water absorption. The use of thin-layer cells allows an expansion of the wavelength region well into the far infrared regime (26).

As mentioned above, the commonly studied wavelength range of 220 to 1400 nm is well covered by a Xe arc lamp and a W filament lamp. Photomultiplier tubes with quartz windows and S20 characteristics (e.g., RCA C31034) and PbS photoresistors (for $\lambda > 500$ nm) used in this laboratory were found to work sufficiently well.

4. ELECTROREFLECTANCE OF METAL ELECTRODES

When a potential is applied to the metal electrode in the so-called double-layer charging region, where no electrochemical reaction takes place, the metal-electrolyte interface behaves like a capacitor. Because of the large double-layer capacity, C_{DL}, of such an interface (typically ranging between 20 and 100 $\mu F \cdot cm^{-2}$; see Fig. 2) large surface charges can be induced (up to about 40 $\mu C \cdot cm^{-2}$ which corresponds to roughly 0.2 electrons per surface atom) and extremely high electric fields are obtained (of the order of 10^7 $V \cdot cm^{-1}$) by a simple potential variation. It has been found that the reflectivity of the metal-electrolyte interface changes with electrode potential. This effect is called electroreflectance (ER) (27-32). Unlike the ER effect in semiconductors where the static electric field penetrates several hundred nanometers into the bulk, the static or low-frequency field applied with the electrode potential is screened within the first atomic layer of the metal because of the high charge-carrier density. Thus, metallic ER is an extremely surface-sensitive probe. This is demonstrated in Fig. 6, where for a jellium metal the surface charge density induced by an external

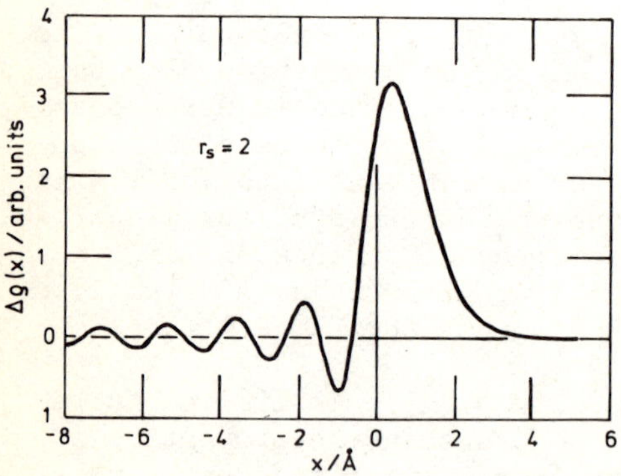

Fig. 6
Surface charge density $\Delta\rho$ of a metal, induced by a weak external electric field. x = 0: edge of the positive-charge background. After ref. (33).

electric field is shown (33). A normal-incidence ER spectrum of Cu(hkl) is shown in Fig. 7 (34). The gross features of this effect, which for Cu is of the order of some 10^{-4} in $\Delta R/R$ for a 100 mV potential modulation, is most easily rationalized by a potential-induced change of the metal surface optical constants (29). As the electrode potential is driven more anodic (cathodic), the free-electron concentration at the metal surface is lowered (increased) with respect to the bulk value, causing the surface optical constants to change. This free-electron effect is easily visualized by separating the metal free- and bound-electron contribution to the dielectric constant, using the Drude expression for the free-electron part (see eq.(5)):

$$\hat{\varepsilon} = \hat{\varepsilon}_f + \hat{\varepsilon}_b = 1 - \frac{\omega_p^2}{\omega^2} + \hat{\varepsilon}_b. \qquad (12)$$

In this approach, the influence of the applied electric field on the bound electrons is neglected ($\hat{\varepsilon}_{2b} = \hat{\varepsilon}_{3b}$) and the change in the surface dielectric constant is solely determined by the potential-induced change Δn in the free-electron concentration at the surface, yielding a potential-dependent ω_p. We can write for the metal surface (18,29):

$$\omega_p = [4\pi(n + \Delta n)e^2/m]^{1/2} \qquad (13a)$$

with

$$\Delta n = C_{DL} \cdot \Delta U/(e_o \cdot d) \qquad (13b)$$

(n: free-electron concentration in the bulk metal; d: penetration depth of the static electric field).

Thus, in the free-electron model, the difference in surface and bulk dielectric constants is given by (18,29):

$$\hat{\varepsilon}_2 - \hat{\varepsilon}_3 = (\hat{\varepsilon}_{3,f} - 1) \cdot (\Delta n/n). \qquad (14)$$

Although this simple model can account for the gross features observed in ER (especially when polycrystalline surfaces are used), it fails to reproduce most of the structural details in spectra obtained with single crystal surfaces. This means that besides the "free-electron" contribution to the ER effect, the bound electrons of the metal (d-electrons and core levels) are indeed affected by the applied electric field (30-32) as seen by the marked dependence of the ER effect on the surface crystallographic orientation. The "bound-electron" contribution contains information on the surface band structure of the metal and its change with electrode potential. E.g., the peak at 2.2 eV in the ER spectrum of Cu arises from an energy modulation of the $3d \rightarrow 4s(E_F)$ interband transition by the potential (31). It is interesting to note that the ER effect of the Cu(113) electrode reveals clearly the two-

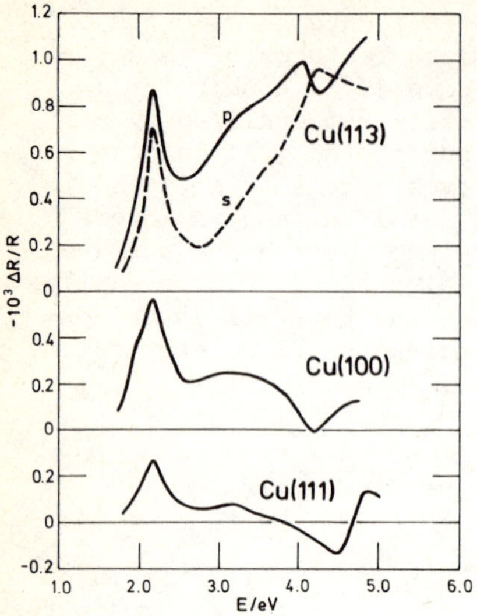

Fig. 7
Normal-incidence ER spectra of Cu single crystal surfaces in 1 N H_2SO_4. U_0 = -0.3 V vs SCE. ΔU_{pp} = 100 mV. p:$\vec{e}$||[001], s:$\vec{e}$||[1̄10]. After ref. (34).

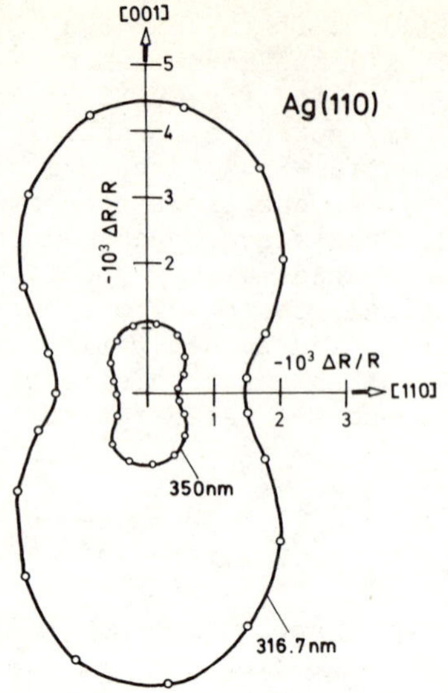

Fig. 8
Normal-incidence ER for Ag(110) as a function of surface crystallographic direction; two different wavelengths. 0.1 N $HClO_4$. U_o = -0.4 V vs SCE.

fold symmetry of the surface. This is demonstrated even better with Ag(110) in Fig. 8. A different optical response $\Delta R/R$ is obtained when the electric field vector of the normal-incident and linearly polarized light is rotated into the [1̄10] direction or into the [001] direction. This has been tentatively explained by an anisotropy in the optical polarizability for the (110) surface parallel to the surface plane (31,35).

Under favourable circumstances the ER spectra of metals can contain information on surface roughness. This is certainly the case for Ag, where surface plasmon (SP) excitation in the large wave-vector limit due to roughness is seen around 3.5 eV. The surface roughness acts as a momentum source which facilitates SP excitation by light. Because of the SP excitation energy being slightly potential-dependent (30), the SP excitation manifests itself in the ER spectra as a derivative-like structure at about 3.5 eV, where the necessary condition $\varepsilon'(Ag) = -\varepsilon(H_2O)$ is fullfilled. In Fig. 9, a sequence of ER spectra is shown for Ag(111) which has been electrochemically roughened by oxidation-reduction cycles up to various degrees. The increasing roughness of the Ag surface is clearly reflected in the SP excitation feature with its center at 360 nm, which is completely absent for the flat Ag(111) surface (see dotted line in Fig. 9).

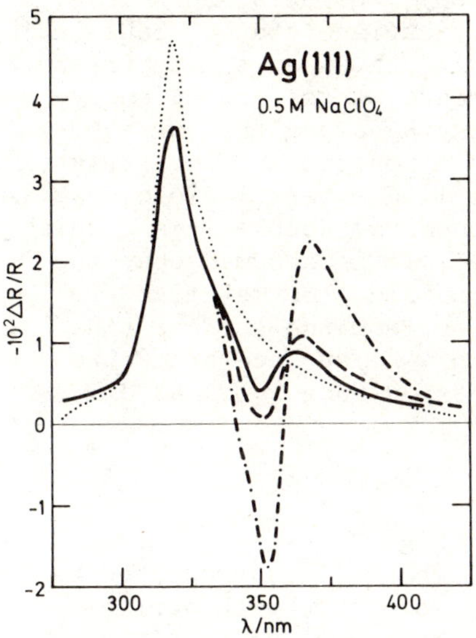

Fig. 9
Normal-incidence ER spectra of Ag(111) in 0.5 M $NaClO_4$ for various degrees of surface roughness. ΔU=500 mV. (•••) flat surface. (——), (---), (—•—) oxidation and reduction of the equivalent amount of 17,32 and 100 monolayers of Ag.

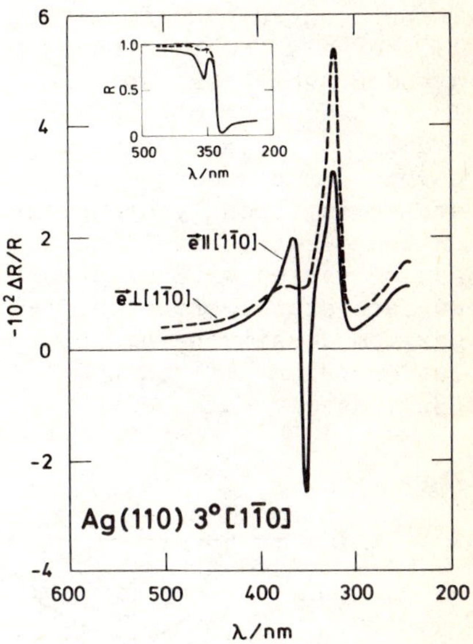

Fig. 10
Normal-incidence ER spectra and static reflectance spectra (insert) of a stepped Ag(110) surface. 0.5 M $NaClO_4$. ΔU = 500 mV. The steps lead in [110] direction. After ref. (31).

In Fig. 10 the normal-incidence ER spectra for a stepped Ag(110) surface are shown for two different directions of the electric vector (31). The surface was prepared to have steps with about 6 nm terrace width, leading into [110] direction. Such a surface represents a very well defined momentum source. Consequently, SP excitation is extremely pronounced (see Fig. 10), however, only when the electric vector points in the direction of [110] where the periodicity of the steps becomes evident. In [001] direction the light hardly sees the steps, and the surface as a whole acts as if flat.

5. SURFACE STATES AT THE METAL-ELECTROLYTE INTERFACE

In recent surface band structure calculations of Ho et al. (36, 37) it has been postulated that optical transitions into empty surface states should be observed for Ag single crystal surfaces. The projected band structure for Ag(100), which is shown in Fig. 11, reveals the existence of two empty surface states, labelled A and B, which are split off the volume states at the upper and lower band edge of the gap. These surface states are strongly localized at the surface, rapidly decaying along the surface normal. As a consequence, the energetic position of the surface states is strongly dependent on an applied external field, e.g., on the electric field in the electrochemical double layer, which can be varied with electrode potential. It is assumed that the surface states probe the region just outside the metal, their maximum density being about 0.1 nm away from the surface (the latter being defined by the position of the surface atom cores) (36). Hence, bulk and surface states experience somewhat dif-

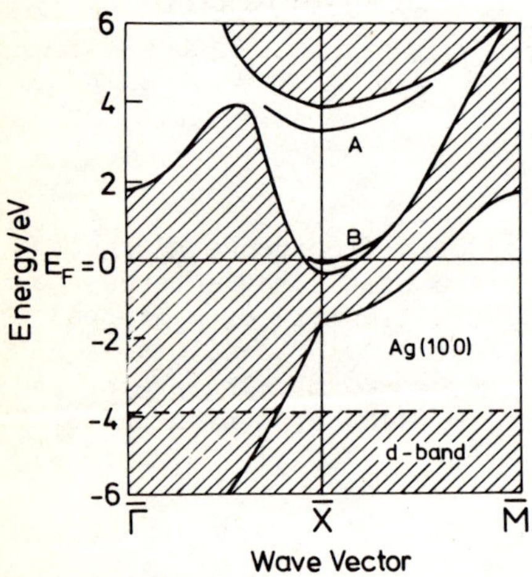

Fig. 11
The Ag(100) projected band structure with surface states A and B. After ref. (37).

ferent electrostatic potentials. E.g., when we apply an electrode potential which is 1 V positive of the respective potential of zero charge, all bulk states of the metal, including the Fermi level, are shifted to higher binding energies (downward in Fig. 11) by 1 eV. The surface states, however, experience an electrostatic potential which is somewhat smaller, as they see only a fraction of the total 1 V potential drop across the double layer. Hence they are shifted to higher binding energies by less than 1 V. As a consequence, bulk and surface states shift relative to each other with electrode potential, and the energy of the optical transition between both states becomes potential-dependent. Such potential-dependent structures in the ER spectra of Ag and Au single crystal surfaces, which we can ascribe to surface states, have indeed been observed (37-41). The results for Ag(100) and Au(100) are shown in Figs. 12 and 13. We note that because of the easy potential variation at the metal-electrolyte interface, this interface is particularly well suited for the detection and study of metallic surface states. In return, from the shift of the energy positions of surface states with electrode potential

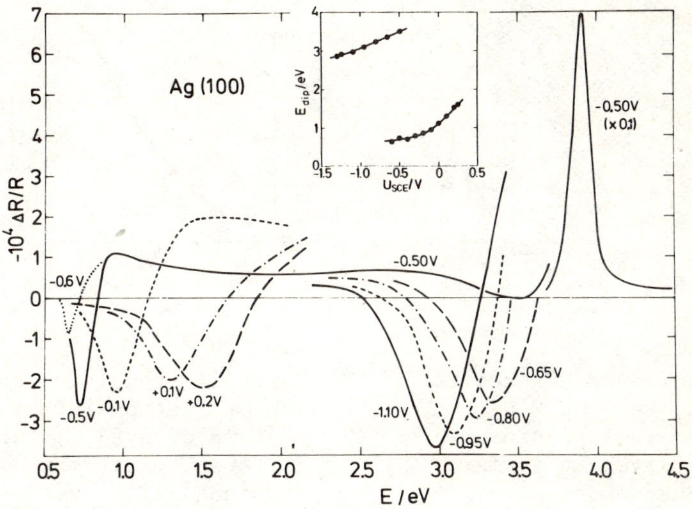

Fig. 12
Normal-incidence ER spectra of Ag(100) in 0.5 M NaF for various bias potentials. After ref. (38). Insert: transition energies as a function of electrode potential.

we obtain information about the microscopic distribution of the electric field in the electrochemical double layer. E.g., in the classical picture of the Helmholtz (double) layer (see Fig. 1b) a linear potential drop is assumed across a distance of about 0.3 nm, the thickness of the compact double layer. If we assume that the surface states probe the region of about 0.1 nm outside the metal surface, we expect a shift of the transition energy of roughly 0.3 eV/V. We observe, however, a much larger value of the order of 1 eV/V (see insert in Fig. 12; see also ref. (38)). This indicates that the potential difference between bulk and surface states is much larger than derived from a linear potential drop across the interface. This may be explained by a non-linear, strongly varying potential distribution in the Helmholtz layer, caused by the microscopic structure of the interfacial water. E.g., the finite size of the water dipoles may lead to regions of overscreening and underscreening, causing an oscillatory potential distribution normal to the surface (37,38).

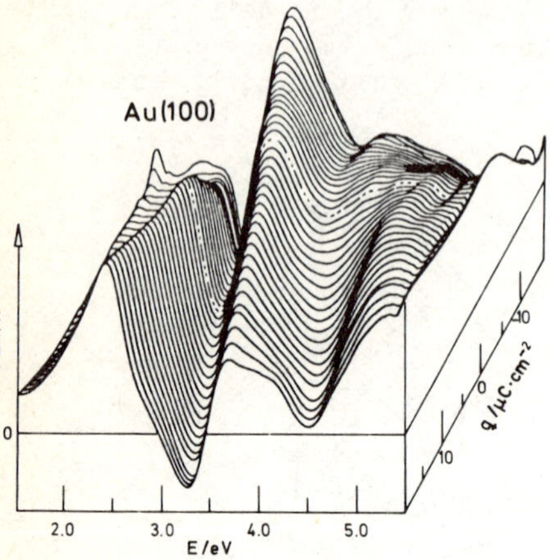

Fig. 13
Normal-incidence ER spectra of Au(100) in 0.02 M NaF (pH 5) for various bias potentials (given in terms of surface charge q). After ref. (39).

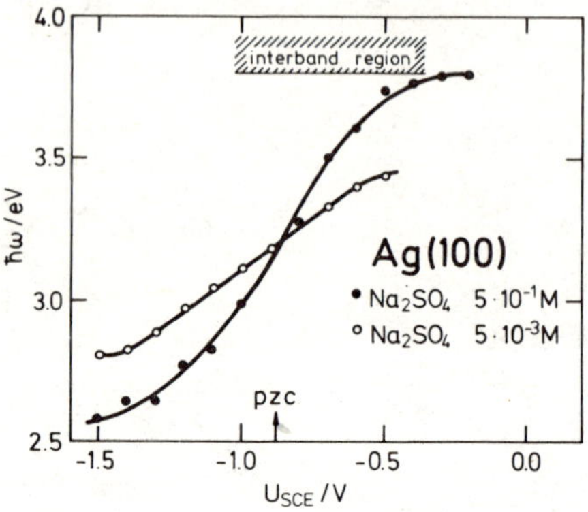

Fig. 14
Shift in transition energy for excitation into surface state A of Ag(100) for two different electrolyte concentrations. After ref. (42).

In Fig. 14, the shift in transition energy for excitation into surface state A of Ag(100) (see Fig. 11) is shown for two different electrolyte concentrations (42). We note that the slope of the shift is clearly dependent on the potential drop across the inner Helmholtz layer, which is reduced in more dilute solutions as the diffuse part of the double layer comes into play. As one would expect, there is no concentration dependence of the surface state position just at the potential of zero charge (-0.9 V vs. SCE for Ag(100)) where the potential drop across the Helmholtz layer vanishes (at least for metals like Ag and Au).

Finally, it may be interesting to note that the above mentioned empty surface states of Ag(100) and Ag(110) have recently been observed at the metal-vacuum interface by inverse photoemission (43,44). The agreement in energy position for both interfaces is excellent. It is worthwhile mentioning that the surface states are readily quenched in UHV by the adsorption of a monolayer of water. The latter has to be condensed onto the surface at about 150 K because of the low interaction energy of water and the metal (see introduction). This apparent difference for the two interfaces in the effect of water on the empty surface states could emphasize the role of temperature which is substantially different in both cases.

6. SURFACE PLASMON EXCITATION AT ELECTRODE SURFACES

Surface plasmons (SP) are collective propagating charge density waves, which are confined to the surface or interfacial region of a solid (45,46). The transverse electric fields coupled with the charge perturbation represent a harmonic wave along the surface with frequency ω and wave vector k_x, parallel to the surface, while perpendicular to the surface the fields decay exponentially. Because of lack of momentum, SP are usually not excited by light in external reflection (if surface roughness is excluded; see Sec. 4). To meet the necessary requirements of momentum conservation, special attenuated total reflection arrangements are necessary (e.g., see insert of Fig. 15), which are usually referred to as Otto or Kretschmann configurations (47,48). The resonance character of this phenomenon makes SP a sensitive probe for studying optical properties of surfaces. In Fig. 15 we show the dispersion curves for various faces of Ag, from which the real part of the complex dielectric function is readily evaluated (35,49). Not only do we see a clear dependence of the surface optical properties on the crystallographic orientation of the Ag surface, but we also notice again a marked anisotropy of the optical response for Ag(110), which reflects the twofold symmetry of that surface. The detection of such differences in the optical constants with normal-incidence reflectance spectroscopy would be nearly impossible since measurements of absolute reflectance values would be required and the differences in these values are less than 1% (49).

The potential dependence of the SP excitation energy is shown in Fig. 16 for silver-electrolyte interfaces. The marked influence of the pzc becomes obvious when comparing the results for various crystal faces. Charging the electrode surface positively causes $\hbar\omega_{sp}$ to shift to lower energies, in agreement with the predictions of classical optics, where ω_{sp} and electron density are connected with each other in much the same way as shown for ω_p in eq.(13a). Charging the surface negatively, however, does not result in a shift of $\hbar\omega_{sp}$ to higher energies, as one would expect, but leaves the SP excitation energy rather unaffected. This asymmetric behaviour, which is not yet fully understood (49-53), is certainly an interesting result as it contains information on the electron density distribution at the metal-electrolyte interface and its change with potential.

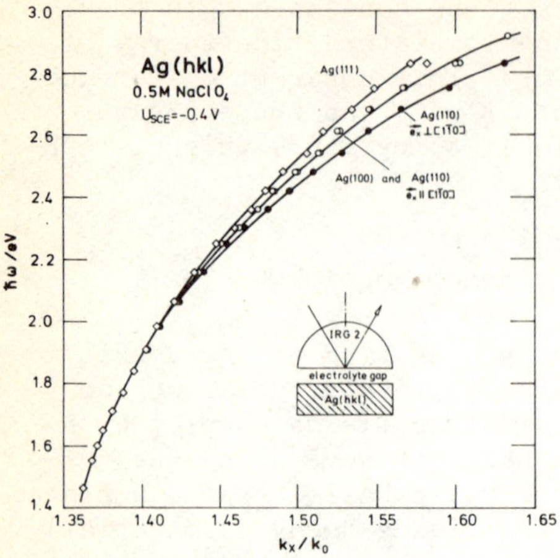

Fig. 15
Surface plasmon dispersion curves for Ag single crystal electrodes in 0.5 M $NaClO_4$. Note the anisotropy in the optical response for Ag(110). After ref. (50).

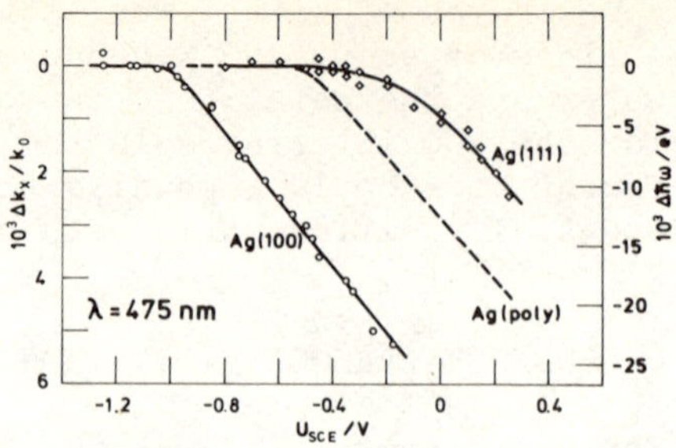

Fig. 16
Surface plasmon excitation energy for various Ag surfaces as a function of electrode potential. 0.5 M $NaClO_4$. IRG 2. After ref. (51).

7. THE STUDY OF ADSORBATES BY DIFFERENTIAL REFLECTANCE SPECTROSCOPY

The optical properties of adsorbates at electrode surfaces are conveniently studied by differential reflectance spectroscopy (DRS), where the relative reflectance change $\Delta R/R$ due to the adsorption process is measured. It is (see Sec. 2)

$$\Delta R/R = [R(\theta) - R(0)]/R(0) \qquad (15)$$

where $R(\theta)$ and $R(0)$ are, respectively, the reflectivities of the adsorbate-covered and the bare surface. In a simple stratified three-phase model with sharp boundaries the observable, $\Delta R/R$, can be correlated with the optical properties of all three phases (substrate, adsorbate, electrolyte), the optical properties of the adsorbate being the only unknown parameters (17). Measurements of $\Delta R/R$ at an oblique angle of incidence and with s- and p-polarized light allow eqs.(9),(10) to be solved for the two optical constants of the adsorbate (17,18,20). Such a treatment, however, assumes the adsorbate optical constants to be isotropic. Attempts have been made to describe the dielectric function of submonolayer adsorbates by an uniaxial tensor (54,55). It remains to be seen, however, whether such a refinement, which is still based on a rather simple 3-phase model (see Fig. 4), meets the expectations.

In the following we consider the formation of metal adsorbates at the metal-electrolyte interface (56). These adsorbates are often formed up to one monolayer at underpotentials, that is, the deposition from the corresponding metal ions in solution occurs at potentials which are positive of the respective Nernst potential for bulk deposition. This is simply a consequence of the fact that the substrate-adsorbate bond exceeds in strength the adsorbate-adsorbate bond (5,56). In Fig. 17, the electrochemical desorption spectra (anodic scans of cyclic current-potential curves) are shown for a Cu monolayer on the three low-index faces of Pt. Distinct differences in the bond energies are readily seen. We note that for Pt(110) monolayer formation occurs in two, energetically quite different, steps.

The $\Delta R/R$ spectra for a Cu monolayer on Pt single crystal surfaces are shown in Fig. 18 (57). They reveal that the optical properties of the Cu monolayer differ markedly from those of bulk Cu, and that the spectra even depend on the crystallographic orientation of the substrate. It was further noted that for Cu

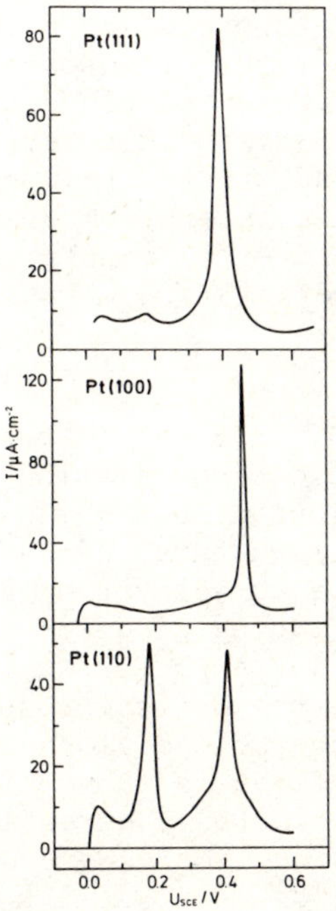

Fig. 17
Electrochemical desorption spectra for Cu on Pt(hkl). Scan rate: 10 mV/s. 1 N H_2SO_4 + 1mM $CuSO_4$.

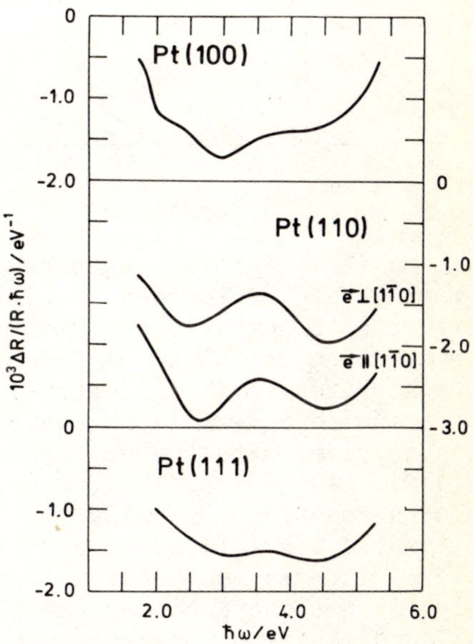

Fig. 18
Normalized reflectance change $\Delta R/R$ for a Cu monolayer on Pt(hkl). Normal incidence. After ref. (57).

on Pt(110) the adsorbate's optical properties show a pronounced anisotropy parallel to the surface, which again reflects the two-fold symmetry of the substrate.

In studying the coverage dependence of $\Delta R/R$ at constant photon energy, we can obtain some information about lateral interactions in the adsorbate layer, and hence we may obtain some clues to the geometric structure of the adlayer. In a case study it was shown (57) (see also eq.11 and substitute d by $\theta \cdot d_0$) that $\Delta R/R(\theta)$ should be a straight line through the origin for θ-independent optical constants of the adsorbate. A reconstruction of the adlayer at a certain coverage should show up in $\Delta R/R(\theta)$ as a change from one straight line through the origin to another one with different slope, while the occupation of new sites at a certain coverage is expected to simply cause a slope change in $\Delta R/R(\theta)$. These three cases are depicted schematically in Fig. 19.

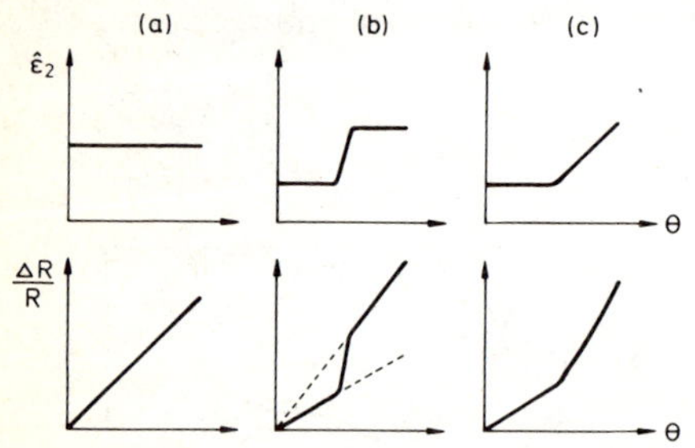

Fig. 19
Schematic diagram of the coverage dependence of $\Delta R/R$ for various types of adsorbates. (a) Coverage-independent optical properties. (b) Reconstruction in the adsorbate. (c) Gradual change in lateral interaction as monolayer is completed. After ref. (57).

Such an investigation was performed for Cu on Pt(hkl), the results for which are reproduced in Fig. 20. The close correlation of experimental observations and model predictions allows some straightforward conclusions to be drawn about the structural behaviour of Cu adsorbates. E.g., a restructuring of the Cu adlayer on Pt(111) at medium coverages seems obvious. An intriguing piece of information is given with the coverage dependence of $\Delta R/R$ for Cu on Pt(110). Besides the lateral anisotropy in the optical response which we mentioned already (see Fig. 18), we note a slope change in $\Delta R/R(\theta)$ for $\vec{e}\perp[110]$ at $\theta = 0.5$. This reflects a change in the lateral interaction between the Cu adatoms in [001] direction, which is not the case for the [110] direction. For a more detailed discussion, however, see refs. (57,58).

Next we briefly discuss the optical properties of thin Ag overlayers and address the question of the minimum thickness of the overlayer needed to acquire bulk optical properties. In order to do so, we recorded the electroreflectance (ER) spectra of a Cu(111) electrode onto which Ag was deposited from solution in various thicknesses (59). As a criterion for the presence of bulk optical constants of Ag, we looked for the appearance of

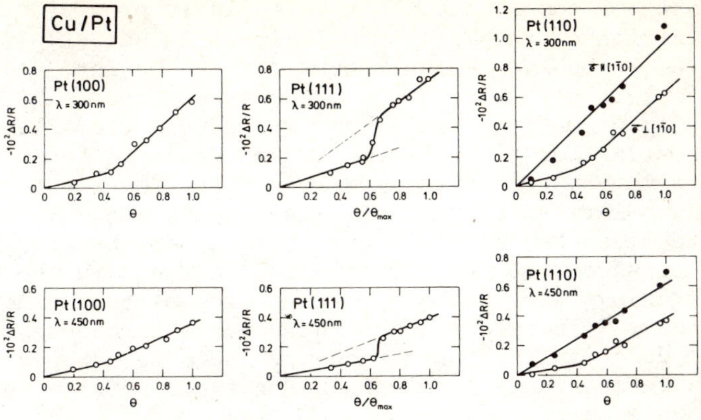

Fig. 20
Normalized reflectance change ΔR/R due to Cu deposition onto Pt(hkl) as a function of Cu coverage. Two different wave lengths. Note the anisotropy in the optical response for Cu on Pt(110). After ref. (57).

the volume plasmon excitation, which for bulk Ag occurs around 3.8 eV and which is a characteristic feature for bulk Ag. Volume plasmon excitation is recognized in the ER spectra of Ag for p-polarized light by a dip in -ΔR/R at about 3.8 eV. The result of our study is shown in Fig. 21. We realize that the volume plasmon structure begins to appear in the spectra already at a thickness of only 2-3 monolayers! We note that the deposition of a monolayer amount of Ag onto Cu(111) causes the ER spectrum of

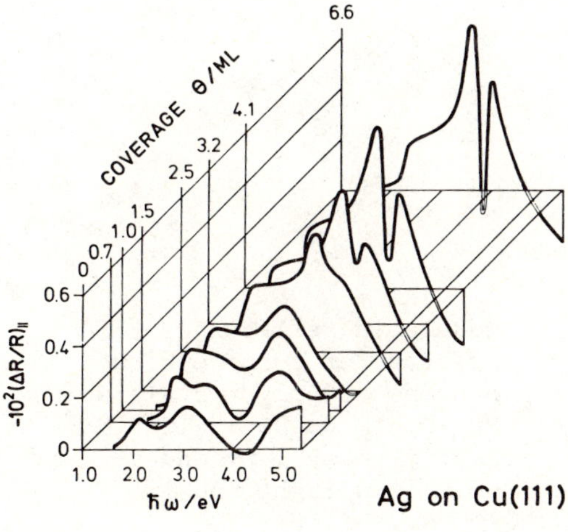

Fig. 21
Electroreflectance (ER) spectra for Cu(111) covered with various amounts of Ag. The Ag coverage is given in units of monolayers (ML). 1 N H_2SO_4. ΔU_{pp} = 300 mV. p-polarization at 45° angle of incidence. After ref. (59).

Cu to disappear completely. This is definite proof for the uniform deposition of the first monolayer of Ag on Cu(111), which seems to be a necessary requirement for the uniform deposition of the next layers also.

Finally we mention the study of adsorbed dye molecules which are interesting for two reasons. Dye molecules often have a transition moment with a well-defined direction in space, and hence by a polarization and angle-dependence study the orientation of the adsorbed molecule with respect to the electrode surface can be assessed. The strong electric field in the double layer may change the absorption properties of the molecule (electrochromism). This effect can be studied rather accurately by potential modulation (ER) spectroscopy. It is then hoped that the dye molecule acts as a local probe for the electric field strength in the double layer.

Electrochromism studies of dye molecules on electrode surfaces have been pioneered by Plieth et al. (60,61). One of their results is reproduced in Fig. 22, which shows the reflectance modulation spectrum of p-amino-nitrobenzene on a polycrystalline Pt electrode. The bands at 436 and 620 nm have been attributed to monomer and dimer absorption, the monomer band shifting to shorter wavelengths with increasingly positive potential. The polarization dependence is striking; an effect is only observable for p-polarized light. It suggests that the molecule is oriented with its dipole moment perpendicular to the Pt surface. We have to bear in mind, however, that two requirements must be met in order to obtain a potential-induced reflectance change. Firstly, the scalar product of light electric vector and transition moment has to be non-zero; and secondly, the transition moment has to have a component in direction of the modulating electric field, which is always normal to the surface.

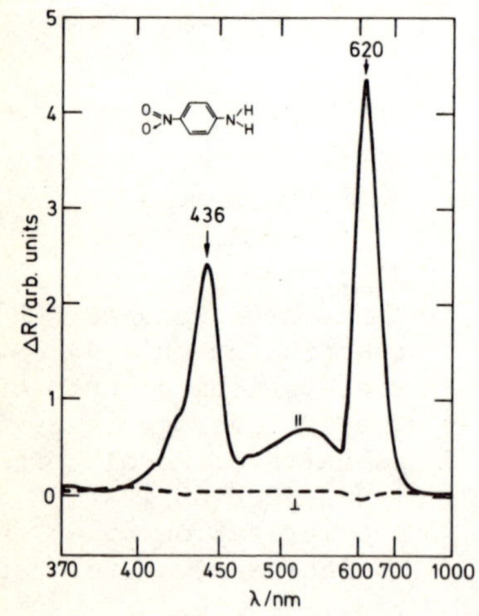

Fig. 22
Modulated reflectance (ER) spectra for p-amino-nitrobenzene on polycrystalline Pt. 0.5 M Na_2SO_4+5·10^{-5} M dye. φ_1 = 65°. p- and s-polarized light. ΔU_{pp} = 100 mV. After ref. (61).

8. THE ROLE OF NON-LOCAL OPTICS IN REFLECTANCE SPECTROSCOPY OF METAL SURFACES

Standard optics is based on the assumption that only transverse electromagnetic plane waves can propagate inside a solid, where the net charge is set zero because of instantaneous screening. This is equivalent to div $\vec{E} = 0$ or $\vec{k} \cdot \vec{E} = 0$ (wave vector and wave amplitude are normal to each other). At high enough frequencies, however, the inertia of the conduction electrons in metals prevents such instantaneous screening and hence charge density waves (plasma waves) can be built up. For metals like gold or silver this plasma frequency lies in the visible or near-UV (e.g., 3.8 eV for Ag). In this frequency regime, the longitudinal plasma waves are eigenmodes of the metal ($\vec{k} \times \vec{E} = 0$), and they are optically excited with p-polarized light where the electric field component normal to the surface induces periodic charge-density fluctuations (62,63). The longitudinal waves propagate into the metal, in addition to the transverse waves, and therefore contribute to the reflectance and transmittance of the metal surface. They have to be taken into account in proper metal optics. E.g., it has been shown in a number of publications (59,64,65) that the ER spectra of Ag around 3.8 eV for p-polarized light could not be reproduced by standard optics, unless one were willing to assume rather peculiar optical properties for the surface layer. This discrepancy was removed by using the proper metal optics, i.e., by including plasma waves in the calculation (non-local optics) (62). In the following we shall demonstrate the influence of plasma waves on ER spectra for thin Ag overlayers on Au(111).

In Fig. 23, $\Delta R/R$ is shown for several wavelengths around the plasma frequency of Ag as a function of film thickness D. The Ag films was deposited onto Au(111) under diffusion-controlled conditions (66). We note pronounced oscillations in $\Delta R/R$ for p-polarized light, while for s-polarization a smooth curve is found (we ignore for this consideration the thickness region below 0.5 nm because the optical properties of the first monolayer are strongly coverage dependent and hence cause structures in $\Delta R/R$ for <u>both</u> polarizations). It can be shown (66) that the oscillations in $\Delta R/R$ for p-polarized light are caused by standing plasma waves, for which the Ag film acts as a resonator at certain thickness values. From the D-values of two succeeding maxima in $(-\Delta R/R)$, the plasma wavelength is readily obtained as some 4 nm (compared to about 320 nm for the transverse wave in this energy region). We mention in passing that reflectance spectroscopy including plasma waves (non-local optics) is much more sensitive to structural effects at the surface or interface (e.g., to electron density profiles) than local optics.

The oscillations in $\Delta R/R$ (Fig. 23) have been reproduced by non-local optics assuming (thickness-independent) bulk optical properties for the Ag-film (66). An evaluation of the curves in Fig. 23 by classical optics clearly would yield wrong results,

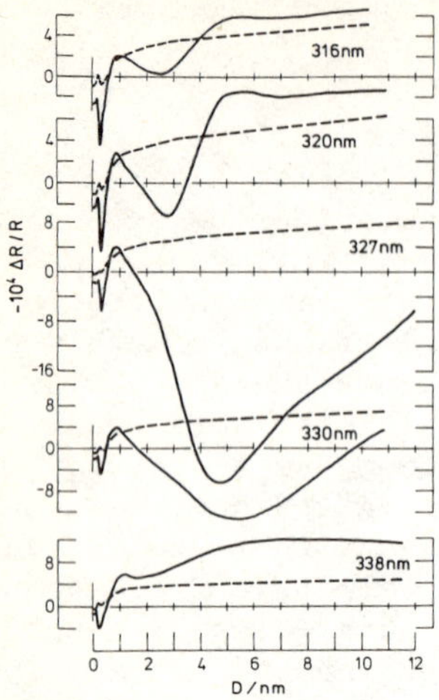

Fig. 23
ER effect for Au(111) during deposition of an Ag overlayer of thickness D. φ_1 = 45°. (——) p- and (---) s-polarization. ΔU_{pp} = 100 mV. After ref. (66).

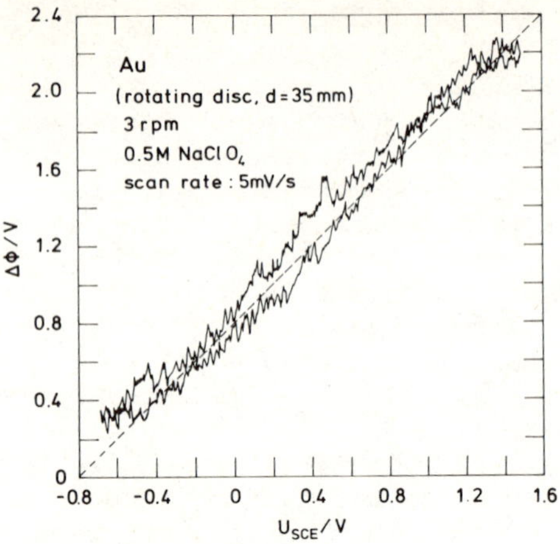

Fig. 24
Change in work function of an emersed gold electrode with emersion potential. $\Delta\phi$ was determined by the Kelvin method. After ref. (71).

namely oscillating optical properties of the Ag film, which is certainly not the case.

9. THE ELECTROCHEMICAL DOUBLE LAYER IN THE EMERSED STATE: AN OUTLOOK

The presence of the bulk electrolyte usually prevents the spectroscopic characterization of electrode surfaces over a wide photon energy range. Extension into the infrared as well as the far-ultraviolet regime should yield valuable new information. One way to overcome these limitations is, as mentioned already above, the use of thin-layer cells to reduced light absorption in the electrolyte. This has been proven to work for in-situ infrared studies (26). An alternative route is the emersion of the electrode from the electrochemical cell. Such a precedure would not only remove the wavelength restrictions for reflectance spectroscopy, it would also - after transfer of the electrode into an UHV chamber - allow the use of all the surface-sensitive methods such as LEED or ESCA which are commonly employed in UHV studies. The obvious drawback of the electrode emersion procedure seems to be the loss of the electrochemical environment and the loss of potentiostatic control. Despite these problems, electrochemists since long try to make use of UHV techniques in interfacial electrochemistry (see, e.g., ref. (67-69)).

The crucial point is whether or not the electrochemical double layer is retained during and after the emersion process. In a joint effort with W. N. Hansen we provided evidence that under certain conditions the electrode can be removed from the electrolyte into an inert atmosphere or even into vacuum with the double layer apparently intact and the bulk electrolyte completely absent (70-74). For example, we measured the work function of a gold electrode after emersion as a function of potential in order to proove that the potential drop across the double layer remains unchanged. The result is shown in Fig. 24 (71). Since work function and electrode potential are conceptually the same, namely the electrochemical potential of the electrons in the metal, we expect a strict linear correlation of unity slope for both quantities, if the double layer remains unchanged during and after emersion. The result of Fig. 24 where the work function of a continuously emersing electrode was determined by the Kelvin method, convincingly shows that indeed the double layer stays intact after emersion.

The chemical composition of the emersed double layer was tested by ESCA. The result for non-specific adsorbing ions is given in Fig. 25, where the surface concentration of cations and anions on a polycrystalline gold electrode is shown as a function of emersion potential (72,73). The ex-situ obtained data are in

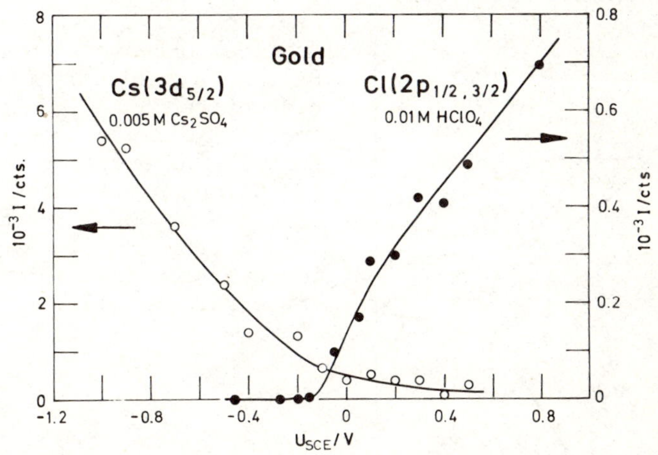

Fig. 25
Surface concentration of cations (Cs^+) and anions (ClO_4^-) on polycrystalline Au as derived from ESCA measurements on emersed gold electrodes as a function of emersion potential. The ions were selected for maximum ESCA cross section.

very good agreement with predictions from classical electrochemistry. When charging the electrode positively (negatively), the anion (cation) builds up on the electrolyte side of the double layer with almost no trace of the counter ion. The pzc of polycrystalline gold is expected to be around -0.05 V vs. SCE, where indeed the surface concentrations of cation and anion are about equal. We mention in passing that sensitivity problems arise in ESCA with ions of low cross sections, e.g. with sulfate. Since cesium perchlorate was not available during our studies, we performed the measurement with two different electrolytes, using Cs_2SO_4 for the cathodic region and perchloric acid for the anodic region.

In Fig. 26, we finally show some results for specifically adsorbing ions. ESCA has been used to determine the adsorption isotherm of Br^- on Au (Fig. 26b), which was found to be in excellent agreement with isotherms determined by in-situ reflectance spectroscopy (73). The effect of specifically adsorbed anions on the counter ion is also clearly seen in the ESCA data of Fig. 26a. When bromide adsorbs, the amount of cations in the double layer increases with respect to the case of non-specific (purely electrostatic) adsorption (dashed curve). The condition of electroneutrality at the electrochemical interface requires that in the case of specific (or superequivalent) adsorption the counter ion must

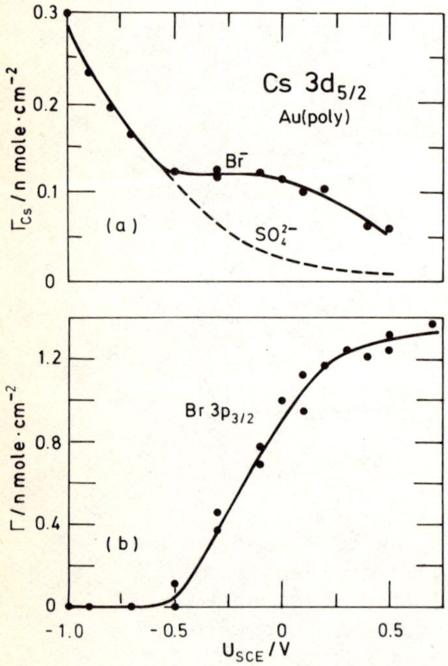

Fig. 26
Surface concentration of Cs^+ and Br^- on gold derived from the respective ESCA signals for emersed gold electrodes as a function of emersion potential. (a) Cs on Au for emersion from $5x10^{-3}$ M Cs_2SO_4 (---) and from 10^{-2} M CsBr (—●—). (b) Br on Au for emersion from 10^{-2} M CsBr. After ref. (72,73).

be present in the double layer to counterbalance the excess charge. Again, the ex-situ results agree perfectly with what has been predicted by electrochemical thermodynamics (12).

The above mentioned results of ex-situ studies on emersed electrodes strongly suggest that it is indeed possible to remove an electrode with its interfacial region intact from an electrolyte. It is obvious that a successful emersion of electrodes makes available a wealth of surface-sensitive techniques for electrode studies, which are not applicable in-situ.

10. CONCLUSION

In-situ reflectance spectroscopy in the visible and near-UV range can yield valuable pieces of information about electronic and structural aspects of electrode surfaces and the electrochemical interface in general, which complement the thermodynamic description obtained by classical electrochemical techniques. To combine these pieces into a complete picture we still must rely to a great extent upon chemical intuition. In comparison to the techniques available for surface studies in ultra-high vacuum, the optical methods are rather unrefined tools ($d/\lambda \ll 1$). Nevertheless, by using an even wider variety of in-situ and ex-situ techniques for the study of electrode surfaces, the hope seems justified that we will be able to derive a realistic microscopic model of the electrochemical interface.

REFERENCES

1. Bockris, J.O'M., Devanathan, M. A. and Müller, K. 1963, Proc. R. Soc. A274, 55.
2. Parsons, R., in: Comprehensive Treatise of Electrochemistry, Vol. 1, ed. by J.O'M. Bockris, B. E. Conway and E. Yeager (Plenum Press, New York, 1980) p. 1.
3. Trasatti, S., in ref. 2, p. 45.
4. Reeves, R., in ref. 2, p. 83.
5. Gerischer, H., Kolb, D. M. and Sass, J. K. 1978, Adv. Phys. 27, 437.
6. McDonald, J. R. and Liu, S. H. 1983, Surface Sci. 125, 653.
7. Valette, G. and Hamelin, A. 1973, J. Electroanal. Chem. 45, 301.
8. Valette, G., Hamelin, A. and Parsons, R. 1978, Z. Phys. Chem. N.F. 113, 71.
9. Klaua, M. and Madey, T. E. 1984, Surface Sci. 136, L42.
10. Madey, T. E. and Netzer, F. P. 1982, Surface Sci. 117, 549.
11. Bange, K., Döhl, R., Grider, D. E. and Sass, J. K. 1983, Vacuum 33, 757.

12. Mohilner, D. M., in: Electroanal. Chem., Vol. 1, ed. by A. J. Bard (Dekker, New York, 1966) p. 241.
13. See, e.g., Cardona, M., in: Optical Properties of Solids, ed. by S. Nudelman and S. S. Mitra (Plenum Press, New York, 1969) p. 137.
14. Wooten, F. 1972, Optical Properties of Solids (Academic Press, New York).
15. Hummel, R. E. 1971, Optische Eigenschaften von Metallen und Legierungen (Springer, Berlin-Heidelberg).
16. McIntyre, J. D. E., in: Adv. Electrochem. Electrochem. Engin., Vol. 9, ed. by R. H. Muller (Wiley, New York, 1973) p. 61.
17. McIntyre, J. D. E. and Aspnes, D. E. 1971, Surface Sci. 24, 417.
18. Kolb, D. M. 1977, J. Physique (Paris) 38, C5-167.
19. Sass, J. K. and Laucht, H. 1976, J. Electroanal. Chem. 67, 260.
20. McIntyre, J. D. E. and Kolb, D. M. 1970, Symp. Faraday Soc. 4, 99.
21. Bewick, A. and Tuxford, A. M. 1970, Symp. Faraday Soc. 4, 114.
22. Wagner, F. T. and Ross, P. N. 1983, J. Electroanal. Chem. 150, 141.
23. Kolb, D. M. and Kötz, R. 1977, Surface Sci. 64, 698.
24. McIntyre, J. D. E., in: Optical Properties of Solids - New Developments, ed. by B. O. Seraphin (North-Holland, Amsterdam, 1976) p. 555.
25. Kolb, D. M. and Gerischer, H. 1973, Electrochim. Acta 18, 987.
26. Russell, J. W., Severson, M., Scanlon, K., Overend, J. and Bewick, A. 1983, J. Phys. Chem. 87, 293.
27. Feinleib, J. 1966, Phys. Rev. Lett. 16, 1200.
28. Hansen, W. N. and Prostak, A. 1968, Phys. Rev. 174, 500.
29. McIntyre, J. D. E. 1973, Surface Sci. 37, 658.
30. Kolb, D. M. and Kötz, R. 1977, Surface Sci. 64, 96.
31. Kötz, R. and Kolb, D. M. 1978, Z. Phys. Chem. N.F. 112, 69.
32. Furtak, T. E. and Lynch, D. W. 1977, J. Electroanal. Chem. 79, 1.
33. Lang, N. D. and Kohn, W. 1973, Phys. Rev. B7, 3541.
34. Kötz, R. and Kolb, D. M., unpublished.
35. Tadjeddine, A., Kolb, D. M. and Kötz, R. 1980, Surface Sci. 101, 277.
36. Ho, K. M., Harmon, B. N. and Liu, S. H. 1980, Phys. Rev. Lett. 44, 1531.
37. Kolb, D. M., Boeck, W., Ho, K. M. and Liu, S. H. 1981, Phys. Rev. Lett. 47, 1921.
38. Boeck, W. and Kolb, D. M. 1982, Surface Sci. 118, 613.
39. Kofman, R., Garrigos, R. and Cheyssac, P. 1981, Thin Solid Films 82, 73.
40. Liu, S. H., Hinnen, C., Huong, C. N. V., de Tacconi, N. R. and Ho, K. M. 1984, J. Electroanal. Chem., in press.
41. Ho, K. M., Fu, C. L., Liu, S. H., Kolb, D. M. and Piazza, G. 1983, J. Electroanal. Chem. 150, 235.

42. Piazza, G. and Kolb, D. M., to be published.
43. Reihl, B., Schlittler, R. R. and Neff, H. 1984, Phys. Rev. Lett. 52, 1826.
44. Reihl, B., Frank, K. H. and Schlittler, R. R. 1984, Phys. Rev. B, in press.
45. Otto, A., in: Optical Properties of Solids - New Developments, ed. by B. O. Seraphin (North-Holland, Amsterdam, 1976) p. 677.
46. Raether, H., in: Physics of Thin Films, Vol. 9, ed. by G. Hass (Academic Press, New York, 1977) p. 145.
47. Otto, A. 1968, Z. Physik 216, 398.
48. Kretschmann, E. 1971, Z. Physik 241, 313.
49. Kolb, D. M., in: Surface Polaritons, ed. by V. M. Agranovich and D. L. Mills (North-Holland, Amsterdam, 1982) p. 299.
50. Tadjeddine, A. and Kolb, D. M., to be published.
51. Tadjeddine, A. and Kolb, D. M. 1980, Proc. 4th Int. Conf. on Solid Surfaces (Cannes) Vol. 1, p. 615.
52. Kötz, R., Kolb, D. M. and Sass, J. K. 1977, Surface Sci. 69, 359.
53. Tadjeddine, A. 1984, J. Electroanal. Chem. 169, 129.
54. Plieth, W., Bruckner, H. and Hensel, H. J. 1980, Surface Sci. 101, 261.
55. Plieth, W. J. 1979, Isr. J. Chem. 18, 105.
56. Kolb, D. M., in: Adv. Electrochem. Electrochem. Engin., Vol. 11, ed. by H. Gerischer and C. W. Tobias (Wiley, New York, 1978) p. 125.
57. Kolb, D. M., Kötz, R. and Yamamoto, K. 1979, Surface Sci. 87, 20.
58. Kolb, D. M., Kötz, R. and Rath, D. L. 1980, Surface Sci. 101, 490.
59. Kötz, R. and Kolb, D. M. 1980, Surface Sci. 97, 575.
60. Plieth, W. J., Gruschinske, P. and Hensel, H. J. 1978, Ber. Bunsenges. Phys. Chem. 82, 615.
61. Schmidt, P. and Plieth, W. 1983, J. Physique (Paris) 44, C10-175.
62. Forstmann, F. and Gerhardts, R. R., in: Festkörperprobleme/ Adv. Solid State Phys., Vol. 22 (Vieweg, Braunschweig, 1982) p. 291.
63. Feibelman, P. J. 1982, Progr. Surface Sci. 12, 287.
64. Kötz, R., Kolb, D. M. and Forstmann, F. 1980, Surface Sci. 91, 489.
65. Forstmann, F., Kempa, K. and Kolb, D. M. 1983, J. Electroanal. Chem. 150, 241.
66. Piazza, G., Kolb, D. M., Kempa, K. and Forstmann, F. 1984, Solid State Commun., in press.
67. Hubbard, A. T. 1980, J. Vac. Sci. Technol. 17, 49.
68. Homa, A. S., Yeager, E. and Cahan, B. D. 1983, J. Electroanal. Chem. 150, 181.
69. Ross, P. N. 1981, Surface Sci. 102, 463.
70. Hansen, W. N. and Kolb, D. M. 1979, J. Electroanal. Chem. 100, 493.

71. Rath, D. L. and Kolb, D. M. 1981, Surface Sci. 109, 641.
72. Hansen, W. N., Kolb, D. M., Rath, D. L. and Wille, R. 1980, J. Electroanal. Chem. 110, 369.
73. Kolb, D. M., Rath, D. L., Wille, R. and Hansen, W. N. 1983, Ber. Bunsenges. Phys. Chem. 87, 1108.
74. Nakai, Y., Zei, M. S., Kolb, D. M. and Lehmpfuhl, G. 1984, Ber. Bunsenges. Phys. Chem. 88, 340.

MOLECULAR STRUCTURE AND ORIENTATION IN THE ELECTRODE/ELECTROLYTE SOLUTION INTERFACE - IN SITU IR SPECTROSCOPY.

Alan Bewick

Department of Chemistry, The University,
Southampton SO9 5NH, England.

Techniques are described which obtain the IR absorption spectra of species, either adsorbed or free in the electrode/ electrolyte solution interphase. A comprehensive selection of applications are discussed to illustrate the capabilities of the methods. These cover both aqueous and non-aqueous systems and the use of single crystal surfaces as well as polycrystalline electrodes.

INTRODUCTION

Data obtained using spectroscopic methods have had a profound influence on the development of many areas of chemistry. In most cases the value of these methods stems from their high sensitivity and molecular specificity which can lead directly to a quantitative assessment and a structural characterisation of the species present in a chemical system and possibly also to information on their interactions and reactions. High vacuum surface science provides a particularly spectacular example of the value of spectroscopic methods and the rapid progress than can be engendered by their intelligent deployment; thus the use of electron diffraction techniques to determine the long range atomic or molecular order on solid surfaces and adlayers in conjunction with UPS, XPS, EELS and IR reflection spectroscopy to characterise molecular structure and orientation have produced rapid advances in the understanding of surface chemistry and surface physics at the fundamental level. There has also been a direct technological benefit in the use of this information to improve catalytic processes and to develop new catalysts. It is a relatively simple step to remove the gas in a solid surface/gas phase system

A. F. Silva (ed.), Trends in Interfacial Electrochemistry, 331–358.

without seriously violating the structural integrity of the interface which can then be characterised using the high vacuum techniques. It would be expected, however, that only in very few cases will it be possible to remove the electrolyte from the electrode/electrolyte solution interface without producing major changes; thus the high vacuum methods have only limited applicability as ex-situ probes for the structure of the interface in electrochemical systems. Fortunately, the ease with which the electrode potential can be used to control and change the electrochemical potential of species in the electrical double layer, the free energy of adsorption/desorption processes and the rate constants for electron transfer processes, allows the small number of species in the electrode/electrolyte solution interphase to be distinguished from the overwhelming number in the bulk electrolyte in a manner which has enabled in-situ spectroscopic methods of investigation to be developed. These methods now include ESR spectroscopy, [1] uv/visible reflection [2], transmission [3] and Raman scattering spectroscopies [4] and IR reflection spectroscopy [5,6,7,8,9]. This paper will be restricted to a discussion of the latter.

IR REFLECTION METHODS

During the last six years in-situ methods have been developed which employ IR radiation specularly reflected from the front surface of a flat electrode (external reflection) to obtain the vibrational spectra of species on the surface or in the solution near the surface. Internal IR reflection techniques have also been devised [10]; these are particularly useful for certain applications but they lack general applicability and place considerable restrictions on the nature of the electrode that can be employed. The external reflection methods require that the radiation passes through the electrolyte solution twice at some angle, θ, away from the surface normal. Most of the solvents commonly employed in electrochemistry absorb more or less strongly in the IR and the most important, water, has broad, strong absorption bands covering extensive regions of the spectrum. Thus it is necessary to employ high intensity spectrometers (high energy throughput) and to use a spectroelectrochemical cell which restricts the radiation to a short pathlength through the electrolyte solution.

CELL DESIGN

Figure 1 illustrates a typical cell design containing the three electrodes required for control of electrode potential and in which the thin layer of electrolyte, typically 1 μm to 50 μm, is produced by pushing the working electrode surface up to the surface of the IR window into the cell [5]. The window can be a disc with parallel working surfaces or it can be prismatic; the latter design minimises reflection losses at the air/window interfaces but it requires a longer pathlength in the window. The thin cell thus formed necessarily has a large value of uncompensated resistance due to the electrolyte layer. Typically, this leads to a time constant of the order of 10^{-2}s for the charging of the double layer in response to a potential step. This factor together with the potential drop across the uncompensated resistance due to faradaic currents needs to be taken into account when designing experiments.

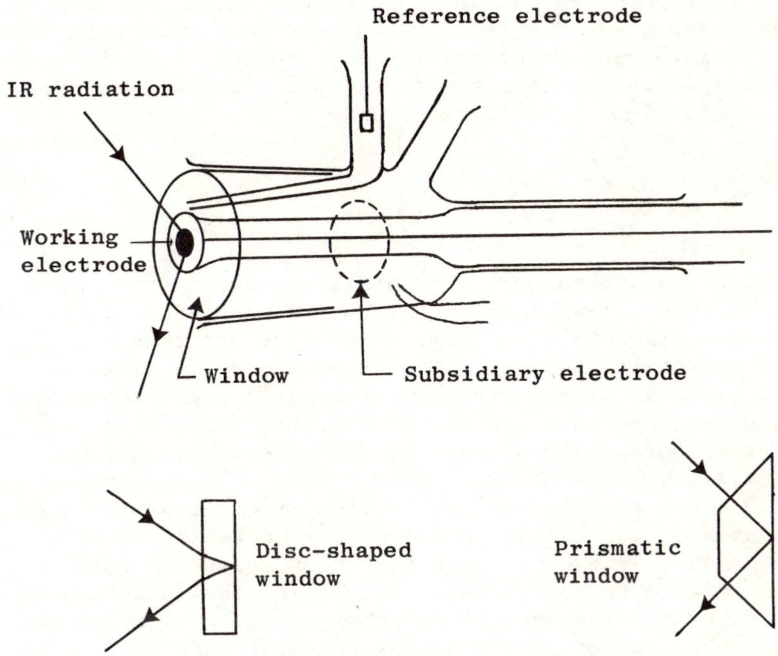

Figure 1 The IR spectroelectrochemical cell.

THE SPECTROMETER

Either a grating monochromator [5] or a Fourier transform spectrometer [6] can be used to obtain the spectra and in each case the paramount characteristic determining the performance is the energy throughput of the instrument. High sensitivity is achieved by alternating the electrode potential between two values, E_1 and E_2, such that the electrochemistry is modulated to produce a change in the IR energy absorbed. This is measured as a relative change in the reflectivity of the electrode $\Delta R/R$ which is the same as the absorbance change so long as $\Delta R/R \ll 1$. When a grating monochromator is employed, the modulation frequency can be made sufficiently high (in the range 1 Hz to 100 Hz) that a phase sensitive detection system can be used to extract the fluctuating component, ΔR, from the output of the IR detector, thus optimising the signal to noise ratio. The technique is then called Electrochemically Modulated Infra Red Spectroscopy (EMIRS). Modulation frequencies in this range cannot be used with an FTIR spectrometer because of the mirror velocities employed in the interferometer and the consequent amplitude modulation frequencies of the components of the radiation. Instead a very low frequency is employed, the electrode potential being held at each potential for 10s to 100s during which time several interferograms are recorded and coadded. In this case, ΔR cannot be obtained directly using a phase sensitive detector; instead the spectra at the two potentials are ratioed to give R_2/R_1 from which $\Delta R/R$ is calculated using:

$$\frac{\Delta R}{R} = \frac{R_2 - R_1}{R_1} = \left(\frac{R_2}{R_1} - 1\right)$$

The technique has, therefore, been distinguished by the name Subtractively Normalised Interfacial Fourier Transform Infra Red Spectroscopy (SNIFTIRS). EMIRS and SNIFTIRS are thus, in essence, very similar techniques but they differ in their use of experiment time, in signal recovery techniques and in the effective frequency range in which the IR detector is operated. SNIFTIRS makes more efficient use of the experiment time by measuring all wavelengths simultaneously and it operates the detector in a better frequency range (the frequencies at which the components of the radiation are modulated by the interferometer). However, these are more than compensated in EMIRS by the immense signal to noise discrimination of the phase sensitive detector and by the improved IR detector performance which results from the measurement of only a narrow spectral range at a time as selected by the exit slit. Thus with existing equipment EMIRS provides a significantly greater sensitivity. The sensitivity attained depends upon a number of factors but, in favourable circumstances, absorbance changes as low as 10^{-6} can be measured. A monolayer of adsorbate on a smooth surface will typically give spectra with absorption bands in the range 10^{-5} to 10^{-2} absorbance and thus the technique readily yields submonolayer sensitivity on low surface area electrodes (even single crystal surfaces). The wavelength range currently accessible extends all the way from the near IR out to about 80 cm^{-1}. Details of the techniques are given in references 5, 6 and 9.

REFLECTION/ABSORPTION AT A METAL SURFACE

For a vibrating molecule to absorb radiation from an incident

IR beam at the frequency of a particular normal mode it must be situated at a position of finite intensity and with an orientation such that there is a finite component of the dipole derivative du_i/dQ_i in the direction of the electric vector of the radiation field, where du_i is the change of dipole for the change of normal mode coordinate dQ_i. At a metal surface there is always a zero intensity for radiation having the electric vector parallel to the surface. As a result s-polarised radiation cannot interact with a molecule on a metal surface; p-polarised radiation has a finite intensity and can interact but the intensity of the resultant absorption band will depend upon the orientation of the molecule with respect to the surface, i.e. there is an apparent selection rule, the surface selection rule [11], which operates in addition to the normal selection rules for IR absorption. In EMIRS and SNIFTIRS measurements the "inactive" s-polarised radiation is prevented from reaching the detector and the relative intensities of the vibrational bands observed in the spectra from the remaining p-polarised radiation are used to deduce the orientation of adsorbed molecules. It should be pointed out, however, that vibrational coupling to adsorbate/adsorbent charge transfer [12] and also an electrochemically activated Stark Effect [7,13,14] can lead to apparent violations of the surface selection rule which can invalidate simple deductions of orientation.

The surface active/surface inactive difference between p-polarised/s-polarised radiation has enabled an alternative modulation technique, polarisation modulation, to be developed [15,16]. In electrochemical applications, it allows surface specificity to be achieved whilst working at fixed potential and without electrochemical modulation of the interface. It can be implemented either on EMIRS or on SNIFTIRS spectrometers and can be very valuable in dealing with electrochemically irreversible systems; however, the achievable sensitivity falls well short of that obtained with electrochemical modulation. It should be noted that its "surface specificity" is not truly surface but extends out into the electrolyte with decreasing specificity to about half a wavelength.

INFORMATION ACCESSIBLE

The EMIRS and SNIFTIRS methods provide the IR vibrational spectra (really the difference spectra - see later) of all species whose population changes either on the electrode surface or in the electrical double layer or in the diffusion layer in response to changing the electrode potential. Spectra will also be obtained for adsorbed species whose population does not change but which undergo a change in orientation or for which the electrode potential alters the intensity, the position or the shape of IR absorption bands. Shifts in band maxima with potential at constant coverage

$(\partial\bar{\nu}_{max}/\partial E)_\theta$ are very common for adsorbed species and they provide valuable information on the nature of adsorbate/absorbent bonding and hence also additional data on adsorbate orientation.

In principle, therefore, these valuable techniques can provide all of the information needed to specify the molecular structure of the electrode/electrolyte solution interphase, the dynamics of adsorption/desorption processes and the nature of reaction intermediates. In many cases, it will not be easy unambiguously to deduce quantitative molecular populations; this aspect has been neglected to date but will certainly command increasing attention in future.

It will be clear that EMIRS and SNIFTIRS spectra are difference spectra and can be somewhat complex [5]. Typically they will contain positive absorption bands from species present in excess at potential E_1 compared to potential E_2 and negative absorption bands from species whose population changes oppositely with potential. In addition, bands which shift with potential will appear as a single bipolar band either with one lobe of each sign (or even more complex structures with three lobes) figure 2.

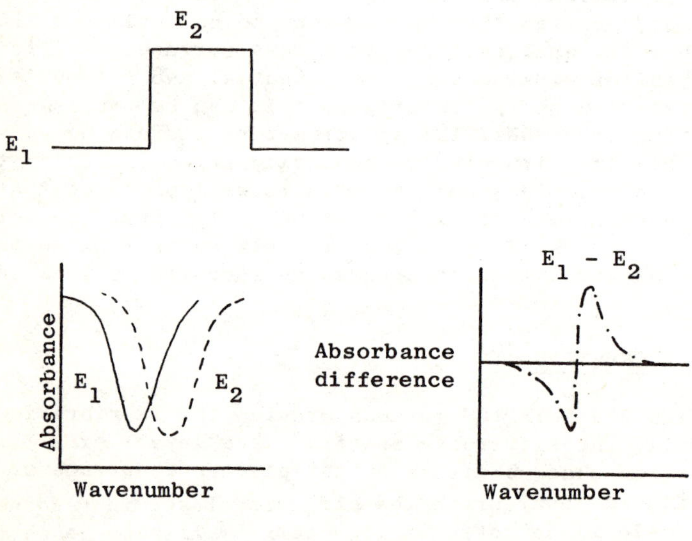

Figure 2 Origin of a bipolar difference band.

APPLICATIONS OF EMIRS AND SNIFTIRS

The Structure of the Double Layer

An exciting objective for the spectroscopic investigation of an electrochemical system would be to identify clearly the different species in the double layer, to measure their changing populations as the electrode potential is varied and to determine which of them are bonded to the surface. Such information would engender a quantum leap in our knowledge of the structure of the double layer which is constrained currently by the limitations of conventional methods of investigation.

An early investigation using SNIFTIRS demonstrated that this objective is not unrealistic [17]. The spectra shown in figures 3 and 4 were obtained for a platinum electrode in contact with a solution of tetrabutylammonium tetrafluoroborate in dry acetonitrile (figure 3) and with the addition of 0.1M water (figure 4). In each case, a base potential of -0.5V ($Ag/0.01M\ Ag^+$) was chosen and spectra were measured showing the difference between this potential and a set of other defined potentials as shown on the diagrams. The large band at 1060 cm^{-1}, exhibiting an increasingly large negative amplitude as the potential is made pro-

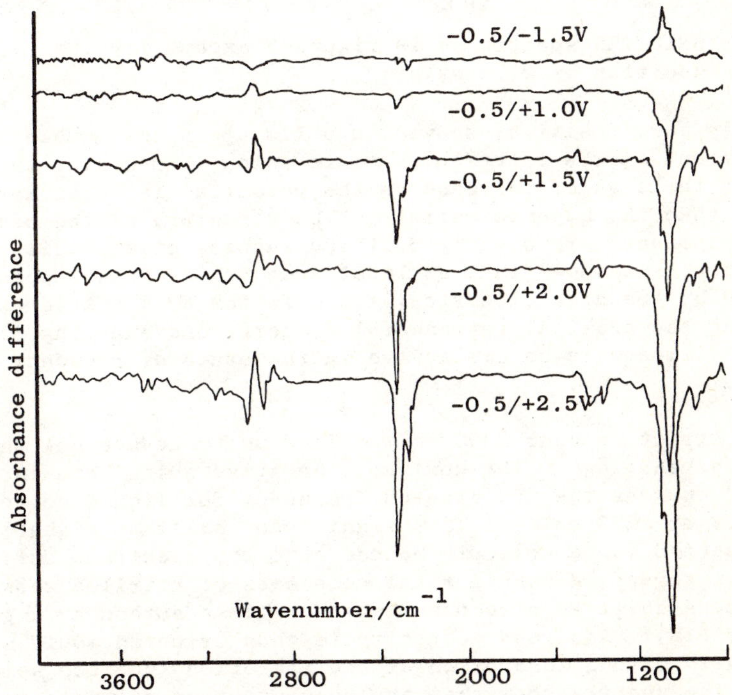

Figure 3 SNIFTIRS difference spectra between the potentials shown for a Pt electrode in 0.1M nBu_4NBF_4/CH_3CN.

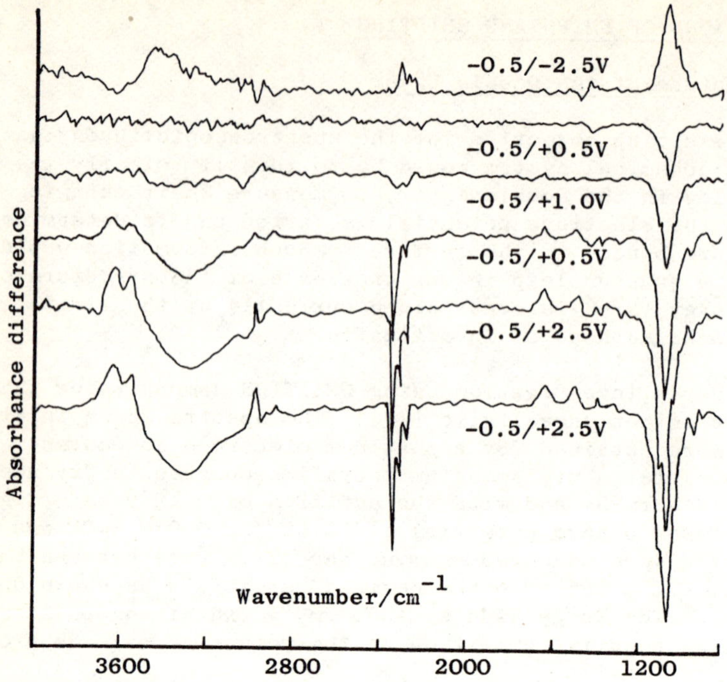

Figure 4 SNIFTIRS spectra as in figure 3 except for the addition of 0.1M water.

gressively more positive, shows the build up in the double layer of the amount of the tetrafluoroborate anion; note also the change in the sign of the band as the potential is taken more negative than the base potential. The structure of the band (three components are clearly distinguishable) gives additional information; it shows that at least some of the excess anion is distorted by the high electrical field in the double layer thus perturbing the original tetrahedral symmetry and enabling other vibrational modes to become active in the state of reduced symmetry.

The structure near 2300 cm^{-1} allows us to deduce how the solvent is behaving. The dominant, negative going band is at 2350 cm^{-1} whereas the C≡N stretch frequency for liquid acetonitrile is at 2220 cm^{-1}. This significant shift to higher energy is as expected for a molecule bonded onto the electrode surface via the nitrogen; N-bonded metal complexes of nitriles show a similar blue shift as also does benzonitrile adsorbed at a gold electrode [18]. Adsorbed acetonitrile thus oriented would show particularly strong absorption of the radiation (surface selection rule). The spectra show that the adsorption of solvent increases strongly as the potential is made increasingly positive. There is also a band at 2220 cm^{-1} from non-adsorbed acetonitrile; it is seen as a positive going band (although still negative of the baseline) superimposed on the low energy side of the larger 2350 cm^{-1} band. This shows that the total amount of free acetonitrile in the double layer decreases with increasing positive potential; presumably it is excluded by the adsorbed solvent and the increasing population of anion.

The addition of water to the system produces some interesting additions to the spectrum, figure 4, particularly in the OH stretch region near 3400 cm^{-1}. The broad negative going band at 3350 cm^{-1} must arise from an extensive hydrogen bonded water structure; this is believed to be associated with the anion and it can be seen that the potential dependence of the band mirrors that of the 1060 cm^{-1} band. There are, however, two additional bands with positive sign at 3625 cm^{-1} and 3550 cm^{-1}, together with another positive band at 1625 cm^{-1} in the water bending region. These values fit closely to the known vibrational frequencies for a symmetric hydrogen bonded complex between one water molecule and two molecules of acetonitrile [19]. The population of this species seems at first to increase and then to decrease as the potential is taken progressively more positive.

A second example of simple double layer changes is of the specific adsorption and potential dependent orientation of bisulphate ion at a gold electrode in aqueous sulphuric acid; this was studied with the EMIRS technique [9,20]. A set of spectra illustrating the effect of increasing positive potential on the adsorption of the anion is shown in figure 5. At the lowest

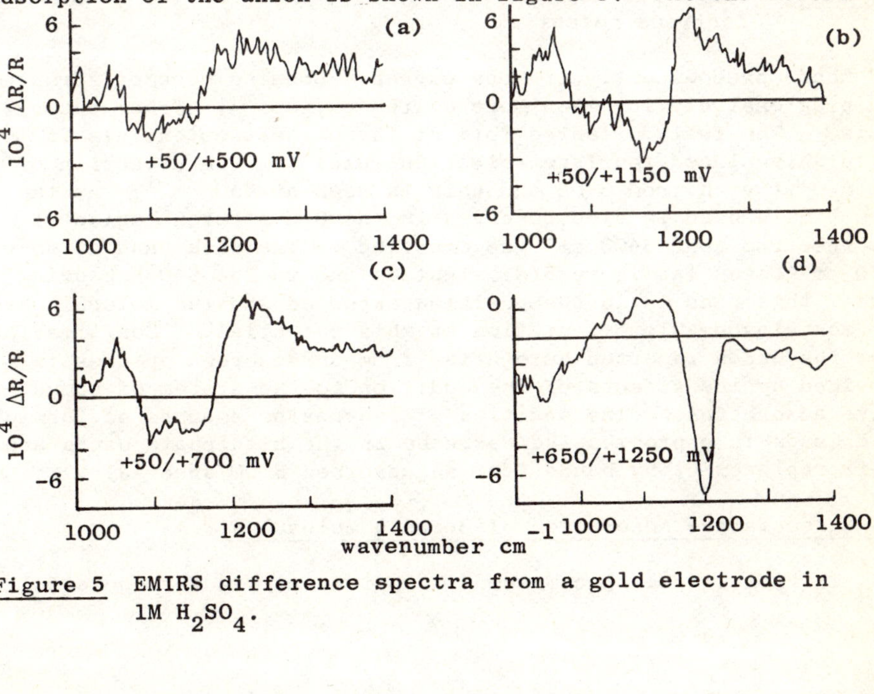

Figure 5 EMIRS difference spectra from a gold electrode in 1M H_2SO_4.

potential, figure 5(a), a small amplitude, negative going band centred on 1100 cm^{-1} is observed. This arises from the stretch modes of the equivalent and non-equivalent S-O bonds in a comparatively weakly adsorbed bisulphate ion present at higher coverage at the more positive potential. Progressive increases in the more positive potential produce an increase in the amplitude of the band, a splitting into two recognisable components and a shift to higher wavenumber. The higher wavenumber component becomes increasingly sharp and dominant until, at very high values of the electrode potential, figure 5(d) only a single narrow band (∿ 40 cm^{-1} at half height) remains together with a smaller band at 950 cm^{-1}. These spectra fit well to an evolutionary scheme for the adsorption as outlined in figure 6. At lower values of the potential, the ion is less strongly adsorbed

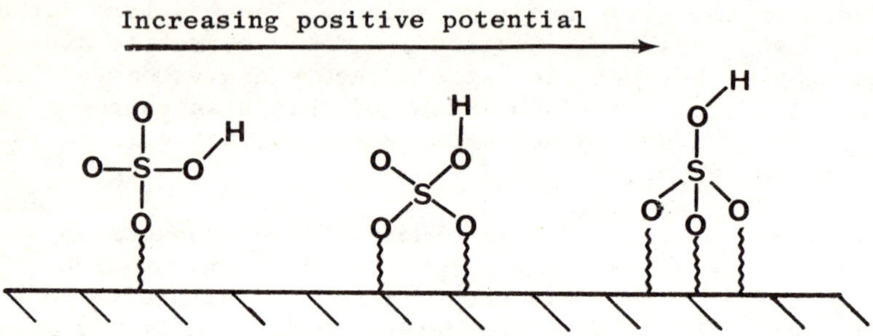

Figure 6 Models for the adsorption of HSO_4^- as a function of electrode potential.

and bonds predominantly via one oxygen. Bonding becomes stronger and progressively involves more of the oxygens at higher potentials. The fully oriented form at the highest potentials is favourably placed (surface selection rule) for the detection of the S-O stretch from S-OH and this is seen at 950 cm^{-1}; at the same time a band is also observed in the O-H stretch region of the spectrum near 3400 cm^{-1} as expected. The weak band at about 1070 cm^{-1} seen in figure 5(d) might be due to the S-O-H bending mode; this band is in the position expected and the molecule has the most favourable orientation at this potential. Confirmation that the bands reported here arise from an adsorbed species is provided by the effects of the addition to the system of competitive adsorbates. The addition of increasing amounts of formic acid leads to a progressive decrease in the bisulphate bands and their replacement by bands from an adsorbed HCOO species.

Non-dissociative Adsorption of Organic Molecules

IR vibrational spectroscopy is potentially a very powerful

tool for investigating the way in which organic molecules interact with a metal surface. Studies of this kind will be particularly interesting for molecules capable of bonding to the metal in alternative orientations by utilising different molecular orbitals. The relative energy levels of these orbitals can be readily varied in a series of related compounds containing different substituents and the electronic energy levels can be varied in the electrode using the electrode potential for fine control and by changing the metal for more major shifts. Investigations of this sort have been started using EMIRS and SNIFTIRS measurements on the adsorption of acrylonitrile, benzonitrile and substituted benzonitriles and on fluorobenzenes.

EMIRS spectra for the adsorption of benzonitrile on a gold electrode from an aqueous sulphuric acid electrolyte [9,18] contain strong bands in the C≡N stretch region indicating increasing coverage of the adsorbate, the negative band at 2250 cm^{-1} in figure 7,

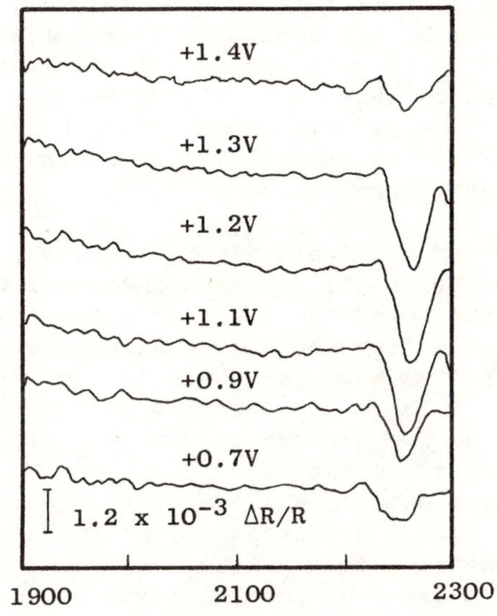

Figure 7. EMIRS difference spectra between OV and the potentials shown for Au/1M H_2SO_4 + 0.02M benzonitrile.

as the potential becomes more positive; at even higher potentials both the electrode surface and the organic become oxidised. The band is at a higher wavenumber than the C≡N stretch of the free molecule, the shift corresponding to that expected for the formation of an N-bonded complex with the metal. This would have an orientation allowing the maximum observable intensity for the C≡N vibration. This negative band partially overlaps a much smaller positive band at the wavenumber corresponding to the free molecule, 2229 cm^{-1}, which can just be discerned on the spectra illustrated. The positive band is the decreased absorbance resulting from the depletion near the electrode of free molecules as they become adsorbed at the positive potential. The disparity in the amplitude of the positive and negative bands results partly from the change from random to fully favourable orientation on adsorption and partly from an increase in the absorption coefficient for the adsorbed species. The orientation of the adsorbed molecule is confirmed by observation of the ring modes. Only the two modes with a symmetry (at 1598 cm^{-1} and 1492 cm^{-1} for the free molecule and shifted to 1595 cm^{-1} and 1480 cm^{-1} in the adsorbed state) are observed. These a_1 modes have the dipole derivative in the same direction as the C≡N stretch whereas the b_1 modes, which are not seen, have the dipole derivative parallel to the metal surface for N-bonded adsorption and are forbidden by the surface selection rule. Thus it appears that benzonitrile adsorbed on gold in a Lewis base/Lewis acid interaction. It is interesting to note that recent experiments show that p-fluorobenzonitrile lies flat on the electrode at low potentials but at higher potentials, where the electrode acts as a stronger Lewis acid, the molecule reorients to the N-bonded configuration [21].

The adsorption of acrylonitrile in identical conditions is remarkably different [7,9,14]. Measurements of double layer capacity indicate that over the range of potential used for the spectroscopic investigation there is a high coverage of the adsorbate. EMIRS spectra show two bands, one at 2125 cm^{-1} and the other at 1525 cm^{-1}. These are each 100 cm^{-1} towards lower energy than respectively the C≡N stretch and the C=C stretch of the free molecule. This indicates that the molecule is lying flat on the electrode surface and bonding via the π-electronic system. The complete absence of bands at the positions for the free molecule show that the potential modulation is not inducing adsorption/desorption, with depletion of acrylonitrile in the solution during the adsorption cycle; this also agrees with the differential capacity data which indicate strong adsorption over the whole range of potential. A major question arises: why are the two bands observed from the adsorbed molecule when it is oriented such that the C≡N and C=C stretch should be forbidden by the surface selection rule? Furthermore, why are these bands single sided? Normally an adsorbed species whose coverage is independent of potential should give rise only to bipolar bands caused by the wavenumber shift $(\partial\bar{\nu}/\partial E)_\theta$. All of these unusual features are explicable in terms of the electrochemically activated Stark effect (TEASE) [7,13,14] in which there is coupling between the high electrical field at the electrode surface and the vibrational motion of the molecule.

Depending upon the magnitude of the electrode potential, a field, up to 10^8 V cm^{-1} in magnitude, will exist in the inner region of the double layer. This will induce in an adsorbed molecule a dipole in a direction perpendicular to the electrode surface and of a magnitude determined by the field strength and the polarisability of the adsorbate. The polarisability of the molecule is dependent upon the internuclear separations and, since these vary during the vibrations of the molecule, the induced dipole will contain fluctuating components at the frequencies of the normal modes. Thus we have a mechanism which can, in principle, enable all the vibrational modes of an adsorbed molecule on an electrode surface to be observed in an IR reflectance measurement regardless of adsorbate orientation. However, in very many cases the intensities of the bands resulting from this mechanism will be very low; see reference [13] for some sample calculations. It is interesting to note that these intensities will be determined by the elements of the polarisability matrix and thus will be governed by the Raman selection rules and not the normal rules for IR spectroscopy.

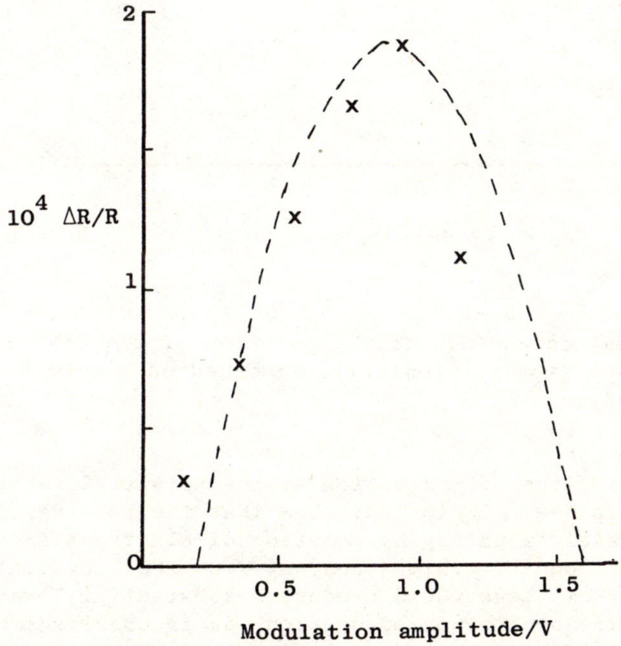

Figure 8 Potential dependence of the amplitude of the C≡N stretch band from acrylonitrile adsorbed on a gold electrode. The dashed curve is a calculated square low field dependence.

Bands produced in this way should have intensities proportional to the square of the electrical field strength; a feature which could be used as a diagnostic factor. Figure 8 is a plot of the amplitude of the C≡N stretch band as a function of potential for adsorbed acrylontrile. The dotted line illustrates the expected dependence calculated crudely by assuming the field to be proportional to the electrode potential but with zero field located at +0.7V (SCE).

Another interesting feature of the spectra is the shift of the band maximum with potential, figure 9. There is a shift to

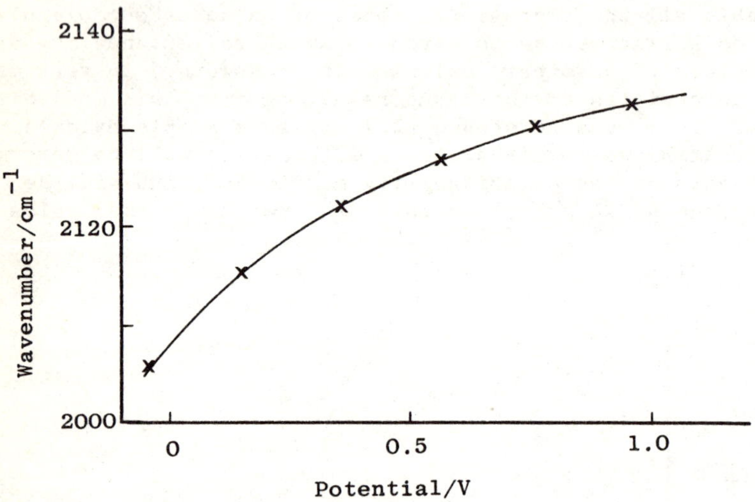

Figure 9 Potential dependence of the position of the C≡N stretch frequency from acrylonitrile adsorbed on a gold electrode.

higher frequency for the C≡N vibration as the potential is taken to more positive values. This indicates that the bonding of the molecule to the metal is mainly by donation of electrons from the gold to the π^* antibonding orbital and not from the π orbital to the metal; either of these would produce a substantial lowering of the vibration frequencies on adsorption, as is observed, but only the former would be expected to lead to the observed direction of the potential dependence.

The investigation, using the SNIFTIRS method, of the adsorption of difluorobenzene on a platinum electrode [22] provides a third interesting example of the interaction of organic adsorbates with metal electrodes. Using modulation potentials to produce adsorption/desorption cycles, spectra were obtained containing ten bands from the major IR active vibrational modes of the molecule in the range 4500 cm^{-1} to 450 cm^{-1} figure 10. The sign of these

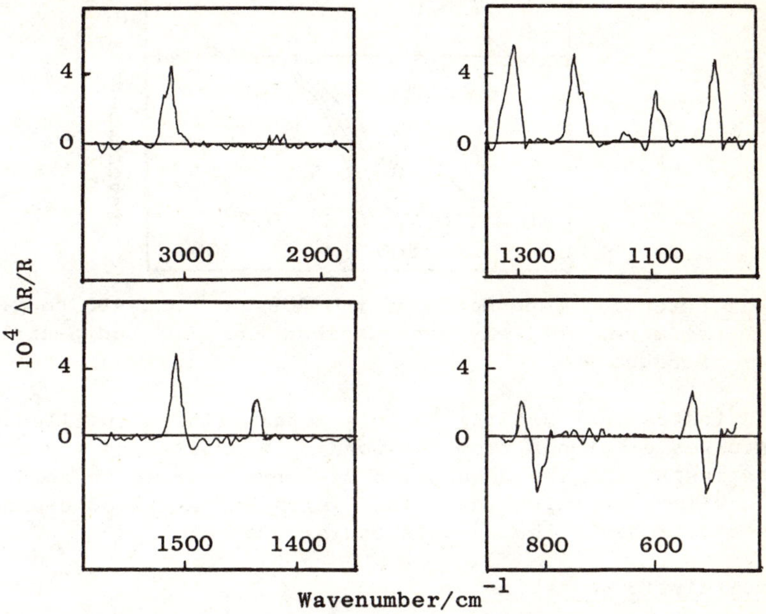

Figure 10 SNIFTIRS spectra from a Pt electrode in 1M $HClO_4$ + 0.5 mM p-difluorobenzene. Modulation from +0.2V to +0.4V.

bands, positive in the figure, show that they originate from difluorobenzene in solution and present in greater concentration at +0.2V than at +0.4V, adsorption taking place at the higher potential and the solution becoming partially depleted. Two of the bands, those at 833 and 509 cm^{-1}, also show negative lobes at lower wavenumber. These two are the only bands from vibrational modes which are allowed for an adsorbed molecule lying flat with the aromatic ring parallel to the electrode, the direction of the dipole derivative then being perpendicular to the plane of the ring. The shift to lower wavenumber on adsorption is as expected for a reduction in effective bond order in the ring. It is clear from the spectroscopic data that this system is a simple case of adsorption/desorption in which the aromatic molecule lies flat on the platinum surface. The potential dependence of the bands, figure 11, confirms the interpretation. The positive bands arising from the molecule in solution are potential independent, as expected, whereas the negative going bands attributed to the adsorbed state are seen to shift to lower wavenumber with increasing positive potential of the platinum. The direction of this

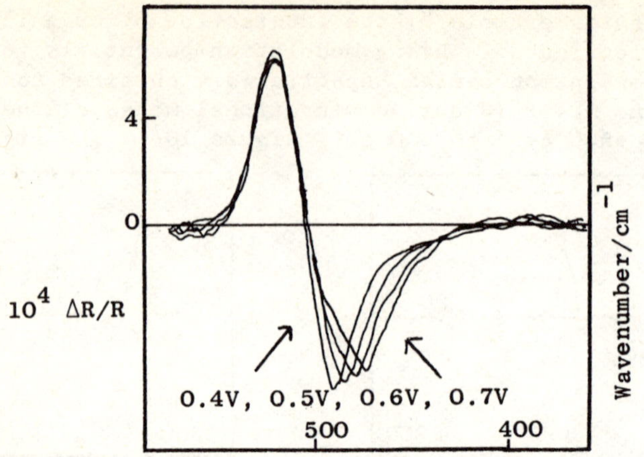

Figure 11 Potential dependence of the 509 cm^{-1} and 495 cm^{-1} bands from respectively non-adsorbed and adsorbed p-difluorobenzene.

shift indicates that bonding to this metal with an unfilled d-band involves electrons from the bonding π orbital of the aromatic. This contrasts with the adsorption of acrylonitrile on gold, discussed earlier, in which the metal, which has a filled d-band, donates electrons to the π* antibonding orbital.

Electrocatalytic Reactions

Electrocatalytic reactions usually produce adsorbed molecular fragments as reaction intermediates or as poisons; thus they provide a particular challenge to the in-situ, spectroscopic methods. In particular, many of them show considerable irreversibility thereby restricting the straightforward application of a technique such as EMIRS relying on continuous modulation of the electrode potential. However, special techniques, such as those developed by Kunimatsu, can be of considerable help in these cases and the use of polarisation modulation adds welcome versatility. The SNIFTIRS method also possesses an advantage in these situations because of the flexibility it allows in the choice of the reference state of the electrode and the periods during the experiment when this reference state is characterised. These difficulties have meant that, to date, for irreversible systems such as fuel cell type oxidations, only poisoned electrodes, which are rapidly formed during continuous pulsing, have been characterised and the detection of the major reaction intermediates has been neglected. This situation should rapidly change as the techniques are increasingly exploited by a growing group of electrochemists. In reversible electrocatalytic reactions, such as hydrogen evolution on platinum or rhodium, the reaction intermediates can be easily characterised [23].

A major early success of the EMIRS method was the identification of the strongly adsorbed poison produced in the oxidation of organic fuels, such as methanol, formic acid, formaldehyde, ethylene glycol etc, on Pt group metallic anodes [7,24,25,26]. The poison was shown to be a $(CO)_{ads}$ species. Before this, the accepted wisdom, accumulated from decades of conventional measurements, was that it was the very different species: $(COH)_{ads}$. Some of the evidence for $(CO)_{ads}$ is given in figures 12, 13 and 14. All of the bands on these spectra are from $(CO)_{ads}$ species, the precise nature of which depend upon the metal used but not upon the organic fuel. The higher wavenumber bond seen on platinum near 2070 cm^{-1} is from CO adsorbed directly on top of a single metal atom (linearly adsorbed); the lower frequency band, near 1850 cm^{-1}, is from CO sitting in between the metal atoms in a site allowing interaction with about three of them. On rhodium both linearly bonded and bridge bonded CO can be observed and, on palladium, three-coordinated CO is seen in addition to these two, the relative amounts of these depending on the total coverage. The same species are formed by direct adsorption of CO from solution instead of oxidation of a fuel [27]. The pattern of CO adsorption on these electrodes very closely parallels that observed on the same metals in gas phase studies and characterised by high vacuum spectroscopic and diffraction methods [28]. On a fully poisoned platinum electrode the hydrogen evolution reaction is sufficiently inhibited to allow polarisation to very negative potentials, e.g. -700 mV (NHE). Even in these extreme conditions the adsorbed CO shows no sign of being reduced to COH. It is, however, oxidised off the surface as CO_2 at potentials above +650 mV where sustained oxidation of most fuels can also be observed. It is also important to note that $(CO)_{ads}$ is the species formed when rigorously deoxygenated conditions are employed; there is no evidence for the initial formation of $(COH)_{ads}$ followed by oxidation to $(CO)_{ads}$ by oxygen.

The mechanism of formation of the poison may vary depending upon the nature of the fuel. One important route may be via the reduction of CO_2. A study of CO_2 reduction on platinum [29] clearly showed the rapid formation of the CO poison, figure 15. Note however that the absorption bands from $(CO)_{ads}$ are only seen when the reduction potential is in the region of formation of adsorbed hydrogen; this suggests that CO_2 reduction occurs via H_{ads}. It is interesting also that both linearly adsorbed CO and the three coordinated form are produced immediately upon reduction of the CO_2. It is difficult to produce the latter form directly from CO in solution; usually the electrode potential has to be cycled to produce an observable quantity. This suggests that the mechanism of formation may involve oxidation of CO to CO_2 followed by reduction via adsorbed hydrogen. When assessing the possible involvement of CO_2 in the poisoning process, it should be remembered that many fuels electrosorb to produce some H_{ads} and also

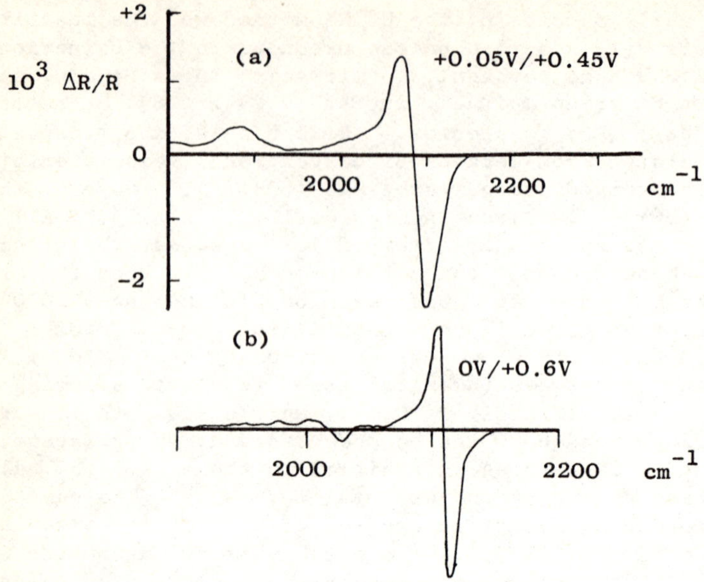

Figure 12 EMIRS difference spectra from (a) Pt in 1M H_2SO_4 + 0.5M CH_3OH and (b) Pt in 0.25M H_2SO_4 + 0.25M HCOOH.

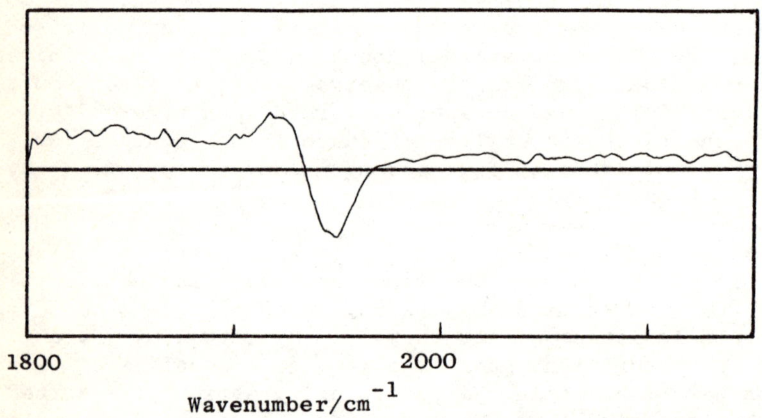

Figure 13 EMIRS difference spectrum from Pd in 1M H_2SO_4 + 0.1M HCHO. Modulation between +0.25V and +0.65V.

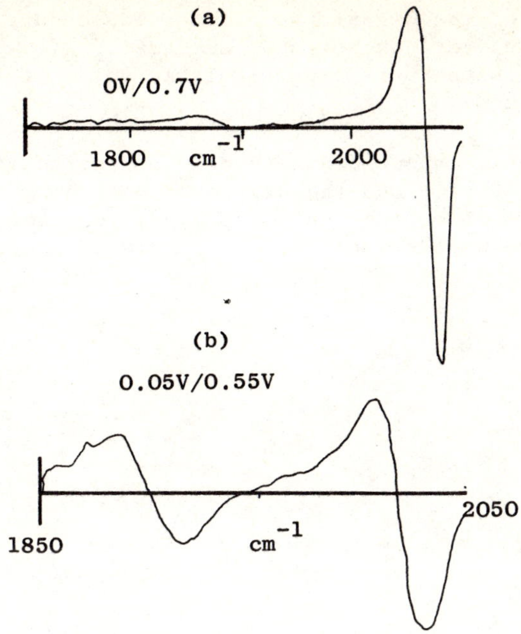

Figure 14 EMIRS difference spectra from (a) Pt in 0.5M $HClO_4$ + 0.1M HCHO and (b) Rh in 0.5M $HClO_4$ + 0.1M HCHO.

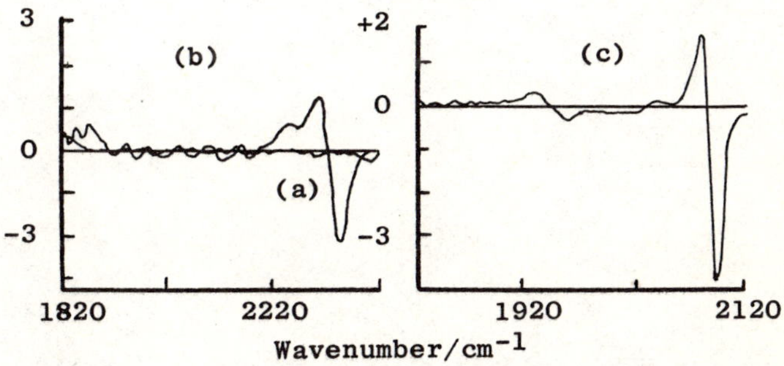

Figure 15 EMIRS difference spectra for reduction of CO_2 at Pt/1M H_2SO_4. Modulation between (a) 0.25V and 0.55V (b) 0.15V and 0.55V (c) 0.05V and 0.45V.

that oxidation of the organic to produce CO_2 occurs readily at low potentials on active, non-poisoned surfaces. It appears also that the three-coordinate CO is a particularly potent form of the poison on platinum.

Spectroscopic measurements show a good correlation between the catalytic activity and the degree of poisoning of single crystal platinum electrodes and of polycrystalline electrodes with partial coverage by adsorbed lead. Figure 16 shows the linear sweep

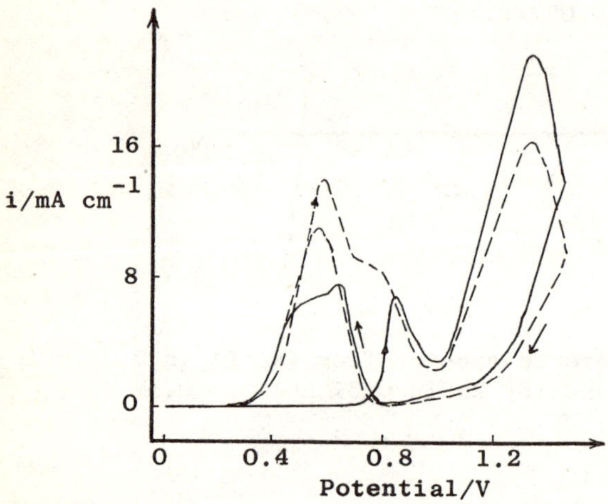

Figure 16 Linear sweep voltammogram for HCHO oxidation on Pt in 1M $HClO_4$ + 0.265M HCHO in the absence (———) and the presence (-----) of lead.

voltammetry for HCHO oxidation on platinum in the absence of (full line) and presence of lead (broken line) [30]. In the first case there is powerful poisoning of the surface on the positive going sweep and OCHO oxidation is fully inhibited until the onset of platinum oxidation but, in the presence of lead, a substantial oxidation wave is seen at low potentials. The spectra corresponding to these cases is shown in figure 17 for two ranges of potential;

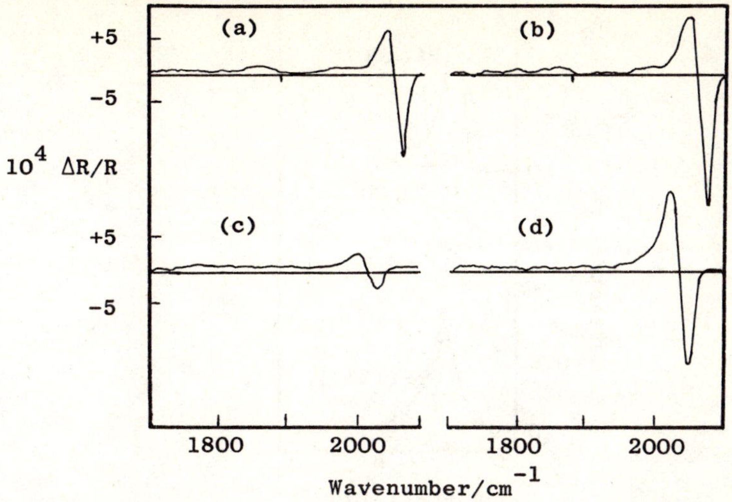

Figure 17 EMIRS difference spectra from Pt in 0.5M $HClO_4$ + 0.1M HCHO in the absence (a) and (b) and the presence (c) and (d) of deposited lead. Modulation between 0V and 0.25V, (a) and (c), and between 0V and 0.4V, (b) and (d).

note that lead adsorption substantially reduces the amount of each form of the poison but that the more highly coordinated form is particularly sensitive [26,31].

It is well known that the crystallographic structure of the catalyst surface has a major effect on the electrocatalytic activity. For example, the oxidation of formic acid on a platinum (111) electrode is relatively slow but the forward and reverse sweeps in linear sweep voltammetry are almost identical, indicating very little poisoning whereas the (100) surface shows heavy poisoning but a higher activity for the non-poisoned surface; some data from Adzic et al [32] is given in figure 18; note the large peak on the reverse scan for the (100) electrode, the peaks for the (111) case being only one sixth as large. The EMIRS spectra for CO adsorption, figure 19, correlate well with the difference in poisoning deduced from the voltammetry [33]. The (100) surface develops a large amount of each kind of adsorbed CO, somewhat more than a polycrystalline electrode. The (111) surface shows none of the more highly coordinated $(CO)_{ads}$ (at 1850 cm^{-1}) and the band at 2050 cm^{-1} is only one seventh of that for the (100) case. The broad absorption between 1700 and 1800 cm^{-1} is in the carbonyl region and is thought to arise from an adsorbed reaction intermediate, possibly $(CHO)_{ads}$.

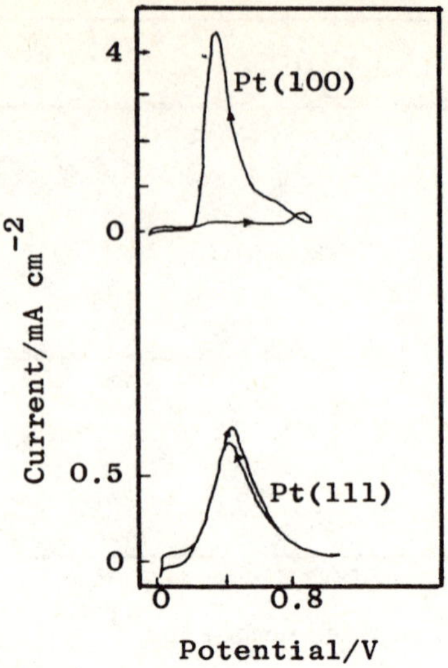

Figure 18 Linear sweep voltammograms for the oxidation of HCOOH on single crystal Pt electrodes.

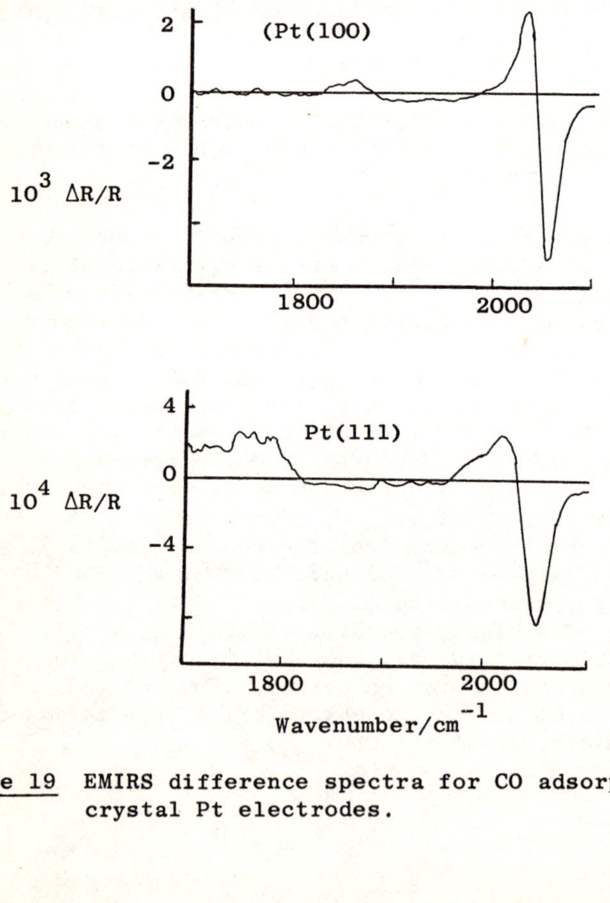

Figure 19 EMIRS difference spectra for CO adsorption on single crystal Pt electrodes.

In almost all of the EMIRS spectra, the $(CO)_{ads}$ species produce bipolar bands. These are formed by the subtraction, in the detection process, of two normal, single-sided absorption bands; these are the C≡O stretch band from $(CO)_{ads}$ at one potential and the same band shifted a little in frequency at the second potential, see figure 2. This can readily be seen using data from polarisation measurements (IRRAS) which allow the simple band at each potential to be obtained, figure 20. This figure also

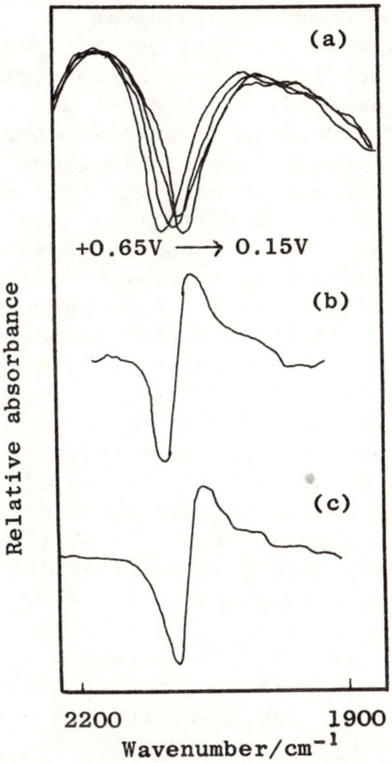

Figure 20

Spectra from Pt/1M $HClO_4$ saturated with CO. (a) IRRAS spectra at 0.65V, 0.45V, 0.25V and 0.15V; (b) difference spectrum from subtracting the +0.05V from the 0.45V spectrum in (a); (c) the EMIRS difference spectrum for modulation between 0.05V and 0.45V.

illustrates the subtraction of two of the IRRAS bands to yield a bipolar band almost identical to the corresponding EMIRS band. The bipolar bands show directly the large shift of band position with potential at constant coverage, $(\partial\bar{\nu}/\partial E)$; for linearly adsorbed CO on platinum gives a value of +35 cm^{-1} V^{-1} for this coefficient. The positive sign indicates that the frequency shift arises from a decrease in the backbonding of electrons from the metal to the π^* antibonding orbital of the CO adsorbate as the electrode potential is made more positive i.e. there is a decrease in the effective electronegativity of the metal. The ability in electrochemistry readily to control and fine tune the chemical nature of the metal and then observe the spectral consequences of the changes in bonding so produced gives a powerful advantage over high vacuum spectroscopic methods. Other examples were given in the preceding discussion on the adsorption of organic molecules.

It has also been exploited in examining the nature of adsorbate/adsorbate interactions [7,35].

DETECTION OF REACTIVE INTERMEDIATES

The EMIRS and SNIFTIRS methods are well suited to the detection and identification of intermediates produced directly by an electron transfer reaction at an electrode or by subsequent reaction of the primary product in a homogeneous process. In these applications, they closely resemble the well established reflectance methods employing radiation in the uv-visible region. For example, the reaction can be alternately switched on and off using the electrode potential thus producing a fluctuating population of the intermediates in the diffusion layer. In these applications the thin-layer cell arrangement used by the IR techniques can be usefully exploited to limit the diffusional process and the amount of reactant accessible. In some cases, it is useful to be able to alter the width of the thin layer to assist in the discrimination between absorption bands from adsorbed and non-adsorbed species. Although the IR methods cannot approach the limiting sensitivity of their uv-visible counterparts, they are sufficiently sensitive to detect short lived intermediates and to give the spectra of such species as radicals, radical ions, carbenium ions and carbanions produced in electroorganic reactions. Studies of irreversible processes will be complicated by the rapid depletion of reactant in the thin layer electrolyte due to its non regeneration at the base potential. It should be possible to devise special techniques to overcome this problem, e.g. continuous flow of fresh solution through the thin layer.

Some data are given in figure 21 for the EMIRS spectra obtained during the reduction/oxidation cycling of benzophenone in acetonitrile/0.1M nBu_4BF_4 electrolyte, the anion radical being alternately produced then oxidised back to the ketone [35,36]. As mentioned earlier, the bands extending downwards correspond to the absorbances from species at the reduction potential, while the positive bands are related to the starting material. In the case of benzophenone the main band to disappear (i.e. positive band) in the spectral region 1500 cm^{-1} to 1800 cm^{-1} is the absorption centred around 1660 cm^{-1} which corresponds to the disappearance of the carbonyl frequency (νC=O) [37,38]. This is accompanied by the appearance of a new band (negative band) centred around 1557 cm^{-1}. This band is most likely to be due to the ketyl formed on reduction ($\nu\dot{C}-\bar{O}$) and appearing at lower frequency due to the decrease in bond order. The aryl stretching bands (C-C stretches) have weaker absorbances than νC=O in the IR spectrum of benzophenone; hence the corresponding bands in the EMIRS spectrum are very weak and are rather difficult to distinguish from the noise (1600 cm^{-1} and 1580 cm^{-1}).

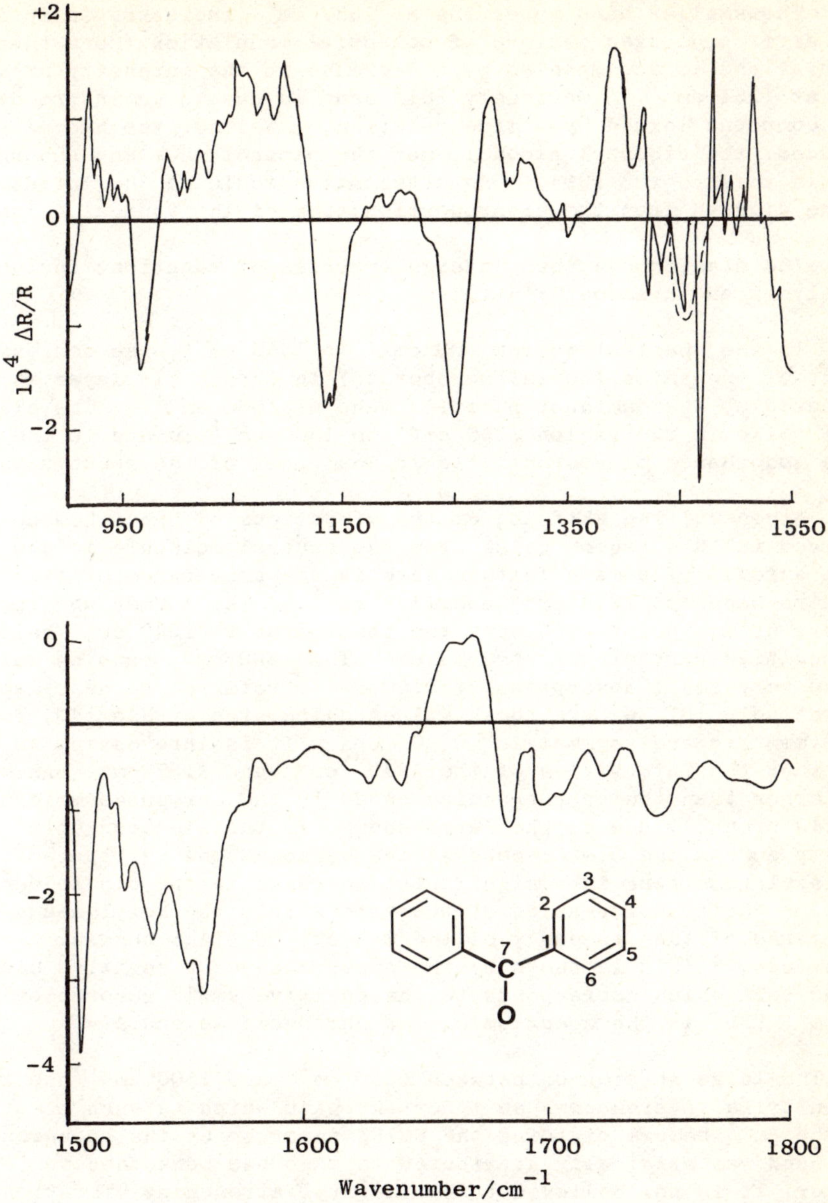

Figure 21 EMIRS spectra from Pt/0.1M nBu_4NBF_4/10 mM benzophenone. Potential modulation from -1.75V to -2.5V at 6.25 Hz.

The smaller band appearing at 1537 cm^{-1} increases in intensity after prolonged periods of potential modulation (more than 6 hours) and is accompanied by a decrease in the intensity of the band at 1557 cm^{-1}. Obviously this band is due to an intermediate or a compound formed by a side reaction. Neither the hydrolysis produced, the dibenzyl alcohol, nor the pinacol show any absorbance in this region [39]. An alternative would be the formation of the dianion from the disproportionation of the radical anion.

The dianion can then undergo a series of reactions including hydrolysis and cyanomethylation.

In the spectral region 900 cm^{-1} to 1550 cm^{-1}, the decreasing amount of the anion (tetrafluoroborate) in the double layer is indicated by the dominant positive band at 1060 cm^{-1}. The high level noise in the region 1350 cm^{-1} to 1500 cm^{-1} is due to the large absorbance of acetonitrile in that part of the spectrum.

A general red shifting of the vibrations of benzophenone is observed in this region going from the neutral molecule to the ketyl anion. The main feature here is the appearance of two negative bands at 1249 cm^{-1} and 1137 cm^{-1}. The former absorption is part of a bipolar band with the absorption at 1280 cm^{-1} being the positive part of the structure. The 1280 cm^{-1} band as well as the very small absorption at 1150 cm^{-1}, referred to as e_2 and e_1 respectively [37,38] are those values calculated by Rao [40] for the symmetric and asymmetric $\nu C_{ar}C_{al}C_{ar}$. It is interesting to note that the intensities of the 1249 cm^{-1} and 1137 cm^{-1} bands are larger than the corresponding bands in the unreduced molecule. This is probably due to the large change in the electron spin density across the C_1- C_7 bond at the benzophenone is reduced to the ketyl resulting in a significant increase in the dipole derivative. The transformation of the ketone into the ketyl leads to a decrease of the frequency of the C-H out-of-plane bending vibrations. This is shown by the appearance of a negative band at 965 cm^{-1} which corresponds to the positive small absorption at 990 cm^{-1} [38] in the spectrum of the unreduced molecule.

The large absorbance between 1350 cm^{-1} and 1500 cm^{-1} due to acetonitrile, overshadows an important band which is seen clearly at 1458 cm^{-1} before dividing the EMIRS spectrum by the background. This band was originally attributed to adsorbed benzophenone [35]. However, it is now believed to be the aryl stretching vibration, shifted from 1600 cm^{-1} by 142 cm^{-1}. Further evidence for this is provided by the EMIRS spectra of substituted benzophenones [36].

ACKNOWLEDGEMENTS

Much of the work discussed was supported in part by the Office of Naval Research, Washington, and by the Science and

Engineering Research Council of the U.K. Collaborative research was facilitated by a Nato Research Grant.

REFERENCES

1. I.B. Goldberg and A.J. Bard, in "Magnetic Resonance in Chemistry and Biology, J.N. Herak and K.J. Adamic (Eds), M. Dekker, New York, 1975.
2. J.D.E. McIntyre, in R.H. Muller (Ed.), Advances in Electrochemistry and Electrochemical Engineering, Vol. 9, Wiley, 1973, p.61.
3. W. Heineman, F. Hawkridge and H. Blount, in "Electroanalytical Chemistry", A.J. Bard (Ed.), M. Dekker. New York, 1984, Vol. 13, p.1.
4. M. Fleischmann and I.R. Hill, in R.K. Chang and T.E. Furtak (Eds.), Electrochemical Effects in Surface Enhanced Raman Scattering, Plenum Press, New York, 1982. R.P. Cooney, M.R. Mahoney and A.J. McQuillan, in R.J.H. Clark and R.E. Hester (Eds), Advances in Infrared and Raman Spectroscopy, Vol. 9, Heyden, London, 1982, p.188.
5. A. Bewick, K. Kunimatsu, B.S. Pons and J.W. Russel, J. Electroanal. Chem., 160 (1984) 47.
6. S. Pons, T. Davidson and A. Bewick, J. Electroanal. Chem., 160 (1984) 63.
7. A. Bewick, J. Electroanal. Chem., 150 (1983) 481.
8. S. Pons, J. Electroanal. Chem., 150 (1983) 495.
9. A. Bewick and S. Pons, in R.J.H. Clark and R.E. Hester (Eds), Advances in Infrared and Raman Spectroscopy, Vol. 12, Wiley Heyden, London, 1985, p.1.
10. H.B. Mark and B.S. Pons, Anal. Chem., 38 (1966) 119. H. Neugebauer, G. Nauer, A. Neckel, G. Tourillon, F. Garnier and P. Lang, J. Phys. Chem., 88 (1984) 652.
11. R.G. Greenler, J. Chem. Phys., 44 (1966) 310.
12. J.P. Devlin and K. Consani, J. Phys. Chem., 86 (1981) 2597.
13. S. Pons, C. Karzeniewski and A. Bewick, J. Phys. Chem., in press.
14. A. Bewick, C. Gibilaro and S. Pons, Langmuir, in press.
15. J.W. Russell, M. Severson, K. Scanlon, J. Overend and A. Bewick, J. Phys. Chem., 87 (1983) 293.
16. H. Seki, K. Kunimatsu and W.G. Golden, Appl. Spectroscopy, in press.
17. S. Pons, T. Davidson and A. Bewick, J. Electroanal. Chem., 140 (1982) 211.
18. A. Bewick, J. Hudec and M. Razaq, in preparation.
19. P. Saumagne and M.L. Josien, Bull. Soc. Chim. France (1958) 813.
20. A. Bewick, M. Razaq, J.W. Russell and T. Solomun, in preparation.
21. A. Bewick and C. Pender, to be published.
22. S. Pons and A. Bewick, Langmuir, 1 (1985) 141.

23. A. Bewick and J.W. Russell, J. Electroanal. Chem., 132 (1982) 329 and 142 (1982) 337.
24. B. Beden, C. Lamy, A. Bewick and K. Kunimatsu, J. Electroanal. Chem., 121 (1981) 343.
25. B. Beden, A. Bewick and C. Lamy, J. Electroanal. Chem., 148 (1983) 147.
26. A. Bewick, in "The Chemistry and Physics of Electrocatalysis", J.D.E. McIntyre, M.J. Weaver and E.B. Yeager (Eds), The Electrochem. Soc. Inc., Proc. Vol. 84-12, 1984, p.301.
27. B. Beden, A. Bewick, K. Kunimatsi and C. Lamy, J. Electroanal. Chem., 142 (1982) 345.
28. N, Sheppard and T.T. Nguyen, in R.J.H. Clark and R.E. Hester (Eds), Advances in Infrared and Raman Spectroscopy, Vol. 5, Heyden, London, 1978.
29. A. Bewick, B. Beden, M. Razaq and J. Weber, J. Electroanal. Chem., 139 (1982) 203.
30. M.D. Spasojevic, R.R. Adzic and A.R. Despic, J. Electroanal. Chem., 139 (1982) 203.
31. A. Bewick and T. Solomun, to be published.
32. R.R. Adzic, A.V. Tripkovic and W.E. O'Grady, Nature, 296 (1982) 137.
33. R.R. Adzic, A. Bewick and M. Razaq, to be published.
34. A. Bewick, M. Razaq and J.W. Russell, Surf. Sci. in press.
35. S. Pons, T. Davidson and A. Bewick, J. Electroanal. Chem., 140 (1982) 211.
36. M. Kalaji, Ph.D. Thesis, University of Southampton, 1985. A. Bewick and M. Kalaji, to be published.
37. A. Perrier-Datin and J. Lebas, Spectrochim. Acta, 25A (1969) 169.
38. G. Menard and A. Mellier, Spectrochim. Acta, 29A (1973) 1273.
39. The Aldrich Library of Infrared Spectra, 2nd edition, 1978.
40. K. Rao and C. Narayanaswamy, Ind. J. Pure Appl. Phys., 8 (1970) 100.

STRUCTURE,THERMODYNAMICS AND KINETICS OF WATER AT INTERFACES

N. G. Parsonage

Imperial College, London, England

ABSTRACT

Monte Carlo and molecular dynamics methods as applied to the study of water at interfaces are briefly reviewed and the results obtained so far are discussed. Structural quantities considered include the O-O distribution function, the distribution of dipole orientations, and the "layering" of the water structure brought about by the interface. The effect of the interface on hydrogen-bond formation is also examined. The time-dependent behaviour of these systems is considered in terms of the self-diffusion coefficients and dipolar relaxation times of the water molecules. Throughout, comparison is made between water near to the interface and bulk (or uniform) water.

1. INTRODUCTION

In most discussions in which electrochemists engage the water is treated as a continuum,with certain properties, and notably the dielectric constant (or relative permittivity), being assigned values near to a surface which are often different from those for for real bulk water. The advantage of this simplification is that it is then possible to concentrate attention on either the metal of the electrode or the ions in solution,or both. This is, of course, a compromise, for electrochemists have always been worried by the necessity of treating water in this way. In recent years, rather more attention has been given to the behaviour of the water molecules themselves. In particular,during the last three years there have been several attacks using simulation methods on the problem of water, treated as discrete molecules, rather than as a

A. F. Silva (ed.), Trends in Interfacial Electrochemistry, 359–371.

continuum, next to a surface. In these studies three different models for water and at least five different types of solid wall have been examined, and the simulation techniques which have been employed have also varied. Not surprisingly, the results do not agree in all respects, but there are a number of questions for which these studies have provided, I think, an agreed answer, and I shall spend most of this lecture on these points. I shall,of course, also endeavour to explain the reasons for the differences in results which do occur. First, however, let me describe the models and techniques which have been used.

2. MODELS FOR WATER

The three models used all combine a distribution of point charges to represent the electrostatic fields of the molecules and a repulsion and/or dispersion interaction of the Lennard-Jones or related type. All three models have been successful in explaining many of the bulk properties of water at normal temperatures and densities. The models which have been used are as follows.

a) The Rowlinson Model (1)

This is based primarily on virial coefficient and dipole moment data. The lone-pair negative charges of value 0.3278 e are placed directly above and below the oxygen atom at a distance from it of 0.2539 Å, the line between these charges being perpendicular to the HOH plane. The hydrogen atoms which carry the same charge as above are a distance of 0.96 A from the oxygen atom.(Figure 1a)

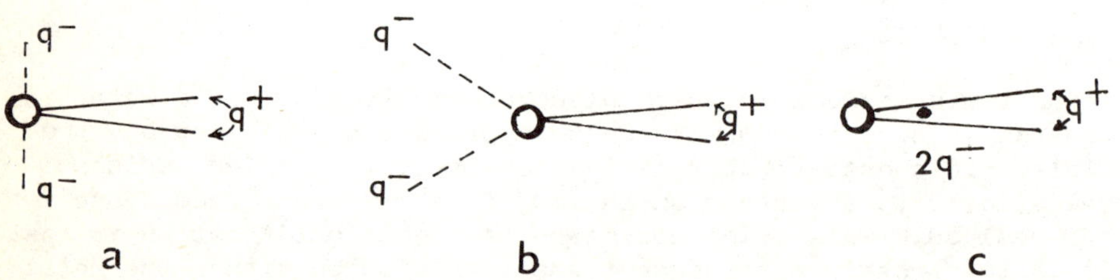

Figure 1. Models for water.

b) The ST2 Model (2)

This model also has two partial positive charges, of value 0.2357e, and two balancing negative charges, but in this case they are distributed at the normal tetrahedral angles around the oxygen atom. The distances from the positive and negative charges to the oxygen atom are 1.0 and 0.8 Å, respectively. As with the Rowlinson potential there is also a Lennard-Jones 12-6 interaction, which acts between the oxygen atoms. An attenuation factor which reduces the interaction at short distances is also incorporated.

c) The MCY Model (3)

This is based on quantum-mechanical results for the interaction of pairs of water molecules. The model involves charges of 0.7174e on each hydrogen atom and a negative charge of twice that amount situated on the C_{2v} axis within the HOH triangle. For a pair of water molecules there is, in addition to the coulombic terms, a repulsive contribution of the form $a_k \exp(-b_k r_{ij})$ between all possible pairs of atomic centres. This model has the computational advantage that, being a 3-centre model (as far as charges are concerned), only 9 coulombic interactions need to be calculated, as against 16 for the other two models discussed here.

It is perhaps surprising that this model, which is planar, has been so successful with respect to the "structure" (radial distribution functions, etc.) of water. It has even been shown to give quite a good account of the phase behaviour of the ices (4).

Table 1 presents a comparison of some of the thermodynamic properties of bulk water as predicted by these models and as found experimentally.

	Expt.	Rowlinson	MCY	ST2
$-U/(\text{kJ mol}^{-1})$	34	33-35	28-29	33-35
$C_v/(\text{J K}^{-1}\text{mol}^{-1})$	75	63-92	75-84	71
$PV/NkT(\rho=1\text{ g cm}^{-3})$	0.05	-4	2.2	0.09(10°C) -0.01(41°C)
μ/ Debye unit	1.86(gas)	1.84	2.19	2.35

Table 1. Experimental and predicted values of properties of bulk water(5).

Apart from the pressure, the models reproduce the experimental results very well. Figure 2 shows the computed and experimental radial distribution functions for water.

3. SIMULATION TECHNIQUES

Both Monte Carlo(MC) and Molecular Dynamics(MD) methods have been used. The sudies so far have, with one exception, considered simulation boxes containing 150 or 216 molecules with the system

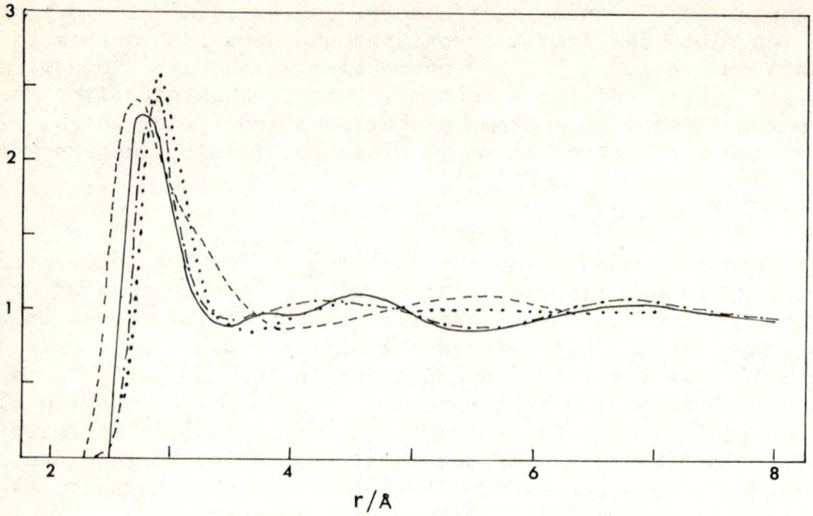

Figure 2. The radial distribution function(oxygen-oxygen) for bulk water. Full line,expt.;broken line,Rowlinson model;chain-line,MCY; dotted line,ST2 (6,7,8).

confined in the $\pm z$ directions by the solid walls which are under investigation. The distance between these walls has been in

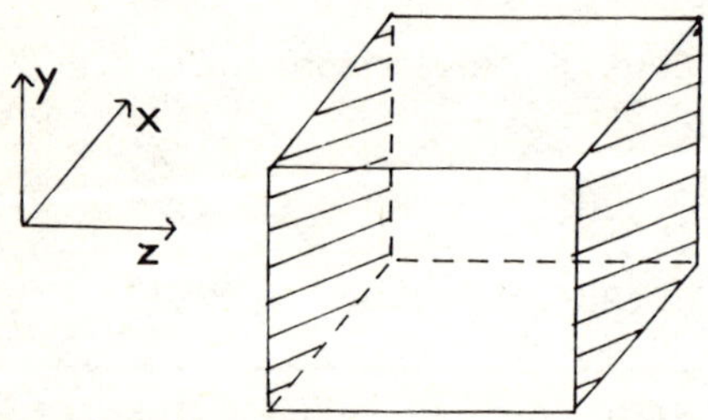

Figure 3. The simulation box

almost every cases ~20 Å, and in deducing the properties of a single solid-fluid interface from the results it is assumed that these walls are sufficiently apart for there to be a negligible amount of interference between the effects of the two walls. For simulations of water this assumption is only approximately true (as will be seen by some of the data shown later). In the x- and y-directions periodic boundary conditions are applied,that is,it is supposed that replicas of the simulation box exist adjacent to it in those directions and that molecules near the edge of the real box can interact with molecules in these replicas. In this way,the system is made pseudo-infinite in these directions. The fact that a spurious periodicity is imposed on the system is probably not a very serious problem for simulation of the pure liquid, since molecular correlations will usually die out within the x- and y-dimensions of the box.

For the Monte Carlo method, which generates a sample of configurations with the probability which is to be found in nature, it is the simplest type, that based on the canonical ensemble (constant T,V,N) which has been used in all the cases we shall discuss. A disadvantage of Monte Carlo in this ensemble is that, as a result of attraction to or retraction from the walls, it quite frequently happens that the density at the middle of the box is not that which had been intended, and adjustments to the dimensions of the box may have to be made to compensate for this (9). Useful as it is,the Monte Carlo method only gives information on static properties. To study time-dependent properties it is necessary to turn to molecular dynamics.

Molecular dynamics, in which by solution of the equations of motion the trajectories (translational and rotational) of all the particles are followed, gives a true representation of the time-development of the model system(10,11). Integration of the differential equations is usually done in time-steps of $\sim 10^{-15}$ s, so that a run of 10 000 time-steps(or passes) would correspond to $\sim$10 ps of laboratory time. Limitations on computing time usually restrict the length of a run to less than 100 000 time-steps or 100 ps, thereby preventing study of processes for which the characteristic time is longer than this. Some success has been had in devising ways of circumventing this problem, but these techniques have not been applied to the systems under consideration here.

Energy, rather than temperature, is a constant in molecular dynamics (at least in its normal form),so a first requirement of any run is to make it refer to a temperature which is of interest. Usually this is done by making periodic adjustments to the kinetic energy of the particles by multiplying them by a scaling factor, so as to bring the kinetic energy to that appropriate to the temperature desired. The procedure is clearly acceptable when applied during the equilibration period at the commencement of a run when the system is engaged in re-distributing its energy and particles as compared with the arbitrary starting configuration; it is much less clear that it is permissible for the remainder of the run, which is to be used for providing the results. Some workers have made very frequent (say every 200 time-steps) adjustments and it is probable that this means that their results do not refer to an equilibrium system, even to the extent that the values of static quantities will be in error (12).

Both the MC and MD techniques can yield values for the dielectric constant (ϵ_r). For a uniform system the equation usually employed is

$$\epsilon_r - 1 = \rho \langle M^2 \rangle / 3NkT\epsilon_o \qquad (1)$$

where ρ is the number density, N is the number of particles in the simulation box, and k, T and ϵ_o have their usual meanings. $\langle M^2 \rangle$ is the mean square polarisation of the simulation system or, less often, for a sphere inscribed in the simulation box. Since $\langle M \rangle$ is zero in zero field, what is required is evaluation of the fluctuation quantity $\langle M^2 \rangle - \langle M \rangle^2$. The convergence of fluctuation quantities in simulation is notoriously bad, and this is the first difficulty in evaluating dielectric constants.

There is another hazard in determining ϵ_r by simulation and for this we need to consider the subject of long-range corrections. Taking account of replicating procedures we have an infinite system, and so each molecule should interact with an infinite number of others. In order to render the computation of the energy of each configuration finite it is necessary to impose a cut-off distance beyond which a molecule does not directly interact with any other particles. This cut-off distance is often one-half of the smallest linear dimension of the box, but it may be such that all molecules and images within a box centred on the molecule in question interact with it. Either way, it is desirable to make a correction for the neglect of longer-range interactions. Simulations of uniform systems usually incorporate such corrections, common methods being the reaction field method (6,7), derived from the theory of Onsager, and, especially for ionic and other polar systems, the Ewald method (13). Inclusion of such corrections is especially important for properties such as ϵ_r which depend upon orientational distributions, since these can be strongly affected by the long-range correction. Equation (1) is, in fact, appropriate where the corrections are made by the Ewald method. Since there is no completely satisfactory way of applying long-range corrections values of ϵ_r suffer another large uncertainty from this cause. Computed values for ϵ_r, even for uniform systems, are thus much less certain than those for many other properties. This topic will be discussed again later in the context of non-uniform systems.

Table 2 summarises the simulations which have been carried out on water-solid interfaces. Because of difficulties in implementing long-range corrections for non-uniform systems these have usually been omitted; only the two studies by Christou et al., which employ an adaptation of the reaction field method of correction, appear to include such a correction.

4. STRUCTURAL RESULTS

Figure 4 shows the singlet distribution functions for oxygen atoms.

Walls	Dimensions/Å z	x,y	Water model	T/K	Moves/particle or time-steps (000s)	Type (nos. particles)	Ref.
Hard wall	20.6	18.6	Row	298	9.4	MC(216)	14
9-3	20.6	18.6	Row	298	11.6	MC(216)	15
Hard wall	5.33,10 20	14.97	MCY	300	10	MC(150)	16
Exp. rep.	20	14.972	ST2	302	54	MD(150)	17
12-6	25.36	18.62	ST2	287	20 ps	MD(216)	9
Hard wall	9.33	18.66	Row	298	9.2	MC(108)	18
9-3	23.62	18.62	ST2	290	36	MD(216)	19
Avge LJ	23.1	14.97	MCY	298	12,18	MD(150)	20

Table 2. Summary of simulations on water-solid interfaces. 9-3 and 12-6 refer to Lennard-Jones type walls; exp.rep. refers to a hard,exponential repulsion; avge.LJ refers to Boltzmann averaging over a particulate LJ wall.

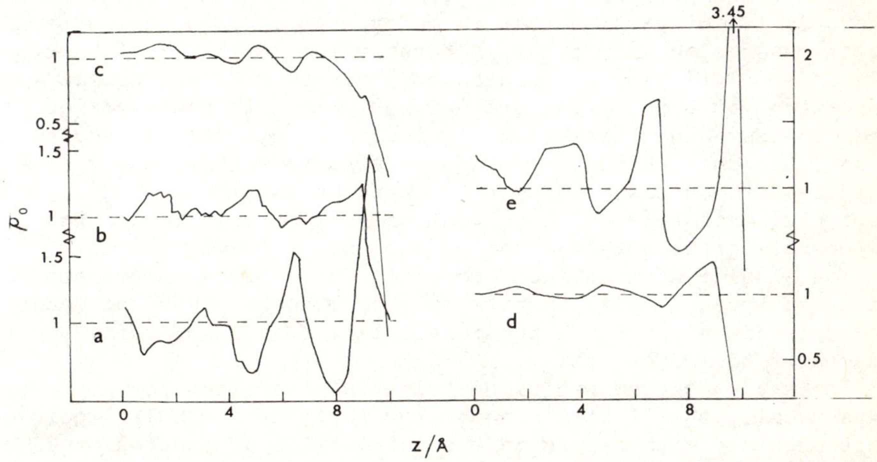

Figure 4. Singlet distribution functions for oxygen atoms. Centre of box is at 0. Function is in units such that bulk water at 1 g cm^{-3} is unity. a)Ref.16; b)Ref.17; c)Ref.9; d)Ref.19; e)Ref.20.

All the simulations show oscillations persisting across the box. Although in most cases these oscillations become quite small near the centre of the box they do not entirely die out. It is for this reason that workers feel that this is the shortest wall-wall distance at which it is at all possible to say that the two halves of the box represent single wall-liquid interfaces. It will be an early priority to increase this distance by,say,~50%.

Whereas some of the simulations show a high first peak followed by marked damping(16, 20), others show of fairly uniform height(14,15,9,19). In systems of simple fluids (noble gases rather than water) the first type of behaviour is common for hard walls provided that the fluid is at sufficient pressure; at lower pressures the fluid tends to retreat from the wall as in the simulations of Christou et al.(14,15) and Sonnenschein and Heinzinger(9).

A question which may be asked of each simulation is,Are the oxygen or the hydrogen atoms nearer to the walls?(We are thinking here only of those molecules which are next to the surface) Three of the four simulations which produced distribution functions without large first peaks show the oxygen atoms to be nearest to the surface(14,15,19). Jonsson's simulation found the hydrogen atoms to approach nearest to the wall(16). In reference (9) in which no large first peak was found there was a strong preference for the dipole to be parallel to the wall, as was also found in references (17,20,18) (see Figure 5). In references (17,20) there was a small preference for orientations with hydrogen closer to the wall than oxygen. All in all,the preference for molecules to be parallel to the wall seems to be a strong one, with preferences between hydrogen and oxygen for proximity to the wall being much weaker. It is interesting that in references (17,20) the central slab of water also showed the preference for parallel orientation. Christou et al.(14,15) who presented their results for dipole orientation on a more fine-grained basis than is done in Figure 5 found considerable alternation of dipole direction right across the box. All this evidence is in general agreement with the conclusion of Henderson et al.(21) that the thickness of the interfacial region in which solvent molecules are oriented is considerably greater than is often supposed.

The singlet distribution functions for hydrogen atoms are similar to those for oxygen but are very much smoother, as would be expected in view of the possible rotations of the hydrogen atoms about the oxygen atom.

The oxygen-oxygen radial distribution functions for surface layers usually show little change from that for the bulk liquid, except possibly with regard to the intensities of the features. Thus,Marchesi(17),who considered a two-dimensionally adapted function, found that the magnitudes of the first peak and trough were enhanced by the wall by ~35%, although the distances at which these features occurred was unchanged. Christou et al.(15) also found no change in the positions of maxima and minima, but in their case there was also no significant change in the intensities.

These curves would normally lead on to a consideration of the dielectric constant. However, in addition to the difficulties mentioned earlier, for non-uniform systems such as these the dielectric constant should be considered as a tensor with four independent elements: $\epsilon_{xx}=\epsilon_{yy}$, ϵ_{zz}, ϵ_{xy} and $\epsilon_{xz}=\epsilon_{yz}$, each of which would need to be evaluated. Furthermore, I am in agreement with Vericat et al.(23) on "the difficulties and ambiguities associated with attempts to define a local dielectric constant". The difficulties extend to the use of the quantities even if they can be determined. The answer, surely, is to avoid introducing a local dielectric constant and instead to determine and discuss quantities more directly concerned with measurable properties.

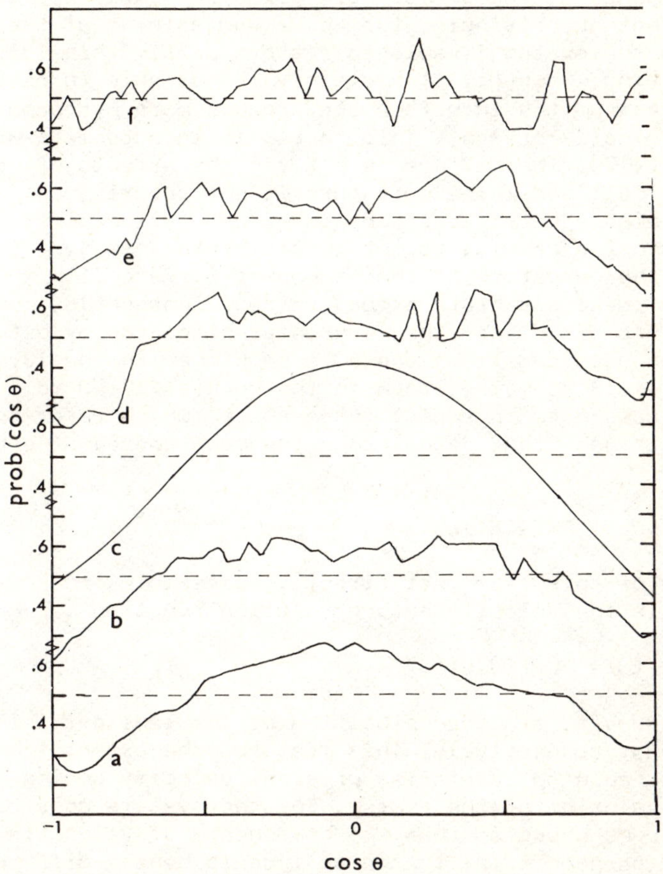

Figure 5. Angular distribution functions for water near the wall. θ is the angle between the z-axis and the direction of the water dipole. a) Ref. 16; b) Ref. 17; c) Ref. 9; d) Ref. 20, average LJ, no image-forces; e) Ref. 20, average LJ, with image-forces; f) Ref. 20, exp.

5. HYDROGEN-BONDING

It is often assumed that the structure-making effect of small non-polar molecules, originally proposed by Frank and co-workers(24) and subsequently supported by computer simulation(25), can be carried over to large non-polar molecules and even to planar surfaces. Christou et al.(15) found that the number of hydrogen-bonds per molecule in a layer which extended from 0.9 Å to 4.0 Å below the interface was only 81% of the bulk value. Most of this reduction is attributable to the purely geometric reason that molecules near the surface would often not have suitable hydrogen-bonding partners available on the side on which the interface occurs. Simple calculation shows that this would reduce the number of hydrogen-bonds to 78% of its bulk value for their system. Thus, it appears, not surprisingly, that by re-adjustment of the structure of the fluid the system is able to restore partly this loss of hydrogen-bonding. However, it certainly is not able to reverse the main effect of the interface and produce more hydrogen-bonding than was originally present. This result is in agreement with that of Lee et al.(19). The difference between the effects of non-polar surfaces of small and large curvature is not surprising if the process for highly curves surfaces is viewed as incipient clathrate formation, since a crucial factor in clathrate formation is matching of the curvature of the non-polar surface with that of a cavity in an ideal clathrate structure(26). Jonsson(16), on the other hand, has calculated the mean interaction energy between nearest neighbours and finds these to be higher in the layer next to the wall(-13.8 kJ mol^{-1}) than in the bulk(-12.7 kJ mol^{-1}). This result does not contradict those above, which refer to the <u>number</u> of hydrogen-bonds rather than the mean <u>energy</u> of a bond.

6. TIME-DEPENDENT PROPERTIES

The quantity of this type most directly accessible from MD simulations is the velocity autocorrelation function

$$z(t) = \langle v(0).v(t) \rangle \qquad (2)$$

where $v(0)$ and $v(t)$ are the velocities of the same particle at time zero and t,respectively. This measures the extent to which a particle is able to retain its original velocity during subsequent evolution of the system. For the systems considered here it would be expected that the components of velocity parallel and perpendicular to the wall would behave differently in this regard, so that it is usual to define modified auto-correlation functions $z_{\parallel}(t)$ and $z_{\perp}(t)$, respectively, referring to components in these directions. Figure 6 shows results from reference (9) for water near to the wall and in the bulk. As with other dense fluids the curve is not a simple monotonic decay. The oscillatory behaviour at <0.3 ps corresponds to the reversal of the velocity at each collision with its immediate neighbours: these represent the oscillations within its cell in the model of that name. The water near the wall shows a greater ability to retain its initial velocity, and this has been attributed to the less complete hydrogen-bonding between the particles. It would be expected, and

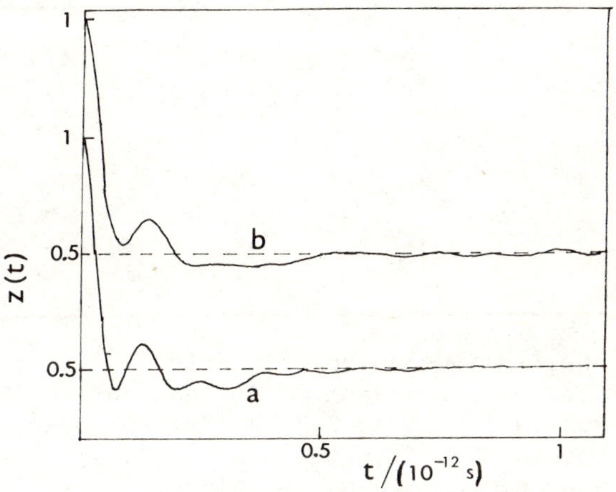

Figure 6. Velocity autocorrelation functions for a)bulk and b)wall water. The curve for wall water refers to $z_{\parallel}(t)$.

is indeed so, that z(t) is related to the self-diffusion coefficient. For a bulk, three-dimensional system the relationship is

$$D = \frac{1}{3} \lim_{t\to\infty} \int_0^t \langle v(0).v(t) \rangle \; dt \tag{3}$$

and for the component parallel to the wall it is

$$D = \frac{1}{2} \lim_{t\to\infty} \int_0^t \langle v_x(0).v_x(t) + v_y(0).v_y(t) \rangle \; dt \tag{4}$$

An alternative, and usually computationally better, way of obtaining diffusion coefficients is by means of the equations

$$(\Delta x(t))^2 + (\Delta y(t))^2 = 4D_{\parallel}t \tag{5}$$

and

$$(\Delta z(t))^2 = 2D_{\perp}t \tag{6}$$

In fact, when the left-hand sides of these last equations are plotted against t the first part of the graph is curved. This is ignored, and only the linear part is used to obtain the diffusion coefficient. Because of the limited length of the box in the z-direction equation (6) is not strictly applicable. However, since during the time represented by the simulation the particles do not travel a large fraction of the length of the box the objection to its use is not too strong. Table 3 shows the results which have been obtained so far. Thus it is not possible to decide

		Ref.9	Ref.17	Ref.20	Expt(27)
Bulk water		2.9	4.2	3.7	1.55(283K) 3.32(314K)
Wall water					
	$D_{\parallel}$	4.8±0.4	1.9±0.4	-	-
	$D_{\perp}$	4.7±0.3	4.1±0.7	-	-
	D	4.7	3.3	3.1,3.1,2.8	-
T/K		287	302	304	-

Table 3. Self-diffusion coefficients(10^{-5} cm^2 s^{-1}). Wall water in references 17 and 20 is that for which the oxygen atom is less than 3.75 A from the averaged LJ wall or less than 2.20 A from the hard wall. In reference 12 the corresponding distance is 5.18 A.

whether diffusion parallel to the wall is faster than diffusion perpendicular to the wall. In considering the differences in the data account should be taken of the fact that Marchesi's run was at a temperature 15 K higher than that of reference 12, which explains the high value which he found for the bulk fluid. Also, he used the suspect procedure of re-scaling the particle velocities at frequent intervals.

Another property of interest which can be calculated is the dipole autocorrelation function

$$\Gamma_1(t) = \langle \mu(0).\mu(t) \rangle \;/\mu(0)^2 \qquad (7)$$

The decay of this function with time may be fitted to an exponential function, thereby giving a dipolar relaxation time. References 17 and 20 indicated a marked slowing down(as compared with bulk water) resulting from the presence of hard,repulsive exponential walls(3.1 and 3.3 as against 2.1 and 2.0 ps for bulk water), but virtually no change arising from the softer averaged LJ wall(2.1 and 2.3,as against 2.1 and 2.0 ps). $\Gamma_1(t)$ is a single-particle correlation and so is not necessarily as vulnerable to error from long-range effects as the two-particle correlations which determine the dielectric constant. Nevertheless, it would be advisable to examine the effect of such corrections on the results obtained.

REFERENCES

(1) Rowlinson,J.S. 1951, Trans. Faraday Soc. 47,pp.120-129.
(2) Stillinger,F.H. and Rahman,A. 1974,J.Chem.Phys. 60,pp.1545-1557.
(3) Matsuoka,O.,Clementi,E. and Yoshimine,M. 1976,J.Chem.Phys. 64, pp.1351-1361.
(4) Morse,M.D. and Rice,S.A. 1982,J.Chem.Phys. 76,pp.650-660.
(5) Nicholson,D. and Parsonage,N.G. 1982,"Computer simulation and the statistical mechanics of adsorption". Academic Press,London, New York,Paris,San Diego,San Francisco,Sao Paolo,Sydney,Tokyo and Toronto,p.353.
(6) Watts,R.O. 1974,Mol.Phys. 28,pp.1069-1083.
(7) Steinhauser,O. 1982,Mol.Phys. 45,pp.335-348.
(8) Lie,G.C.,Clementi,E. and Yoshimine,M. 1976,J.Chem.Phys. 64, pp.2314-2323.
(9) Sonnenschein,R. and Heinzinger,K. 1983,Chem.Phys.Lett. 102, pp.550-554.
(10) Alder,B.J. and Wainwright,T.E. 1959,J.Chem.Phys. 31,pp.459-466.
(11) Rahman,A. 1964,Phys.Rev. 136,pp.A405-411.
(12) Nose,S. 1984,Mol.Phys. 52,pp.255-268.
(13) Adams,D.J. and McDonald,I.R. 1976,Mol.Phys. 32,pp.931-947.
(14) Christou,N.I.,Whitehouse,J.S.,Nicholson,D. and Parsonage,N.G. unpublished work.
(15) Christou,N.I.,Whitehouse,J.S.,Nicholson,D. and Parsonage,N.G. 1981,Roy.Soc.Chem.,Faraday Symp. 16,pp.139-149.
(16) Jonsson,B. 1981,Chem.Phys.Lett. 82,pp.520-525.
(17) Marchesi,M. 1983,ChemPhys.Lett. 97,pp.224-230.
(18) Antonchenko,V.Ya.,Ilyin,V.V. and Makovsky,N.N. 1983,Doklady Acad.Nauk Ukrainshoy SSR,ser.A. 1,pp.53-55.
(19) Lee,C.Y.,McCammon,J.A. and Rossky,P.J. 1984,J.Chem.Phys. 80, pp.4448-4455.
(20) Barabino,G.,Gavotti,C. and Marchesi,M. 1984,Chem.Phys.Lett. 104,pp.478-484.
(21) Henderson,D.,Blum.L. and Lozada-Cassou,M. 1983,J.Elec.Chem. 150,pp.291-303.
(22) Barker,J.A. and Watts,R.O. 1973,Mol.Phys. 26,pp.789-792.
(23) Vericat,F.,Blum,L. and Henderson,D. 1983,J.Elec.Chem. 150, pp.315-324.
(24) Frank,H.S. and Evans,M.W. 1945,J.Chem.Phys. 13,pp.507-532.
(25) Geiger,A.,Rahman,A. and Stillinger,F.H. 1979,J.Chem.Phys. 70,pp.263-276.
(26) Parsonage,N.G. and Staveley,L.A.K. 1978,"Disorder in crystals". Clarendon Press,Oxford. Ch.11.
(27) Mills, 1973,J.Phys.Chem. 77,pp.685.

MODELS OF THE ELECTRICAL DOUBLE LAYER:
An Overview of the Present Situation

Roger Parsons*

Laboratoire d'Electrochimie Interfaciale des CNRS
1 Place Aristide Briand
92195 France

1 - INTRODUCTION

Although attempts to model the electrical double layer are more than a century old there is still no satisfactory explanation for the variety of phenomena observed. In the last few years, as a result of advances in experimental techniques as well as the development of theoretical interest in interphases, there has been an attempt to construct a model of this region based more firmly on molecular concepts. The present paper surveys some of the progress made in this direction towards the solution of seven problems which arise in the consideration of the electrical double layer.

2 - THE DIFFUSE DOUBLE LAYER

It was realised in 1910 and 1913 by Gouy [1] and Champman [2] respectively that thermal motion of the ions constituting the solution side of a metal/electrolyte double layer would result in a distribution of these ions in the direction normal to the interface. Their description of this distribution is based on the same postulates as the later Debye-Hückel theory of electrolyte interaction [3]. It postulates point charge ions in a structureless dielectric medium. The non-linearized Poisson--Boltzmann equation can be readily solved for a plane interface

* Present Adress:
Department of Chemistry
The University Southampton SO95NH

A. F. Silva (ed.), Trends in Interfacial Electrochemistry, 373–385.

and leads to explicit expressions for such quantities as the solution charge, the component charges due to each ionic species, the interfacial capacity as a function of the potential difference across this region [4].

The model is so crude that it has been criticized frequently and many attempts have been made to improve it by introducing for example the finite size of the ions [5], change in the dielectric properties of the medium (6) non-ideality of the ionic solution [7] etc.. Such effects tend to produce deviations from the G-C theory only under conditions of high charge and/or high concentration where other effects mask the behaviour of the diffuse layer. Nevertheless a substantial part of the phenomena observed are interpreted with the aid of G-C theory; many of the experimental "facts" depend on this interpretation. It is therefore important to know whether deviations from expected behaviours should be attibuted to the unreliability of the G-C model itself or to other factors such as the presence of the "inner layer" discussed below.

The most satisfactory tests of the G-C model seem to be a comparison with experiment [8] but the validity of the analysis of the model can also be checked by a Monte Carlo evaluation [9]. The latter confirms the conclusion drawn from previous tests of model, that the G-C solution of the Poisson-Boltzmann equations is remarkably good for 1:1 electrolytes under not too extreme conditions. The same is true if the ions are given finite sizes. Deviations occur first with the potential-distribution or with the relation between total potential drops and quantities like charge or surface excess, i.e., quantities depending on an integral across the difuse layer. Relations amongst these integral quantities retain good agreement with G-C predictions to conditions where this would be unexpected on general grounds [10] since a great many of the G-C equations depend on relations between these integral properties it is difficult to devise a satisfactory experimental test which is sufficiently sensitive.

One such test was attempted about 20 years ago [8]. The basis is the expectation that in a solution containing two cations of different charge, the surface excesses of these will be sensitive to the potential distribution. Experimental measurements of the capacity of an electrode in solutions of $KCl + MgCl_2$ were analysed thermodynamically to give the surface excess of three constituent ions. The relative amounts of K^+ ang Mg^{2+} were found in satisfactory agreement with G-C predictions and though the observation of an increasing surface excess of Cl^- as the negative charge on the metal increases cannot yet be explained satisfactorily.

It may be concluded that the G-C theory is certainly the correct low concentration limit (like the Debye-Hückel theory) and is probably usable well into practical useful regions of concentration and potential although at present quantitative estimates of the deviation from an exact model are not available.

The dependence of much subsequent analysis on the G-C equations means that further developments of this part of the double layer model are of great interest.

3 - THE INNER LAYER

The Gouy-Champmann theory leads to a capacity which is a cosh function of the potential drop across the diffuse layer. If the diffuse layer is assumed to extend right up to a structureless metal surface, this prediction is in violent contrast with experiment. This was already realised by Gouy [11] and more formally expressed by Stern [12]. The simplest modification is that of attributing a finite size to the ions but only with their interaction with the metal surface. Thus the inner boundary of the diffuse layer becomes the plane containing the centres of ions which are at there distance of closest approach to it. This plane has come to be known [4] as the "outer Helmholtz plane" (OHP). The total potential drop between metal and electrolyte can then be divided into two parts: - one which occurs between the metal and the OHP and one which occurs across the difuse layer. The capacity of the interphase then behaves formally as a series combination of two capacities: that of the "inner layer" and that of the diffuse layer.

It has become customary, following Grahame [4] to calculate the inner layer capacity by subtracting (reciprocal by) the calculated G-C capacity from the experimental capacity. In the absence os specific adsorption (see below) this leads to an inner layer capacity which is a function of interfacial charge but not of electrolyte concentration. However, it is clear that this conclusion depends on the validity of the assumption of a "one-dimensional double layer" in which the properties are uniform in a plane parallel to the interface and also on the validity of G-C theory.

Recently, Henderson et al [13] have used an improved model of electrolyte to calculate the double layer capacity. The ions and the solvent are represented as hard spheres carrying respectively a charge and a dipole. Using the hipernetted chain theory and the mean sherical approximation, they found that the reciprocal capacity at the point of zero charge could be expressed in the form of two terms: one arises from the ions and is equivalent to the G-C term described above while the other arises from the solvent. However, there is no spatial separation as in the earlier model, both terms represent the distribution of particles in the whole double layer. Nevertheless the solvent term, like the "inner layer capacity" dominates at high concentrations. It is also very insensitive to the ionic size which could eventually provide an explanation for the observed insensitivity of the capacity at negative charges to the nature of the cation [14, 15].

This model, which is undoubtedly closer to reality than the G-C theory, casts doubt on the existence of an inner layer as usually concieved. It is therefore worth considering whether there is more direct evidence for the existence of such a layer. One experiment seems to provide this. Is was observed by Gouy [16] that for some electrolytes which may be expected to interact weaklywith mercury, the interfacial tension of the mercury/ /electrolyte interface increases with increasing electrolyte concentration at the potential zero charge (electrocapillary maximum). Following the Gibbs equation

$$(\partial\gamma/\partial\mu_{salt})_{\sigma=0,T,P} = - \Gamma^{(w)}_{salt} \quad (1)$$

This implies a negative surface excess of the salt. The surface excess is normally calculated relative to the solvent, that is, it can be expressed in terms of the actual surface concentrations by

$$\Gamma^{(w)}_{salt} = \Gamma_{salt} - \Gamma_{w} (x_{salt}/x_{w}) \quad (2)$$

where x_{salt} and x_{w} at the mole fractions of salt and solvent respectively, in the bulk of the solution. The experimental observation has been interpreted by several subsequent groups [17-20] in terms of concentration profile of the salt and the solvent which are both uniform but that of the salt drops to zero further from the metal surface than that of the solvent. If the drop is vertical in both cases, the difference in distance of closest approach is found to be about 0.3 nm for aqueous solutions or about, the diameter of one water molecule. This suggests that there is some basis for the physical existence of the inner layer. Similar experiments have been done for the air/electrolyte [21] and oil/electrolyte [22] interphases with similar results.

4 - THE CONTRIBUTION OF THE METAL

Most of the discussions of double layer structure at the metal/electrolyte interphase up to the present time have treated the metal as a plane polarizable structureless wall incapable of maintaining a field beyond the surface. The possibility of a diffuse layer of charge within the metal was considered by Rice [23] in 1926, assuming a Gouy-Champman like model. However, it has usually been considered that because of the very short Thomas-Fermi screening length (corresponding to a very short Debye lenght) that this model reduces to the conventional model for pratical purposes (see by [24]). Recently, however, prompted by the work on electron distribution at the free surface of metals by Lang and Kohn [25] the concept has been reviewed. It is evidently essential to consider a substantial surface dipolar layer arising from the overshoot of electrons to account for the electronic work function of metals. The presence of an electrolyte would modify this distribution but it seems unlikely that it would eliminate it. Badiali [26], Kornyshev [27], Schmickler [28] among others have suggested that in fact the surface electronic structure of a metal does make a substantial contribution to the observed behaviour in particular to the capacitance.

This arises not so much from the electron distribution as such but from its modification as the state of charge changes. In further work, attempts have been made [29, 30] to treat the interface as a whole by combining this "jellium" model for the metal with the hard sphere model for the electrolyte. This appears to give a quantitive explanation for the effect of the nature of the metal, in terms of the free electron density of sp metal, although at present only at the point of zero charge.

This type of model although certainly very simplified, does have the advantage of being based on clear physical principles, whereas previous explanations of the nature of the metal make use of rather vague concepts of hydrophilicity or other types of chemical analogy. On the other hand it seems premature to conclude that the major features of the real system are satisfactorily included in the physical model. One aspect which needs to be considered, for example, is the observation that the capacity of the double layer at a non-metallic solid like AgI seems to be rather similar to that on mercury [31]. This needs both experimental and theoretical exploration. Other aspects are considered in the next section.

5 - SOLVENT STRUCTURE IN THE INTERPHASE

The form of the inner layer capacity/charge curve [32] calculated as described above has usually been ascribed to the orientation of a monolayer of dipolar solvent molecules in contact with the metal surface. A variety of detailed models has been proposed since the original two-position model of Watts-Tobin [33]. These have usually aimed to explain the inner layer capacity curve for mercury in water as a function of charge and temperature obtained by Grahame [32] although considerable effort has been expended in applying such models to systems with non-aqueous solvents [34, 35]. Although these observations are to some extent dependent on the G-C theory, the variety of behaviour and the characteristic temperature dependence argue that the properties of the solvent play a major role in this behaviour. It remains to be seen whether a combination of the solvent orientation model with the electron overlap from the metal can account for the totality of the observations. Work in this direction has begun [29, 30].

The solvent orientation models in themselves are open to a number of criticisms despite their ability to reproduce the experimental results: the parameters particularly those representing the energy of the different states are arbitrary and have no satisfactory physical interpretation. The model can yield non-physical solutions like negative capacities for apparently reasonable physical parameters [36]. An attempt to overcome some of these problems by simulation to find minimum energy configuration [37] has not yet let to an improvemed representation of experiment.

One result which may provide a clue to the setting up of a better model is the isotope effect. Experimental data [38] show that the capacity is about 5% lower in D_2O in the region of the "hump" than is the H_2O. Models depending on dipole and polarization effects are incapable of explaning such a change. Guidelli's detailed quasi-chemical model incorporating hydrogen bonding however gives an explanation in terms of the change in hydrogen bond strength with the replacement of, H, the by D [39].

Most of the comparison of these models with the experiment has been done with mercury electrodes and so the effect of crystal structure has not been considered much. The inner layer capacity for Au [40] and Ag [41] seems to be quantitatively different from that on mercury, the "hump" is larger and more symmetrically placed at the pcz. It is not clear which is the more typical behaviour, nor have the differences yet found a satisfactory explanation.

It is not known yet whether the charge and temperature dependence of the 'inner layer' capacity can be explained by the hard sphere models because these are at present available only for the potential of zero charge. It is obviously to be hoped that these properties will emerge 'naturally' from a model firmly based on molecular properties and using more exact statistical procedures.

6 - SPECIFIC ADSORPTION

The orignial definition of specific adsorption, proposed by Gouy [16], was adsorption which occured at the potential of zero charge, where there was clearly no adsorption due to the charge on the electrode. However, the observation of different types of behaviour at positively charged electrodes even amongst salts which show no positive adsorption at the pzc indicates that there are specific effects on anion adsorption which call for an extension of this definition. Although different definitions and names for this phenomenon have been proposed, all evaluations of specific adsorption use the operational definition that specific adsorption is adsorbed that deviates from that predicted from G-C theory.

The evaluation of specific adsorption then depends on the presence of one type of ion in the electrolyte which can reasonably be assumed not to be specifically adsorption. Typically anionic specific adsorption is calculated assuming that the cation (often an alkali ion) is not specifically adsorbed. Thermodynamic analysis of the experimented capacity or interfacial tension curves yields the surface excesses of cation Γ_+ and anion Γ_- which, in sufficiently dilute solution can be put equal to the surface concentrations. If the cation is presented only in the diffuse layer then G-C theory may be used [4] to calculate from Γ_+ the surface concentration of anion in the diffuse layer (containing a single 2-2 valent electrolyte).

$$\Gamma^{2-s} = 2\ n\ L_D\ [(\Gamma_+/2\ n\ L_D + 1)^{-1} - 1] \qquad (3)$$

L_D the Debye lenght given by

$$L_D = (\varepsilon kT/z^2\ \varepsilon^2\ n)^{1/2} \qquad (4)$$

where n is the concentration of ions and ε is the permittivity.

The amount of anion specifically adsorbed is then obtained by difference.

$$\Gamma_- = \Gamma_- - \Gamma_-^{2-s} \qquad (5)$$

It is clear that such a definition depends on the validity of the G-C theory. However, as noted above, relations such as (3) seem to be relatively accurately given by G-C theory so that the results may not be unduly sensitive to deficiencies in G-C theory. On the other hand the form of equation (3) shows that the errors in Γ_-^{2-s} are the greatest when specific adsorption is weak and the solution is dilute: these errors are particularly troublesome [19] when the ion may be specifically adsorbed on an electrode carrying a charge of opposite sign.

These problems arise for example when the attempt is made to establish whether F^- is adsorbed specifically. Such adsorption is certainly weak and occurs only on positively charged electrodes if at all. Increase of the concentration of the alkali fluoride in the solution would increase the amount of specific adsorption but at the same time it increases the errors in G-C and also the difference between the surface excess and the surface concentration and consequently the inaccuracy in estimating the latter. Various othe techniques have been used to establish the existence of specific adsorption bul all quantitative estimates depend on G-C theory in one form or another.

The discussions in sections 3, 4 and 5 depend on the assumption that the calculated inner layer properties contain no contribution from specific adsorption. For mercury electrodes, Grahame gave arguments for the absence of specific adsorption of fluoride [42]. The problem was discussed in detail by Schiffrin [43] who was able to suggest more quantitative limits to the specific adsorption of F^- in terms of concentration, temperature and charge on the electrode. It may be concluded from his work taht specific adsorption of F^- plays no role in the regions usually considered when.models of the inner layer are fitted to experiment in aqueous solution. The situation is less clear in non-aqueous solution because no such thorough analysis has been carried out.

That this conclusion for mercury electrodes connot automatically be carried over to other metals has been shown recently by Valette in his studies of Ag single crystal electrodes [41]. From these, it appears that F^- is more strongly adsorbed on a positively charged Ag electrode than ClO_4^-, BF_4^- or PF_6^-, the last being strongly adsorbed, probably not specifically adsorbed. Valette was able to use the method of mixed electrolytes [44] to determinate the specific adsorption of F^- from solutions containing PF_6^-. This method much reduces the dependence of the results on the errors in G-C theory. However, as in the case of the study of Cs^+ adsorption from solutions containing Li^+ [45] it is difficult, if not impossible to distinguish weak specific adsorption form an effect due to the closer approach of the ion under study to the electrode compared to the reference ion [46].

Despite the weaknesses of the methods fos establishing the absence of specific adsorption, it is essential that they be fully used before an attempt to fit a model which assumes no specific adsorption. Both in this case and in the study of specific adsorption itself it is necessary to keat in mind the assumptions which have been used to obtain the properties under discussion. The use of models replacing the G-C model, is clearly desirable but as yet it has not been attemped.

7 - ADSORPTION OF ORGANIC MOLECULES

The adsorption of neutral species usually organic molecules has been studied for a variety of reasons. Here the discussion will be limited to the attempt to use them as a probe for inner layer structure especially solvent structure. The absence of strong electrostatic interaction with the electrodes means that the molecule/electrode interaction energy falls off rapidly with distante. Hence adsorption can usually be considered to occur only in the first layer next to the metal (monolayer adsorption) and this is borne out by the observed amounts adsorbed. Such adsorption is conveniently described by an adsorption isotherm like that used for gas adsorption, although at an electrode the dependence on the electrical state of the interphase must also be taken into account. Also it must be noted that unlike single gas adsorption, adsorption from solution is always a replacement process. In the simplest case where a molecule replaces a solvent molecule of the same size and with the same interactions with its neighbours the adsorption obeys the Langmuir isotherm

$$\theta/(1 - \theta) = \beta a \qquad (6)$$

where θ is the fraction of adsorbate molecules in the adsorbed layer, a is the bulk activity of these molecules and β is the adsorption coefficient (dependent on the electrical state of the interphase). It should be noted that in contrast with the system of gas adsorption on a solid, the Laugmvir isotherm does not imply localised adsorption on sites.

Since organic molecules have quite different interactions from water, it is not surprising that a few organic adsorptions follow equation (6) and by analogy with the van der Waals equation, Frumkin proposed a modified isotherm

$$(\theta/(1 - \theta) \exp A\theta = \beta a \tag{7}$$

where the coefficient A expresses this difference of interactions. Frumkin's isotherm has been found to fit a wide variety of systems, perhaps because one freely adjustable parameter (A) and one adjustable to some extent (the saturation coverage Γ_s) are sufficient to express results of limited accuracy.

In the last ten years data of considerably increased precision have become available notably those obtained by Mohilner et al [48] using a high precision computer-controlled pressure gauge in the measurement of the mercury/electrolyte interfacial tension. Those data for the adsorption of 2-butanol from aqueous Na_2SO_4 have been analysed under precise thermodynamic conditions. The interphase was treated as a two-dimensional solution, which involves arbitrary choice of its thielmers, and the resulting excess Gibbs energy of mixing was expressed as a power series of the surface mole fraction. Seven terms were required but the physical significance of these is not yet clear. Perhaps of greater significance is the analysis of the temperature dependence which shows that the entropy of mixing in the simple interpretation of Frumkin's isotherm in which the entropy therm $\theta/(1-\theta)$ is based on the concept of random mixing and is of course positive. Thus it seems necessary, as might be expected from the behaviour of the corresponding bulk mixtures, that the nature of the molecular interactions needs to be taken account in the form of incipient de-mixing. However, it must be noted that the behaviour of bulk alcohol + water mixtures remains imperfectly understood.

The statistical basis of Frumkin's isotherm is not very rigorous ans Sangaranarayanan and Rangarajan have recently started a programme [49] of examining the consequences of simple models of the adsorbed layer allowing up to two positions for the solvent or the adsorbate or the solvent. They conclude that not only Frumkin isotherm but most of its competitors including the Mohilner expansions are in fact approximations. It was concluded that most of the deductions about surface layer structure from the form of the adsorption isotherms and, a portion from the values of its constants, are unjustified and probably misleading.

Nevertheless more qualitative conclusions may be more

reliable. For example, the evidence for the orientation of water at the mercury electrode from experiments in which particular molecules are adsorbed seems to point clearly to a very small net orientation: the replacement of water by pyrazine [50] which is adsorbed flat and has no net dipole confirmed by the replacement of water by the hydrocarbon chains of alcohols etc. [51]. Similarly the expectation that highly polar molecules will tend to align with their dipoles parallel to the interface [52] is confirmed by the experiments on glycine and glycylglycine in their zwitterionic forms [53]. Deductions about the relative strengths of adsorption of water on different metallic surfaces by observation of the relative strengths of adsorption of a non--polar molecule like di-ethyl ether seem less certain and not always in accord with other estimates of the rather vague concept of hydrophilicity [54].

The great range of types of organic compound provides the possibility of a variety of very subtle differences of interaction. In particular small modifications of structure can lead to quite large changes of observed properties. Adsorption of the geometric isomers. the maleate and fumarate ions, seems to be equally strong but the former almost suppresses the characteristic "hump" of the mercury/aqueous solution interphase while the latter enhances it [55]. It has been supposed that the two ions interact differently with the water in the inner layer. An even smaller difference of structure between mannitol and sorbitol provokes an apparently greater degree of adsorption of the latter, despite the fact that the former is the less soluble [56].

It is evident that the explanation of systems of these types will require a very considerable advance in the sophistication of models whose exact statistical treatment already represents a formidable problem.

8 - PARTIAL CHARGE TRANSFER

The idea that the "chemical" interaction of an ion or molecule with an electrode would involve some redistribution of the electrons between the two partners was introduced into electrochemistry by Lorenz and Salié [57] and it is obviously to be expected from the general knowledge of chemical bonds. Lorenz's arguments started from a thermodynamic coeficient

$$\ell = - F^{-1} (\partial\sigma/\partial\Gamma)_E = (\partial\mu/E)_\Gamma$$

which was identified clearly by Barker [58] as the avarage number of unit charges supplied to the electrode from the external circuit when one molecule of a species is adsorbed at constant potential. The interpretation of this experimentally observable coefficient is not straightforward. Vetter and Schultze [59] pointed out that it contains in general contributions not only from partial charge transfer, but also from the movement of charge through the double layer field (the geometrical effect expressed by g), the replacement of solvent dipoles by the dipole of the adsorbed species as well as polarizability effects not included in the partial charge transfer. Thus they expressed the coefficient ℓ in the form

$$= gz - \lambda(1 - g) + k_{ad} - \nu k_{w} \tag{9}$$

where λ is the fraction of charge lost by the ion of charge originally zc, g is a geometric factor expressing the fractional of the double layer potential drop crossed by the ion, k_{ad} the contribution of the dipole of the adsorbing species and k_w that of each of the ν water molecules it displaces. The various contributions to ℓ cannot be disentangled experimentally at present although in some cases limiting values suggest that one overhelms the rest. For example in many systems of submonolayer metal deposition (UDP) the fact that ℓ is equal to the valence of the metallic ion in solution suggests strongly that the ion has completly lost its charge [60]. Schultze and Koppitz [61] have attempted to use correlations between the behaviour of simple cations and anions to deduce the contribution of partial charge transfer, but as yet there is no independent confirmation of their conclusions. Spectroscopic methods seem a promising source of such information but so far this has not been forthcoming.

9 - CONCLUSION

The large body of experimental information on the metal/ /electrolyte interphase has been predominantly obtained by thermodynamic or similar methods which do not give a occess to molecular properties. It is therefore difficult to construct statistical theories based on molecular and atomic properties which give a unique fit to experiment. Nevertheless the current interest of theorists has lead to some clarification of the issues involved and hope for further progress. The recent development of in situ spectroscopic techniques notably Raman and IR also provides the possibility for more specific information about interfacial structures which will lead to the construction of better models.

REFERENCES

1. G. Gouy, J. Phys. Chim., 9 (1910) 457.
2. D.L. Chapman, Phil. Mag., 25 (1913) 475
3. P. Debye and E. Mückel, Physikal Z., 24 (1923) 185.
4. D.C. Grahame, Chem. Reviews, 41 (1947) 441
5. V. Freiser, Z. Elektrochem., 56 (1952) 822
6. M.J. Sparnaay, Rec. Trav. Chim., 77 (1958) 382

7. A.L. Loeb - J. Colloid Sci., 6 (1951) 75
8. R. Parsons and S. Trasatti, Trans. Faraday Soc., 65 (1969) 3314.
9. G.M. Torrie and J.P. Valleau, J. Chem. Phys., 73 (1980) 5807.
10. H.D. Hurwitz, A. Sanfeld and A. Steinchen-Sanfeld, Electrochim. Acta, 9 (1964) 929.
11. G. Gouy, Ann. Phys., (9) 9 (1917) 129
12. O. Stern, Z. Elektrochem., 30 (1924) 508
13. D. Henderson, L. Blum and M. Lozada-Cassou, J. Electroanal. Chem., 150 (1983) 291.
14. D.C. Grahame, J. Electrochem. Soc., 98 (1951) 343
15. N.F. Mott, R. Parsons and R.J. Watts-Tobin, Phil. Mag., 7 (1962) 483.
16. G. Gouy, Ann. Chem. Phys., (7) 29 (1903) 161.
17. Z.A. Iofa and A.N. Frumkin, Zhur. Fiz. Khim., 18 (1944) 268.
18. D.C. Grahame and R. Parsons, J. Amer. Chem. Soc., 83 (1961) 1291
19. R. Parsons and F.G.R. Zobel, J. Electroanal. Chem., 9 (1965) 333.
20. J.A. Harrison, J.E.B. Randles and D.J. Schffrin, J. Electroanal. Chem., 25 (1970) 197.
21. J.E.B. Randles, Advances in Electrochemistry and Electrochemical Engineering Vol. 3 1963 Ed. P. Delahay and C.W. Tobias, Wiley, New York. pl.
22. H.H. Girault and D.J. Schiffrin, J. Electroanal. Chem. 150 (1983) 43.
23. O.K. Rice, J. Phys. Chem., 30 (1926) 1501.
24. R. Parsons, Modern Aspects of Electrochemistry Vol. 1 (1954) Ed. J.O M. Bockris and B.E. Conway, Butterworths, London, p. 103.
25. N.D. Lang and W. Kohn, Phys. Rev., B 1 (1970) 4555, B 3 (1971) 1215.
26. J.P. Badiali, M.L. Rosinberg and J. Goodisman, J. Electroanal. Chem., 143 (1983); 150 (1983 25.
27. A. Kornsyshev, W. Schmickler and M. Vorotyntsev, Phys. Rev. B 25 (1982) 5244.
28. W. Schmickler, J. Electroanal. Chem., 150 (1983) 19.
29. W. Schmickler and D. Henderson, J. Chem, Physics, 80 (1984) 3381.
30. J.P.Badiali, M.L.Rosinberg, F. Vericat and L. Blum, J. Electroanal. Chem., 158 (1983) 253.
31. J. Lyklema, Trans. Faraday Soc., 59 (1963) 418.
32. D.C, Grahame, J. Amer. Chem. Soc., 79 (1957) 2093.
33. R.J. Watts-Tobin, Phil. Mag., 6 (1961) 133.
34. R. Parsons, Electrochimica Acta, 21 (1976) 681.
35. W.R. Fawcett, Israel J. Chesmistry 18 (1979) 3.
36. I.L. Cooper and J.A. Harrison, J. Electroanal. Chem., 66 (1975) 85.
37. R. Parsons and R.M. Reeves, J. Electroanal. Chem. 123 (1981) 141.

38. R. Parsons, R.M. Revees and P.N. Taylor, J. Electroanal. Chem., 50 (1974) 149.
39. R. Guidelli, J. Electroanal. Chem., 110 (1980) 205
40. A. Hamelin, J. Electroanal. Chem., 138 (1982) 395.
41. G. Valette, J. Electroanal. Chem., 122 (1981) 285.
42. D.C. Grahame, J. Amer. Chem. Soc., 76 (1954) 4815
43. D.J. Schiffrin, Trans. Faraday Soc., 67 (1971) 3318.
44. G. Valette, J. Electroanal. Chem., 138 (1982) 37.
45. R. Parsons and A. Stockton, J. Electroanal. Chem., 25 (1970) App 10.
46. W.R. Fawcett, J. Electroanal. Chem., 39 (1972) 474.
47. A.N. Frumkin, Z. Physikal. Chem., 116 (1925) 46.
48. D.M. Mohilner, J. Physical Chem., 81 (1977) 244 and Private Communication.
49. M.V. Sangaranarayanan and S.K. Rangarajan, Can. J. Chem., 59 (1981) 2072); J. Electroanal. Chem., 176 (1984) 1, 29, 45, 65, 99, 119.
50. B.E. Conway, J.G, Matheson and H.P. Dkar, J. Phys. Chem., 78 (1974) 1226.
51. J.O'M. Bockris, M.A.V. Devanathan and K. Müller, Proc. Roy. Soc., A 274 (1963) 55.
52. R. Parsons, J. Electroanal. Chem., 109 (1980) 369.
53. L.M. Baugh and R. Parsons, J. Electroanal. Chem., 41 (1973) 311; Croat. Chim. Acta, 45 (1973) 127.
54. J. Lipkowski, C. Hinnen, C. Nguyen Van Huong, R. Parsons and J. Chevalet, J. Electroanal. Chem., 143 (1983) 375.
55. R. Parsons and J.T, Reilly; J. Electroanal. Chem., 24 (1970) App 23.
56. R. Peat and S. Shannon, J. Electroanal. Chem., 159 (1983) 229.
57. W. Lorenz and G. Salié, Z. Physikal Chem., 218 (1961) 259.
58. G.C. Barker, Pure Appl. Chem., 15 (1967) 239.
59. K.J. Vetter and J.W. Schultze, Ber. Bungsenge, 76 (1972) 920, 927.
60. D.M. Kolb, Advances in Electrochemisty and Electrochemical Engineering Ed. H. Gerischer and C.W. Tobias, Vol II (1978) Wiley, New York p. 125.
61. J.W. Schultze and F.D. Koppitz, Electrochim. Acta, 21 (1976) 327, 337.

MONOLAYER MODELS OF METAL-WATER INTERPHASES AND THEIR USE IN THE INTERPRETATION OF DIFFERENTIAL CAPACITY CURVES AND OF ORGANIC ADSORPTION

Rolando Guidelli

Dipartimento di Chimica, Università di Firenze,
Firenze, Italy

Monolayer models developed to account for the properties of metal-water interphases are reviewed. The concepts of nearest-neighbour lattice statistics commonly employed in molecular models of the adsorbed monolayer are examined, and the main physical and mathematical approximations involved in these models are pointed out. The various mathematical approximations are critically compared. Experimental data obtained for aqueous systems in the absence of specific adsorption are examined within the context of different monolayer models. Upon generalizing monolayer models so as to include the presence of polymeric neutral solute molecules, theoretical adsorption isotherms are derived and their features are compared with the experimental adsorption behaviour of aliphatic compounds from aqueous solutions.

1. INTRODUCTION

The subject of solvent structure at electrified interphases has been considered in several reviews (1-5). Most of the discussion in these reviews, with the exception of that by Fawcett (5), is directed to metal-water interphases, since they have been the most extensively investigated experimentally. Here, only models based on a molecular approach to interfacial solvent properties will be considered.

Two different approaches to the development of such models can be distinguished. One approach is based on the traditional division of the double layer into two regions: (i) the inner layer, immediately adjacent to the metal surface, whose properties are regarded as independent of electrolyte concentration and exclusively determined by those of the solvent when solute specific

A. F. Silva (ed.), Trends in Interfacial Electrochemistry, 387–452.

adsorption can be excluded; (ii) the diffuse layer, namely the region between the inner layer and the bulk solution, in which ionic concentrations differ fron their bulk values and whose properties are considered to be satisfactorily expressed by the Gouy-Chapman theory. Such a separation relies on Grahame's classical measurements of the differential capacity of the Hg/aqueous NaF interphase as a function of the charge density σ_M on the metal and of electrolyte concentration. Grahame (6) showed that the observed capacity C could be represented as a series combination of an inner-layer capacity C_i , which is independent of NaF concentration at constant σ_M , and of a diffuse-layer capacity C_d , which can be estimated from the Gouy-Chapman theory:

$$1/C = 1/C_i + 1/C_d \tag{1}$$

Strictly speaking, C_i is not a directly measurable quantity. However, it practically coincides with the measured differential capacity in concentrated (∿1M) solutions, where $1/C_d$ is almost negligible at all charge values. Although Grahame's separation in eqn.(1) is formal and does not necessarily imply the existence of a concentration-independent *inner-layer* region physically distinct from a *diffuse-layer* region satisfying the Gouy-Chapman theory, all molecular models based on this separation aim at explaining the behaviour of the inner-layer capacity C_i through a monolayer of solvent molecules lying between the metal surface and the distance of closest approach of nonspecifically adsorbed ions (the so-called outer Helmholtz plane, o.H.p.). This physical situation is simulated by regarding the monolayer of solvent molecules as lying between two structureless, continuous, planar distributions of charge, of opposite sign, on the metal surface and on the o.H.p.. This amounts to assuming that the solvent molecules and the ions outside the inner layer may be approximated as a dielectric continuum. Cooper and Harrison (7) stressed the roughness of the simplifying assumptions involved in monolayer models, and concluded that these models are invalid, also in view of the fact that they cannot coherently explain both the temperature-dependence and the charge-dependence of the differential capacity C_i. Although the crudeness of certain physical approximations introduced in monolayer models can hardly be denied, the latter criticism is unjustified. Thus, at present, only monolayer models can predict correctly a wide variety of interfacial phenomena, though, in many cases, by having recourse to a high number of adjustable parameters.

The other approach to the development of molecular models of electrified interphases is more fundamental in nature, in that it is not restricted to a single monolayer of solvent molecules. The electrolytic solution in the interfacial region is usually approximated as a mixture of charged hard spheres (the model ions) and uncharged hard spheres with imbedded point dipoles (the model solvent molecules). This *dipolium* model (8-10) is a logical extension of the so-called *primitive* model (11), in which the electrolyte is represented by a system of charged hard spheres in a uniform dielectric medium of dielectric constant ε . Just as monolayer models, dipolium models regard the electrode as a planar hard wall, on which a uniform charge density σ_M is distributed. Dipolium models are usually treated on the basis of distribution function theories (12,13). It should be noted that a pair distribution function $g(\vec{r}_1,\vec{r}_2)$ expresses the probability of finding particles 1 and 2 at the positions defined by the radius vectors $\vec{r}_1$ and $\vec{r}_2$, all other particles being distributed statistically around these two fixed particles. The pair distribution function is normalized in such a way that it tends to unity when the distance between the two fixed particles tends to infinity. The total correlation function $h(\vec{r}_1,\vec{r}_2)$ is then defined by

$$h(\vec{r}_1,\vec{r}_2) = g(\vec{r}_1,\vec{r}_2) - 1$$

Its name is justified by the fact that, as the distance between the two particles tends to infinity, namely as the correlation between these particles tends to vanish, $h(\vec{r}_1,\vec{r}_2)$ tends to zero. In practice, all distribution function theories are based on the solution of the Ornstein-Zernike integral equation:

$$h(\vec{r}_1,\vec{r}_2) = c(\vec{r}_1,\vec{r}_2) + \rho\int c(\vec{r}_1,\vec{r}_3)\, h(\vec{r}_3,\vec{r}_2)\, d\vec{r}_3$$

According to this equation, which holds rigorously, the total correlation function between two particles can be separated into two contributions. The first contribution, herein denoted by $c(\vec{r}_1,\vec{r}_2)$, is called the direct correlation function and accounts for the correlations which are transmitted directly between the two particles. The second contribution stems from correlations transmitted through a third particle 3, and is made up of direct correlations between particles 1 and 3 multiplied by the total correlation function between particles 3 and 2. Naturally, the latter contribution is integrated over all possible particles 3 and is multiplied by the number density ρ of the particles. The various distribution-function theories differ by the approximate way in which they express the direct correlation function in the Ornstein-Zernike equation. Thus, for instance, the Percus-Yevick theory sets

$$c(\vec{r}_1,\vec{r}_2) = \{1 - \exp[v(\vec{r}_1,\vec{r}_2)/ kT]\}\, g(\vec{r}_1,\vec{r}_2) \qquad (2)$$

where $v(\vec{r}_1,\vec{r}_2)$ is the interaction potential between particles 1 and 2 . As expected, $c(\vec{r}_1,\vec{r}_2)$ is zero whenever the pair interaction potential vanishes, namely when there are no direct correlations between the two particles. At sufficiently large distances, $g(\vec{r}_1,\vec{r}_2)$ tends to unity and $v(\vec{r}_1,\vec{r}_2)$ becomes $\ll kT$, so that the exponential in eqn.(2) can be replaced by the first two terms in its series expansion, yielding

$$c(\vec{r}_1,\vec{r}_2) \cong 1 - [1 + v(\vec{r}_1,\vec{r}_2)/(kT)] = -v(\vec{r}_1,\vec{r}_2)/(kT)$$

This expression of $c(\vec{r}_1,\vec{r}_2)$, which generalizes the asymptotic behaviour of the direct correlation function to all intermolecular spacings, is employed in the so called *mean spherical approximation*. This is a very popular theory, mainly because it permits the obtainment of analytic solutions even for rather complicated models, such as the dipolium models previously outlined. Naturally, in this case, the distribution functions of interest are those between the hard wall and the hard spheres, and express the probability of finding a sphere at a given distance from the wall. Unfortunately, the mean spherical approximation is a linearized theory which holds only for weak interaction potentials, and hence for small electrode charges σ_M (8-10) and large ion-ion average distances (i.e., low electrolyte concentrations). Moreover, only solvent-solvent interactions of the dipole-dipole type are considered, which limits the applicability of the results to organic solvents. In fact, the hydrogen bonds between water molecules are too strong to be approximated by dipole-dipole interactions. On account of these mathematical approximations, the penalty which one pays in adopting dipolium models rather than the simpler monolayer models is that one is less able to correlate experimental data.

Both monolayer and dipolium models regard the electrode surface as a structureless planar distribution of charge. It is evident that the simulation of the structured metal surface by a planar distribution of charge can only be a first approximation to the true physical situation. Thus, it is well known that a metal can be regarded as a lattice of ion cores immersed in a sea of conduction electrons, which are free to move within the metal. If we assume that the dielectric constant of the metal is infinitely high, then, as soon as a monolayer of solvent dipoles approaches the metal, the electron gas within the metal will distribute itself in such a way as to maintain the electric field in the metal equal to zero even in the presence of the dipoles. This redistribution of the electron gas, usually referred to as *charge induction* , is expected to occur even if the metal surface is approximated as a structureless plane; in this case, the charge induction produces on the dipoles the same electrostatic effect which would be produced if a perfect mirror image of the dipoles, albeit with an inversion in the sign of the poles, existed at an equivalent position behind the metal surface plane. It should be kept in mind that images do not actually exist, and hence there is no potential difference between the images and the metal surface plane, as would be if images were made of real charges.

More fundamental approaches have been recently adopted to express the behaviour of the electron gas at metal surfaces, by using the so called *jellium model*(14,15). In the jellium model, the charge of the ion cores of the metal lattice is considered to be smeared out uniformly throughout space up to the metal surface, where it drops abruptly to zero. Conversely, the conduction electrons in the metal are allowed to spill over beyond the metal surface , as a consequence of the quantum-mechanical requirement that the electron wave functions extend, as exponentially decaying tails, into regions of negative total energy. The jellium model has been recently combined with a monolayer model (16) as well as with dipolium models (17,18), in order to provide a more complete picture of the metal-solution interphase.

In the following we will confine ourselves to an examination of monolayer models and of their potentialities in explaining the experimental behaviour of metal-water interphases.

2. EXPERIMENTAL BEHAVIOUR OF METAL-WATER INTERPHASES

Let us briefly summarize the salient properties of metal-water interphases in the absence of specific adsorption.

I. The curve of the inner-layer differential capacity C_i against the charge σ_M for the Hg/H_2O interphase at 0° C has a central maximum at +2 $\mu C\ cm^{-2}$ with two minima at +10 $\mu C\ cm^{-2}$ and -13 $\mu C\ cm^{-2}$ (see Fig.1) ;at extreme positive charges C_i rises sharply, whereas only a moderate increase is observed at extreme negative

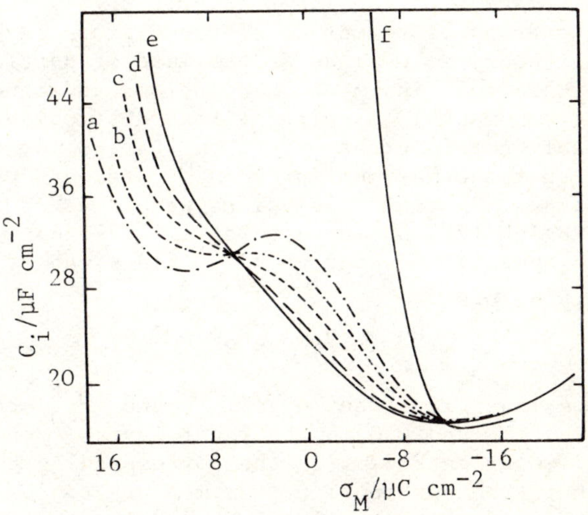

Figure 1. Curves a to e are C_i vs. σ_M plots at the Hg/ aqueous NaF interphase at 0, 25, 45, 65 and 85°C(19). Curve f is a C_i vs. σ_M plot at the Ga/aqueous 0.5M Na_2SO_4 interphase (Fig. 12 of Ref.4).

charges. As temperature increases, the central maximum decreases in height and ultimately disappears at 85°C, where the capacity curve shows a single minimum. The curves in Fig.1 were obtained by Grahame (19) at the interphase between Hg and aqueous solutions of NaF. Although there is some evidence (20-22) that F^- ion is specifically adsorbed at very positive charges, this adsorption does not become significant until ~6 μC cm^{-2}. Hence, it is widely accepted that the central hump in aqueous fluoride systems is to be ascribed mainly to water properties. A different explanation is provided by Bockris and coworkers (23,24), who ascribe the hump to a progressive increase of anion adsorption with increasing positive charge density. Now, it is certainly true that the capacity hump observed at mercury in aqueous systems is notably affected by the presence and nature of specifically adsorbed electrolyte anions, but F^- specific adsorption seems too small to influence the shape of the inner-layer capacity curves of Fig.1 at σ_M values less positive than +6 μC cm^{-2}.

II. The entropy of formation of the inner layer at Hg for aqueous fluoride systems (namely in the absence of ionic specific adsorption), as calculated by Harrison et al. (25) on the basis of capacity data, shows a maximum for charge densities in the range from -4 to -6 μC cm^{-2}. The entropy of formation of the interphase is defined as the difference in entropy of the components of the interphase when they are forming part of it, and when they are present in the bulk of the corresponding phases. The estimate of the entropy of formation of the inner layer, ΔS^{M-2}, requires an estimate of the entropy of formation of the diffuse layer via the Gouy-Chapman theory as well as an estimate of partial molar entropies for single ions in the bulk solution, and hence involves nonthermodynamic arguments. In spite of this difficulty, the ΔS^{M-2} values are probably not in great error, so that we may state with some confidence that the maximum in the ΔS^{M-2} vs. σ_M curve lies at negative charges. Similar conclusions were reached by Reeves (1), Hills and Hsieh (26) and Hurwitz et al. (27,28). In the absence of ionic specific adsorption, ΔS^{M-2} is given by (25)

$$\Delta S^{M-2} = S^{M-2} - m^{M-2}_{Hg}\,\bar{S}_{Hg} + m^{M-2}_{H_2O}\,\bar{S}_{H_2O} ,$$

where S^{M-2} is the inner-layer entropy, m^{M-2}_{Hg} and $m^{M-2}_{H_2O}$ are the numbers of moles of Hg and H_2O in the inner layer, and $\bar{S}_{Hg}$ and $\bar{S}_{H_2O}$ are partial molar entropies in the corresponding bulk phases. If we assume that the contribution from mercury to ΔS^{M-2} is negligible and that $m^{M-2}_{H_2O}$ varies with σ_M only slightly, then the entropy variations observed have to be ascribed to changes in the entropy of the adsorbed water molecules due to changes in the orientation of water dipoles by the electric field. This implies that the maximum state of disorder of water in the inner layer is attained at a negative charge, probably in the range from -4 μC cm^{-2} to -6 μC cm^{-2}. It should be noted that ΔS^{M-2} is obtained by double integration of $[\partial(1/C_i)/\partial T]_{\sigma_M}$ over charge. In this connection it is useful to note that the experimental plot of $[\partial(1/C_i)/\partial T]_{\sigma_M}$

vs. σ_M at the Hg/aqueous NaF interphase from which the ΔS^{M-2} vs. σ_M plot was derived shows a maximum at a negative charge which is close to the charge of maximum ΔS^{M-2} (25).

III. Several pieces of experimental evidence indicate that at the potential of zero charge (p.z.c.) the water molecules in the inner layer have a net preferential orientation with the negative end of the dipole pointing towards the metal; differently stated, the water molecules at the p.z.c. make a negative contribution to the inner-potential difference, $\Delta\phi$, across the inner layer. Thus, the temperature coefficient of $\Delta\phi$ at the p.z.c., $(\partial\Delta\phi/\partial T)_{\sigma_M=0}$, is positive (29). Since an increase in temperature is expected to have a randomizing effect upon dipole orientation, such a positive temperature coefficient suggests that $\Delta\phi$ becomes progressively less negative with increasing T. Another experimental observation pointing to a negative value of $(\Delta\phi)_{\sigma_M=0}$ is represented by the positive shift in the p.z.c. following adsorption of non-polar organic molecules such as pyrazine (30) at the Hg/H_2O interphase. Since adsorption consists in a displacement of solvent molecules by organic solute molecules from the inner layer, the above positive shift in the p.z.c. indicates that the displaced water molecules have a net negative orientation towards the metal. The occurrence of the maximum state of disorder of the solvent in the inner layer at a negative charge density (see point II) leads to the same conclusion, since a positive shift from this charge to the p.z.c. will produce a net alignment of the water dipoles with the negative end towards the metal.

IV. Inner-layer capacity measurements have been carried out at liquid metals other than mercury (gallium, indium-gallium alloys, etc.) as well as at solid electrodes. All capacity data for polycrystalline electrodes have been obtained with surfaces of unknown topography. Since differential-capacity curves depend on the crystallographic orientation, more significant results have been obtained at solid single-crystal electrodes (31). Particular attention has been devoted to sp metals. In most cases, these metals cannot be polarized to potentials positive with respect to the p.z.c., owing to oxidation of the electrode. In general, with low-melting sp metals the inner-layer capacity increases more rapidly when proceeding towards less negative charges, and attains a higher value at the p.z.c., the higher is the hydrophilicity of the metal (4). Incidentally, a measure of the surface hydrophilicity of metals, namely of their ability to bind water molecules through the oxygen atoms, can be provided by the enthalpy of formation of the corresponding metal oxides. Curve f in Fig.1 shows the C vs σ_M plot for the interphase between a very hydrophilic metal such as gallium and an aqueous solution of 0.5M Na_2SO_4.

3. GENERAL APPROACH TO MONOLAYER MODELS

In general all monolayer models assume that the monolayer

consists of a regular two-dimensional lattice of sites in which each site is occupied by a solvent molecule in a given orientation. There is a potential energy of interaction between nearest-neighbour (n.n.) solvent molecules, which depends on their relative orientations, but no interaction between higher-neighbour molecules is usually allowed for. In the presence of specific adsorption, solute molecules are also allowed to occupy the lattice (see Fig.2).

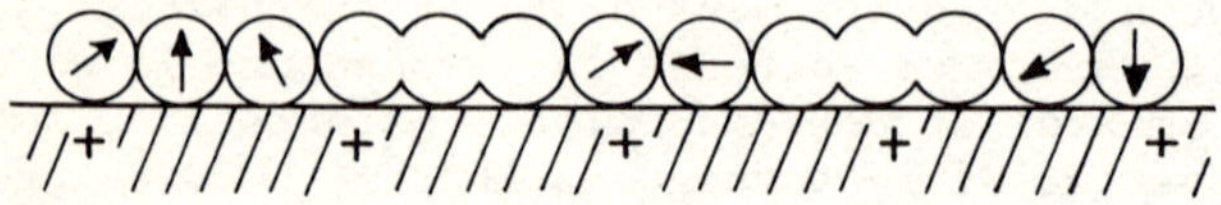

Figure 2. Schematic picture of a monolayer of differently oriented solvent dipoles and of trimeric solute molecules.

If their size is larger than that of solvent molecules, then they will occupy $\underline{n}$ neighbouring sites on the lattice, with $\underline{n} > 1$. Each *n-meric* solute molecule consists of $\underline{n}$ segments, each occupying one site on the lattice. In this case the potential energy of the monolayer is expressed as the sum of the energies of all n.n. pairs of particles (by the term *particles* we shall denote all allowed orientations of the solvent molecules and of the solute segments), with the obvious exclusion of the pairs of chemically bonded segments belonging to the same solute molecule. Let us denote by N_i the number of particles of the i-th type per unit surface of the monolayer, and by ${}^{\chi}N_{jk}$ the number of n.n. pairs of particles of the j-th and k-th type which follow each other in the positive direction of a given axis χ. Incidentally, the axes along which the n.n. sites of a given adsorption site are aligned are two at 90° angles for a square lattice, and three at 60° angles for a hexagonal lattice.

The canonical partition function for a monolayer characterized by a given set $\{N_i\}$ of numbers of particles, namely by a given *distribution* $\{N_i\}$, is expressed by the equation

$$Q = (\prod_i q_i^{N_i})\{\sum_{\{{}^{\chi}N_{jk}\}} g(\{N_i\},\{{}^{\chi}N_{jk}\}) \exp[-(\sum_{j,k,\chi} {}^{\chi}N_{jk}\, {}^{\chi}w_{jk})/kT]\} \qquad (3)$$

with

$$q_i = \exp[-(U_i^{el} + U_i^{c})/kT] \qquad (4)$$

Here ${}^{\chi}w_{jk}$ is the energy of lateral interaction between a $\underline{j}$ and a $\underline{k}$ particle following each other along the positive direction of the χ-axis. The triple summation $\Sigma_{j,k,\chi}$ is such that $\underline{j}$ and $\underline{k}$ extend over all particles and χ extends over all axes of the lattice. The quantity $g(\{N_i\},\{^XN_{jk}\})$ expresses the number of all distinct configurations of the monolayer which are characterized by the same set $\{^XN_{jk}\}$ of numbers of pairs. Hence, the summation between braces in eqn.(3), which extends over all possible sets $\{^XN_{jk}\}$ compatibly with the given distribution $\{N_i\}$, expresses the contribution to the partition function Q from all lateral interactions within the monolayer.

The further contribution to Q from the interactions of the monolayer with the metal and the solution phases is expressed by the factor $(\Pi_i\, q_i^{N_i})$ in eqn.(3), where q_i is the *single particle partition function* for the i particle and the product Π_i extends over all particles. Equation (4) shows that q_i has the form of a Boltzmann exponential factor whose exponent contains the energy of interaction of the i particle with the two contacting phases. This energy is considered to be made up of a contribution U_i^{el} due to coulombic interactions and of a further contribution U_i^c due to *chemical interactions*, namely interactions which cannot be directly explained on the basis of coulombic forces. If the monolayer is regarded as a planar array of molecules sitting in a vacuum under the action of a uniform external field $\vec{X}$ due to the charge σ_M on the metal, then the only nonzero interaction energy is U_i^{el}. This energy is usually estimated approximating the solvent molecules (or the solute segments) by point dipoles, thus setting $U_i^{el} = -X\mu_i$, where X and μ_i are the components of $\vec{X}$ and of the dipole moment $\vec{\mu}_i$ of the i particle along the axis normal to the monolayer and pointing from the metal towards the solution. Naturally, the magnitude of $\vec{X}$ is given by $|X|$, since $\vec{X}$ is normal to the monolayer. If the metal surface is regarded as a plane of dielectric discontinuity in passing from the monolayer to the metal phase, then a further contribution to U_i^{el} stems from the electrostatic interaction of each dipolar molecule with its own image in the metal-surface plane. It must be noted, however, that the use of images across distances comparable with molecular dimensions is only a first approximation to the actual behaviour of the electron gas in the metal (32). A still cruder approximation is involved in simulating the screening effect which the mobile ions in the diffuse layer exert on the electric field resulting from the charge density σ_M by means of a further plane of dielectric discontinuity, placed on the solution side of the monolayer. In this case a real dipole is imaged an infinite number of times, as in a hall of mirrors.

The presence of the U_i^c term in eqn.(4) permits us to account for any specific chemical interactions of the monolayer particles with the metal and solution phases, such as hydrogen bonding with the contacting water molecules of the second monolayer and chemisorption interactions with the surface metal atoms. Since the magnitude of these interactions depends not only on the nature of the adsorbed molecules but also on their orientation, each particle of the monolayer, namely each orientation of solvent monomers or of solute segments, can be given a different U_i^c value. However, one should avoid introducing too many different U_i^c values, since they have the character of adjustable parameters; now, it is evident that too many adjustable parameters may ensure a satisfactory agreement between experimental behaviour and the predictions of a given model, independently of the actual validity of the physical assumptions made in the model. The chemical interaction expressed by the U_i^c term does not depend on the position of the $\underline{i}$ particle of the monolayer with respect to that of the particle (or particles) of one of the two (or both) contacting phases with which the $\underline{i}$ particle interacts. In other words, this interaction must be regarded as smeared out uniformly throughout the boundaries of the monolayer with the two contacting phases. It is therefore clear that the approach based on the use of U_i^c values has not the general validity of a statistical mechanical theory dealing with a system of metal ion cores, electrons, and solvent and solute molecules extending from the bulk metal to the bulk solution, in which reliable pairwise interaction potentials between the various particles are employed. Unfortunately, such a theory is a formidable task, especially for a hydrogen-bonded polar liquid like water.

3.1. Constraints and equilibrium conditions

The sets $\{N_i\}$ and $\{{}^{\chi}N_{jk}\}$ are subject to a number of quite intuitive constraints which depend on the geometry of the lattice. Thus, if the monolayer consists of solvent monomers in an unspecified number of orientations and of a n-meric solute molecule S (which for simplicity will be regarded as consisting of $\underline{n}$ identical segments in a fixed orientation), then we can write:

$$N_T = \sum_{i \neq S} N_i + nN_S \tag{5}$$

where the summation extends over all solvent orientations, N_S is the number of solute molecules per unit surface, and N_T is the constant number of adsorption sites per unit surface of the lattice. The further constraint (33-35)

$$cN_k = \sum_{j,\chi} ({}^{\chi}N_{kj} + {}^{\chi}N_{jk}) \tag{6}$$

states that the sum, extended over all axes χ, of all n.n. pairs having the $\underline{k}$ particle as one of the two partners, is simply equal to the number N_k of $\underline{k}$ particles times the number $\underline{c}$ of n.n. sites of any given site; $\underline{c}$, which is called the *coordination number of the lattice*, equals 6 for a hexagonal lattice and 4 for a square lattice. Equation (6) holds only for monomeric molecules, and must be modified in order to be applied to n-meric molecules, with n>1. In this case $\underline{c}$ in eqn.(6) must be replaced by the total number of n.n. pairs having each segment of a single n-meric molecule at one of the two ends, i.e. cn, minus the number of those pairs which consist of two chemically bonded intramolecular segments (see Fig.3). For linear-chain and branched-chain

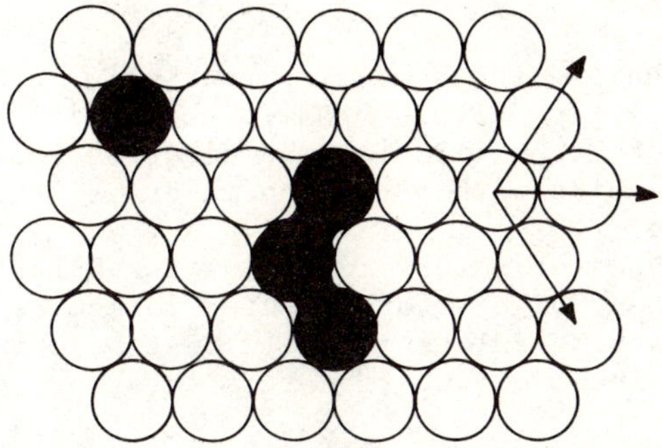

Figure 3. Top view of a hexagonal lattice in which a monomer and a trimer are evidenced.

molecules, which are the only types of n-meric molecules considered here, the latter number equals 2(n-1). (The number of intramolecular pairs in a n-meric molecule is actually (n-1), but it is counted twice in the above computation.) Hence, for a n-meric solute molecule S , we will write:

$$cpN_S = \sum_{j,\chi} ({}^{\chi}N_{Sj} + {}^{\chi}N_{jS}) \quad \text{with } p \equiv \left[cn-2(n-1)\right]/c \tag{7}$$

Besides the constraints of eqns.(5) to (7), the equilibrium conditions for the passage between any two different orientations of the same molecule must be accounted for. These conditions must be written both for solvent molecules and for solute segments. Here, having assumed for simplicity that solute segments are adsorbed in a fixed orientation, we shall confine ourselves to solvent molecules. At equilibrium, the Helmholtz free energy A of the monolayer is invariant with respect to any changes in orientation. We recall that A is related to the canonical partition function by the simple equation

$$A = -kT \ln Q \tag{8}$$

Let the solvent orientations introduced into the monolayer model range from 1 to $\underline{m}$. If we select a reference orientation (say, orientation 1) in such a way that all reorientations involve the passage from orientation $\underline{j}$, with $1<j\leqslant m$, to orientation 1, then the invariance of A is expressed by the (m-1) equations:

$$\left(\frac{\partial \ln Q}{\partial N_j}\right)_{N_i \neq N_1} = \left(\frac{\partial \ln Q}{\partial N_j}\right)_{N_i} + \left(\frac{\partial \ln Q}{\partial N_1}\right)_{N_i} \frac{dN_1}{dN_j} = 0 \text{ with } 1<j\leqslant m \tag{9}$$

where the subscript N_i refers to all particles, including any solute segments. Noting that $dN_1/dN_j=-1$ in view of the constraint of eqn.(5), the relationships of eqn.(9) take the form:

$$(\partial \ln Q / \partial N_j)_{N_i} = (\partial \ln Q / \partial N_1)_{N_i} \tag{10}$$

Recalling that the electrochemical potential $\bar{\mu}_j$ of any particle $\underline{j}$ equals $(\partial A/\partial N_j)_{N_i} = -kT(\partial \ln Q/\partial N_j)_{N_i}$ by definition, eqn.(10) simply states that at equilibrium the electrochemical potential of the water molecules within the monolayer is independent of their orientation.

In the presence of solute molecules S , a further equilibrium for the displacement of $\underline{n}$ adsorbed water molecules from the monolayer by one adsorbing solute molecule must be considered. Choosing once again the solvent orientation 1 as reference, this equilibrium can be written

$$nW_1 + {}^{b}S \rightleftarrows n\,{}^{b}W + S \tag{11}$$

where W_1 denotes a water molecule in the orientation 1 and the superscript $\underline{b}$ denotes the bulk solution phase. The choice of the reference solvent orientation is clearly immaterial once the equilibrium conditions of eqn.(10) are accounted for. At equilibrium, the Helmholtz free energy change involved in the reaction of eqn.(11) equals zero. In view of eqn.(8), this statement is written

$$\left(\frac{\partial \ln Q}{\partial N_S}\right)_{N_i \neq N_1} = \left(\frac{\partial \ln Q}{\partial N_S}\right)_{N_i} + \left(\frac{\partial \ln Q}{\partial N_1}\right)_{N_i} \frac{dN_1}{dN_S} = \left(\frac{\partial \ln {}^{b}Q}{\partial {}^{b}N_S}\right) \tag{12}$$

where the quantities ${}^{b}Q$ and ${}^{b}N_S$ refer to the bulk solution phase. If the solute concentration c_S in the bulk solution is relatively low, then the corresponding activity coefficient can be regarded as constant, especially if the solute is neutral. Hence we can write

$$-kT(\partial \ln {}^{b}Q/ \partial {}^{b}N_S) \equiv {}^{b}\bar{\mu}_S - n\,{}^{b}\bar{\mu}_W \cong \text{const.} + kT \ln c_S \tag{13}$$

where ${}^{b}\bar{\mu}_W$ and ${}^{b}\bar{\mu}_S$ are the electrochemical potentials of water and solute in the bulk solution. Noting that $(dN_1/dN_S)=-n$ in view of eqn.(5), combination of eqns.(12) and (13) yields

$$(\partial \ln Q/\partial N_S)_{N_i} - n(\partial \ln Q/\partial N_1)_{N_i} \cong -\text{const.}/kT - \ln c_S \tag{14}$$

The presence of the unspecified constant *const.* in eqn.(14) does not permit us to obtain the constant of the adsorption equilibrium of eqn.(11), and hence the adsorption coefficient of the solute S.

Nonetheless, the usefulness of eqn.(14) is justified by the fact that it permits us to derive the shape of the adsorption isotherm for the solute S at constant charge as well as its charge dependence.

3.2. Physical approximations in monolayer models

The use of monolayer models involves a number of approximations. Some of them aim at simplifying the physical picture before treating the model from a statistical-mechanical point of view. In addition to these *physical* approximations, a number of *mathematical* approximations is usually made in order to simplify the statistical-mechanical treatment, which would be prohibitively involved if the general expression of eqn.(3) for the canonical partition function were directly employed.

The very use of a *lattice model* has the character of a physical approximation. This model assumes that the equilibrium positions of the adsorbed molecules are discrete and regularly distributed within the monolayer region. In this respect *continuum models*, which allow the equilibrium positions to occupy any point in space, seem more realistic. It is evident that lattice models underestimate the configurational entropy with respect to continuum models, although they may well predict the correct dependence of this entropy upon the electrical state. At any rate, on solid metals it is possible that the structure of the metal substrate may induce an analogous lattice-like structure in the adsorbed monolayer.

Apart from the physical approximations involved in the use of any lattice model, further physical approximations are required at a later stage. They concern (i) the choice of the lattice geometry (square lattice, hexagonal lattice, etc.), (ii) the choice of the orientations of solvent and solute molecules to be introduced into the model, and (iii) the choice of the pairwise interactions between the monolayer particles (Van der Waals interactions, dipole-dipole interactions, H-bonds, and so on). As concerns the choice of the orientations, a judicious choice should include those orientations which make the largest contribution to the canonical partition function of eqn.(3). Unfortunately, the task of singling out these orientations is far from trivial. It should be noted that those interactions, $\underline{w}$, which depend neither on the orientations of the interacting molecules nor on their nature (when in the presence of different molecular species), are practically non-operative in lattice models. This is apparent from eqn.(3), where the contribution to the exponent ($\Sigma_{j,k,\chi}{}^{\chi}N_{jk}{}^{\chi}w_{jk}$) from these particular interactions can be set equal to $w\Sigma_{j,k,\chi}{}^{\chi}N_{jk}=wcN_T/2$ in view of the constraints of eqns.(5) and (6). Therefore $\underline{w}$ contributes a constant multiplying factor to the partition function, and hence affects neither orientation nor adsorption equilibria (see eqns.10 and 14).

A physical approximation which is frequently adopted in

monolayer models consists in separating the effects of orientation polarization from those of distortional polarization in the following formal way. Orientation polarization is accounted for by approximating solvent and solute molecules by permanent point dipoles, which are allowed to assume a given number of orientations. In estimating dipole-dipole interactions as well as the interactions between the adsorbed dipoles and the external field due to σ_M, distortional polarization is accounted for by assuming that the dipoles are embedded in a fictituous dielectric medium having a dielectric constant ε_∞ related to the sole distortional polarization. Relatively recent measurements of high-frequency dielectric constant of water (36,37)give for ε_∞ a value close to 4.5. On the other hand, the ε_∞ values which provide the best fit between the theoretical predictions of monolayer models and experimantal results usually range from 5.3 to 6. In this respect ε_∞ must be regarded as an adjustable parameter, albiet within a relatively narrow range of values. In the framework of this physical picture, the potential drop $\Delta\phi$ across the monolayer is considered to be the sum of two contributions:

$$\Delta\phi = 4\pi\sigma_M d/\varepsilon_\infty - 4\pi \sum_i \mu_i N_i/\varepsilon_\infty \tag{15}$$

where d is the monolayer thickness and μ_i is the normal component of the dipole moment of the i particle in the direction from the metal towards the solution. The first contribution expresses the potential drop, due to the charge σ_M on the metal, across a continuum of dielectric constant ε_∞. The second contribution, frequently referred to as the *surface dipole potential* and denoted by $\Delta\chi$, stems from the net orientation of the permanent dipoles. Upon differentiating eqn.(15) with respect to σ_M, we obtain the reciprocal of the inner-layer differential capacity:

$$C_i^{-1} = d(\Delta\phi)/d\sigma_M = 4\pi d/\varepsilon_\infty - (4\pi/\varepsilon_\infty) \sum_i \mu_i \, dN_i/d\sigma_M \tag{16}$$

While the first contribution to C_i^{-1} in eqn.(16) is independent of σ_M, the second contribution varies with charge and hence accounts for the charge dependence of C_i.

A more realistic physical picture of the adsorbed molecules consists in regarding them as polarizable point dipoles characterized by a permanent dipole moment $\vec{\mu}$ and a distortional polarizability α. An approach used by Levine, Bell and Smith (38) and ,subsequently, by Fawcett et al. (39-42) is as follows. For simplicity, we shall consider a monolayer exclusively populated by solvent molecules. All solvent molecules are ascribed an average value $\bar{\mu} \equiv (\Sigma_i \mu_i N_i)/N_T$ of the normal component of the permanent dipole moment. The average value $\bar{\mu}_{tot}$ of the total dipole-moment normal component is then set equal to

$$\bar{\mu}_{tot} = \bar{\mu} + \alpha\,\xi$$

where ξ is the average electric field acting on a solvent molecule and $\alpha\xi$ is the average value of the induced dipole moment. ξ is expressed in implicit form by the relationship

$$\xi = 4\pi\sigma_M + c\bar{\mu}_{tot}/d^3 = 4\pi\sigma_M + c(\bar{\mu}+\alpha\xi)/d^3 \qquad (17)$$

The first contribution to ξ in eqn.(17) is the electric field created in a vacuum by a uniform charge distribution of density σ_M on the metal and by the corresponding charge distribution of density $-\sigma_M$ due to the diffuse-layer ions. Noting that the field created in a vacuum by a dipole of dipole moment μ_{tot} in a direction normal to the dipole (and hence parallel to the electrode surface plane) at a distance d equals μ_{tot}/d^3, the second contribution to ξ expresses the field created at any adsorption site by the c nearest-neighbour solvent dipoles. (In a regular lattice of spherical molecules, the inner-layer thickness d coincides with the distance between any two n.n. molecules.) A limitation of this approach is that it neglects the *local order* which any particle tends to create in the n.n. sites by favouring their occupation by those particles which interact with it more attractively(see Sec. 3.3). In this respect the solvent molecules immediately adjacent to a given solvent molecule have an average orientation which depends on that of the *central* solvent molecule and which may differ appreciably from that expressed by $\bar{\mu}_{tot}$. Hence the electric field ξ expressed by eqn.(17) may be notably in error.

3.3. Mathematical approximations in monolayer models

Mathematical approximations aim at simplifying the general expression of eqn.(3) for the canonical partition function. A familiar procedure in statistical mechanics consists in replacing the summation $\Sigma\{N_{jk}\}$ over all possible sets $\{^{\chi}N_{jk}\}$ by its maximum term. This procedure is justified by the fact that in general this term makes by far the most significant contribution to the summation . Equation (3) is therefore simplified as follows:

$$Q = (\prod_i q_i^{N_i})\, W(\{N_i\}) \exp[-(\sum_{j,k,\chi} {}^{\chi}N_{jk}^{\dagger}\, {}^{\chi}w_{jk})/kT] \qquad (18)$$

with

$$g(\{N_i\},\{^{\chi}N_{jk}^{\dagger}\}) \simeq \sum_{\{^{\chi}N_{jk}\}} g(\{N_i\},\{^{\chi}N_{jk}\}) \equiv W(\{N_i\}) \qquad (19)$$

Here the $^{\chi}N_{jk}^{\dagger}$'s are the numbers of pairs which maximize Q, namely those which satisfy the equation

$$(\partial \ln Q/\partial\, {}^{\chi}N_{jk})_{N_i} = 0 \qquad (20)$$

with Q expressed by eqn.(3), compatibly with the constraints of eqn.(6). In writing eqn.(18) we have also assumed that the $g(\{N_i\},\{^{\chi}N^{\dagger}_{jk}\})$ factor for the maximizing set $\{^{\chi}N^{\dagger}_{jk}\}$ is practically equal to the total number, $W(\{N_i\})$, of lattice configurations corresponding to the distribution $\{N_i\}$. This further assumption is more drastic than those adopted in first-order mathematical approximations , such as the quasi-chemical approximation (see further), but it is convenient to retain it for the present purposes.

If all particles in the monolayer are monomeric, then $W(\{N_i\})$ is given by the equation

$$W(\{N_i\}) = N_T! \ / \ (\prod_i N_i! \ ) \tag{21}$$

This expression is readily obtained by considering that $W(\{N_i\})$ is the number of ways of assigning N_T particles into N_T distinguishable adsorption sites. If the particles were all distinguishable, then there would be N_T choices for filling the first site, (N_T-1) for the second site, and so on. Hence W would be equal to $N_T!$. Since the particles of a given type $\underline{i}$ (namely the molecules of a given species in a given orientation) are actually indistinguishable, in order to cancel the originally assumed distinguishability from $W=N_T!$, this expression must be divided by $N_i!$. Following the same procedure for all types of particles, eqn.(21) is obtained. In the presence of a n-meric solute molecule S , only approximate expressions of W are available. A popular expression due to Flory (43) and valid for chain molecules is the following:

$$W(\{N_i\}) = 2^{-N_S} \left[(c-1)/N_T\right]^{(n-1)N_S} \ N_T!/(\prod_i N_i!) \tag{22}$$

where the product $\prod_i$ extends over all monomeric particles and over the solute molecule S. The relation of the various factors composing the partition function of eqn.(18) to the thermodynamic variables of the monolayer appears in a particularly clear way by substituting Q from eqn.(18) into eqn.(8) while taking eqn.(4) into account:

$$A \equiv U-TS = -kT \ln Q = \sum_i N_i(U_i^{el}+U_i^{c}) + \sum_{j,k,\chi} {}^{\chi}N^{\dagger}_{jk}\, {}^{\chi}w_{jk} - kT \ln W(\{N_i\}) \tag{23}$$

A comparison between the second and the fourth member of eqn.(23) shows that the sum of the interaction energies between all n.n. pairs in the monolayer and of the interaction energies of all monolayer particles with the two contacting phases is a measure of the internal energy U of the monolayer. Moreover, the configurational entropy S of the monolayer is measured by k ln W,and hence increases with the number of possible configurations on the lattice sites.

The various mathematical approximations differ by the way

in which the maximization of the Q expression in eqn.(3) is carried out in order to obtain the ${}^{\chi}N^{\dagger}_{jk}$ values. Henceforth the dagger will be omitted to simplify notations, and the symbol ${}^{\chi}N_{jk}$ will be used exclusively to denote the numbers of pairs which maximize the partition function. The simplest and most frequently adopted approximation consists in carrying out the maximization procedure upon disregarding the exponential factor in eqn.(3), namely upon regarding all interaction energies ${}^{\chi}w_{jk}$ either as equal to zero or (which has the same effect) as equal between themselves. This amounts to assuming that the various particles are distributed randomly among the sites. Differently stated, the presence of a particle at a given site is assumed not to favour the occupation of the n.n. sites by particles which interact attractively with it over those which interact repulsively. This approximation is referred to as *random-mixing* , *random*, *Bragg-Williams*, or *zero-order* approximation. According to such an approximation, the total number of pairs between two monomeric particles j and k which follow each other along both directions of all axes of the lattice is given by the number of sites surrounding all j particles per unit surface (i.e. , cN_j) , times the probability of any of these sites being occupied by the k particle; the latter probability is then equated to the ratio of the number of sites surrounding all k particles (i.e., cN_k) to the total number of sites surrounding any given particle per unit surface of the monolayer (i.e., cN_T). Whence:

$$\sum_{\chi} ({}^{\chi}N_{jk} + {}^{\chi}N_{kj}) = cN_jN_k / N_T \tag{24}$$

and, for j=k :

$$\sum_{\chi} {}^{\chi}N_{jj} = (c/2)\, N_j^2/N_T \tag{25}$$

In the random approximation the number of pairs between any two particles j and k is clearly the same independently of the order in which they follow each other along a given axis χ (i.e., ${}^{\chi}N_{jk} = {}^{\chi}N_{kj}$) and independently of the choice of the axis. Hence, when dealing with this approximation, we shall denote by N_{jk} the total number of pairs of j and k particles :

$$N_{jk} \equiv \sum_{\chi}({}^{\chi}N_{jk} + {}^{\chi}N_{kj}) \quad ; \quad N_{jj} \equiv \sum_{\chi} {}^{\chi}N_{jj} \tag{26}$$

In the presence of n-meric solute molecules S, eqn.(24) must be modified by considering that the number of sites surrounding a solute molecule equals $cp = [cn-2(n-1)]$ (see eqn.7). Moreover, in the case of a mixture of monomeric particles and of a n-meric solute molecule S , the total number of sites surrounding all particles is given by

$$c \sum_{i \neq S} N_i + cpN_S \equiv cN_p \tag{27}$$

Hence, by the same reasoning leading to eqn.(24), the numbers of pairs between the various particles will now be given by

$$N_{jk} = cN_j N_k / N_p \qquad \text{for } \underline{j} \text{ and } \underline{k} \neq S \tag{28}$$

$$N_{jj} = cN_j^2 \,/\, 2N_p \quad ; \quad N_{jS} = cN_j pN_S \,/\, N_p \quad \text{for } j \neq S \tag{29}$$

$$N_{SS} = cp^2 N_S^2 \,/\, 2N_p \tag{30}$$

It should be noted that the expressions of eqns.(24),(25) and (28)-(30) for the numbers of pairs satisfy directly the constraints of eqns.(6) and (7), which are therefore redundant in the random approximation.

By its very nature, the random approximation cannot account for the interactions between the *horizontal components* of the dipole moments of n.n. molecules, namely the components parallel to the electrode-surface plane. In the absence of an external field parallel to the electrode-surface plane, the number of dipolar molecules with a given dipole-moment normal component and a given horizontal component is certainly equal to the number of molecules with the same normal component and an apposite horizontal component, for obvious symmetry reasons. Hence, if these dipolar molecules are distributed randomly among the sites as assumed in the random approximation, the overall effect of their horizontal components at any given site will vanish. In this respect, horizontal dipoles in the random approximation behave as nothing but holes in the monolayer, as far as their energies of lateral interaction are concerned. Differently stated, in the random approximation the probability of occupation of any given site by a particle $\underline{j}$ is not affected by the interaction energies of this particle with the n.n. particles, and hence it is merely proportional to the number of $\underline{j}$ particles, N_j. Hence, if a given molecular species is assumed to be adsorbed in $\underline{m}$ different orientations, the monolayer behaves as though all molecules of this species were adsorbed in the same orientation; the dipole moment $\bar{\vec{\mu}}$ of this orientation is simply the weighted average of the dipole moments $\vec{\mu}_j$ of all allowed orientations, the weighting factor of the $\underline{j}$ orientation being equal to $N_j / \Sigma_{j=1}^{m} N_j$:

$$\bar{\vec{\mu}} = \sum_{j=1}^{m} \vec{\mu}_j N_j \Big/ \sum_{j=1}^{m} N_j \tag{31}$$

Clearly, in view of the preceding considerations, the only nonzero component of the $\bar{\vec{\mu}}$ vector is in the direction normal to the electrode surface plane. Hence, the interactions between dipole-moment horizontal components are non-operative. Note that, in view of eqn.(31), Levine's procedure (38) of eqn.(17) for taking the distortional polarizability α of adsorbed molecules into account is perfectly consistent with the random approximation. This is no longer true when local order is accounted for.

When the lateral interaction energies ${}^{X}w_{jk}$'s are comparable with kT, which is a measure of the energy of thermal agitation, the random approximation may turn out to be seriously in error. In fact, under these conditions, a particle at a given site will tend to favour the occupation of the n.n. sites by those particles which interact with it more attractively, compatibly with the randomizing forces of thermal agitation. In other words a particle accupying a given site will tend to create a *local order* about this site. A first-order mathematical approximation which accounts for local order and yields an analytic solution of the statistical-mechanical problem is the *quasi-chemical approximation* (33,44). It is perfectly equivalent to a procedure introduced by Bethe (33, 45). According to the quasi-chemical approximation the number, $g(\{N_i\},\{{}^{X}N_{jk}\})$ of configurations in eqn.(3) is first estimated as if the various numbers of pairs ${}^{X}N_{jk}$'s did not interphere with one another. Since the ${}^{X}N_{jk}$'s are actually interconnected, such a procedure overestimates $g(\{N_i\},\{{}^{X}N_{jk}\})$; this quantity is then approximately corrected via a correction factor which is arbitrarily assumed to be only a function of $\{N_i\}$. The numbers of pairs which maximize the partition function are ultimately found to satisfy the equation

$$ {}^{X}N_{jk}\,{}^{X}N_{lm} / ({}^{X}N_{jm}\,{}^{X}N_{lk}) = \exp\left[-({}^{X}w_{jk}+{}^{X}w_{lm}-{}^{X}w_{jm}-{}^{X}w_{lk})/kT\right] \quad (32) $$

Since the overall energy in the exponent of the Boltzmann factor of eqn.(32) is that required to convert a (jm) pair and a (lk) pair into a (jk) pair and a (lm) pair, this equation has exactly the same form as the well known law of mass action for chemical reactions. This is the reason why this method is called the quasi-chemical approximation.

More accurate mathematical approximations, some of which also account for second and higher neighbour interactions, are reported in Domb's review (46). Usually, however, a moderate increase in accuracy is attained at the expense of a notable increase in the compexity of the calculations. Rigorous results for any well-defined model can be obtained by computer simulation techniques, such as the Monte Carlo method, which do not involve mathematical approximations. Application of Monte Carlo computations (47) to monolayer models requires the generation of a sequence of configurations of the particles in the lattice by successive random dispacements. Not all configurations which occur are accepted, the decision whether to accept or reject a particular configuration being made in a way which ensures , asymptotically, that the sampled configurations make a significant contribution to the canonical partition function. Any function of the lattice configuration is then obtained as an unweighted average over the resulting set of configurations. Typically, between 10^5 and 10^6 configurations are generated. The main drawback of this essentially exact method is that the computations are fairly expensive to carry out, so that one is limited to a rather small number N_T of particles in the sample studied.

3.4. Comparison between different mathematical approximations

The predictions of the rigorous Monte Carlo method, the random approximation and the quasi-chemical approximation are compared in Figs. 4 and 5 for a model of a square lattice of point dipoles (48), whose orientations are represented by the vectors in Fig. 6. The colatitude θ is given the values 0 (one vertical orientation directed towards the solution), $\pi/4$, $\pi/2$, $3\pi/4$ and 2π (one vertical orientation directed towards the metal). Each of the θ values $\pi/4$, $\pi/2$ and $3\pi/4$ is associated with eight values of the longitude, which increases from 0 to 2π by $\pi/4$ steps. The values of ε_∞=5.5, d=3Å, T=298°K and of the magnitude μ_T=1.8D of the dipole moment which are employed in the calculations are close to those commonly used in monolayer models of water molecules. Figure 4 shows the average value $\bar{\mu}=\Sigma\mu_i N_i/N_T$ of the dipole-moment normal component as a function of charge σ_M. Imaging in the metal surface plane was considered in calculating curve a, whereas it was ignored in calculating curves b and c. The solid curves a and b were calculated by the quasi-chemical approximation, whereas the solid curve c was obtained by the random approximation. Points were obtained by the Monte Carlo method. Agreement between the Monte Carlo method and the quasi-chemical approximation is excellent, the differences between the corresponding $\bar{\mu}$ values being usually less than 1%. Conversely, the curve calculated by the random approximation (curve c) is decidedly at variance with the corresponding curve b, calculated by the two other methods. Figure 5 shows how the numbers of molecules with the same colatitude θ , divided by N_T, vary with charge. Calculations were carried out for the same parameters as in Fig. 4 upon ignoring images. Numbers on each curve refer to the θ value. Even the more detailed picture of the monolayer behaviour in Fig. 5 shows the excellent agreement between the quasi-chemical approximation (solid curves) and the Monte Carlo method (points). It is interesting to note that the orientations which predominate at zero charge are the horizontal ones ($\theta=\pi/2$). This is due to the fact that the interaction energy between two point dipoles, a distance d apart, attains a maximum negative value, $w=-2\mu_T^2/(d^3\varepsilon_\infty)$, when they are in line, head to tail. With the parameters employed in the calculation, this interaction energy equals -1.1kT, and hence is comparable with the energy of thermal agitation. While the head-to-tail configuration corresponds to maximum attraction, the configurations of maximum repulsion are those with the dipoles

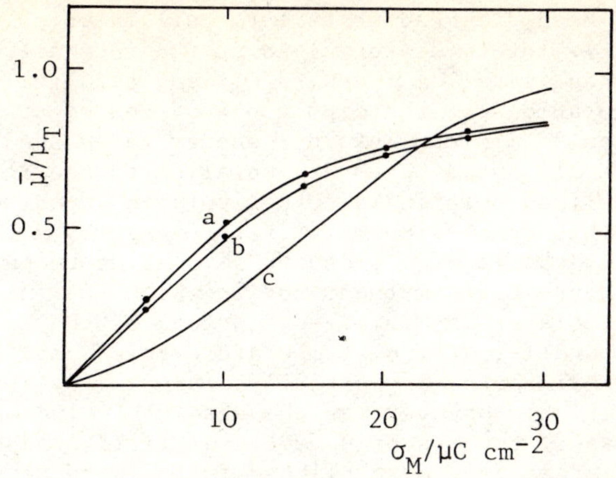

Figure 4. $(\bar{\mu}/\mu_T)$ vs. σ_M plots calculated as described in the text.

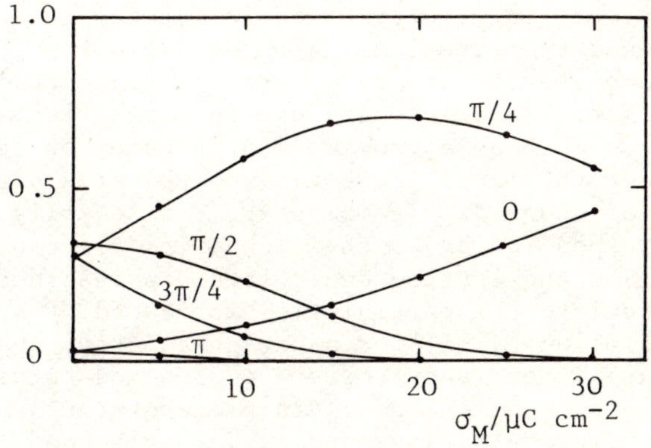

Figure 5. Distribution of the various orientations as a function of σ_M.

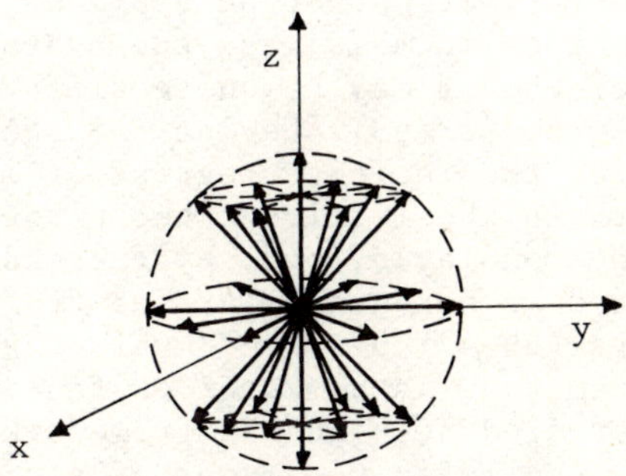

Figure 6. Orientations introduced into the model. The metal-surface plane is defined by the (x,y)plane.

in line, either head to head or tail to tail, in which case w equals $2\mu_T^2/(d^3\varepsilon_\infty)$. Intermediate values of the interaction energy are attained by antiparallel ($w=-\mu_T^2/d^3\varepsilon_\infty$) and parallel ($w = \mu_T^2/d^3\varepsilon_\infty$) configurations of the dipole pairs. Hence, at zero charge, the configuration of minimum internal energy for a square lattice is that in which all the dipoles are parallel to the interface and oriented in lines pointing alternatively in opposite directions (49,50). This conclusion was first reached by Parsons and Reeves (50) on the basis of a computer simulation technique similar to the Monte Carlo method, but in which the internal energy of the monolayer is minimized, instead of the Helmholtz free energy. Naturally, at room temperature this configuration of minimum internal energy is partially disrupted by the forces of thermal agitation. Moreover, as the absolute value of the charge σ_M increases, the resulting external field tends to favour orientations progressively more aligned in the direction of the field over the horizontal ones (see Fig. 5).

3.5. Approximations concerning the coordination number $\underline{c}$

We have seen that in monolayer models lateral interactions are usually confined to nearest neighbours. A simple approximation has been employed by Levine and coworkers (38,51) as well as by Fawcett and coworkers (39-42) to account for the effect of non-nearest neighbours. Such an approximation is based on Topping's calculation (52) of the mutual potential energy of a plane array of point dipoles of equal dipole moment μ_T, and equally oriented with their axes perpendicular to the plane. The electrostatic field at any site of the lattice due to the dipoles on all the other sites is equal to the partial field produced by a single n.n. dipole, multiplied by ≃11 for a hexagonal array and by ≃9 for a square array. It has been therefore proposed to use an *effective coordination number* $c_e=11$ in place of the *geometrical* number c=6 for a hexagonal array, and $c_e=9$ in place of c=4 for a square array. This approximation is consistent with the random approximation, according to which all molecules of the same species behave as though they had the same vertical average orientation $\underline{\mu}$ (see eqn. 31). However, it cannot be applied when local order is accounted for, because in this case the average orientation of second nearest neighbours may be quite different from that of first nearest neighbours (53). Levine and coworkers (38,51) propose to account for the infinite regress of images of the adsorbed dipoles both in the metal-surface plane, z=0, and in the outer boundary of the monolayer, z=d, by increasing the effective coordination number of a hexagonal lattice from 11 to 15.2. Fawcett and coworkers (39-42) use a c_e value greater than 11 as an adjustable parameter. As a matter of fact, an increase in c_e above 11 is unjustified even in the framework of the random approximation. Thus, the infinite images of a normal dipole are aligned with it, head to tail, in the direction normal to the monolayer. Consequently they damp the electric field created by the *real dipole* within the monolayer, rather than enhancing it. In this respect c_e should be decreased with respect to 11 to account for images. At any rate, the effect of images can be easily embodied in any statistical-mechanical treatment by modifying the lateral interaction energies (53). Thus, the interaction energy w_{jk} between two dipoles j and k in the presence of images is given by $-\vec{\mu}_j \cdot \vec{F}_k$, where $\vec{\mu}_j$ is the dipole moment of the j dipole and $\vec{F}_k$ is the electric field created by the other dipole, k, and by its infinite regress of images at the position occupied by the j dipole. For two normal dipoles with dipole moments of magnitude μ_T, placed a distance d from each other and midway between the z=0 and z=d planes, w_{jk} equals $\mu_T^2/(d^3\varepsilon_\infty)$ in the absence of images, and $\mu_T^2(1-2^{-5/2})/(d^3\varepsilon_\infty)$ in the presence of a single image in the metal-surface plane, z=0. In the presence of an infinite image system in the two boundaries z=0 and z=d, we have

$$w_{jk} = \frac{\mu_T^2}{d^3\varepsilon_\infty}\left[1 - 2\sum_{l=1}^{+\infty} \frac{2l^2-1}{(1+l^2)^{5/2}}\right] \tag{33}$$

where the summation extends over all positive integer values of l. The preceding equation clearly shows that images decrease the absolute value of lateral interaction energies.

4. USE OF MONOLAYER MODELS IN INTERPRETING THE DIFFERENTIAL-CAPACITY BEHAVIOUR

The actual application of a monolayer model of solvent molecules to the interpretation of experimental differential-capacity curves requires precise assumptions to be made as to the number and nature of the solvent orientations to be introduced into the model. The most satisfactory approach would consist in generating a very high number of discrete orientations (or, even better, a continuous distribution of orientations) characterized by the same number of orientations per unit solid angle along all directions of three-dimensional space. Just as in Monte Carlo techniques, the choice of the most favourable orientations at any given charge σ_M would then be left to the statistical-mechanical treatment, which will automatically privilege those orientations which occur most frequently in the lattice configurations of higher probability density. Unfortunately this approach is rather time-consuming (48), and a very limited number of solvent orientations is generally adopted, the choice of these orientations being frequently made on the basis of rather arbitrary criteria.

4.1. Two-state models

The simplest choice of solvent orientations is made in the so called *two-state models*. Here solvent dipoles are allowed to assume only two normal orientations, with the negative end pointing to the metal (*up* orientation) or away from it (*down* orientation). These models originate from the Ising model employed to explain ferromagnetism and antiferromagnetism by interactions between spins of certain electrons in neighbouring atoms making up a pure crystal. While the assumption that electrons can be in either one of two spin states is quite reasonable, the corresponding assumption that solvent molecules can be in either one of two opposite normal orientations is much less realistic, as we already pointed out. The absence of horizontal orientations is particularly disappointing, although the use of the random approximation, which is insensitive to interactions between the horizontal components of the dipoles (see Sec. 3.3.), may justify such an absence.

For the sake of generality, let us assume that the components of the *up* (↑) and *down* (↓) dipoles in the direction towards the solution are different in magnitude:

$$\mu_\uparrow \neq -\mu_\downarrow \tag{34}$$

Note that $\mu_\downarrow$ is negative since the down dipole is directed towards the metal. In the random approximation, application of eqn.(23) to the present model yields:

$$-kT \ln Q = N_\uparrow U_\uparrow^c + N_\downarrow U_\downarrow^c - (4\pi\sigma_M/\varepsilon_\infty)(\mu_\uparrow N_\uparrow + \mu_\downarrow N_\downarrow) + cN_\uparrow N_\downarrow w_{\uparrow\downarrow}/N_T + cN_\uparrow^2 w_{\uparrow\uparrow}/(2N_T) + cN_\downarrow^2 w_{\downarrow\downarrow}/(2N_T) - kT \ln [N_T!/(N_\uparrow! N_\downarrow!)] \tag{35}$$

In writing eqn.(35) use was made of eqns.(24)-(26) to express the numbers of pairs and of eqn.(21) to express the total number of configurations. Moreover, the energies $U_\uparrow^{el}$ and $U_\downarrow^{el}$ were expressed by setting the external electric field due to σ_M equal to $4\pi\sigma_M/\varepsilon_\infty$ (see eqn. 15). Upon substituting ln Q from eqn.(35) into eqn.(10) and rearranging terms, we obtain the following equilibrium condition for the reorientation between up and down dipoles:

$$\frac{N_\uparrow}{N_\downarrow} \equiv \frac{\theta_\uparrow}{\theta_\downarrow} = \frac{\exp(-U_\uparrow/kT)}{\exp(-U_\downarrow/kT)} \tag{36}$$

with

$$\theta_\uparrow \equiv N_\uparrow/(N_\uparrow+N_\downarrow) \quad ; \quad \theta_\downarrow \equiv N_\downarrow/(N_\downarrow+N_\uparrow) \quad ;$$

$$U_\uparrow \equiv U_\uparrow^c - 4\pi\sigma_M\mu_\uparrow/\varepsilon_\infty + c\theta_\uparrow w_{\uparrow\uparrow} + c\theta_\downarrow w_{\uparrow\downarrow} \quad ; \tag{37}$$

$$U_\downarrow \equiv U_\downarrow^c - 4\pi\sigma_M\mu_\downarrow/\varepsilon_\infty + c\theta_\downarrow w_{\downarrow\downarrow} + c\theta_\uparrow w_{\uparrow\downarrow} \quad ;$$

In deriving eqn.(36), use was made of Stirling's approximation in the form $d(\ln N!)/dN \simeq \ln N$. Noting that $N_\uparrow+N_\downarrow=N_T$ in view of eqn.(5), eqn.(36) can also be written in the alternative form:

$$N_\uparrow/N_T = \exp(-U_\uparrow/kT)/[\exp(-U_\uparrow/kT) + \exp(-U_\downarrow/kT)] \tag{38}$$

together with an analogous expression for the $(N_\downarrow/N_T)$ ratio. Equation (38) can be immediately generalized to a monolayer of solvent dipoles in any given number, m, of orientations:

$$N_i/N_T = \exp(-U_i/kT) / \sum_{j=1}^{m} \exp(-U_j/kT) \qquad \text{for } 1 \leqslant i \leqslant m \qquad (39)$$

with $N_T = \sum_j N_j$. Just as in eqn.(37), U_i expresses the overall contribution of the $\underline{i}$ orientation to the internal energy of the system (i.e., the chemical and electrostatic energies of interaction with the two contacting phases and the energies of lateral interaction with all possible orientations). If the monolayer contains two or more different molecular species, it is still possible to use for each of these species an expression analogous to eqn.(39):

$$N_i / \sum_{j=1}^{m} N_j = \exp(-U_i/kT) / \sum_{j=1}^{m} \exp(-U_j/kT) \qquad \text{for } 1 \leqslant i \leqslant m \qquad (40)$$

In this case, however, the summation extends over all orientations of the chosen molecular species, and hence $\sum_j N_j$ is $<N_T$. Moreover, the energy U_i must include the energies of lateral interaction of the $\underline{i}$ orientation not only with all orientations of the same molecular species, but also with those of all other species populating the monolayer. Equations (39) and (40) have been frequently misused in the literature. Thus, eqn.(39) has been applied so as to embody orientations of different molecular species (see Sec. 4.3.); this procedure is clearly inappropriate since a mere reorientation cannot convert a molecular species into a different one. In applying eqn.(40) to all orientations of a given species, lateral interactions between this species and the remaining species of the monolayer have been neglected in the U_i terms; moreover, the surface coverages $\theta_i = N_i/N_T$ included in these terms (see eqn. 37) have been erroneously identified with $N_i/\sum_{j=1}^{m} N_j$, in spite of the fact that $\sum_{j=1}^{m} N_j$ is $<N_T$ (23,24).

Returning to two-state models, inner-layer capacity curves are calculated by combining the implicit expressions of $N_\uparrow$ and $N_\downarrow$ as a function of charge in eqn.(38) with the expression of eqn.(16) for C_i, which in the present case takes the form

$$1/C_i = 4\pi d/\varepsilon_\infty - (4\pi/\varepsilon_\infty)(\mu_\uparrow dN_\uparrow/d\sigma_M + \mu_\downarrow dN_\downarrow/d\sigma_M) \qquad (41)$$

Equations (38) and (41) predict a C_i vs. σ_M curve with a rounded maximum at intermediate charges. This behaviour is readily visualized by considering that the $dN_\uparrow/d\sigma_M$ and $dN_\downarrow/d\sigma_M$ derivatives make a maximum contribution to the charge-dependence of C_i at intermediate charges, whereas they tend to vanish at extreme charges, where all solvent dipoles are almost completely aligned in the direction of the external field. From eqns.(38) and (41) it can be readily seen that the capacity maximum is attained for $\theta_\uparrow = \theta_\downarrow = 1/2$. The charge σ_m of maximum capacity is therefore obtained from eqns. (36) and (37) by setting $U_\uparrow = U_\downarrow$ (53):

$$4\pi\sigma_m(\mu_\uparrow-\mu_\downarrow)/\varepsilon_\infty = (U^c_\uparrow - U^c_\downarrow) + c(w_{\uparrow\uparrow}-w_{\downarrow\downarrow})/2 \tag{42}$$

The lateral interaction energies $w_{\uparrow\uparrow}$ and $w_{\downarrow\downarrow}$ in eqn.(42) are clearly equal to:

$$w_{\uparrow\uparrow} = \mu_\uparrow^2 \;/\; (d^3\varepsilon_\infty) \quad ; \quad w_{\downarrow\downarrow} = \mu_\downarrow^2 \;/\; (d^3\varepsilon_\infty) \tag{43}$$

where $\underline{d}$ is the distance between n.n. dipoles. If we set $\mu_\uparrow = -\mu_\downarrow$ (and hence $w_{\uparrow\uparrow}=w_{\downarrow\downarrow}$) as well as $U^c_\uparrow = U^c_\downarrow$, then eqn.(42) yields $\sigma_m=0$. The prediction that the capacity maximum lies at zero charge is, however, at variance with the experimental behaviour of the Hg/NaF interphase (see point I of Sec.2.). In priciple,a way to predict a positive σ_m value, in accordance with experiment, consists in setting $U^c_\uparrow > U^c_\downarrow$ while retaining $\mu_\uparrow=-\mu_\downarrow$ (see eqn.42). This amounts to assuming that down molecules interact with the two contacting phases more attractively than up molecules. Since, however, the capacity maximum is attained for $\theta_\uparrow=\theta_\downarrow=1/2$, a negative shift in the charge density from $\sigma_M=\sigma_m$ to $\sigma_M=0$ will cause $\theta_\downarrow$ to become greater than $\theta_\uparrow$. Hence the surface dipole potential $\Delta\chi = -(4\pi/\varepsilon_\infty)\mu_\uparrow N_T(\theta_\uparrow-\theta_\downarrow)$ will be positive at zero charge (see Fig.7),

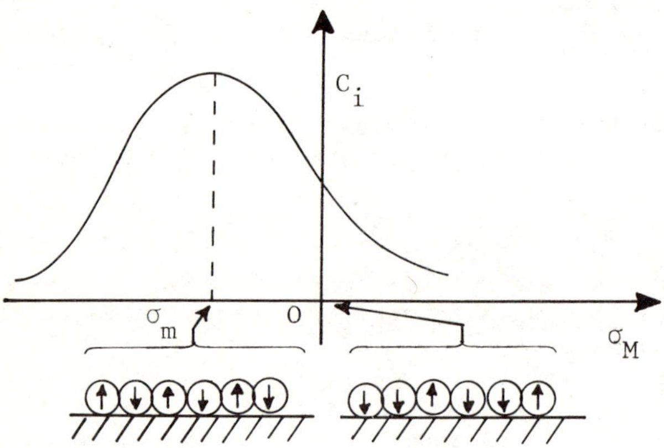

Figure 7. Schematic picture of the C_i vs σ_M curve and of the distribution of dipoles at $\sigma_M=\sigma_m$ and at $\sigma_M=0$ for $\mu_\uparrow=-\mu_\downarrow$ and $U^c_\uparrow>U^c_\downarrow$

in contrast with several pieces of experimental evidence which point to a negative $\Delta\chi$ value (see point III of Sec. 2.). To account simultaneously for a positive σ_m value and a negative $\Delta\chi$ value at zero charge, we must assume that $\mu_\uparrow$ is $> -\mu_\downarrow$, and hence that $w_{\uparrow\uparrow}$ is $> w_{\downarrow\downarrow}$. In this case σ_m is positive in view of eqn.(42), while at the same time $\Delta\chi=-(4\pi/\varepsilon_\infty)N_T(\mu_\uparrow\theta_\uparrow+\mu_\downarrow\theta_\downarrow)$ is surely negative at the capacity hump, where $\theta_\uparrow=\theta_\downarrow=\frac{1}{2}$, and may well remain negative at zero charge (38,53).

The first application of the two-state model to metal-solution interphases is due to Watts-Tobin (54). His approach differs from the general approach previously outlined by two main points. In applying eqn.(36), lateral-interaction energies are not introduced explicitly; rather, the solvent dipoles are regarded as subject to a total electric field

$$X_{tot} = \frac{\Delta\phi}{d} = \frac{4\pi\sigma_M}{\varepsilon_\infty} - \frac{4\pi\mu_\uparrow}{d\varepsilon_\infty} N_T(\theta_\uparrow - \theta_\downarrow) \tag{44}$$

as obtained by applying eqn.(15) to the present model, with $\mu_\uparrow = -\mu_\downarrow$. Clearly X_{tot} includes not only the external field ($4\pi\sigma_M/\varepsilon_\infty$) due to σ_M, but also the mean field due to all partially aligned dipoles. In applying eqn.(36), the energies $U_\uparrow$ and $U_\downarrow$ are therefore set equal to

$$U_\uparrow = U^c_\uparrow - \mu_\uparrow X_{tot} \quad ; \quad U_\downarrow = U^c_\downarrow - \mu_\downarrow X_{tot} \tag{45}$$

with X_{tot} given by eqn.(44). This *mean field approximation* is conceptually equivalent to the random approximation, though it may lead to somewhat different results. It must also be stressed that Watts-Tobin's approach is *potential-centric* rather than *charge-centric*. Thus, the inner-layer capacity C_i is obtained by differentiating the charge σ_M, as derived from eqn.(44), with respect to $\Delta\phi$:

$$C_i = \frac{d\sigma_M}{d(\Delta\phi)} = \frac{\varepsilon_\infty}{4\pi d} + \frac{2\mu_\uparrow N_T}{d} \frac{d\theta_\uparrow}{d(\Delta\phi)} \tag{46}$$

In writing eqn.(46) use has been made of the constraint of eqn.(5), $\theta_\uparrow + \theta_\downarrow = 1$. The derivative $d\theta_\uparrow/d(\Delta\phi)$ is then obtained from eqns.(36), (44) and (45). Equation (46) yields calculated curves of C_i vs $\Delta\phi$.

All treatments of two-state models which have followed the pioneering work by Watts-Tobin are *charge-centric* in that they use equations of the type of eqn.(41) to derive C_i as a function of σ_M. Incidentally, calculated C_i vs σ_M curves can be more directly compared with experiment than calculated C_i vs $\Delta\phi$ curves, because σ_M is a thermodynamic quantity, whereas the estimate of $\Delta\phi$ requires modelistic assumptions. The charge-centric approach was first proposed by Bockris, Devanathan and Müller (23) for the particular case of $U^c_\uparrow = U^c_\downarrow$ and $\mu_\uparrow = -\mu_\downarrow$. Curiously enough, these authors did not use such an approach to explain the experimental capacity hump at the Hg/H_2O interphase, but rather to discard such an explanation. According to these authors, eqn.(41) yields C_i values higher than those found experimentally. Hence they conclude that the capacity hump is not due to water reorientation, but rather to anion specific adsorption. As a matter of fact, in employing eqn.(41) Bockris and coworkers (23) used a factor 2π instead of 4π in the second term of this equation. Once the correct expression is employed, the calculated C_i values become comparable with the observed ones (3). An essentially analogous two-state model, apart from the use of a somewhat different formalism, was subsequently presented by Cooper and Harrison (55). Levine, Bell and Smith (38) presented a two-state model in which distortional polarization is accounted for as outlined in connection with eqn.(17). They also allowed for the possibility that $\mu_\uparrow$ is different from $-\mu_\downarrow$.

All two-state models so far mentioned make use of the random approximation. The present author (53) examined a two-state model on the basis of the quasi-chemical approximation to account for local order. It was thus shown that as the energies $w_{\uparrow\uparrow}$, $w_{\uparrow\downarrow}$ and $w_{\downarrow\downarrow}$ of lateral interaction become increasingly greater than kT, the single hump in the C_i vs σ_M curve splits into two symmetrical peaks whose separation increases progressively. This is because an increase in the magnitude of lateral interactions tends to favour highly ordered configurations in which any up dipole tends to be surrounded by the maximum number of down dipoles. These ordered configurations, which are particularly stable in the proximity of $\theta_\uparrow=\theta_\downarrow=\frac{1}{2}$, tend to resist an increase in the external field, giving rise to a capacity minimum for $\theta_\uparrow=\frac{1}{2}$. However, at sufficiently high $|\sigma_M|$ values, the $\theta_\uparrow$ value will start to deviate appreciably from $\frac{1}{2}$ in any of the two directions, until ordered configurations will ultimately collapse. It is at this point that $d\theta_\uparrow/d\sigma_M$, and hence the capacity C_i, attains a maximum value.

We have seen that the best agreement between the predictions of two-state models and experiment is attained by setting $\mu_\uparrow>-\mu_\downarrow$ (38,53). However, it is difficult to attach a physical significance to this assumption, which merely reflects the inadequacy of two-state models. At most, we can state that the position $\mu_\uparrow>-\mu_\downarrow$ suggests an asymmetric behaviour of interfacial water with respect to charge, and a greater tendency of water molecules to align along the external field at positive than at negative charges.

4.2. A three-state model with a horizontal orientation

An attempt to improve the two-state model by introducing a third orientation parallel to the electrode was made by Fawcett (39). In the framework of the random approximation, this author uses eqn.(39) to express the $N_\uparrow/N_T$, $N_\downarrow/N_T$ and $N_\cdot/N_T$ ratios, where $N_\cdot$ denotes the number of horizontal dipoles per unit surface and $N_T=N_\uparrow+N_\downarrow+N_\cdot$. The expressions for $U_\uparrow$, $U_\downarrow$ and $U_\cdot$ account for distortional polarization by the use of the polarizability α instead of the dielectric constant ε_∞ (see eqn.17). Dropping this complication here, these expressions become

$$U_\uparrow = U_\uparrow^c - 4\pi\sigma_M\mu_\uparrow + c\theta_\uparrow w_{\uparrow\uparrow} + c\theta_\downarrow w_{\uparrow\downarrow} ;$$
$$U_\downarrow = U_\downarrow^c - 4\pi\sigma_M\mu_\downarrow + c\theta_\downarrow w_{\downarrow\downarrow} + c\theta_\uparrow w_{\uparrow\downarrow} ; \quad U_\cdot = U_\cdot^c \qquad (47)$$

Lateral interactions of up or down dipoles with the horizontal one do not appear in eq.(47) because the electrostatic interaction between two dipoles so oriented as to form a right angle is zero.

Lateral interactions between horizontal dipoles are also lacking, in spite of the fact that the strongest electrostatic interactions are established between two horizontal dipoles in line, in a head-to-tail configuration (see Sec. 3.4.). This is a direct consequence of the random approximation being blind to lateral interactions between horizontal dipoles. Horizontal dipoles do not even interact with the normal external field $4\pi\sigma_M$, and hence they actually behave as holes in the monolayer (see Fig.8). Since the external

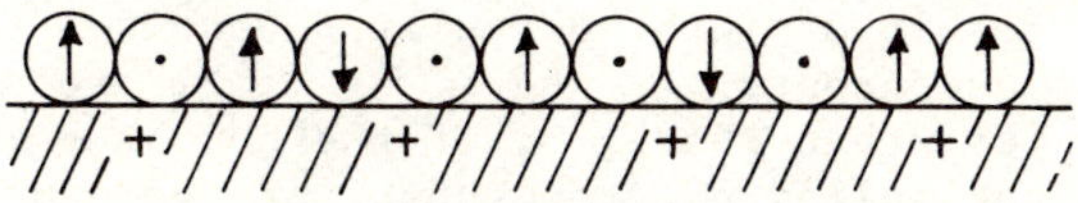

Figure 8. Schematic picture of Fawcett's three-state model (39).

field stabilizes the up or down dipoles, depending on whether σ_M is positive or negative, Fawcett is forced to set the chemical interaction energy $U^c_\cdot$ more negative (namely more attractive) than $U^c_\uparrow$ or $U^c_\downarrow$, in order to make $N_\cdot/N_T$ appreciably different from zero. If $U^c_\cdot$ is made sufficiently negative with respect to $U^c_\uparrow$ and $U^c_\downarrow$, then horizontal dipoles are *chemically* stabilized to such an extent with respect to up and down dipoles in the proximity of zero charge, so as to give rise to a capacity minimum over this σ_M range. In a sense, setting $U^c_\cdot$ less than $U^c_\uparrow$ and $U^c_\downarrow$ amounts to simulating the strong lateral interactions between horizontal dipoles. However, two limitations of this procedure are evident: (i) while lateral interactions between horizontal dipoles can be treated on the same footing as those between up and down dipoles, $U^c_\cdot$ is an adjustable parameter; (ii) the relative effects of the chemical energies $U^c_\uparrow$, $U^c_\downarrow$ and $U^c_\cdot$ depend exclusively upon the distribution $\{N_i\}$ of the dipoles, whereas those of lateral interactions also depend on the lattice configuration, at least provided local order is accounted for. Fawcett and coworkers (39,40,56,57) have extensively used this model to interpret the differential capacity curves at the interphase between mercury and a number of aprotic solvents. A multistate model for unassociated solvents recently presented by Fawcett and de Nobriga (42) suffers from the same limitations proper to the random approximation.

4.3. Cluster models

The suggestion that clusters of water molecules play an important role in determining the behaviour of the inner-layer capacity at metal-water interphases was first put forward by Damaskin and Frumkin (58). These authors assume that the monolayer is populated by *up* and *down water clusters*, as well as by *up water monomers*. Figure 9 shows a schematic picture of the model. The clusters are aggregates of hydrogen-bonded water

Figure 9. Schematic picture of Damaskin and Frumkin's monolaver model (58)

molecules which are considered to be adsorbed with the net dipole moment directed either towards the solution (*up clusters*) or in the opposite direction (*down clusters*). Since the constituent molecules of the up and down clusters are oriented with respect to one another as a result of intermolecular H-bonding, their dipole-moment normal components, $\mu_{\uparrow c}$ and $\mu_{\downarrow c} = -\mu_{\uparrow c}$, are assumed to have a magnitude less than that, $\mu_{\uparrow}$, of the up water monomers. Here, as well as in the following, the subscripts $\uparrow c$ and $\downarrow c$ will be used to denote up and down clusters, respectively. In Damaskin and Frumkin's model the role of up monomers is no less important than that of clusters, since for the first time these monomers are assumed to be chemisorbed to a different extent depending on the nature of the metal.

Equation (40) is correctly applied to the two cluster orientations, upon assuming that *all* lateral interactions are entirely negligible:

$$\left.\begin{aligned} \frac{N_{\uparrow c}}{N_{\uparrow c}+N_{\downarrow c}} &= \frac{\exp(-U_{\uparrow c}/kT)}{\exp(-U_{\uparrow c}/kT)+\exp(-U_{\downarrow c}/kT)} \\ \frac{N_{\downarrow c}}{N_{\uparrow c}+N_{\downarrow c}} &= \frac{\exp(-U_{\downarrow c}/kT)}{\exp(-U_{\uparrow c}/kT)+\exp(-U_{\downarrow c}/kT)} \end{aligned}\right\} \qquad (48)$$

with

$$U_{\uparrow c}=-4\pi\sigma_M\mu_{\uparrow c}/\varepsilon_\infty \quad ; \quad U_{\downarrow c}= -4\pi\sigma_M\mu_{\downarrow c}/\varepsilon_\infty \qquad (49)$$

However, the sum $(N_{\uparrow c}+N_{\downarrow c})$ is set equal to the constant number of adsorption sites per unit surface, N_T, as if the up monomers did not occupy any space. This is a hardly justifiable assumption, since the correct application of the constraint of eqn.(5) to the present situation yields

$$N_T = N_{\uparrow} + nN_{\uparrow c} + nN_{\downarrow c} \qquad (50)$$

where <u>n</u> is the number of adjacent adsorption sites occupied by one water cluster. The chemisorbed water monomers, with the oxygen atom pointing towards the metal, are allowed to increase freely with increasing positive charge according to the exponential relationship $N_{\uparrow} = N^{\circ}_{\uparrow}\exp(\gamma_{\uparrow}\sigma_M)$, where $N^{\circ}_{\uparrow}$ and $\gamma_{\uparrow}$ are adjustable parameters depending on the nature of the metal. The potential difference $\Delta\phi$ across the monolayer is obtained by applying eqn.(15) to the present case:

$$\Delta\phi = 4\pi\sigma_M d/\varepsilon_\infty - (4\pi/\varepsilon_\infty)(\mu_\uparrow N_\uparrow + \mu_{\uparrow c}N_{\uparrow c} + \mu_{\downarrow c}N_{\downarrow c}) \tag{51}$$

and $1/C_i$ is finally obtained by differentiating eqn.(51) with respect to σ_M while taking eqns.(48) and (49) into account. Damaskin and Frumkin's model contains five adjustable parameters. In the absence of water chemisorption (i.e., for $N_\uparrow^\circ=0$) the model reduces to a two-state model, and hence predicts the usual bell-shaped curve with a maximum at zero charge (see curve 1 in Fig.10).

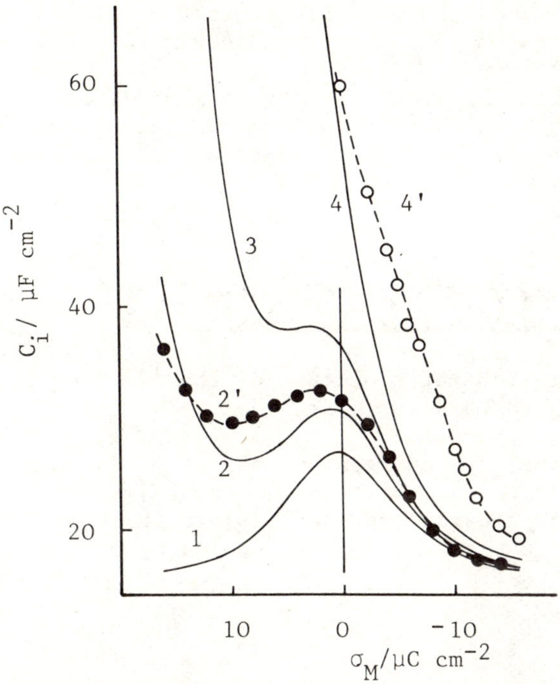

Figure 10. C_i vs. σ_M curves according to Damaskin and Frumkin's model (58).

Naturally, in this case the hump is due to reorientation of clusters from the up orientation at more positive charges to the down orientation at more negative charges. When $N_\uparrow^\circ$ is >0, the chemisorbed monomers tend to produce a roughly exponential increase in C_i towards more positive charges, which is steeper the higher is $N_\uparrow^\circ$. Curves 2 to 4 in Fig. 10 refer to increasing $N_\uparrow^\circ$ values at constant $\gamma_\uparrow$. It is evident that such an exponential increase in capacity has the effect of shifting the capacity hump from zero charge towards positive charges, thus simulating the experimental C_i vs. σ_M curve for the Hg/H_2O interphase at 0°C (dashed curve 2' in Fig.10). For sufficiently high $N^\circ_\uparrow$ values, the exponential increase in capacity due to up monomers starts to be felt at sufficiently negative charges so as to obliterate the hump due to reorientation of clusters. The resulting calculated curves simulate the differential capacity curves on hydrophilic metals (see point IV in Sec.2.). For comparison, Fig.10 shows the C_i vs. σ_M curve at the interphase between Cd and aqueous KF at 0°C (curve 4'). Naturally,when $N^\circ_\uparrow$ is >0, the presence of up monomers at zero charge causes $\Delta\chi$ to be negative at this charge, in agreement with experiment (point III in Sec.2.). The physical modelling in Damaskin and Frumkin's approach is appealing, but the corresponding statistical-mechanical treatment is definitely incorrect. This explains why this cluster model was subsequently improved by Parsons (59), by Damaskin himself (60) and by Fawcett et al.(41).

Parsons slightly modifies the physical model by the further introduction of a *down chemisorbed monomer* (see Fig.11), so as to allow for the possibility that clusters break up into monomers

Figure 11. Schematic picture of the adsorbed monolayer according to Parsons' model (59).

not only towards positive charges but also towards negative ones. To determine the dependence of $N_\uparrow, N_\downarrow, N_{\uparrow c}$ and $N_{\downarrow c}$ upon charge, Parsons applies the general equation (39) to the $N_\uparrow/N_T, N_\downarrow/N_T$, $N_{\uparrow c}/N_T$ and $N_{\downarrow c}/N_T$ ratios , using the following expressions for the various energies U_i:

$$U_{\uparrow c}=-X\mu_{\uparrow c} \; ; \; U_{\downarrow c} = -X\mu_{\downarrow c} \; ; \; U_\uparrow = U^c_\uparrow - X\mu_\uparrow \; ; \; U_\downarrow = U^c_\downarrow - X\mu_\downarrow \quad (52)$$

where $\mu_\uparrow=-\mu_\downarrow$, $\mu_{\uparrow c}=-\mu_{\downarrow c}$ and $X\equiv 4\pi\sigma_M/\varepsilon_\infty$ is the external field ; $\mu_{\uparrow c}$ is regarded as much less than $\mu_\uparrow$. Upon comparing eqn.(52) with eqn.(37), it is apparent that the energies in eqn.(52) do not include the lateral interactions between the four different particles populating the monolayer, which interactions are therefore regarded as negligible. Moreover, the chemical energies $U^c_{\uparrow c}$ and $U^c_{\downarrow c}$ of up and down clusters are equated to zero. This simply means that $U^c_{\uparrow c}=U^c_{\downarrow c}$ is taken for convenience as the ground state for the energies of chemical interaction of all particles with the two contacting phases. Thus, from eqn.(39) it is evident that the N_i/N_T ratios, and hence the inner-layer capacity C_i, do not depend on all U^c_i values but rather on the differences between these values, which reduces the number of these adjustable parameters by one unit with respect to the number of particles. The application of eqn.(39) to Parsons' model and, more generally, to all cluster models is inappropriate because this equation expresses a reorientation equilibrium. Conversely, an interchange between monomers and clusters does not involve a mere reorientation, but also, and mainly, a breaking or forming of H-bonds. Even passing over this inconsistency, clusters, by their very nature, are expected to occupy a number n of lattice sites greater than unity, once water monomers are assumed to occupy a single site. Hence, application of the constraint of eqn.(5) to the present model yields:

$$N_T = N_{\uparrow} + N_{\downarrow} + nN_{\uparrow c} + nN_{\downarrow c} \qquad (53)$$

Upon postulating a fictituous reorientation equilibrium between a water monomer and a cluster, the condition for such an equilibrium must be therefore modified with respect to eqn.(10) in order to account for the difference in size between monomers and clusters. This is readily done by applying the general equation (9) to the interchange between a monomer and a cluster (e.g., between the up monomer and the up cluster). Setting $N_1 \equiv N_{\uparrow}$ and $N_j \equiv N_{\uparrow c}$ in eqn.(9) and noting from the constraint of eqn.(53) that $dN_{\uparrow}/dN_{\uparrow c} = -n$, the following equilibrium condition:

$$(\partial \ln Q/\partial N_{\uparrow c})_{N_i} = n\,(\partial \ln Q/\partial N_{\uparrow})_{N_i} \qquad (54)$$

is obtained in place of eqn.(10). The use made by Parsons of eqn.(38), which stems directly from eqn.(10) (or, which is the same, from eqn.(54) with n=1), clearly implies the use of the constraint (53) with n=1, namely:

$$N_T = N_{\uparrow} + N_{\downarrow} + N_{\uparrow c} + N_{\downarrow c}$$

Differently stated, Parsons' approach is formally correct, provided the up (down) clusters are regarded as consisting of a single water molecule with a dipole-moment normal component much less than that of the up (down) chemisorbed monomers. This situation is schematically depicted in Fig. 12.

In spite of the dubious physical significance of clusters consisting of a single water molecule, Parsons' model accounts

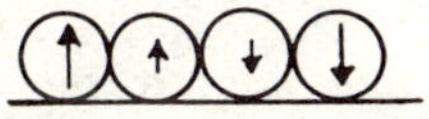

Figure 12.

quite satisfactorily for the shape of the differential capacity curves at the interphase between Hg and aqueous NaF over the whole temperature range from 0 to 85°C. Thus, the calculated curves in Fig.13 reproduce the main features of the experimental curves in

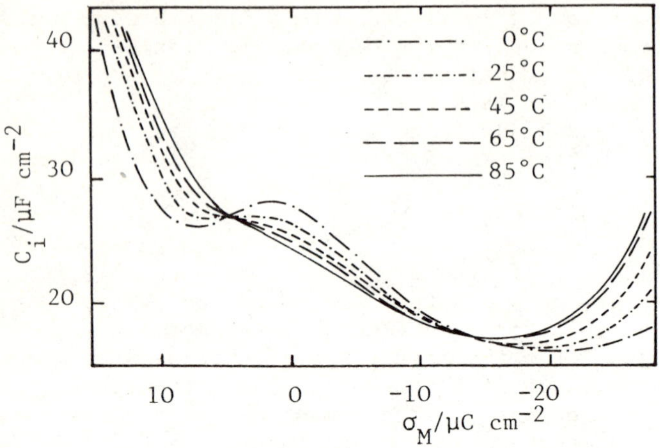

Figure 13. C_i vs σ_M curves calculated on the basis of Parsons' model (59) at different temperatures for ε_∞ = 5.3, d=3.1Å, $\mu_{\uparrow c}$=0.53 D, $U^c_\uparrow$= 4.6 kT and $U^c_\downarrow$= 9.2 kT, with T=273°K.

Fig.(1) quite well, notably the negative temperature coefficient in the region of the hump and the positive temperature coefficient at the most positive charges. Since the entropy of formation of the inner layer, ΔS^{M-2}, is directly obtained from the temperature dependence of C_i (see point II in Sec.2.), it is possible that Parsons' model also accounts satisfactorily for the experimental behaviour of this quantity (61). The central maximum in the calculated curves of Fig.13 occurs in a charge range where clusters predominate and corresponds to reorientation of clusters. Since $\mu_{\uparrow c}$ is $<\mu_\uparrow$, when the charge density becomes more positive the electrostatic energy of up clusters decreases less rapidly than that of up monomers. Thus, up clusters eventually convert into up monomers at sufficiently positive charges, giving rise to an increase in C_i. An analogous conversion of down clusters into down monomers gives rise to an increase in C_i at extreme negative charges. Such an increase is less pronounced than that at positive charges because $U^c_\uparrow$ is set $<U^c_\downarrow$. Parsons' approach involves five adjustable parameters, namely ε_∞, d, $\mu_{\uparrow c}$, $U^c_\uparrow$ and $U^c_\downarrow$, since $\mu_\uparrow$ is equated to the experimental value of the dipole moment of water in the gas phase, μ_T=1.84 D, and hence cannot be regarded as truly adjustable. It should also be noted that the ε_∞ and d values can be varied only within relatively narrow ranges. In fact, physically significant values of d and ε_∞ cannot differ too much from the minimum O-O distance in ice (d= 3.78 Å) and from the

square of the refractive index of water (ε_∞=4.5), respectively.

Damaskin (60) tried to improve Parsons' model by introducing explicitly the number n of water molecules composing a cluster and by taking lateral interactions into account. In his attempt he disregards down chemisorbed monomers, just as in the original model proposed by Damaskin and Frumkin (see Fig. 9). As a matter of fact, Damaskin continues making an improper use of eqn.(39), by estimating the $N_\uparrow/N_\Sigma$, $N_{\uparrow c}/N_\Sigma$ and $N_{\downarrow c}/N_\Sigma$ ratios through this equation, where N_Σ is defined by

$$N_\Sigma \equiv N_\uparrow + N_{\uparrow c} + N_{\downarrow c}$$

The procedure is entirely analogous to that employed by Parsons and implies that all particles involved have the same size, namely that n=1 and $N_\Sigma=N_T$=constant. Conversely, Damaskin allows N_Σ to vary with σ_M, while keeping constant $N_T \equiv N_\uparrow + nN_{\uparrow c} + nN_{\downarrow c}$ (see eqn. 50), with n>1. It is evident that this way of introducing n in explicit form contrasts with the use of eqn.(39). Damaskin accounts for lateral interactions by a procedure analogous to that employed by Watts-Tobin (see eqn. 44). Thus, he writes the energies U_i in eqn.(39) in the form:

$$U_\uparrow = U^c_\uparrow - \mu_\uparrow X_{tot} \quad ; \quad U_{\uparrow c} = -\mu_{\uparrow c} X_{tot} \quad ; \quad U_{\downarrow c} = -\mu_{\downarrow c} X_{tot}$$

where

$$X_{tot} \equiv 4\pi\sigma_M/\varepsilon_\infty + \lambda(\Delta\chi/d)$$

and the surface dipole potential $\Delta\chi$ is correctly expressed by the equation

$$\Delta\chi = -(4\pi/\varepsilon_\infty)(N_\uparrow\mu_\uparrow + N_{\uparrow c}\mu_{\uparrow c} + N_{\downarrow c}\mu_{\downarrow c})$$

Here, however, the mean field ($\Delta\chi$/d) created by the partially aligned dipoles is attenuated at will with respect to the external field, $4\pi\sigma_M/\varepsilon_\infty$, by means of the adjustable parameter λ, which is given values between 0 and 1. Damaskin's treatment involves seven adjustable parameters, namely ε_∞, d, $\mu_\uparrow$, $\mu_{\uparrow c}$, $U^c_\uparrow$, λ and n. In this case $\mu_\uparrow$ must also be regarded as adjustable, since the best fit between the predictions of the model and experiment is attained for a $\mu_\uparrow$ value which is twice the experimental value of the dipole moment of water in the gas phase. To obtain a reasonable agreement between calculated and experimental values of $\partial(1/C_i)/\partial T$ (see point II of Sec. 2.), Damaskin assumes that n decreases linearly from 3 to 2 as the temperature increases from 0 to 85°C. On the whole, Damaskin obtains a somewhat better fit than Parsons. However, in view of the higher number of adjustable parameters in Damaskin's approach, the better fit does not necessarily reflect a better physical modelling.

Parsons' four-state model consisting of up and down monomers and of up and down clusters was later reexamined by Fawcett, Levine et al. (41). These authors continue using the random approximation, but they directly account for the distortional polarizalibity α of water on the basis of eqn.(17) and use an effective coordination number $c_e > 11$. Moreover, they take into proper account the difference in size between clusters and monomers both in the constraints and in the equilibrium conditions for reorientation. Thus, the constraint of eqn.(5) is written in the form of eqn.(53) and the constraint of eqn.(7) is used instead of that of eqn.(6), with the result that the appropriate expressions of eqns.(28)-(30) for the numbers of n.n. pairs are obtained. In addition, the fictituous conditions for monomer-to-cluster reorientation are correctly derived from eqn.(54), with n>1, rather than from eqn.(10),used in all preceding cluster models. The present model is more successful in describing the experimental behaviour of the inner-layer capacity at the Hg/H_2O interphase than those proposed by Parsons (59) and Damaskin (60). Thus, besides predicting the correct temperature dependence of C_i, it also predicts a positive temperature coefficient of $\Delta\phi$ at zero charge, in accordance with experiment (see point III of Sec. 2). In this connection it should be noted that all preceding cluster models predict a negative value of $(\partial\Delta\phi/\partial T)_{\sigma=0}$. The adjustable parameters are seven in number, namely d, c_e, $\mu_{\uparrow c}=-\mu_{\downarrow c}$, $\alpha_{\uparrow c}=\alpha_{\downarrow c}$, $U^c_\uparrow$, $U^c_\downarrow$ and $U^c_{\downarrow c}$, where $\alpha_{\uparrow c}=\alpha_{\downarrow c}$ is the distortional polarizability of clusters. The dipole-moment normal components $\mu_\uparrow$ and $\mu_\downarrow=-\mu_\uparrow$ of the fully aligned monomers and their distortional polarizability $\alpha_\uparrow=\alpha_\downarrow$ are given the experimental values of water in the gas phase. The chemical energy $U^c_{\uparrow c}$ of up clusters is taken as reference, and hence is equated to zero. To account for the temperature dependence of C_i, $\mu_{\uparrow c}=-\mu_{\downarrow c}$ and $U^c_{\downarrow c}$ are allowed to vary linearly with temperature by the introduction of two adjustable temperature coefficients. It is therefore difficult to establish how much of the improvements in describing the experimental data is to be ascribed to a better modelling, and how much to the increased number of adjustable parameters.

In general, the weakness of all cluster models consists in assuming fictituous *reorientation equilibria* between monomers and clusters. A direct consequence of this assumption is that the energies $U^c_{\uparrow c}$ and $U^c_{\downarrow c}$ of chemical interaction of clusters with the two contacting phases must be made more negative, namely more attractive, than the corresponding energies $U^c_\uparrow$ and $U^c_\downarrow$ of chemisorbed monomers, in order to describe the experimental behaviour. It is evident that the assumption that clusters interact with the metal and solution phases more attractively than chemisorbed monomers apparently contrasts with the concept of chemisorption. The practical reason for setting $U^c_{\uparrow c}$ and $U^c_{\downarrow c}$ less than $U^c_\uparrow$ and $U^c_\downarrow$ is that the absolute value of the dipole-moment normal component of chemisorbed monomers is higher than that of clusters. Hence up and down monomers are electrostatically more stable than the corresponding clusters at all charge densities σ_M, unless clusters are assumed to be more strongly anchored to the two contacting phases. Only by the latter assumption will clusters be more stable than chemisorbed monomers at not too high $|\sigma_M|$ values, as required in order to account for experimental behaviour. From a conceptual point of view, the necessity of setting $U^{c}_{\uparrow c}$ and $U^{c}_{\downarrow c}$ less than $U^{c}_{\uparrow}$ and $U^{c}_{\downarrow}$ can be visualized by the following thought experiment. Imagine passing from an adsorbed up cluster made up of three water molecules to three chemisorbed up monomers at an uncharged electrode through the following steps, schematically depicted in Fig. 14. The adsorbed cluster is brought from the monolayer to a charge-free vacuum (step I), then the two H-bonds within the cluster are broken and the three resulting monomers are arranged in a parallel configuration (step II) and, finally, the three dipoles are brought back from the vacuum into the monolayer (step III). If lateral interactions of the cluster and of the three resulting up monomers with the surrounding particles in the monolayer are disregarded, the opposite of the energy U_I involved in step I is the *true* energy of chemical interaction of the up cluster with the two contacting phases, whereas U_{III} is three times the energy of chemical interaction of an up monomer. Now, even though

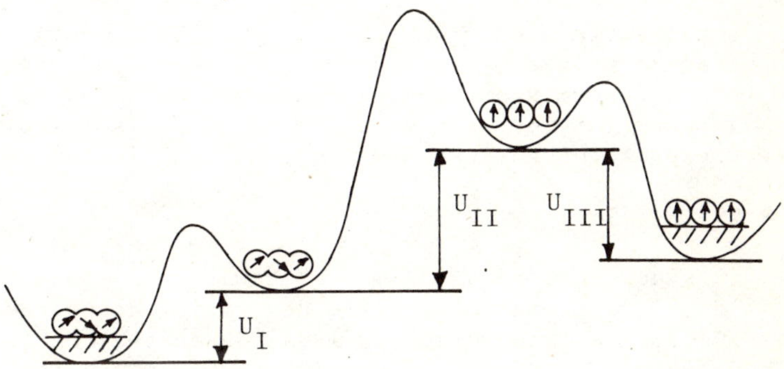

Figure 14

U_{III} is certainly $<-U_I$ if monomers are chemisorbed and clusters are only physically adsorbed, the energy U_{II} involved in step II, which is mainly spent in H-bond breaking, causes the energy level of the final state in Fig. 14 to be higher than that of the initial state. In cluster models this situation is formally accounted for by setting $U^{c}_{\uparrow}$ and $U^{c}_{\downarrow}$ greater than $U^{c}_{\uparrow c}$ and $U^{c}_{\downarrow c}$. Differently stated, the chemical energies $U^{c}_{\uparrow}$ and $U^{c}_{\downarrow}$ of monomers relative to those, $U^{c}_{\uparrow c}$ and $U^{c}_{\downarrow c}$, of clusters embody the positive energy spent in H-bond breaking. This formal approach is somewhat questionable from a statistical-mechanical point of view because the contribution to the internal energy of the monolayer from the H-bonds between the adsorbed water molecules depends on the *configuration* of the two-dimensional lattice, whereas that from the chemical interactions of the adsorbed molecules with the two contacting phases depends exclusively upon the *distribution* of the different water orientations.

4.4. Lattice models of H-bonded water molecules

A *chemical approach* which can be used to overcome the above intrinsic limitations of all cluster models consists in treating the formation of water clusters exactly by the same methods as the formation of ordinary molecules by introducing the idea of a *chemical equilibrium* between water monomers and clusters. Such an approach was used by Bockris and Habib (3,24), who consider an up monomer and a down monomer to be in equilibrium with a dimer resulting from H-bond formation between the two oppositely oriented water molecules. An equilibrium constant K_d for this dimerization reaction is calculated from the H-bond energy and from the energies of other interactions. Since the external field exerts opposite effects on the two oppositely oriented monomers and the dimer has no net dipole moment, K_d is independent of σ_M. In applying the law of mass action to the dimerization equilibrium Bockris and Habib write:

$$K_d = \theta_d/(1 - \theta_d)^2$$

where θ_d is the surface coverage by the dimer and $(1-\theta_d)$ is the overall surface coverage by up and down monomers, and hence they conclude that θ_d assumes a charge-independent value of 0.68. As a matter of fact, in view of the way in which K_d is calculated, the correct expression for the dimerization equilibrium is

$$K_d = \theta_d/(\theta_\uparrow\theta_\downarrow)$$

and hence θ_d should attain a maximum value at intermediate charges, where $\theta_\uparrow$ and $\theta_\downarrow$ are comparable, while decreasing at extreme charges. The above fundamental inconsistency invalidates the conclusions of Bockris and Habib's approach. It should also be noted that there is no apparent reason for the water molecules composing the dimer to have opposite orientations, in view of the possibility of these molecules rotating freely around the H-bond.

A more convenient way to overcome the limitations of cluster models consists in treating H-bonds between the variously oriented water molecules of the monolayer on the same footing as all other lateral interactions, such as Van der Waals forces or dipole-dipole interactions. More precisely, the monolayer can be treated as a two-dimensional lattice of single water molecules distributed over different orientations, and such that two n.n. molecules may interact via H-bond formation provided they are conveniently oriented with respect to each other. It must be stressed that such a monolayer model of H-bonded water molecules requires a statistical-mechanical treatment which takes local order into adequate account. In fact, if a number of water molecules with several orientations were distributed randomly among an equal number of lattice sites in accordance with the random approximation, the probability of H-bond formation would be unreasonably low, in view of the strong directional character of H-bonds.

Parsons and Reeves (50) examined two-dimensional hexagonal and square lattices of water molecules. Lateral interactions were estimated by approximating a water molecule by a tetrahedral arrangement of two positive and two negative charges according to a modified Stillinger-Rahman interaction potential. Equilibrium configurations of the lattice were obtained by a computer simulation technique which minimizes the internal energy of the monolayer. The calculations, which were confined to an uncharged electrode, indicate the strong preference for water molecules to assume orientations with an almost horizontal dipole moment (see Sec.3.4.).

The present author examined a two-dimensional hexagonal lattice of water molecules adsorbed in several orientations, by taking local order into account on the basis of the quasi-chemical approximation (62). All possible orientations (36 in number) having two H-bonding directions aligned along two of the three axes of the hexagonal lattice were included in the model. Of the infinite orientations having only one H-bonding direction aligned along any of these three axes , those with maximum alignment of their dipole moments in the directions towards or away from the metal surface (24 in number) were also included in the model. To simulate *up chemisorbed water molecules* a further water orientation was considered, with the dipole moment fully aligned in the direction away from the metal surface and , hence, with no H-bonding directions parallel to the electrode surface plane . Nearest-neighbour water molecules following each other along a given axis were considered to form an H-bond only when they have a proton-accepting direction and a proton-donating direction pointing exactly towards each other. Hence chemisorbed water molecules cannot form H-bonds within the monolayer. The energy w_H of H-bond formation was set equal to its experimental value, 19 kJ mol^{-1}. Besides H-bonds, lateral interactions between n.n. molecules were assumed to include the electrostatic interactions between the normal components of their dipole moments; the latter interactions were estimated by regarding water molecules as unpolarizable point dipoles embedded in a medium of dielectric constant ε_∞=5.7. The infinite regress of images of the normal components of the adsorbed water dipoles in the metal surface plane and in the outer boundary of the monolayer, z=d, was directly accounted for in estimating both dipole-dipole lateral interactions (see eqn. 33) and the electrostatic energies U_i^{el} in the exponent of the single-particle partition functions q_i (see eqn. 4). The electrostatic interactions of the water molecules with the external field $4\pi\sigma_M/\varepsilon_\infty$ were estimated by approximating these molecules by point dipoles. All water orientations were assumed to have the same energy of chemical interaction with the two contacting phases except for the chemisorbed orientation, whose chemical energy was set more negative (namely, more attractive) than that of the remaining orientations by an amount ΔU^c. This model involves the three adjustable parameters ε_∞ , d and ΔU^c, the first two of which are allowed to vary only within narrow ranges of values.

Agreement between the predictions of this model and experimental behaviour was found to be unsatisfactory for all reasonable values of the parameters ε_∞, d and ΔU^c. The reason for this disagreement is to be found in the tendency of the great majority of water molecules to be oriented so as to form two H-bonds per molecule with the surrounding molecules, giving rise to an exceedingly H-bonded two-dimensional lattice. A satisfactory agreement between theoretical predictions and experiment was found only upon discarding the 36 orientations with two H-bonding directions parallel to the electrode surface plane from the treatment, so as to permit the formation of no more than a single H-bond per adsorbed molecule within the monolayer. Such an elimination can be formally justified by postulating that the 24 orientations with a single horizontal H-bonding direction have an energy of chemical interaction with the two contacting phases which is much more negative than that for the 36 orientations with two horizontal H-bonding directions. This amounts to assuming that the singly H-bonding orientations are notably stabilized with respect to the doubly H-bonding ones by chemical interactions with the two contacting phases. From a physical point of view, this assumption can be readily explained by considering that the water molecules with a single horizontal H-bonding direction are free to rotate around the resulting horizontal H-bond so as to direct the other H-bonding directions in the most favourable way for the formation of further H-bonds with the water molecules of the second layer.

Here we shall be mainly concerned with an improved version of the above multi-state model of H-bonded water molecules (63). Improvements concern both the mathematical approximations and the physical modelling. Thus, dipole-dipole interactions between n.n. water molecules are now calculated exactly, without disregarding the contribution from the horizontal components of the dipoles. The water molecules are considered to occupy a square lattice. We shall assume that the water orientations which make the most important contribution to the canonical partition function are obtained, via a slight rotation, starting from the *ideal orientations* having the four H-bonding directions pointing towards four of the eight vertices of the elementary cubic cell (see Fig. 15). Clearly, the ideal orientations have the dipole moment pointing from the centre of the cubic cell towards the centre of any of the six faces, depending on the positions of the H atoms; all these ideal orientations have the same a priori probability of forming two H-bonds with the adjacent water molecules of the solution phase, provided we assume that the latter molecules have analogous orientations within the ideal cubic lattice representing the natural extension towards the bulk solution of the square lattice in direct contact with the electrode. The slight rotations imparted to the above ideal orientations are such as to lift one of the two H-bonding directions pointing towards the metal so as to permit the formation of a bent H-bond between two adsorbed water molecules occupying cubic cells having a lateral edge in common. Naturally, such a

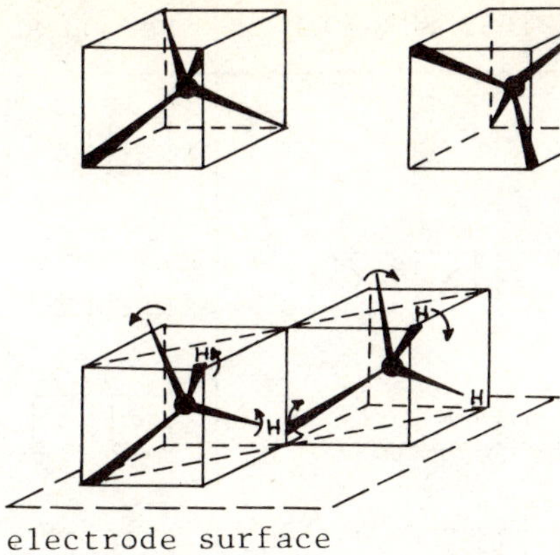

Figure 15. (a) Two *ideal orientations* in which the positions of H atoms and electron pairs are unspecified. (b) rotations imparted to two ideal orientations in order to form an H-bond within the monolayer.

rotation also produces a bending of any H-bonds which the adsorbed water molecules may form with the adjacent water molecules of the solution phase. Ab initio calculations on water dimers (64,65) show that there is little or no difference in the total dimer energy for bending angles $\leqslant 25°$. Hence, the moderate bending of H-bonds which is required to allow the formation of H-bonds within the adsorbed monolayer without producing a breaking of H-bonds between the first and the second layer of water molecules, can be reasonably assumed to take place with no appreciable decrease in the energy of H-bonds, at least as a first approximation. It is evident that, with this model, the assumption of a maximum of one H-bond per adsorbed water molecule within the monolayer as well as the assumption that all adsorbed orientations have the same energy of chemical interaction with the water molecules of the solution phase, become entirely natural.

Figure 16 summarizes all orientations included in the statistical-mechanical treatment. The first column shows a front view of all *groups* of water orientations, each group being characterized by a given orientation relative to the electrode-surface plane. For symmetry reasons, each orientation $\underline{i}$ of a given group is characterized by the same number N_i of water molecules adsorbed in that orientation per unit surface of the monolayer. All orientations are depicted schematically in the second and third columns through a top view of their dipole moment and of their *lifted* H-bonding direction. Arrows denote dipole moments parallel to the electrode surface plane, whereas triangles and circles at the centre of the square cell denote normal dipole moments directed towards the metal and towards the bulk solution, respectively.

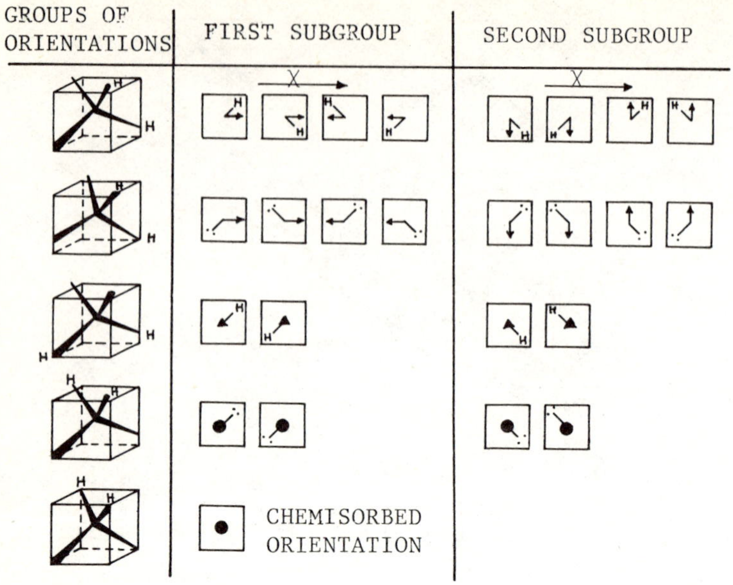

Figure 16

The H letters and colons at the end of the diagonal segments representing the lifted H-bonding directions denote hydrogen atoms and electron lone pairs respectively. The orientations of each group are subdivided into two subgroups. For symmetry reasons, each orientation of a given subgroup is characterized by the same number $^{X}N_{ii}$ of pairs of water molecules which are both adsorbed in that orientation and which follow each other along the X-axis, per unit surface of the monolayer. The use of groups and subgroups notably simplifies numerical calculations, because they can be carried out directly on these groups and subgroups, rather than on the single water orientations. Figure 16 also includes a group (group 5) consisting of a single orientation characterized by full alignment of the dipole moment in the direction away from the electrode. This orientation cannot form H-bonds within the adsorbed monolayer, and simulates a chemisorbed water molecule. All water orientations in Fig. 16, including that of group 5, are characterized by the same energy of chemical interaction with the solution phase. However, by definition, the chemisorbed orientation will interact with the metal phase more attractively than the remaining orientations, and hence its overall energy U_5^c of chemical interaction with the two bulk phases will be more negative than that of the other orientations by an amount which we shall continue denoting by ΔU^c.

For reasonable values of the three adjustable parameters ε_∞, d, and ΔU^c, this molecular model yields C_i vs σ_M curves of the shape shown in Fig. 17. For zero or low negative values of ΔU^c, these curves are bell-shaped, with the hump lying at zero charge. At sufficiently negative ΔU^c values an increase in capacity followed by a rapid drop to the minimum value ($\varepsilon_\infty/4\pi d$) starts to be observed when proceeding towards positive charges, while at the same time

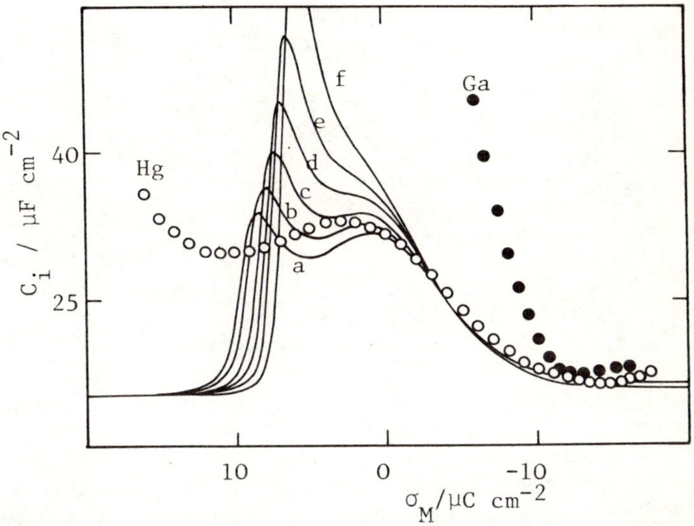

Figure 17. C_i vs σ_M curves calculated for ε_∞=5.5, d=3.2Å, T=273°K and for ΔU^c=-5.7 (a), -5.8 (b), -5.9 (c), -6 (d), -6.1 (e) and -6.2 k·298°K (f). For comparison the experimental curves at the Hg/H_2O and Ga/H_2O interphase are also reported (see Fig. 1).

the hump shifts towards small positive charges. As ΔU^c becomes progressively more negative, the rising section of the C_i vs σ_M curve at positive charges becomes steeper, while both this section and the subsequent capacity drop shift towards less positive σ_M values. In Fig. 18 the numbers of molecules belonging to each of the five groups of orientations of Fig. 16, divided by N_T, are plotted against σ_M for a typical set of parameters. At sufficiently negative charges the water molecules are almost equally distributed between groups 2 and 3; this situation allows the formation of a maximum of H-bonds between pairs of water molecules having an *almost horizontal* and an *almost down* dipole moment respectively. At these negative charges all $dN_i/d\sigma_M$ values are very small, and hence the inner layer capacity C_i practically attains the value $\varepsilon_\infty/(4\pi d)$ corresponding to pure distortional polarization, in view of eqn.(16). As we approach the zero charge there is a gradual passage from the *almost down* orientations of group 3 to the *almost up* orientations of group 4, and a concomitant passage from the almost horizontal orientations of group 2 to the almost horizontal orientations of group 1, which may form H-bonds with the up orientations of group 4. The maximum rate of change

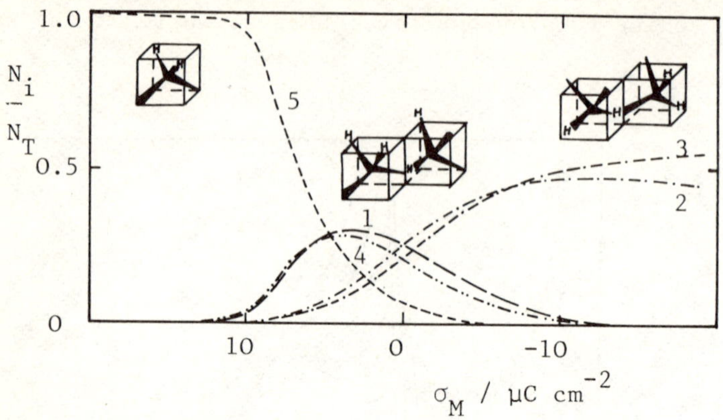

Figure 18. Distribution of water molecules among the five groups of Fig. 16 as a function of σ_M for ε_∞=5.5, d=3.2Å and ΔU^c=-5.85 k·298°K at T=273°K.

from the orientations of groups 2 and 3 to those of groups 1 and 4 occurs in the proximity of zero charge, and gives rise to the capacity hump. As σ_M becomes progressively more positive, the water molecules in the chemisorbed orientation 5 increase in number first slowly, and then gradually more rapidly, until ultimately all water molecules become chemisorbed. The rapid increase in the number of chemisorbed molecules with increasing σ_M is due to the disruptive effect of these molecules (which cannot form H-bonds within the monolayer) upon the strongly H-bonded two-dimensional network. Such a rapid increase is responsible for the rising section of the differential capacity curve at positive charges (see Fig. 17). Naturally, as soon as all water molecules in the monolayer become chemisorbed, all derivatives $dN_i/d\sigma_M$ in eqn.(16) vanish and C_i attains the value $\varepsilon_\infty/(4\pi d)$ corresponding to constant solvent orientation.

The calculated C_i vs σ_M curves in Fig. 17 account for a number of features of the inner-layer capacity at the Hg/H_2O interphase in the absence of ionic adsorption (see point I in Sec. 2). However, by comparing in Fig. 17 the inner-layer capacity curve of the Hg/H_2O interphase at 0°C with the corresponding calculated curves, it is apparent that the experimental rising section occurs at much more positive charges and is not followed by a capacity drop over the range of accessible charges. As ΔU^c is made increasingly negative, namely as the chemical interaction of the chemisorbed orientation with the metal becomes increasingly attractive, the rising section in the calculated C_i vs σ_M curves increases progressively and ultimately obliterates the hump. This accounts qualitatively for the experimental observation that the inner-layer capacity of the interphases between low-melting sp-metals and water increases progressively with an increase in the hydrophilicity of the metal (see point IV of Sec. 2.). However, a comparison between the calculated C_i vs σ_M curves of Fig. 17 and the experimental curve of the Ga/H_2O interphase shows that in the latter curve the capacity starts to increase over its minimum value at more negative charges than in the calculated curves. The present model is also able to account for the decrease in the capacity hump and for its shift towards less positive charges as observed in passing from the Hg/H_2O to the Hg/D_2O interphase (66). Thus, ever though several physical properties, such as bond angle, bond length, dipole moment, dielectric constant and refractive index, are very similar for H_2O and D_2O , there is convincing evidence for the OD...O bond being stronger than the OH...O bond (67-69). Figure 19 shows that the

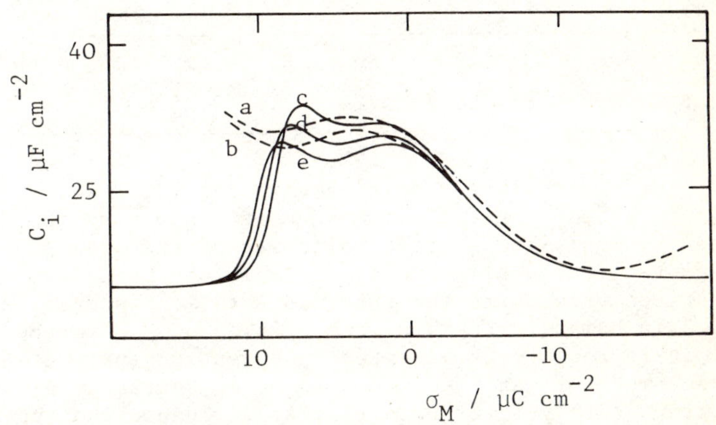

Figure 19. The dashed curves a and b are experimental C_i vs σ_M plots for the Hg/H_2O and Hg/D_2O interphases at 298°K (66). The solid curves were calculated for ε_∞ =5.5, d=3.2Å, ΔU^c=-5.85 kT, T=298°K and for w_H=19 (c), 19.25 (d) and 19.50 kJ mol^{-1}.

main effect of an increase in the value w_H of the H-bond energy upon calculated capacity curves, under otherwise identical conditions, is that of depressing the hump and of shifting it towards less positive values. The capacity along the negative branch of the hump and along the negative minimum remains practically unaltered, once again in agreement with the experimental behaviour of the Hg/D_2O interphase relative to that of the Hg/H_2O interphase. Incidentally, according to the model the capacity hump is higher, the higher the rate of reorientation of the water dipoles with a change in the external field due to σ_M . This explains why an increase in w_H , by favouring H-bond formation over dipole reorientation, produces a decrease in the capacity hump.

The capacity C_i vs σ_M curves of the Ag/H_2O (70) and Au/H_2O (31,71) interphases differ considerably from those of the low-melting sp metals because, even after correction for surface roughness, they show a high peak symmetrical about the p.z.c.. Moreover, the minimum capacity C_{min} corresponding to constant solvent orientation which is attained at far negative charges is appreciably greater than that observed with low-melting sp metals, and depends on the crystal face. Thus, C_{min} ranges from 20 to 25 $\mu F\ cm^{-2}$ on Ag single faces, whereas is close to 17 $\mu F\ cm^{-2}$ on low-melting sp metals. The behaviour of Ag and Au is apparently contradictory (72,73) because the high capacity values at the p.z.c. would seem to indicate a strong water reorientation stemming from water chemisorption (see the curves for high $|\Delta U^c|$ values in Fig. 17), whereas the symmetry of the capacity hump about the p.z.c. suggests that there is no preferential orientation of water at the p.z.c., and hence no water chemisorption. These contradictory aspects can possibly be reconciled if we compare the following values of interatomic distance in crystals of sp metals,

Cu	Ag	Au	Cd	Sn	Bi	In	Tl	Pb	
2.56	2.89	2.88	2.98	3.02	3.11	3.25	3.41	3.50	(in Å)

with the intermolecular O-O distance in ice (2.80 Å). The latter distance is probably the minimum value attainable by the distance $\underline{d}$ between two contiguous water molecules of the adsorbed monolayer. Thus, in practically all monolayer models of the Hg/H_2O interphase $\underline{d}$ is ascribed values greater than 2.8 Å (3.1 Å in Parsons' model (59), 3.3 in Damaskin-Frumkin model (58), and 3.2 in the present model). It is now widely accepted that lattice spacing may affect the arrangement of adsorbed atoms in metal surfaces, giving rise to important catalytic effects (35,74,75). Hence, noting that the interatomic distances in Ag and Au crystals are very close to the minimum distance 2.8 Å between two contiguous water molecules, it is reasonable to postulate that the lattice spacing may modulate the arrangement of adsorbed water molecules on the metal surface. Such a modulation should consist in a penetration of water molecules into the voids left by the regular arrangement of the surface metal atoms (70), regarded as immobile spheres, with the intermolecular distance between adsorbed water molecules matching the interatomic distance between surface metal atoms. The interpenetration of water molecules and metal atoms involves a decrease in the inner-layer thickness, whereas, for the Ag and Au metals, the matching between intermolecular and interatomic distances involves a contraction of the distance between contiguous adsorbed water molecules. To interpret the differential capacity behaviour of Ag and Au, we have slightly modified our model by setting the intermolecular distance between water molecules (henceforth referred to as d_l) generally different from the inner-layer thickness d. Figure 20 shows C_i vs σ_M curves calculated for ΔU^c=0 by decreasing progressively $\underline{d}$ while maintaining d_l constantly equal to the experimental value

2.89 Å of the interatomic distance in Ag crystals. It can be seen that the $\underline{d}$ value, 2.4 Å, which provides the best fit to the minimum inner-layer capacity C_{min}=24 μF cm^{-2} of the (100) silver face (70) at far negative charges, also accounts satisfactorily for the height of the corresponding capacity hump at zero charge. It should be noted that a gradual decrease in $\underline{d}$ under otherwise identical

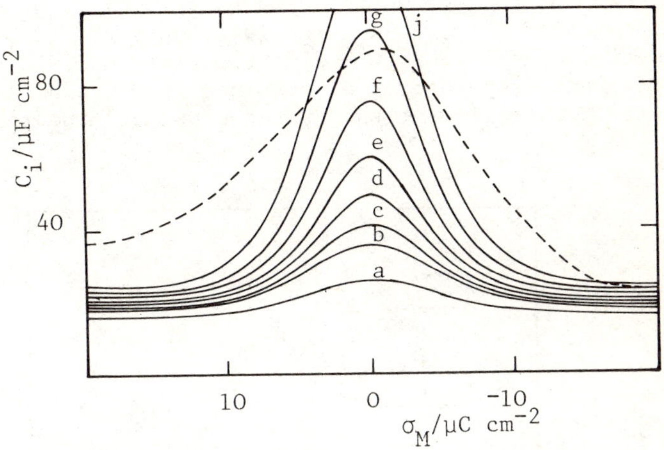

Figure 20. C_i vs σ_M plots calculated for ε_∞=5.5, d_l= 2.89 Å, ΔU^c=0, T=298°K and for $\underline{d}$ equal to 3.2 (a), 2.9 (b), 2.8 (c), 2.7 (d), 2.6 (e), 2.5 (f), 2.4 (g) and 2.3 Å (j). The dashed curve is the experimental C_i vs σ_M plot at the (100) silver face (70).

conditions causes the calculated C_i vs σ_M curves to increase at zero charge more than at far negative charges. The preceding example shows that the high C_i value at zero charge for Ag does not necessarily imply a high hydrophilicity; conversely, the symmetry of the capacity peak about σ_M=0 points to a low hydrophilicity, in agreement with an hydrophilicity scale based on the enthalpy of formation of metal oxides. In other words, if the metal surface has a high degree of hardness so as to be able to modulate the arrangement of the adsorbed water molecules, it will notably affect the inner-layer capacity both at the hump and at the minimum. Under these conditions the C_i value at zero charge cannot be regarded as a measure of the metal hydrophilicity.

We have seen from Fig. 17 that our monolayer model predicts an unavoidable drop of the inner-layer capacity to the value $\varepsilon_\infty/(4\pi d)$ corresponding to constant solvent orientation at positive charges, whereas no such drop is experimentally observed at the Hg/H_2O interphase. It is interesting to observe that the experimental behaviour of the inner-layer capacity of the Hg/H_2O interphase at far positive charges can hardly be explained by any inner-layer model based on the general and widely accepted equation:

$$1/C_i = d(\Delta\phi)/d\sigma_M = 1/K_{ion} + dg^{H2O}(dip)/d\sigma_M \qquad (55)$$

where K_{ion} is a σ_M-independent capacity at constant solvent orientation and $g^{H2O}(dip)$ is the surface dipole potential due to the adsorbed solvent dipoles, so far denoted by $\Delta\chi$. Incidentally, eqn.(16) is a particular case of eqn.(55). If we assume that $dg^{H2O}(dip)/d\sigma_M$ is practically zero along the flat capacity minimum C_{min}= 17 μF cm^{-2} of the Hg/H_2O interphase, then, in view of eqn. (55), the integral $\int(C^{-1}-C_{min}^{-1})d\sigma_m$ should measure $g^{H2O}(dip)$, apart from an additive constant. Figure 21 shows plots of $\int(C^{-1}-C_{min}^{-1})d\sigma_M$ vs σ_M at different temperatures as obtained from Grahame's inner-layer capacity curves (19) of the Hg/(aqueous NaF) interphase

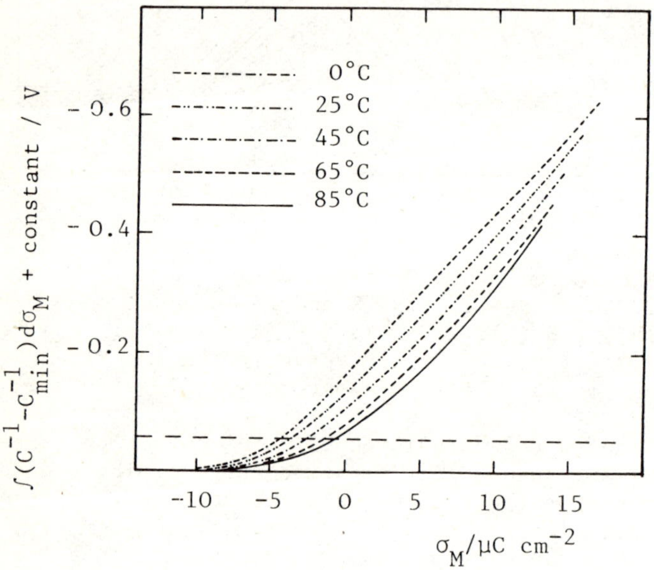

Figure 21.

(see also Fig. 1). This procedure for obtaining $g^{H2O}(dip)$, apart from an additive constant, is entirely analogous to that followed by Trasatti (76), but is also extended to positive charges. Now, several pieces of experimental evidence indicate that the $g^{H2O}(dip)$ value at zero charge, $g^{H2O}(dip)_0$, for Hg at 25°C should be close to -70 mV (4). This permits us to locate the $g^{H2O}(dip)$=0 axis in the proximity of the dashed horizontal straight line of Fig.21. This implies that $g^{H2O}(dip)$ tends asymptotically to a positive saturation value at far negative charges, whereas it reaches negative values more than five times greater at σ_M >10 μC cm^{-2}, without showing any tendency towards saturation. To explain such a behaviour with a model based on eqn.(55), one must assume that the saturation orientation of water molecules at far negative charges is unreasonably low with respect to the maximum orientation corresponding to full alignment of all water molecules in the direction normal to the electrode surface. This is indeed the typical assumption of all cluster models, even though H-bond formation within the monolayer does not prevent adsorbed water molecules from undergoing an appreciable reorientation under the action of the external field via a rotation around the H-bonds. To explain the behaviour in Fig. 21 through eqn.(55) without having to postulate such a strong asymmetry in the limiting orientations of water dipoles at far positive and far negative charges one should assume that the $g^{H_2O}(dip)=0$ axis intersects the curves in Fig. 21 not far from their middle points. This would imply that $g^{H_2O}(dip)$ on mercury is appreciably positive at $\sigma_M=0$, in contrast with several pieces of experimental evidence (see point III in Sec. 2). In conclusion, to achieve a better agreement between the C_i vs σ_M curves of the Hg/H_2O interphase at far positive charges and theoretical predictions, one should use an inner-layer model based on an expression for C_i less approximate than that in eqn.(55). Apart from the rather fictitious separation of the solvent polarizability into an orientational and a distortional contribution which is implicit in eqn.(55), the use of this equation involves two further main simplifying assumptions: (1) the surface dipole potential $g^M(dip)$, due to spill-over of the electrons into the solution, is considered to be σ_M-independent and hence not to contribute to C_i; (2) the thickness of the inner layer and all interactions between this layer and the solution phase (e.g., imaging conditions on the solution side of the inner layer, propagation of the water structure existing in the adsorbed monolayer to water molecules beyond the first monolayer) are assumed to be independent of σ_M , as implied by the σ_M-independence of K_{ion} in eqn.(55).

As concerns assumption (1), recent models of electrified interphases in which the electron gas in the metal is considered to interact both with a uniform background of metal ions (*jellium*) and with the solvent molecules (16-18,77,78), indicate that $g^M(H_2O)$ depends appreciably upon the charge and hence contributes to the inner-layer differential capacity C_i. However, consideration of $dg^M(dip)/d\sigma_M$ does not seem to explain the rising section of the capacity curve of the Hg/H_2O interphase at positive charges. On the contrary, Badiali et al. calculate a C_i vs σ_M curve which decreases towards positive charges (see Fig. 4 in Ref. 78, where $q_S \equiv -\sigma_M$). A modified jellium model proposed by Price and Halley (79) predicts an increase in C_i towards positive charges, but also the temperature independence of C_i. As concerns the simplifying assumption (2), a possible way of removing it would require the statistical-mechanical treatment of a solution layer, adjacent to the electrode, several molecules thick and including non-specifically adsorbed ions. Unfortunately, this is a quite involved statistical-mechanical problem which has not yet been solved. Nonetheless, we may estimate in a very naif way the effect of the propagation of the water structure, as created within the adsorbed monolayer by a strongly hydrophilic metal, into the bulk liquid beyond this monolayer. First of all it must be stressed that any metal surface will act as hydrophilic and structure-making only at not too negative charges, when water molecules start to be chemisorbed to an appreciable extent. According to the present model, chemisorbed water molecules differ from all other adsorbed water molecules because they are unrotated relative to their elementary cubic cells and, hence, cannot form H-bonds within the adsorbed monolayer. In the framework of a cubic lattice, these chemisorbed water molecules will form undistorted, and hence somewhat stronger, H-bonds with the adjacent water molecules of the second monolayer. We may therefore expect that a gradual increase in the number N_5 of chemisorbed water molecules will be accompanied by a progressive propagation of the chemisorbed-water structure beyond the first monolayer. Such a propagation should also produce a gradual *squeezing out* of nonspecifically adsorbed ions from the water monolayers involved in this propagation of structure. Since the mobile ions in the vicinity of the outer boundary z=d of the adsorbed monolayer exert a screening effect of the electric field resulting from the charge density σ_M, the approximate effect of these ions is the same as that of an ideal conductor placed in the vicinity of z=d. In this respect a gradual squeezing out of ions can be roughly accounted for by a gradual shift of the imaging plane on the solution side of the monolayer from the boundary z=d towards the bulk solution. For simplicity, we shall assume that the distance, z=d', of this imaging plane from the electrode surface is expressed by the relation:

$$d' = d(1 + N_5/N_T) \qquad (56)$$

This amounts to stating that the water structuring induced by the metal surface extends only up to the second water monolayer and propagates linearly with the number N_5 of chemisorbed water molecules per unit surface. Figure 22 shows C_i vs σ_M curves calculated on the basis of the present model by allowing d' to vary as in eqn.(56). It is apparent that these curves are in better agreement with experiment than those in Fig. 17. Thus, if $-\Delta U^c$ is low enough for water chemisorption to become appreciable only in the proximity of zero charge, the curves in Fig. 22 show a notable rise in capacity at positive charges, where chemisorbed water molecules definitely predominate. These curves account rather satisfactorily for the experimental behaviour of the Hg/H_2O interphase. Conversely, when $-\Delta U^c$ is so high as to give rise to appreciable water chemisorption even at negative charges, the calculated C_i vs σ_M curves start showing an appreciable increase in capacity at these negative charges, thus accounting satisfactorily for the experimental behaviour of the interphases between water and strongly hydrophilic metals. It might be argued that, as d' approaches 2d, the second water monolayer would also be expected to contribute to $g^{H_2O}(\text{dip})$ and hence to C_i , whereas this contribution is not accounted for in the calculated curves of Fig. 22. It is, however, reasonable to believe that the main contribution to C_i stems from the first monolayer, which is the only one containing chemisorbed water molecules. At any rate, even if the second monolayer were just a replica of the first one, the values of $1/K_{ion}$ and $dg^{H_2O}(dip)/d\sigma_M$ for the whole bilayer would be twice as great as those of the first monolayer, and hence the capacity C_i would be reduced to half in

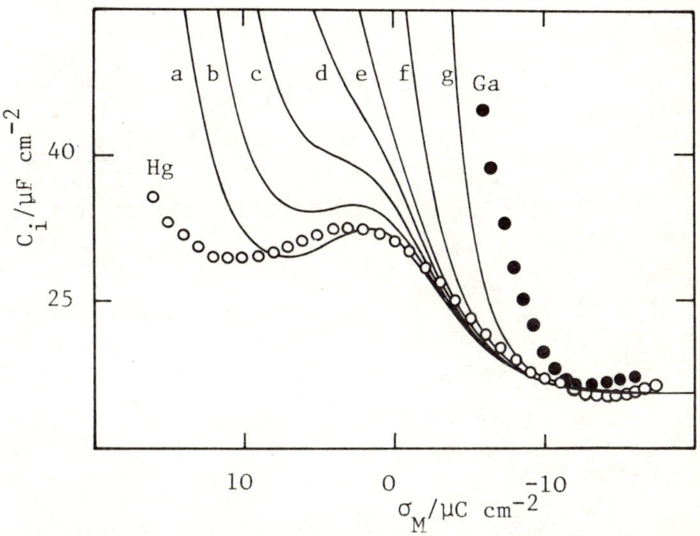

Figure 22. C_i vs σ_M curves calculated for ε_∞=5.5, d=3.2 Å, T=273°K and for ΔU^c=-5.2 (a), -5.4 (b), -5.6 (c), -5.8 (d), -6 (e), -6.5 (f) and -7.5 k·298°K (g), upon assuming that the imaging plane z=d' on the solution side of the monolayer is expressed by eqn.(56). For comparison the experimental curves at the Hg/H_2O and Ga/H_2O interphases are also reported (see Fig. 1).

view of eqn.(55). Conversely, the calculated capacity of the first monolayer increases by a factor greater than 2 when shifting the imaging plane from z=d to z=2d. Figure 23 shows a number of C_i vs σ_M curves calculated at different temperatures on the basis of eqn.(56) using the ΔU^c value which gives the best fit to the experimental C_i vs σ_M curve on Hg at 0°C. Comparison of these curves with the experimental curves in Fig. 1 indicates that the use of an imaging plane z=d' depending upon water chemisorption according to eqn.(56) accounts rather satisfactorily for the temperature dependence of the inner-layer capacity of the Hg/H_2O interphase. We may therefore conclude that the preceding approach, albeit quite crude, yields qualitatively correct results.

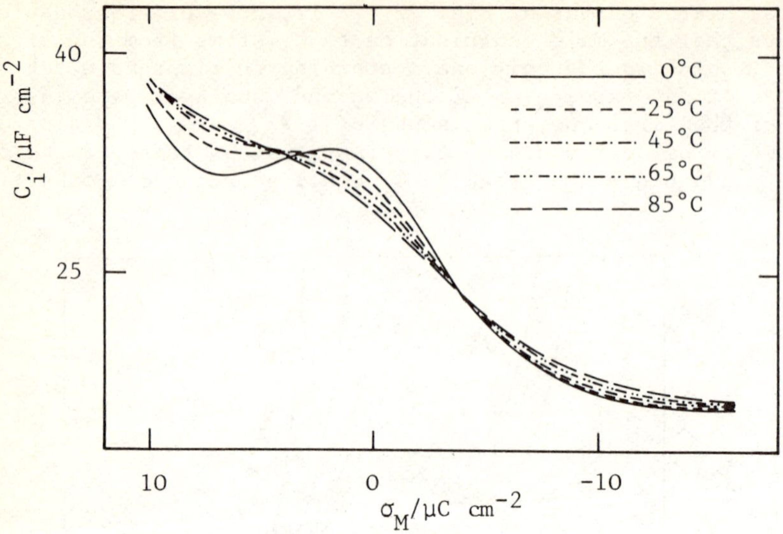

Figure 23. C_i vs σ_M curves calculated at different temperatures on the basis of eqn.(56) for ε_∞=5.5, d=3.2 Å and ΔU^c=-5.3 k·298°K.

5. USE OF MONOLAYER MODELS IN INTERPRETING THE ADSORPTION BEHAVIOUR OF ALIPHATIC COMPOUNDS.

Disturbance of the double layer with neutral aliphatic molecules produces data of diagnostic value for double-layer studies. The adsorption behaviour of a large number of aliphatic compounds shows appreciable similarities in spite of the presence of different functional groups. Some typical features are:
I. Adsorption satisfies the Frumkin isotherm:

$$c_S \exp(-\Delta G^\circ_{ads} / kT) = \theta_S \exp(a\theta_S)/(1-\theta_S) \tag{57}$$

where $\theta_S \equiv nN_S/N_T$ is the surface coverage by the organic surfactant, c_S is its bulk concentration, ΔG°_{ads} is the standard free energy of adsorption at zero coverage and a is the so called *Frumkin interaction factor*. Roughly linear plots of ln $[c_S(1-\theta_S)/\theta_S]$ vs θ_S at constant applied potential E or at constant charge σ_M (henceforth referred to as *Frumkin-isotherm plots*) have been reported countless times in the literature (80-82). Usually the slope of these plots, which measures the interaction factor a, assumes negative values ranging from -1 to -3.
II. The standard free energy of adsorption, ΔG°_{ads}, is a quadratic function of σ_M as well as of E. The maximum value of $-\Delta G^\circ_{ads}$, which corresponds to a maximum in the adsorptivity of the organic surfactant, is attained for a charge value, σ_{max}, which frequently lies

in the range from -2 to -4 μC cm^{-2}.

Frumkin derived the isotherm of eqn.(57) on the basis of a model which lacks a molecular character. However, this isotherm can be readily derived from eqn.(14) in the framework of the random approximation, upon disregarding the presence of water molecules in the monolayer, namely upon regarding adsorption as a random distribution of adsorbing monomeric solute molecules over the initially free adsorption sites of a two-dimensional lattice (83,84). This amounts to regarding the monolayer as consisting of monomeric solute molecules adsorbed in a single orientation and of free adsorption sites which, by their very nature, interact neither between themselves, nor with the solute molecules nor with the external field. Naturally, this implies that the $n(\partial \ln Q/\partial N_1)_{N_i}$ term in eqn. (14) is dropped. By this approach the Frumkin interaction factor a turns out to be equal to (cw_{SS}/kT), where c is the coordination number of the lattice and w_{SS} is the lateral interaction energy between n.n. solute molecules. This is the reason why in the early literature a negative experimental value of a is generally regarded as an indication of strong attractive interactions between solute molecules. Incidentally, this point of view is still retained by several russian researchers (80,81,85). The use of the Frumkin isotherm has been criticized (86) in view of the very rough assumptions involved in its statistical-mechanical derivation. In the present author's opinion, the Frumkin isotherm should be regarded as a convenient formalism to parametrize adsorption data, and any molecular theory of organic adsorption should be able to account for the experimental observation that many adsorption data satisfy this isotherm. Naturally, the physical significance to be attached to the parameters of the Frumkin isotherm should be critically reassessed.

Bockris and coworkers (23,87) were the first to stress the mayor role played by solvent-field and solvent-solvent interactions in organic adsorption. Before them, Kirkwood (88) had already pointed out the importance of the role of the solvent in producing a new force between solute molecules in the bulk other than the direct Van der Waals forces. This concept is now commonly applied to bulk solutions, such that the term *hydrophobic interactions* has been coined to express those forces which are mainly dependent on the properties of water, rather than on those of organic solutes. According to Bockris and coworkers (23,87), the adsorption behaviour of aliphatic compounds can be predicted correctly, largely by considering the reorientation of the adsorbed solvent dipoles with a change in the direction and strength of the external field. At high field strengths, the organic molecules are expelled from the monolayer by the competing solvent dipoles, which interact more favourably with the field.

Application of the random approximation to a monolayer model of water and solute molecules permits us to explain the quadratic dependence of ΔG°_{ads} upon σ_M (23,87), but not the Frumkin-isotherm behaviour (53). Consider a monolayer consisting of up and down solvent dipoles and of the n-meric molecules of a neutral surfactant S, adsorbed in a single orientation. We shall assume for simplicity that the solute molecules interact neither between themselves nor with the solvent molecules by setting $w_{SS}=w_{S\uparrow}=w_{S\downarrow}=0$. Moreover, we shall set $U^c_\uparrow=U^c_\downarrow$ and $\mu_\downarrow=-\mu_\uparrow$; the latter assumption implies that $w_{\uparrow\uparrow}=w_{\downarrow\downarrow}=-w_{\uparrow\downarrow}$ in view of eqn.(43). Upon applying the expression of eqn.(18) for the partition function Q to the three particles of the monolayer, with $W(\{N_i\})$ expressed by eqn.(22) and the numbers of pairs expressed by eqns.(28)-(30), and introducing the resulting expression for Q into the equilibrium condition of eqn.(14) for solvent displacement, we readily obtain the adsorption isotherm (53):

$$c_S \propto \frac{(q_\uparrow q_\downarrow)^{n/2}}{q_S} \quad \frac{\theta_S}{n(1-\theta_S)^n} \quad \frac{(1+\theta_\uparrow/\theta_\downarrow)^n}{(\theta_\uparrow/\theta_\downarrow)^{n/2}} \tag{58}$$

The quantities $q_\uparrow$, $q_\downarrow$ and q_S are single-particle partition functions (see eqn. 4). Having set $\mu_\uparrow=\mu_\downarrow$, the external field exerts exactly opposite effects upon the up and down water dipoles, and hence the $(q_\uparrow q_\downarrow)$ product is independent of σ_M. A charge-independence can also be postulated for q_S, provided the solute molecule is very weakly polar, as already postulated in setting $w_{SS}=w_{S\uparrow}=w_{S\downarrow}=0$. The $\theta_\uparrow/\theta_\downarrow$ ratio in eqn.(58) is obtained from eqn.(36), with $U_\uparrow$ and $U_\downarrow$ expressed by eqn.(37). Incidentally, this equation is also valid in the presence of a non-interacting solute molecule, provided that $\theta_\uparrow$ and $\theta_\downarrow$ are now defined by:

$$\theta_\uparrow = N_\uparrow/N_T \qquad ; \qquad \theta_\downarrow = N_\downarrow/N_T$$

where $N_T=N_\uparrow+N_\downarrow+nN_S$. Rearranging terms and using the preceding constraint in the form $\theta_\uparrow+\theta_\downarrow+\theta_S=1$, we have:

$$\frac{\theta_\uparrow}{\theta_\downarrow} = \exp\left[-\frac{8\pi\sigma_M\mu_\uparrow}{\varepsilon_\infty kT} - \frac{2cw_{\uparrow\uparrow}}{kT}\,\frac{\theta_\uparrow/\theta_\downarrow-1}{\theta_\uparrow/\theta_\downarrow+1}\,(1-\theta_S)\right] \tag{59}$$

which expresses $(\theta_\uparrow/\theta_\downarrow)$ implicitly as a function of σ_M and θ_S. It is interesting to note that in the absence of lateral interactions, namely for $w_{\uparrow\uparrow}=0$, the $\theta_\uparrow/\theta_\downarrow$ ratio in eqn.(59) becomes independent of θ_S, and hence the isotherm of eqn.(58) at constant charge reduces to a Flory-Huggins isotherm (34,43)

$$c_S \exp(-\Delta G^\circ_{ads}/kT) = \theta_S/\left[n(1-\theta_S)^n\right] \tag{60}$$

which was originally derived in lattice statistics to account for the behaviour of a mixture of non-interacting particles of different size. Figure 24 shows a plot of ln $\left[c_S(1-\theta_S)^n/\theta_S\right]$ vs θ_S at different charge values as obtained for T=298°K, $\mu_\uparrow$=1.84 D, ε_∞=6, n=3, c=11 and $w_{\uparrow\uparrow}$=0.2 kT. Numbers on each curve denote the charge σ_M in μC cm^{-2}. An entirely symmetrical behaviour is predicted at negative charges. Extrapolation of these curve to θ_S=0 yields

$\Delta G^{\circ}_{ads}/kT$, apart from an additive constant. The calculated plot of $(\Delta G^{\circ}_{ads}/kT)$ vs σ_M, as shown in Fig. 24a, is almost parabolic in shape, in accordance with experimental behaviour. Apart from this encouraging result, the behaviour expressed by eqns.(58) and (59) is at variance with the Frumkin-isotherm behaviour. Thus, in the absence of lateral interactions eqn.(58) reduces to the Flory-Huggins isotherm of eqn.(60), whereas the Frumkin isotherm reduces to the Langmuir isotherm

$$c_S \exp(-\Delta G^{\circ}_{ads}/kT) = \theta_S/(1 - \theta_S) \qquad (61)$$

which is a particular case of the Flory-Huggins isotherm when solute and solvent molecules are of equal size. The ability of the Frumkin isotherm to account for the adsorption properties of the majority of aliphatic compounds was therefore tentatively explained by assuming that the adsorbing surfactant displaces small groups of H-bonded water molecules (clusters) of the same size as the surfactant (80,86). Incidentally, the concept of H-bonded clusters was subsequently employed in the cluster models described in Sec. 4.3. At any rate, it is difficult to see how water molecules in the interphase would adjust themselves into displaceable groups with the same area as the adsorbing organic molecules, for various sizes of the latter (2). Even confining ourselves to a monomeric solute (n=1), eqns.(58) and (59) lead to calculated ln $[c_S(1-\theta_S)/\theta_S]$ vs θ_S plots of slopes opposite to those observed experimentally, as clearly shown in Fig. 24. More precisely, eqns.(58) and (59) predict an increase in θ_S with increasing c_S at constant charge which is slower than in the absence of lateral interactions, when the Langmuir isotherm applies. This contrasts with the negative

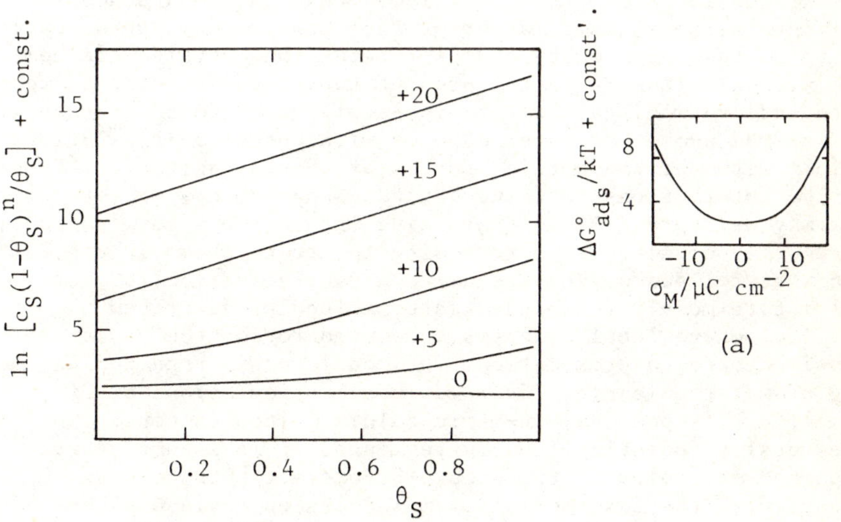

Figure 24.

slope commonly exhibited by experimental Frumkin-isotherm plots. Incidentally, Bockris and coworkers (23) erroneously omit the $(1-\theta_S)$ factor in eqn.(59), which factor accounts for the progressive decrease in solvent-solvent interactions *per given number of adsorbed solvent molecules* with an increase in θ_S , following the gradual random interposition of adsorbed solute molecules between the solvent molecules. The effect of this omission is that the $(\theta_\uparrow/\theta_\downarrow)$ ratio becomes independent of θ_S , and hence the adsorption isotherm at constant σ_M reduces to a Langmuir isotherm at all changes, provided that n=1.

The unsatisfactory predictions of eqns.(58) and (59) are a direct consequence of the random approximation. Thus, we have seen in Sec. 3.3 that in the random approximation all molecules of the same species behave as though they had the same vertically directed average dipole moment $\vec{\bar{\mu}}$, as expressed by eqn.(31). At zero charge, where $\theta_\uparrow$ equals $\theta_\downarrow$ in view of the assumptions $U^c_\uparrow=U^c_\downarrow$ and $\mu_\uparrow=-\mu_\downarrow$, the average dipole moment $\vec{\bar{\mu}}\equiv(\theta_\uparrow\vec{\mu}_\uparrow+\theta_\downarrow\vec{\mu}_\downarrow)$ equals zero, and hence lateral interactions between solvent molecules are vanishingly small. This explains the horizontal plot at zero charge in Fig. 24. However, as $|\sigma_M|$ increases, $|\vec{\bar{\mu}}|$ increases accordingly with a resulting increase in the repulsive forces between n.n. parallel solvent dipoles, until all solvent dipoles are fully aligned along the external field. To a certain extent, these repulsive interactions favour the displacement of adsorbed solvent molecules from the monolayer by the adsorbing solute molecules. However, as θ_S increases, an increasing number of non-interacting adsorbed solute molecules surrounds the residual adsorbed solvent molecules. Hence, the latter molecules are less strongly repelled by their nearest neighbours, which makes their removal by further adsorbing solute molecules more difficult. This explains why θ_S increases with increasing c_S more slowly than in the absence of lateral interactions. As a matter of fact, the random approximation proves to be too rough when lateral interaction energies are comparable with kT. Thus, with the present two-state orientation model for water adsorption, each water dipole will tend to be surrounded by the maximum number of water dipoles with opposite orientation, compatibly with the randomizing forces of thermal agitation. Hence, attractive interactions between contiguous oppositely oriented water dipoles may well prevail over repulsive interactions between equally oriented water dipoles, thus reversing the trend shown in Fig.24(53)

In what follows we shall examine a generalization (89) of the multistate model of H-bonded water molecules described in Sec. 4.4, where local order is accounted for on the basis of the quasi-chemical approximation. The monolayer is regarded as consisting of water molecules adsorbed in all orientations of Fig. 16 as well as of open-chain n-meric solute molecules consisting of structurally indistinguishable segments. It is assumed that the adsorbed molecules of the solute S interact both between themselves and with the adsorbed water molecules only via dipole-dipole interactions. For simplicity, the dipole moment of the solute molecules is assumed to have a charge-independent orientation and to be distributed uniformly over the n segments, (μ_S/n) denoting the normal component of the dipole moment per segment. The adsorption isotherm is obtained by expressing the partition function Q, with $W(\{N_i\})$ given by eqn.(22), in the framework of the quasi-chemical approximation, and by substituting it into the equilibrium condition of eqn.(14) for solvent displacement.

Figure 25 shows a number of isotherms calculated for different σ_M values by setting w_H equal to the experimental value 19 kJ mol^{-1}, T=298°K, n=3, μ_S=-1.2 D, ε_∞=5.5, d=3.2 Å and ΔU^c=-5.85 kT. The latter three values are those providing the best fit between the capacity curve at 0°C calculated on the basis of the same model in the absence of specifically adsorbed solutes and the corresponding experimental curve for the Hg/H_2O interphase (see Fig . 17). The numbers on each curve of Fig. 25 are charge values in μC cm^{-2}. The calculated isotherms are plotted on a ln $[c_S(1-\theta_S)/\theta_S]$ vs θ_S graph, in order to point out any deviations from the Frumkin-isotherm behaviour. The resulting Frumkin-isotherm plots are almost linear over a wide range of θ_S values, with standard deviations from a straight line <2% for $0<\theta_S<0.75$, other than at very positive charges.

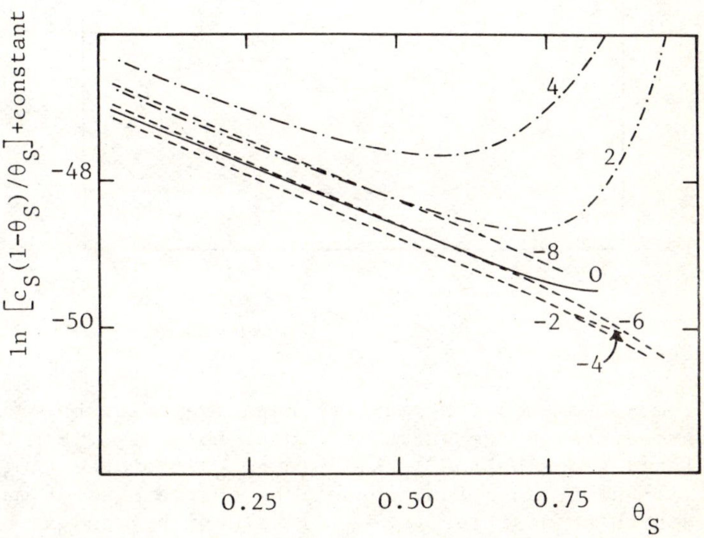

Figure 25.

By least-squares fitting of the Frumkin isotherm plots to a straight line over the θ_S range from 0 to 0.5, values of $\Delta G^\circ_{ads}/kT$ and of $\underline{a}$ are immediately obtained from the corresponding intercepts and slopes. Figure 26 shows ΔG°_{ads} vs σ_M and $\underline{a}$ vs σ_M plots as calculated for μ_S=-0.6 and -1.2 D and for values of the remaining parameters equal to those in Fig. 25. It is apparent that $\Delta G^\circ_{ads}/kT$ depends almost parabolically upon σ_M and that the charge σ_{max} of maximum adsorption shifts towards more negative values with a gradual increase in the negative value of μ_S , namely the more the positive end of the solute dipoles points towards the metal. The calculated Frumkin interaction factor a in Fig. 26 assumes negative values

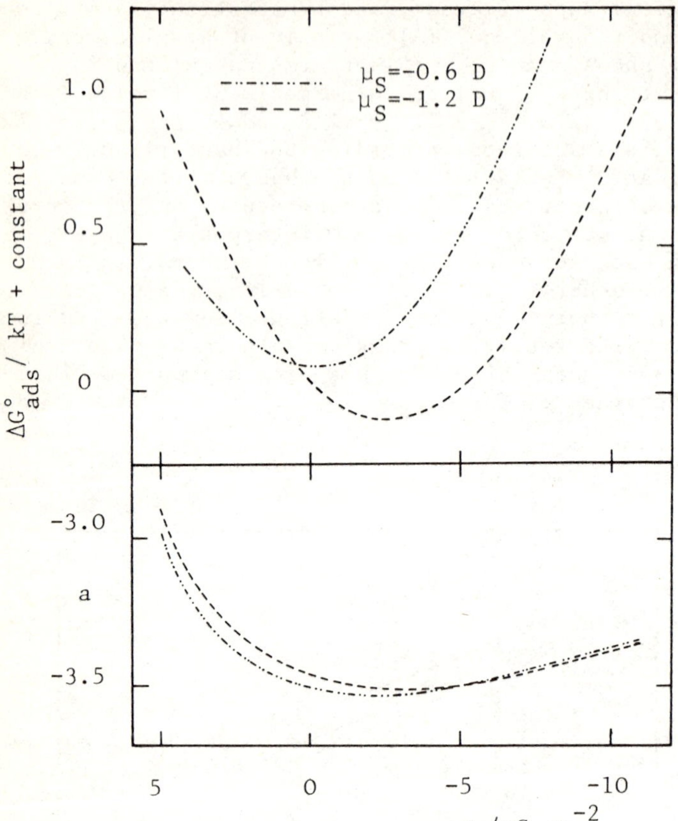

Figure 26.

ranging from -3 to -3.5, the maximum negative value being attained at an intermediate charge which is close to σ_{max}. Figure 27 shows the calculated slope $b=\partial(\Delta G^\circ_{ads}/kT)/\partial(\sigma_M-\sigma_{max})^2$ of the $(\Delta G^\circ_{ads}/kT)$ vs $(\sigma_M-\sigma_{max})^2$ plot, as a function of the charge σ_{max} of maximum adsorption both at positive $(\sigma_M-\sigma_{max})$ values (b_+) and at negative ones (b_-). The different values of σ_{max} in Fig. 27 were obtained by ascribing different values to μ_S , while maintaining the other parameters as in Figs. 25 and 26. It is evident that the b_- value shows a parabolic dependence upon charge, with a flat maximum in the proximity of the zero charge. This is due to the fact that the availability of the adsorbed water molecules to assume electrostatically more favourable orientations with varying charge is a maximum in the proximity of zero charge; hence the rate of solute desorption with increasing $|\sigma_M-\sigma_{max}|$ will be a maximum there. The

b_+ value decreases rather rapidly at positive charges, due to water chemisorption.

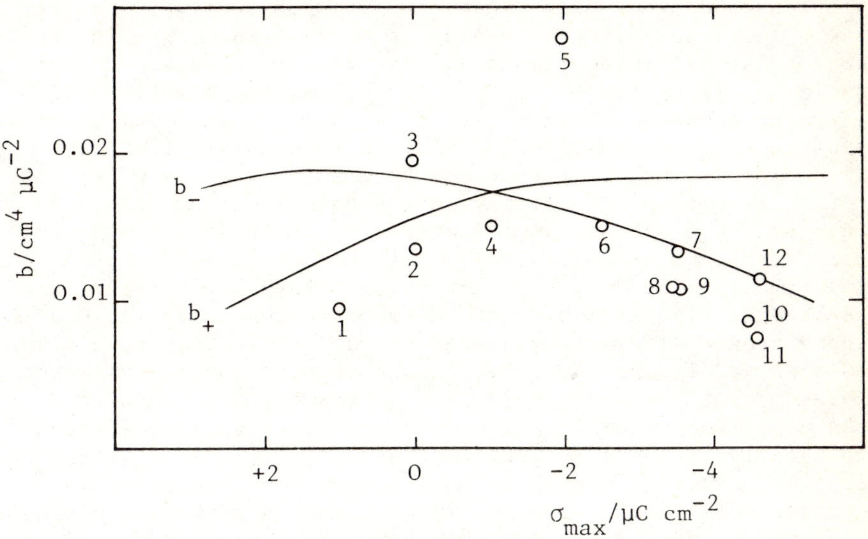

Figure 27. The solid curves are calculated plots of b_+ and b_- vs σ_{max}. Circles are experimental $\underline{b}$ values for 2-methylthioethanol (1), sucrose ($2 \equiv b_+$, $3 \equiv b_-$), 2-butyne-1,4-diol (4), n-butanol (5), butane-1,4-diol (6), ethylene glycol (7), propionitrile (8), acetonitrile (9), diethylether (10), and succinonitrile ($11 \equiv b_-$, $12 \equiv b_+$).

The fact that the calculated Frumkin-isotherm plots are relatively linear can be qualitatively explained by the water-water interactions being more attractive than the remaining lateral interactions. Taking local order into proper account, these attractive interactions cause water molecules to aggregate, namely to occupy a maximum of adjacent sites compatibly with the randomizing forces of thermal agitation. Hence, the probability of an adsorbing n-meric solute molecule finding $\underline{n}$ adjacent water molecules to displace from the adsorbed monolayer is greater than expressed by the $(1-\theta_S)^n$ factor in the Flory-Huggins isotherm of eqn.(60). Loosely speaking, for sufficiently high water-water interaction energies, this probability tends to be expressed by the $(1-\theta_S)$ factor in accordance with the Frumkin isotherm behaviour, other than at very high surface coverages. It should be noted that the present molecular model yields approximately linear Frumkin-isotherm plots for both n=1 and n>1, and hence is much more satisfactory than the macroscopic models having recourse to removable water clusters of definite size (80,86). The fact that the slope $\underline{a}$ of the calculated Frumkin-isotherm plots is negative (see Figs.

25 and 26) is likewise explained by the strongly attractive water-water lateral interactions. As θ_S is gradually increased, the interposition of an increasing number of solute molecules between the residual water molecules decreases their mutual attractive interactions and hence makes their removal progressively easier. Hence θ_S increases with increasing c_S more rapidly than in the absence of lateral interactions, giving rise to a negative slope of the log $[c_S(1-\theta_S)/\theta_S]$ vs θ_S plots. The maximum negative value of a at intermediate charges (see Fig. 26) is mainly due to H-bonds between adsorbed water molecules and, to a minor extent, to attractive dipole-dipole interactions between contiguous water dipoles. As σ_M becomes progressively more negative the amount of H-bonding remains practically unaltered, but the gradual rotation of the adsorbed water molecules around the H-bonds within the monolayer, as produced by the increasing external electric field, causes lateral dipole-dipole interactions between water molecules to become progressively less attractive and ultimately repulsive. Hence, the overall lateral interactions between contiguous water molecules become less attractive as σ_M becomes more negative, and the negative value of a decreases accordingly. The decrease in the negative value of a with increasing $|\sigma_M|$ is more pronounced at positive σ_M values (see Fig. 26), because over this charge range water molecules are gradually chemisorbed with a resulting breaking of H-bonds within the monolayer. This also causes progressive deviations from the Frumkin-isotherm behaviour, as shown in Fig. 25.

The adsorption behaviour of the majority of aliphatic compounds agrees satisfactorily with the predictions of the present model. Thus, the Frumkin isotherm behaviour is generally satisfied, with negative interaction factors a ranging from -1 to -3 (80-82). A roughly parabolic dependence of a upon charge with a maximum negative value in the proximity of σ_{max}, such as that predicted in Fig. 26, was reported for t-butanol (90) and butyronitrile (91), and was also verified with several aliphatic amines (92). The majority of organic surfactants exhibits a parabolic dependence of ΔG°_{ads} upon σ_M analogous to that shown in Fig. 26. Usually, aliphatic compounds are found to exhibit σ_{max} values ranging from -2 to -4 $\mu C\ cm^{-2}$. To explain this feature on the basis of the present model, we must assume that the solute dipoles are adsorbed with the positive end towards the metal, which assumption is frequently made in the literature. It might be shown, however, that negative values of σ_{max} can also be explained, at least partly, by postulating a moderate H-bonding between solute and water molecules, which is usually possible with organic molecules having a hydrophilic functional group. The aliphatic compounds whose adsorption behaviour on mercury has been thoroughly examined at constant charge are relatively limited in number. Nonetheless, if the available b_- values for these compounds are plotted against σ_{max}, as done in Fig. 27, a definite trend is apparent. Thus, b_- increases as σ_M becomes progressively less negative, in agreement with the predictions of the model. The experimental evidence for the predicted decrease of b_- as σ_{max} becomes progressively more positive is more ambiguous, because it is based on a single experimental point. It should also be noted that, even though the experimental b_- values in Fig. 27 refer to organic compounds with different Γ_{max} values, whereas the corresponding expected values were all calculated for n=3, agreement between experimental and calculated b_- values is semiquantitative. In the majority of adsorption data available in the literature, either b_- values are reported or else b_- and b_+ values are reported to be the same. Nonetheless succinonitrile, which is assumed to be adsorbed in a fixed orientation, is characterized by $b_+>b_-$ and $\sigma_{max}=-4.6$ μC cm^{-2} (93), whereas sucrose is characterized by $b_+<b_-$ and $\sigma_{max}=0$ μC cm^{-2} (94). Both experimental findings are in agreement with the predictions of Fig. 27.

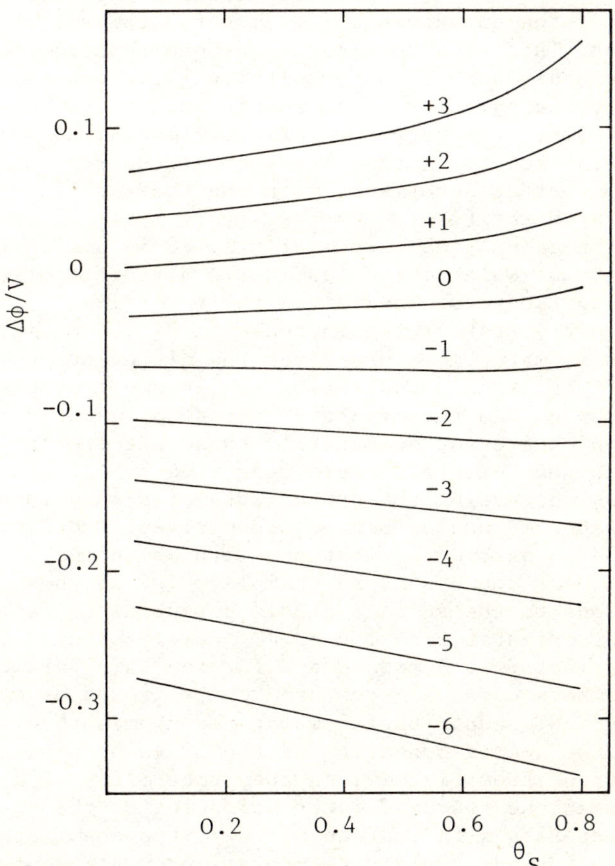

Figure 28.

Figure 28 shows plots of $\Delta\phi$ vs θ_S at several σ_M values, as calculated for μ_S=-0.6 D while maintaining all other parameters as in Figs. 25 to 27. Numbers on each curve denote charge values in $\mu C\ cm^{-2}$. These calculated curves exhibit average slopes which pass from negative to positive values with increasing σ_M , becoming roughly horizontal in the proximity of σ_{max} , which is $\simeq$-1 $\mu C\ cm^{-2}$ for this particular case. This behaviour has been reported for the majority of aliphatic compounds. It was rationalized by Parsons (95) on the basis of the equation

$$(\partial\Delta\phi/\partial\Gamma)_{\sigma_M} = d\Delta G^{\circ}_{ads}/d\sigma_M$$

which stems directly from the electrocapillary equation and from the condition for congruence with respect to charge. It should be noted that the calculated curves of Fig. 28 account satisfactorily for this experimental behaviour, without having to postulate congruence. The plots in Fig.28 show a curvature which increases with increasing $|\sigma_M-\sigma_{max}|$, especially towards positive charges, the convexity of the curvature being directed towards the most horizontal plot. This behaviour has also been reported for several aliphatic compounds, such as butyronitrile (91), t-butanol (90), n-butanol (96), diethylether (96), and several n-aliphatic amines (92). In particular, the curvature of experimental $\Delta\phi$ vs θ_S plots at σ_M=0 has been used to distinguish whether the adsorption behaviour is in better agreement with the two-parallel-capacitors model or else with the two-series-capacitors model (80,97). The curvature in the calculated curves of Fig. 28 is due to the fact that the numbers of water molecules in the different orientations of Fig. 16 do not decrease linearly with increasing θ_S at constant charge. Differently stated, the increase in θ_S is accompanied by a redistribution of water molecules over the different groups of water orientations. Such a redistribution is due to the fact that H-bond formation within the monolayer tends to favour certain water orientations which are not necessarily those electrostatically more stable in the strong electric field generated by σ_M. Hence, as θ_S increases, thus decreasing the probability of H-bond formation of the residual water molecules between themselves, these water molecules will tend to assume orientations with a minimum of electrostatic and *chemical* energy. As an example, Fig. 29 shows plots of the contributions to the surface dipole potential $\Delta\chi$ from all the groups of water orientations with $\mu_i \neq 0$ (namely groups 3, 4 and 5 of Fig. 16) against θ_S , for σ_M=0 and for the same parameters as in Fig. 28. Numbers on each curve denote the groups of water orientations. The notable contribution which the chemisorbed water molecules make to the upward concavity of the $\Delta\chi$ vs θ_S plot at zero charge is apparent from Fig. 29. As the probability of H-bond formation between the residual water molecules of the monolayer necessarily decreases with increasing θ_S , the chemisorbed orientation of group 5 tends rapidly to prevail over all other orientations, because of its stronger chemical interaction with the metal.

Thus, even though the total number of adsorbed water molecules decreases as $(1-\theta_S)$ with increasing θ_S, the number of chemisorbed molecules first increases slowly, and then decreases rapidly so as to attain the zero value at $\theta_S=1$.

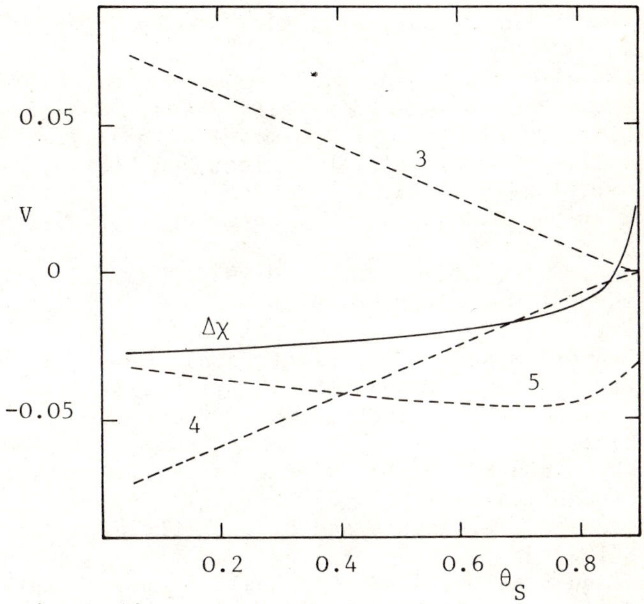

Figure 29.

In conclusion, multistate models of H-bonded water molecules in which local order is accounted for explain several features of inner-layer capacity curves as well as of organic adsorption. A further notable improvement in the physical picture may possibly be gained by including further water monolayers and ionic species into the model.

REFERENCES

1) R.M.Reeves, Mod. Aspects Electrochem., 9(1974)239.
2) B.E.Conway, Adv. Colloid Interface Sci., 8(1977)91.
3) M.A.Habib, Mod. Aspects Electrochem., 12(1977)131.
4) S.Trasatti, Mod. Aspects Electrochem., 13(1979)81.
5) W.R.Fawcett, Israel J. Chem.. 18(1979)3.
6) D.C.Grahame, J. Am. Chem. Soc., 76(1954)4819.
7) I.L.Cooper and J.A.Harrison, J. Electroanal. Chem., 86(1978) 425; Electrochim. Acta, 22(1977)519.
8) L.Blum and D.Henderson, J. Chem. Phys., 74(1981)1902.

9) D.Henderson and L.Blum, J. Electroanal. Chem., 132(1982)1.
10) D.Henderson, L.Blum and M.Lozada-Cassou, J. Electroanal. Chem., 150(1983)291.
11) L.Blum, J.Phys. Chem., 81(1977)136.
12) J.P.Hansen and I.R.McDonald, *Theory of Simple Liquids*, Academic Press, London, 1976.
13) R.O.Watts and I.J.Mc Gee, *Liquid State Chemical Physics*, Wiley, New York, 1976.
14) N.D.Lang and W.Kohn, Phys. Rev. B1 (1970)4555; Phys. Rev. B3 (1971)1215.
15) F.Garcia-Moliner and F.Flores, *Introduction to the Theory of Solid Surfaces*, Cambridge University Press, Cambridge, 1980.
16) W.Schmickler, J. Electroanal. Chem., 150(1983)19.
17) J.P. Badiali, M.L.Rosinberg, F.Vericat and L.Blum, J. Electroanal. Chem., 158(1983)253.
18) W.Schmickler and D.Henderson, J. Chem. Phys., 80(1984) 3381.
19) D.C.Grahame, J. Am. Chem. Soc., 79(1957)2093.
20) A.W.M.Verkroost, M.Sluyters-Rehbach and J.H.Sluyters, J. Electroanal. Chem., 24(1970)1.
21) G.J.Hills and R.M.Reeves, J. Electroanal. Chem., 31(1971)269.
22) D.J.Schiffrin, Trans. Faraday Soc., 67(1971)3318.
23) J.O'M.Bockris, M.A.V.Devanathan and K.Müller, Proc.R.Soc., London, A274(1963)55.
24) J.O'M.Bockris and M.A.Habib, J. Electroanal. Chem., 65(1975) 473; Electrochim.Acta, 22(1977)41.
25) J.A.Harrison, J.E.B.Randles and D.J.Schiffrin, J. Electroanal. Chem., 48(1973)359.
26) G.J.Hills and S.Hsieh, J. Electroanal. Chem., 58(1975)289.
27) H.D.Hurwitz and C.V.D'Alkaine, J. Electroanal. Chem., 42(1973)77.
28) A.Jenard and H.D.Hurwitz, J. Electroanal. Chem., 51(1974)377.
29) J.E.B.Randles and K.S.Whiteley, Trans. Faraday Soc., 52(1956) 1509.
30) B.E.Conway and H.P.Dhar, Croat. Chem. Acta, 45(1973)109.
31) For a review, see A.Hamelin. T.Vitanov, E.Sevastyanov and A.Popov, J. Electroanal. Chem.. 145(1983)225.
32) R.A.Oriani and C.A.Johnson, Mod. Aspects Electrochem., 5(1969)90.
33) R.Fowler and E.A.Guggenheim, *Statistical Thermodynamics*. Cambridge at the University Press. Cambridge, 1965.
34) I.Prigogine, A.Bellemans and V.Mathot, *The Molecular Theory of Solutions*, North-Holland Publishing Company, Amsterdam. 1957.
35) A.Clark, *The Theory of Adsorption and Catalysis*, Academic Press. New York, 1970.
36) J.B.Hasted in W.A.P.Luck (Ed), *Structure of Water and Aqueous Solutions*, Verlag Chemie, Berlin, 1974, p. 377.
37) H.P.Schwan. R.J.Sheppard and E.H.Grant, J.Chem. Phys., 64(1976) 2257.
38) S.Levine, G.M.Bell and A.L.Smith, J. Phys. Chem., 73(1969)3534.
39) W.R.Fawcett, J. Phys. Chem., 82(1978)1385.
40) Z.Borkowska and W.R.Fawcett, Elektrokhimiya, 16(1980)1692.

41) W.R.Fawcett, S.Levine, R.M.de Nobriga and A.C.Mc Donald, J. Electroanal. Chem., 111(1980)163.
42) W.R.Fawcett and R.M.de Nobriga, J. Phys. Chem., 86(1982)371.
43) P.J.Flory, J. Chem. Phys., 10(1942)51.
44) E.A.Guggenheim, *Mixtures*, Oxford University Press, Oxford. 1952.
45) H.A.Bethe, Proc. Roy. Soc. A, 150(1935)552.
46) C.Domb, Adv. Phys., 9(1960)149.
47) J.P.Valleau and S.G Whittington in B.J.Berne (Ed.), *Statistical Mechanics. Part A: Equilibrium Techniques*, Plenum, New York, 1977, pp. 137-194.
48) R.Guidelli and G.Aloisi, in preparation.
49) R.Parsons, J. Electroanal. Chem., 109(1980)369.
50) R.Parsons and R.M.Reeves, J. Electroanal. Chem., 123(1981)141.
51) S.Levine, A.L.Smith and E.Matijevic, J. Coll. Interface Sci., 31(1969)409.
52) J.Topping, Proc. Roy. Soc., A113(1927)67.
53) R.Guidelli, J. Electroanal. Chem., 110(1980)205.
54) R.J.Watts-Tobin, Phil. Mag., 6(1961)133.
55) I.L.Cooper and J.A.Harrison, J. Electroanal. Chem., 66(1975)85.
56) W.R.Fawcett, B.M.Ikeda and J.B.Sellan, Can. J. Chem., 57(1979) 2268.
57) Z.Borkowska and W.R.Fawcett, Can. J. Chem., 59(1981)710.
58) B.B.Damaskin and A.N.Frumkin, Electrochim. Acta, 19(1974)173.
59) R.Parsons, J. Electroanal. Chem., 59(1975)229.
60) B.B.Damaskin, J. Electroanal. Chem., 75(1977)359.
61) R.Parsons, Electrochim. Acta, 21(1976)681.
62) R.Guidelli, J. Electroanal. Chem., 123(1981)59.
63) R.Guidelli, in preparation.
64) P.A. Kollman and L.C.Allen, J. Chem. Phys., 51(1969)3286.
65) J.Del Bene and J.A.Pople, J. Chem. Phys., 52(1970)4858.
66) R.Parsons, R.M.Reeves and P.N.Taylor, J. Electroanal. Chem., 50(1974)149.
67) E.Whalley, Trans. Faraday Soc., 53(1957)1578.
68) J.H.Wang, J. Am Chem. Soc., 73(1951)4181.
69) G.Nemethy and H.A.Scheraga, J. Chem. Phys., 41(1964)680.
70) G.Vallette, J. Electroanal. Chem., 122(1981)285; 138(1982)37.
71) A.Hamelin, J. Electroanal. Chem., 138(1982)395.
72) G.Vallette, J. Electroanal. Chem., 139(1982)285.
73) S.Trasatti, J. Electroanal. Chem., 138(1982)449; 172(1984)27.
74) A.A.Balandin, Russ. Chem. Rev. (English Transl.) 11(1962)589.
75) B.M.W.Trapnell, Advan. Catalysis, 3(1951)1.
76) S. Trasatti, J. Electroanal. Chem., 82(1977)391.
77) J.P.Badiali, M.L.Rosinberg and J.Goodisman, J. Electroanal. Chem., 130(1981)31; 150(1983)25.
78) J.P.Badiali, M.L.Rosinberg and J. Goodisman, J. Electroanal. Chem., 143(1983)73.
79) D.Price and J.W.Halley, J. Electroanal. Chem., 150(1983)347.
80) B.B.Damaskin, O.A.Petrii and V.V.Batrakov, *Adsorption of Organic Compounds on Electrodes*, Plenum, New York, 1971.
81) B.B.Damaskin and V.E.Kazarinov in J.O'M.Bockris, B.E.Conway and

E.Yeager (Eds.), *Comprehensive Treatise of Electrochemistry*, Vol 1, Plenum, New York, 1980, p.353.
82) S.Trasatti, J. Electroanal. Chem., 53(1974) 335.
83) Ref. 33, p.431.
84) Ref. 35, p.74.
85) V.A.Kir'yanov, V.S.Krylov and B.B.Damaskin, Elektrokhimiya, 6(1970)533.
86) R.Parsons, J. Electroanal. Chem., 8(1964)93.
87) J.O'M.Bockris, E.Gileadi and K.Müller, Electrochim. Acta, 12 (1967)1301.
88) J.G.Kirkwood in W.D.Mc Elroy and B.Grass (Eds.), *A Symposium on the Mechanism of Enzyme Action*, John Hopkins University Press, Baltimore, 1954.
89) R. Guidelli and M.L.Foresti, in preparation.
90) A.De Battisti, B.A.Abd-El-Nabey and S.Trasatti, J. Chem. Soc., Faraday Trans. I, 72(1976)2076.
91) R.Amadelli, A.Daghetti, L.Vergano, A.De Battisti and S.Trasatti, J. Electroanal. Chem., 100(1979)379.
92) G.Aloisi, M.Carla', M.L.Foresti and R.Guidelli, unpublished results.
93) A.De Battisti, V.Faggiano and S.Trasatti, J. Electroanal. Chem., 73(1976)327.
94) R.Parsons and R.Peat, J. Electroanal. Chem., 122(1981)299.
95) R. Parsons, Trans. Faraday Soc., 55(1959)999.
96) E.Dutkiewicz, J.D.Garnish and R.Parsons, J. Electroanal. Chem., 16(1968)505.
97) A.N.Frumkin, B.B.Damaskin and A.A.Survila, J. Electroanal. Chem., 16(1968)493.

THE CAPACITY OF THE METAL/LIQUID ELECTROLYTE INTERPHASE: CONTRIBUTION OF THE METAL

Wolfgang Schmickler
Institut für physikalische Chemie
der Universität Düsseldorf
D 4000 Düsseldorf, FRG

ABSTRACT: The jellium model for metals is reviewed. The polarizability of the electronic plasma in the surface region increases the double layer capacity. The model is complemented by a model for the electrolyte solution, which is represented as an ensemble of hard sphere ions and dipoles. The relation of the model to experimental data, and possible extensions, are discussed.

I) INTRODUCTION

The inverse capacity of a metal/electrolyte interphase can be decomposed into a contribution $1/C_{dif}$, which depends on the electrolyte concentration, and a contribution $1/C_i$, which is independent of it:

$$\frac{1}{C} = \frac{1}{C_{dif}} + \frac{1}{C_i} \tag{1}$$

C_{dif} is usually called the "diffuse layer" capacity, and for a 1-1 electrolyte which does not specifically adsorb, and for not too high electrolyte concentrations, it is well represented by the familiar Gouy-Chapman expression C_{GC}. $1/C_i$ is is conveniently obtained from a Parsons and Zobel/1/ plot: experimental values of $1/C$ are plotted against $1/C_{GC}$, and extrapolated to infinite electrolyte concentrations. C_i is generally called the "inner layer" capacity, since it was conventionallly attributed to an "inner layer" of solvent molecules at the metal surface. We shall keep this terminology, though we do not share this interpretation.

A. F. Silva (ed.), Trends in Interfacial Electrochemistry, 453–472.

This inner layer capacitance depends on the nature of the metal and the solvent, but it seems to be independent of the electrolyte, as long as it is not specifically adsorbed. For a given interphase, it is a function of the electrode charge or potential, and temperature. In the conventional view, which was unchallenged till about 1980, C_i was assumed to be due to an "inner layer" of solvent molecules at the metal surface, which was thought to constitute a region into which the electrolyte ions could not penetrate. The metal itself was considered to be a perfect conductor, which does not make a direct contribution to the interfacial capacity, but which affects the dielectric properties of the "inner layer" through chemical interactions.

Metal	$C_i / \mu F cm^{-2}$	ε_i	g / V
Hg	28	9.8	0.08
Tl	33.5	11.6	0.20
Cd	52	18.2	0.23
In(Ga)	60	21.0	0.20
Zn	100	35.0	0.32
Ga	135	47.0	0.33

Tab.1: Double layer capacity and dipole potential for the interphase between water and various sp-metals at the p.z.c.; (after Trasatti/2/)

This view was well summed up in a paper by Trasatti/2/, in which he collected data for the inner layer capacity at the p.z.c., and calculated from them an "effective inner layer dielectric constant" ε_i from the assumption, that the inner layer constitutes a capacitor with a plate separation of one molecular diameter. In tab.1 we reproduce his data for contacts of sp-metals with water, tab.2 gives the data for DMSO. The calculated dielectric constants ε_i vary considerably with the metal; for water they are always smaller than the bulk value, which is attributed to the fact, that chemisorbed water is less free to rotate. For the interphase Ga/DMSO ε_i is larger than the bulk value, which is difficult to explain.

Metal	$C_i/\mu Fcm^{-2}$	ε_i
Hg	23	15.8
In(Ga)	53	36.5
Ga	93	64.5

Tab.2: The double layer capacity for the interphase between DMSO and various sp-metals at the p.z.c.; the bulk dielectric constant of DMSO at room temperature is 46.6; (after Trasatti/2/).

This conventional model does not not offer any prescription how to to estimate the effective dielectric constants ε_i from microscopic parameters such as the strength of the adsorption bond. There is also an apparent inconsistency: the estimated values for the dipole potential g of water at the p.z.c. increase in almost the same order as the inner layer capacities (see tab.1), which indicates an increasing degree of orientation of the solvent due to stronger chemisorption bonds. However, a more strongly bonded molecule, which is less free to rotate, would be expected to give rise to a lower effective dielectric constant.

For a few systems, particularly mercury/water, the capacity/charge characteristics, including the temperature dependence, could be reproduced from molecular models of the inner layer which contained a few adjustable parameters; the pioneering paper by Parsons/3/ sparked off a series of works of similar nature. All these models worked for one particular metal, usually Hg, only; they could not reproduce the characteristics for other metals unless parameters pertaining to the solvent were also changed.

Already in 1928, O.K. Rice/4/ had suggested, that the inner layer capacity is due to the metal, and calculated its value in the Thomas-Fermi model. However, the values obtained in this model are generally too small compared with experimental data/5/. Also, attributing all of the inner layer to the metal, as Rice did, could not explain its dependence on temperature and charge - in the Thomas-Fermi approximation the metal capacity is independent of the charge -, and his work was almost forgotten for the next 70 years.

In the meantime, the theory of metal surfaces in the vacuum had made considerable progress, and effects like field penetration into the metal, electron spill-over at the surface, became well established. If these were important for metals in the vacuum, they should also have relevance to electrochemical systems, and could also affect the double layer. So the question, whether the metal itself makes a contribution to the capacity, was reexamined. Kornyshev et al./5,6/ applied the formalism of nonlocal electrostatics to the double layer problem; this led to a number of formal relations, expressing the double layer capacity through the nonlocal dielectric function of the interphase, and an important qualitative result/5/: any model of the interphase, which does not allow for electron spill-over into the solution, is incompatible with experimental data. Soon afterwards "jellium", a quantum mechanical model for sp-metals, was introduced into double layer theory first by Badiali et al./7/ and by myself/8/, and recently several other papers have appeared or are in press (see also the papers by Halley et al., and the cluster calculation approach by van der Erden, both at this conference).

Simultaneously, considerable progress was made in the theory of electrolyte solutions: a model, in which the solution is represented as an ensemble of hard sphere ions and dipoles, the metal as a charged hard wall, was solved by Carnie and Chan/9/ and Blum and Henderson/10/. These works also cast doubt on the existence of an inner layer in the conventional sense. This subject is reported in the article by D. Henderson in this volume, which is in several ways complementary to mine.

These two new developments - quantum mechanical models for the metal, progress in the statistical mechanics of electrolyte solutions - have been combined/11,12/, and have led to a new view of the double layer. As we shall demonstrate below, the new theories, which are presently emerging, are more explicit than the old models, they can explain a greater variety of facts, and are in line with modern theories of metal surfaces and electrolyte solutions.

Here, we shall concentrate on the effect that the metal has on the inner layer capacity. The plan of the rest of this article is as follows: First, we shall present the jellium model for sp-metals; in section III we shall explain, by what mechanism the metal surface contributes to the double layer capacity, and finally we shall give some results for the inner layer capacity calculated from a model, in which the metal is represented as jellium, the solution as an ensemble of hard sphere ions and dipoles.

II) JELLIUM - A MODEL FOR sp-METALS

A metal consists of positively charged ions and electrons. In the jellium model the ions are represented by a constant positive background charge of value n , which drops abruptly to zero at the surface, the jellium edge (see fig.1), which we take at $x=0$, the metal occupying the region $x<0$. The electrons form a plasma of density $n(r)$, which interacts with itself, with the positive background, and with any external forces, e.g. an external electrostatic field acting on the surface. Due to their small mass the electrons spill over the metal surface, so that a small negative surface charge exists on the vacuum side $x>0$ (the solution side in electrochemical systems), which for an uncharged surface is balanced by a corrresponding positive excess charge on the metal side. Thus a surface dipole is formed at the the jellium edge, it gives rise to a corresponding surface potential drop, which makes an important contribution to the electronic work function. We shall show later, that the variation of this surface potential with the charge density determines the contribution of the metal to the inner layer capacity.

Several excellent review articles on jellium exist (see e.g. refs.13,14, and further references therein), so we shall concentrate on those points which are directly relevant to electrochemical systems. Jellium is usually treated within the Hohenberg - Kohn - Sham formalism/15,16/, in which the energy of the system is written as a functional of the electronic density $n(r)$. Using atomic units ($e_0=\hbar=m_e=1$) throughout, we can write:

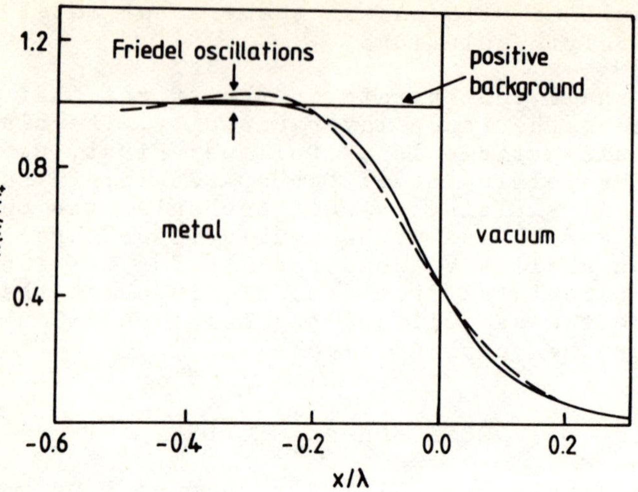

Fig.1: The jellium model for a metal in vacuum; solid line: exact electronic profile, broken line: approximate profile using the trial functions of eq.(11). $\lambda = 2\pi/k_F$ is the Fermi wavelenght.

$$E[n(r)] = \int v(r)\,n(r)\,dr + \frac{1}{2}\int \frac{n(r)\,n(r')}{|r-r'|}\,dr\,dr' - \frac{3}{4}\left(\frac{3}{\pi}\right)^{1/3}\int n(r)^{4/3}dr - 0.056\int \frac{n(r)^{4/3}}{0.079+n(r)^{1/3}}dr + T_{kin} \quad (2)$$

where in the first term v(r) is an external potential, and the other terms denote the electrostatic self-interaction, the exchange energy, the correlation energy in the Wigner approximation, and the kinetic energy, in this order. For the latter, a local expansion is often used:

$$T_{kin}[n(r)] = \frac{3}{10}(3\pi^2)^{2/3}\int n(r)^{5/3}dr + \frac{1}{72}\int \frac{(\vec{\nabla} n(r))^2}{n(r)}dr \quad (3)$$

This is particularly useful, when solutions for n(r) are sought within a family of trial functions.

With v(r) determined by the positive background, eq.(2) is for a metal in the vacuum. When an electrolyte solution is present in the region $x>0$, at least three additional terms arise:
1) The interaction of the plasma with the charge distribution in the solution. In a one-dimensional model, in which the ionic charge is smeared out in the y-z direction, this can be written as:

$$E_{ion} = \int_0^\infty \varphi_{jel}(x)\, \rho_c(x)\, dx \qquad (4)$$

where φ_{jel} is the electrostatic potential generated by the plasma, and ρ_c the ionic charge density in the solution.
2) The interaction of the plasma with the solvent polarization:

$$E_{dip} = -\int_0^\infty E_{jel}(x)\, P(x)\, dx \qquad (5)$$

$E_{jel}(x)$ is the electric field generated by the jellium.
3) A term representing the repulsion of the electrons from the closed shell of electrons on the solvent molecules:

$$E_{rep} = \int V_{rep}(r)\, n(r)\, dr \qquad (6)$$

where V_{rep} is a repulsive potential pertaining to the solvent molecules. Ideally, this potential should come from a quantum mechanical model of the solvent. So far, such expressions are not available, and only simple model potentials of the form:

$$V_{rep}(x) = V_b\, \theta(x)$$

or $\qquad (7)$

$$V_{rep}(x) = U\, \delta(x-a) \quad , a>0$$

have been used in actual calculations. In these model potentials the parameters are chosen on a semi-empirical basis for each particular solvent.

In the ground state of the system, the electronic energy attains its minimum for the true electronic density profile. From this condition a set of self-consistent equations for n(r) can be derived. On this basis Lang and Kohn/17/ have calculated the electronic density profiles over a range of bulk densities exactly. An attractive scheme to obtain approximate solutions is, to choose a suitable family of trial functions for n(x) with several free parameters, calculate the electronic energy as a function of these parameters using the gradient expansion for the kinetic energy, and to minimize the energy within this family of functions. This approach was first taken by Smith/18/: with a simple one-parameter family of trial functions he obtained jellium work functions surprizingly close to the exact values of Lang and Kohn. So far, all calculations on electrochemical systems have used this trial function method, so we shall discuss it in greater detail, and give several examples of suitable sets of functions that can be used for charged surfaces.

The simplest family, due to Partenskii and Smorodinskii/19/ is:

$$n(x) = n_+ - \frac{1}{2}[n_+ + \alpha q]\, e^{\alpha x} \qquad \text{for } x < 0$$

$$n(x) = \frac{1}{2}[n_+ - \alpha q]\, e^{-\alpha x} \qquad \text{for } x > 0$$

q is the surface charge density. The preexponential factor has been chosen in such a way, that the charge balance condition is fulfilled; for an uncharged surface this family of trial functions reduces to that employed by Smith/18/. The only free parameter α has the meaning of an inverse length characterizing the extent of electronic spill-over. The simple form of the preexponential factor limits the application to the region $\alpha q/n_+ \ll 1$.

A more general family has been used by Paasch and Hietschold/20/, and by Badiali et al./7/:

$$n(x) = n_+ (1 - A)\, e^{\alpha(x - x_0)} \qquad \text{for} \quad x < x_0$$
$$n(x) = n_+ B\, e^{-\beta(x - x_0)} \qquad \text{for} \quad x > x_0 \qquad (9)$$

Of the five parameters A, B, x_0, α, β, three are fixed by the conditions, that n(x) and its derivative n'(x) must be continuous, and by charge balance.

The correct electronic density profile exhibits small oscillations of decreasing amplitude, the Friedel oscillations, on the metal side of the surface (see fig.1) Except for very low electronic densities, these oscillations are small, and this justifies their neglect in the two families of trial functions above. There is, however, a useful relation which these families do not fulfill: the Budd and Vannimenus sum rule/21,22/, which gives an exact expression for the electrostatic potential drop $\varphi(0)-\varphi(-\infty)$ between the jellium surface and the bulk:

$$n_+ [\varphi(0) - \varphi(-\infty)] = -\frac{q^2}{2 n_+} - \frac{k_F^2}{2}\left[0.4 - 0.0829\, r_s - \frac{0.0796\, r_s^3}{(r_s + 7.8)^2}\right] \qquad (10)$$

where $k_F = (3\pi^2 n_+)^{1/2}$ is the Fermi momentum, and $r_s = (3/4\pi n_+)^{1/3}$ is the Wigner-Seitz radius. This relation has the nature of a force or pressure balance condition: it can be derived by placing a sheet of charge density q at a position outside the jellium far from the jellium tail, and equilibrating forces. The first term on the r.h.s. is the pressure on the sheet, and the second term the electrostatic pressure inside the bulk of the jellium in the Wigner approximation. In many applications of the jellium model, e.g. for calculations of the work function or of the double layer capacity, the surface potential drop $\varphi(+\infty) - (-\infty)$ generated by the electronic spill-over plays an important role; this sum rule gives an expression for a sizable part of this surface potential drop.

Schmickler and Henderson/23/ have devised a family of trial functions, which incorporates both Friedel oscillations, and fulfills the Budd and Vannimenus sum rule:

$$n(x) = n_+ [1 - A e^{\alpha x} \cos(\gamma x + \delta)] \quad \text{for} \quad x < 0$$
$$n(x) = n_+ B e^{-\beta x} \quad \text{for} \quad x > 0 \tag{1}$$

Of the six parameters four are determined from the continuity of n(x) and n'(x), from the charge balance, and from the sum rule; the other two are determined as usual by energy minimization. Not only the electronic density profiles (see fig.1), but also the work functions calculated from this set of functions agree very well with the exact values of Lang and Kohn/23/.

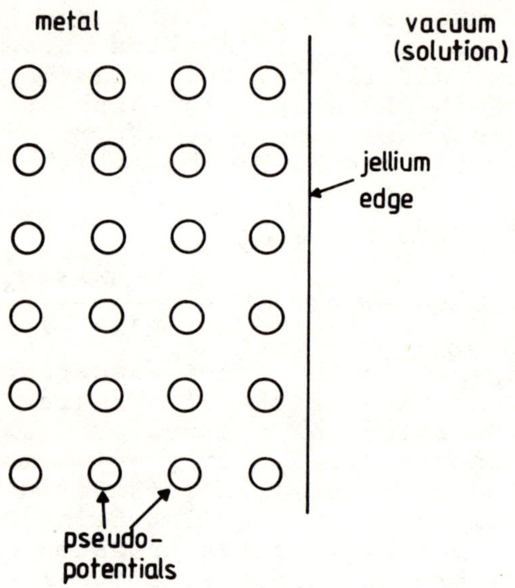

Fig.2: Jellium with a pseudopotential lattice

The "structureless" jellium described above is a reasonable model for polycrystalline sp-metals, and can be used to estimate work functions and similar surface properties. For several applications, particularly for the study of single crystall surfaces, and also for calculating surface energies, the discrete nature of the ionic lattice must be accounted for. The constant positive charge background can be replaced by a lattice of ions, and a pseudopotential centered at each lattice site (see fig.2). The jellium edge is half a lattice spacing in front of the first row of lattice sites. A suitable pseudopotential is, for example, the Ashcroft potential/24/:

$$v_{ps}(r) = 0 \qquad \text{for} \qquad r < r_c$$
$$v_{ps}(r) = -\frac{z}{r} \qquad \text{for} \qquad r > r_c \tag{12}$$

where z is the ionic charge number, and the radius r_c has been fitted for each metal to give good bulk properties. Inclusion of such pseudopotential lattices into the jellium leads to a marked improvement for the surface energies/17/, and makes it possible to calculate surface properties of single crystall surfaces/25/.

III) THE CONTRIBUTION OF THE METAL TO THE DOUBLE LAYER CAPACITY

We now return to the subject of our main interest, the double layer capacity, and consider the interface between a metal occupying the region $x<0$, modelled as jellium, and an electrolyte solution in the region $x>0$. The double layer capacity is determined by the variation of the electrostatic potential drop between the bulk of the solution and the metal with the electrode charge. The electrostatic potential $\varphi(x)$ is governed by the Poisson equation, which we write in the form:

$$\Delta \varphi(x) = -4\pi\left[\rho_m(x) + \rho_s(x) - \frac{d}{dx}P(x)\right] \tag{13}$$

where ρ_m is the charge density of the jellium, consisting of a positive ionic and a negative electronic contribution, ρ_s is the ionic charge density of the solution, and P(x) is its polarization. Integrating twice we obtain for the total potential drop across the interface:

$$\varphi(\infty) - \varphi(-\infty) = 4\pi\int_{-\infty}^{\infty} x\,\rho_m(x)\,dx + 4\pi\int_{0}^{\infty} x\,\rho_s(x)\,dx + 4\pi\int_{0}^{\infty} P(x)\,dx \tag{14}$$

This relation eneables us to write the double layer capacity as a series combination of a metal capacity, defined by:

$$\frac{1}{4\pi C_m} = -\frac{d}{dq}\int_{-\infty}^{\infty} x\,\rho_m(x)\,dx \tag{15}$$

and an electrolyte capacity:

$$\frac{1}{4\pi C_s} = -\frac{d}{dq}\int_{0}^{\infty} [x\,\rho_s(x) + P(x)]\,dx \tag{16}$$

Note, that these two contributions are not independent of each other, since any reasonable theory will contain interaction terms between the metal and the solution such as our eqs.(4)-(6). The metal capacity is thus determined by the variation of the surface dipole potential with the charge density q. It is easily seen, that this capacity is negative for jellium, i.e. the capacity between jellium and an electrolyte solution is greater than that between a perfect conductor and a solution/7,8/. This is illustrated in fig.3: On application of an electric field directed normally to the surface and pointing towards the solution, the electrons are pushed back into the metal, and the magnitude of the surface dipole is decreased. Thus, the change in the dipole moment is such as to oppose the applied field. We may thus consider the jellium surface as a dipole with a fixed orientation, but with a considerable polarizability, and this polarizability makes the interfacial capacity greater than it would be for a perfect conductor.

An alternative way of viewing the metal capacity can be obtained in the following way: We rewrite the metallic charge density in the form: $\rho_m(x)=\rho_m^0(x)+\delta\rho_m(x)$, where the first term corresponds to the uncharged metal surface. Using the Budd-Vannimenus sum rule eq.(10) for the potential drop between the bulk of the jellium and the surface, we obtain:

$$\frac{1}{4\pi C_m} = \frac{q}{n_+} - \frac{d}{dq}\int_{0}^{\infty} x\,\delta\rho_m(x)\,dx \tag{17}$$

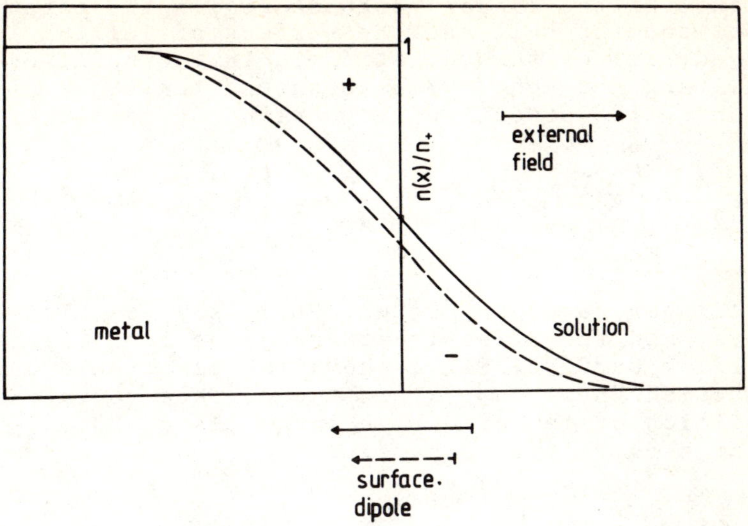

Fig.3: Polarization of the electronic plasma under the influence of an external field.

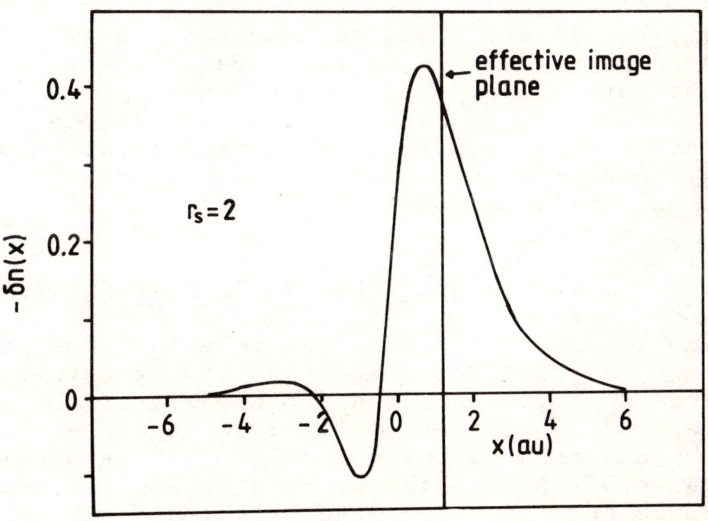

Fig.4: Induced surface charge

Note, that in the linear response region the potential drop between the bulk and the surface of jellium is unchanged, all of the variation is in the spill-over region. Lang and Kohn/17/ have shown, that the position of the effective image plane of jellium is:

$$x_{im} = \frac{d}{dq} \int_{-\infty}^{\infty} x\, \delta \rho_m(x)\, dx = \frac{d}{dx} \int_{0}^{\infty} x\, \delta \rho_m(x)\, dx \qquad (18)$$

in the linear response regime. Obviously, x_{im} is positive, and the effective image plane is in front of the jellium surface. Fig.4 shows the electronic charge distribution $\delta n(x) = -\delta \rho_m(x)$ in the surface region, and the position of the effective image plane. So, at the point of zero charge we have:

$$C_m(q=0) = -\frac{1}{4\pi x_{im}} \qquad (19)$$

The fact, that the effective image plane is in front of the jellium surface means, that a capacitor, in which one of the plates consists of jellium, has an effective plate separation which is smaller by an amount x_{im} than it would be, if this plate were a perfect conductor. Since x_{im} is of the order of 0.5 A, this effect is enetirely negligible for normal capacitors. However, if one thinks of the double layer as a capacitor with an effective plate separation of about 3 Å, then this effect is important, and explains, why the metal capacity is negative.

IV) THE INNER LAYER CAPACITY OF THE JELLIUM/SOLUTION INTERPHASE

The jellium model for the metal must be supplemented by an equally sophisticated model for the electrolyte solution; it is important, that both parts of the system are treated with about the same degree of accuracy, since an unbalanced model may overemphasize the influence of one subsystem. At present, the best model for the solution which can be solved mathematically is a mixture of hard sphere ions and dipoles. So the interphase between jellium and an ensemble of hard sphere ions and dipoles should be a good representation of the double layer. This model has been investigated by Badiali et al./11/ and by

Schmickler and Henderson/12/; both treatments differ in several ways. The latter work is reported in the article by D. Henderson in this volume; we shall briefly summarize the main points which are relevant for the extension of this work, by the same authors, that we shall describe below.

The equations for an ensemble of hard sphere ions and dipoles in contact with a hard wall can be solved for small charge densities on the wall in the mean spherical approximation/9,10/. In our model, the hard wall is the jellium edge. So, analytical expressions exist for the Laplace transforms of the ionic concentration profiles, and the solvent polarization P(x). The one parameter trial functions of eq.(8) were used for the electronic density distribution. The repulsive potential was taken as V   (x)=V   (x); the interaction terms between the jellium and the solution are readily calculated. Since the equations for the hard sphere ions and dipoles have been solved only in the linear response region, the calculations were limited to the p.z.c.. When compared with experimental data, the model gives good results for the inner layer capacity of sp-metals of the second and third column of the periodic table in contact with water, DMSO, and acetonitrile; it also reproduces well the temperature dependence of the capacity of the Hg/H_2O interphase/12/.

We report here on an extension of this model to single crystal surfaces; an extension to higher charges will be given elsewhere/27/. The study of single crystal surfaces has been a major advance in the electrochemistry of the last decade. In particular, Hamelin, Valette,/31,32/ and others have demonstrated the influence of the metal surface structure on double layer properties. For single crystal silver, the inner layer capacity characteristics in aqueous solutions have been determined for the three main faces/29/. We have therefore begun to investigate the capacity of single crystal surfaces in a model of jellium with pseudopotentials. The technique employed was the same as in ref.12 - in particular, the one parameter trial functions of eq.(18) were used - but an additional term, as described by Monnier and Perdew/25/, was introduced into the energy of the electronic plasma. Using Ashcroft pseudopotentials, we obtained for the inner layer capacity of single crystal silver in contact with an aqueous solution the folowing results:

face	C_i^{theo} /μFcm^{-2}	C_i^{exp} /μFcm^{-2}
100	19.3	90
110	25.2	108
111	18.3	69

Tab.3: Inner layer capacities of single crystal surfaces of Ag/H_2O; experimental values after Trasatti/32/

These calculations were performed under the assumption, that one valence electron per silver atom is free (n_+=0.0087 au). The calculated values for the capacity are far lower than the experimental values; this is probably due to the fact, that silver is an sd-metal, and there may well be some influence of the d-electrons. However, the important point is, that our calculations give the correct order for the three faces; even the relative variation $\delta C_i/C_i$ has the right order of magnitude.

But it must be pointed out, that such calculations do not work well for metals with higher electronic densities, for them, negative capacities may result. This is probably due to the fact, that we have used very simple trial functions only, which do not work so well for high densities.

V) OUTLOOK

In the double layer model outlined above, the metal and the solvent contribute equally to the inner layer capacity. The contribution of the metal is due to the polarizability of the electronic plasma in the surface region; the contribution of the solution is a boundary effect: the dielectric properties of the solvent are afffected by the presence of the metal surface acting as a hard wall; dielectric saturation is important at higher electrode charges.

The new theory avoids some of the difficulties of the older models mentioned in the introduction, and is much more explicit about mechanisms and interactions. At the present stage, model calculations involve only one semiempirical parameter, the height V_b of the repulsive barrier; the older models for monolayers of solvent contained between four and eleven such parameters.

Let us summarize what the new theory has achieved so far:

- it gives the correct trend for the inner layer capacity at the p.z.c. for sp-metals of the second and third column in contact with water, DMSO, and acetonitrile -

- it reproduces the temperature dependence of the inner layer capacity of the Hg/H_2O interface/12/ -

- it explains, why the diameter of the ions is not important/12/ -

- for Ag/H_2O it gives the correct order for the inner layer capacity of the three main single crystal surfaces -

- it explains deviations from the Parsons and Zobel plot at high electrolyte concentrations/33/ -

- it gives the right sign and order of magnitude for the dipole potentials at the p.z.c. (see refs. 7 and 28, and the article by S. Trasatti in this volume) -

For a theory that is both relatively simple and new, this is an impressive list. There is, however, also a long list of desirable improvements and extensions:

- A solution of the hypernetted chain equation at high electrode charges should be attempted, so that capacity-charge characteristics can be calculated -

- the potential representing the repulsion of the electrons from the solvent should be based on a quantum mechanical calculation. -

- the study of single crystal surfaces should be persued with more suitable sets of trial functions. -

- the quality of the mathematical approximations - trial function method, mean spherical approximation - should be checked -

One may wonder, that the concept of chemisorption, or strong chemical interactions - usually of the oxygen end with the metal -, which played such an important role in the older models, does not appear explicitly in the new theories. However, due to the electronic spill-over, there is a preferential orientation of the negative end of the solvent dipoles towards the metal at the p.z.c./28/; this could be interpreted as a weak chemical interaction. Strong chemisorption effects do not seem to occur in the systems studied so far. This is in line with recent UHV studies (see the article by J. Sass in this volume), which indicate, that the interaction of a water molecule with a metal is generally of the same order of magnitude as the water-water interaction. There may, however, be systems in which strong chemisorption is important - after all, capacity charge characteristics in non-aqueous solvents may differ widely -, and the concept may have to be reintroduced.

Acknowledgement: I would like to thank Dr. D. Henderson, IBM San Jose, for a fruitful and pleasurable cooperation, and to acknowledge financial support by the Deutsche Forschungsgemeinschaft and the Fond der Chemischen Industrie.

References

1) R. Parsons and F. Zobel, J. Electroanal. Chem. 9(1965)333

2) S.Trasatti, in: Advances in Electrochemistry and Electrochemical Engeneering, ed. by H. Gerischer and C.W. Tobias, Vol.10, Wiley-Interscience, 1977; S.Trasatti, J. Electroanal. Chem., 123(1981)121

3) R. Parsons, J. Electroanal. Chem. 59(1975)229

4) O.K. Rice, Phys. Rev. 31(1928)1051

5) A. Kornyshev, W. Schmickler, and M. Vorotyntsev, Phys. Rev. B 25(1982)5244

6) A. Kornyshev and M. Vorotyntsev, Can. J. Chem. 59(1981)2031

7) J.P. Badiali, J. Goodisman, and M. Rosinberg, J. Electroanal. Chem. 130(1981)31; 143(1983)73; 150(1983)25

8) W. Schmickler, J. Electroanal. Chem. 150(1983)19

9) S.L. Carnie and D.Y.C. Chan, J. Chem. Phys. 73(1980)2949

10) L. Blum and D. Henderson, J. Chem. Phys. 74(1981)1902

11) J.P. Badiali, M.L. Rosinberg, F. Vericat, and L. Blum, J. Electroanal. Chem. 158(1983)253

12) W. Schmickler and D. Henderson, J. Chem. Phys. 80(1984)3381

13) "Theory of the inhomogeneous electron gas", ed. by S. Lundquist and N.H. March, Plenum Press, 1983; see in particular the articles by N.D. Lang and W. Kohn

14) N.D. Lang, in: Solid State Physics, ed. by F. Seitz, D. Turnbull, and H. Ehrenreich, Vol.28, pp.225-300, Academic Press, 1973

15) P. Hohenberg and W. Kohn, Phys. Rev. 136(1964)B864

16) W. Kohn and L.J. Sham, Phys. Rev. 140(1965)A1133

17) N.D. Lang and W. Kohn, Phys. Rev. B1(1970)4555; Phys. Rev. B3(1971)1215; Phys. Rev. B7(1973)2541

18) J.R. Smith, Phys. Rev. 181(1969)522

19) M. Partenskii and Ya. Smorodinskii, Sov. Phys. Solid State 16(1974)423

20) G. Paasch and M. Hietschold, Phys. stat. solid. b) 67(1975)743, ibid. 83 (1977) 209

21) H.F. Budd and Vannimenus, Phys. Rev. Lett. 31(1973)1212; 31(1973)1430(E)

22) J. Heinrichs and N. Kumar, Phys. Rev. B 12(1975)802

23) W. Schmickler and D. Henderson, Phys. Rev. B, in press

24) N.W. Ashcroft, Phys. Lett. 23(1966)48

25) R. Monnier and J.P. Perdew, Phys. Rev. B 17(1978)2595

26) W. Schmickler and D. Henderson, J. Electroanal. Chem., in press

27) W. Schmickler and D. Henderson, to be published

28) W. Schmickler, Chem. Phys. Lett. 99(1983) 135

29) G. Valette and A Hamelin, J. Electroanal. Chem.45(1973)301;
G. Valette, J. Electroanal. Chem. 122(1981)285;
G. Valette, A. Hamelin, and R. Parsons, Z. Phys. Chem. NF 113(1978)71

30) A. Hamelin, T. Titanov, E. Sevastinov, and A. Popov, J. Electroanal. Chem. 145(1983)225

31) D. Henderson and W. Schmickler, to be published.

32) S. Trasatti, J. Electroanal. Chem. 172(1984)27

33) L. Blum, D. Henderson, and R. Parsons, J. Electroanal. Chem. 161(1984)389

RECENT DEVELOPMENTS IN THE THEORY OF ELECTRIFIED INTERFACES

Douglas Henderson

IBM Research Laboratory, San Jose, California 95193

Until recently, all theories of the double layer were based on a model in which the charge on the metal is assumed to be confined to surface layer of zero thickness, the interfacial solvent layer is assumed to be a monolayer, and outside the "inner" layer the solvent is assumed to be a dielectric continuum. It is shown that even within this model the well-known Gouy-Chapman-Stern theory overestimates the potential difference of the interface. Moreover, more plausible models can be formulated which agree with experiment but do not require a large number semiempirical parameters and which account for some of the deviations between the earlier models and experiment. In particular, it is shown that the so called "inner" layer is fictitious. The contribution of the solvent to the interfacial properties arises from a layer which is as thick as the entire double layer. In addition, it is shown that the metal makes a significant contribution to the interfacial properties

1. INTRODUCTION

The electrical double layer is not only a problem of some practical importance but also a rather interesting theoretical problem since it involves a system composed of both a classical fluid (the electrolyte) and a quantum fluid (the electrons in the metal electrode). Furthermore, the classical fluid is itself interesting because of the long range coulombic forces due to the ions and the orientational forces due to the solvent molecules. Even in the bulk, the theory of an electrolyte is only partly developed.

In this paper, the metal will be modelled as a jellium, that is a collection of free electrons with the metallic ions represented by a smeared out background of positive charge which is spatially fixed and the

A. F. Silva (ed.), Trends in Interfacial Electrochemistry, 473–521.

electrolyte will be modelled as a mixture of charged hard spheres (the ions) and hard spheres with embedded point dipoles (the solvent molecules, usually water). This model is displayed in Fig. 1.

The properties of the double layer are conveniently described in terms of *profiles* which give the probability of finding a particle at some point x, normalized so that the profile is the bulk density far from the interface. For the electrolyte the profiles will be denoted by $\rho(x)$ while for the jellium they will be denoted by n(x). Thus, adopting the convention used in Fig. 1 of placing the metal in the left half plane and the electrolyte in the right half plane,

$$\rho(\infty) = \rho \tag{1.1}$$

and

$$n(-\infty) = n\ . \tag{1.2}$$

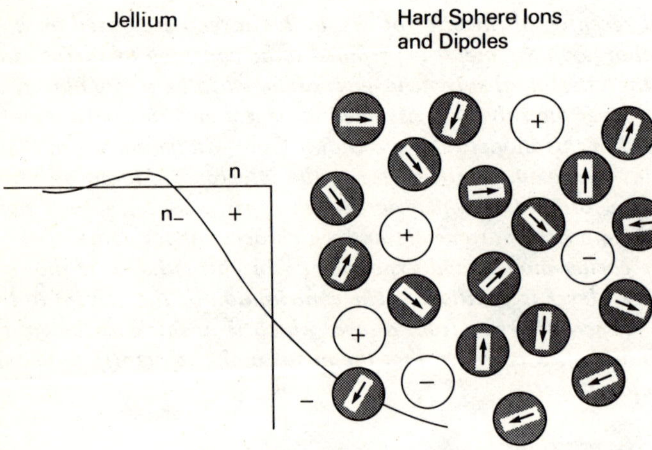

Figure 1. *Ion-dipole-jellium model of the double layer.*

Throughout this article it will be assumed that the interface is perfectly flat and smooth so that the profiles can be assumed to be a function only of the distance x normal to the interface (or in the case of the solvent molecules x and the orientation of the molecules). It is often convenient to divide $\rho(x)$ by $\rho(\infty)$ to obtain a *singlet distribution* function

$$\begin{aligned} g(x) &= 1 + h(x) \\ &= \rho(x)/\rho\ . \end{aligned} \tag{1.3}$$

The profile for the metallic ions is just

$$n_+(x) = \begin{cases} n & , x<0 \\ 0 & , x>0 , \end{cases} \tag{1.4}$$

while that of the electrons in the metal would be of the general form

$$n_-(x) = \begin{cases} n - Ae^{\alpha x}\cos(\beta x + \gamma) & , x<0 \\ Be^{-\delta x} & , x>0 , \end{cases} \tag{1.5}$$

which is also displayed in Fig. 1. Thus, well inside the metal there is no net charge. However, there is a net charge at the metal surface. The charge/unit area on the metal electrode is just

$$-e\int_{-\infty}^{\infty} [n_-(t)-n_+(t)]dt = E/4\pi , \tag{1.6}$$

where e is the magnitude of the electronic charge. The potential drop across the metal part of the double layer is

$$V_m = 4\pi e\int_{-\infty}^{\infty} t[n_-(t)-n_+(t)]dt . \tag{1.7}$$

In the electrolyte, the profile for ions of species i is

$$\rho_i(x) = \rho_i g_i(x) , \tag{1.8}$$

where ρ_i is the bulk density of the ions of species i. Typical profiles for a binary salt are displayed in Fig. 2. The density profile of all the ions is

$$\rho(x) = \sum_{i=1}^{m} \rho_i(x) = \sum_{i=1}^{m} \rho_i g_i(x) , \tag{1.9}$$

where m is the number of species of ions in the solution. The charge profile is

$$e\sum_{i=1}^{m} z_i\rho_i(x) = e\sum_{i=1}^{m} z_i\rho_i g_i(x) , \tag{1.10}$$

where z_i is the valence (including the sign) of an ion of species i.

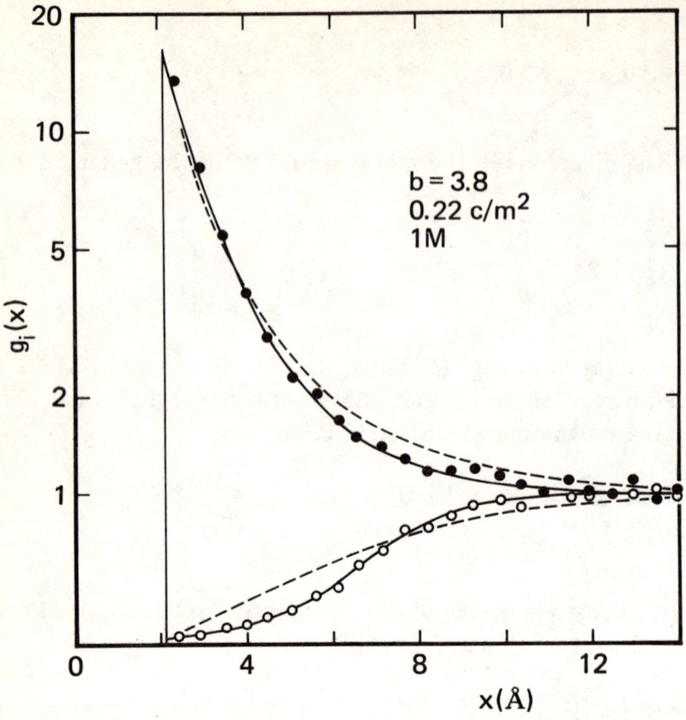

Figure 2. *Ionic profiles for the primitive model of 1M electrolyte near a hard charged electrode (T=298k, σ=425Å, $z_1=z_2=1$, ε=78.5). The points give the computer simulation values. The solid curve gives the HNC/MSA and MPB5 results while the broken curve gives the GCS results.*

In the bulk, charge neutrality means that

$$\sum_{i=1}^{m} z_i \rho_i = 0 \; . \tag{1.11}$$

At the interface, the net charge accumulated in the electrolyte near the interface must be equal in magnitude, but opposite in sign, to the charge on the metal electrode. Thus,

$$e \sum_{i=1}^{m} z_i \rho_i \int_0^\infty g_i(t)dt = -E/4\pi \; . \tag{1.12}$$

The profile, $\rho_s(x,\Omega)$, for the solvent molecules is a function of the position x and orientation Ω. The density profile is then

$$\rho_s(x) = \rho_s g_s^0(x) , \tag{1.13}$$

where ρ_s is the bulk density of the solvent molecules,

$$g_s^0(x) = \frac{1}{4\pi} \int g_s(x,\Omega) d\Omega , \tag{1.14}$$

and

$$g_s(x,\Omega) = \rho_s(x,\Omega)/\rho_s . \tag{1.15}$$

The profile, $g_s(x,\Omega)$, may be expanded in spherical harmonics. For dipoles the result [1] is particularly simple

$$g_s(x,\Omega) = g_s^0(x) + \sqrt{3}\, \Delta h_s(x) \cos\theta + \cdots , \tag{1.16}$$

where

$$\Delta h_s(x) = \frac{\sqrt{3}}{4\pi} \int g_s(x,\Omega) \cos\theta \, d\Omega . \tag{1.17}$$

The potential difference across the electrolyte part of the interface is

$$V_e = 4\pi e \sum_{i=1}^{m} z_i \rho_i \int_0^\infty t g_i(t) dt + \frac{4\pi}{\sqrt{3}} \rho_s \mu \int_0^\infty \Delta h_s(t) dt , \tag{1.18}$$

where μ is the dipole moment of the dipolar hard spheres. The total potential difference is the sum of the metal and electrolyte terms,

$$V = V_m + V_e . \tag{1.19}$$

Until recently, the interfacial region of the metal has been assumed to be of negligible thickness and as a result V_m=O. It will be shown in Section 8 that this is too simple a picture. However, the metallic contributions will not be considered until Section 8. For this reason the electrolyte potential difference V_e will not be subscripted in Sections 2 to 7.

Force balance considerations [1-3] require that

$$\sum_{i=1}^{m} \rho_i(\sigma_i/2) + \rho_s(\sigma_s/2) = \beta\left(p + E^2/8\pi\right) , \tag{1.20}$$

where $\beta=1/kT$ (T is the temperature and k is the Boltzmann constant), p is the pressure of the electrolyte and σ_i and σ_s are the ionic and solvent molecule diameters, respectively.

One last exact theorem [4] which can be mentioned states that the potential difference across the electrolyte is a monotonic function of the charge on the electrode. Although this seems obvious, this theorem does not seem to have been established until recently. Although the total potential difference is monotonic, it turns out that individual components of the potential difference need not be monotonic.

2. Gouy-Chapman-Stern Theory

As is well known, Gouy [5], Chapman[6], and Stern [7] have developed a theory of the double layer which has formed the basis of most subsequent work.

The theory is based upon the *primitive* model, where the solvent is treated as a dielectric continuum whose dielectric constant is ε. Thus, the interaction potential among the ions is

$$u_{ij}(r) = \begin{cases} \infty & , r<\sigma_{ij} \\ \dfrac{z_i z_j e^2}{\varepsilon r} & , r>\sigma_{ij} , \end{cases} \tag{2.1}$$

where

$$\sigma_{ij} = \frac{\sigma_i + \sigma_j}{2} \tag{2.2}$$

and σ_i is the diameter of an ion (charged hard sphere) of species i.

The potential profile of the ions is obtained from Poisson's equation

$$\frac{d^2\phi(x)}{dx^2} = -\frac{4\pi e}{\varepsilon} \sum_{i=1}^{m} z_i \rho_i g_i(x) . \tag{2.3}$$

If the solution is sufficiently dilute, the ions are usually far from each other and the ionic diameter may be neglected, except that it must be remembered that an ion cannot get closer to the electrode than $\sigma_i/2$. Thus,

$$g_i(x) = \begin{cases} 0 & , x<\sigma_i/2 \\ \exp\{\beta z_i e\phi(x)\} & , x>\sigma_i/2 , \end{cases} \tag{2.4}$$

might be a reasonable approximation.

Substitution of Eq. (2.4) into (2.3) gives a differential equation which can be solved analytically, for the following binary salts at least, $|z_1|=|z_2|=z$ and $|z_1|=1$, $|z_2|=2$. For a binary symmetric salt, $|z_i|=z$ and $\sigma_i=\sigma$ for i=1, 2, the result is

$$\phi(x) = \frac{2}{\beta ze} \sinh^{-1}\frac{b}{2} - \frac{E(x-\sigma/2)}{\varepsilon}, \quad 0<x<\sigma/2 , \tag{2.5}$$

where $E/4\pi$ is the charge density on the electrode and

$$\tanh\left[\frac{\beta ze\phi(x)}{4}\right] = \frac{b/2}{1+\sqrt{1+\frac{b^2}{4}}} \exp\{-\kappa(x-\sigma/2)\}, x>\sigma/2 , \tag{2.6}$$

where

$$b = \frac{\beta zeE}{\varepsilon\kappa} \tag{2.7}$$

is the charge on the electrode in dimensionless units and

$$\kappa = \left\{\frac{4\pi\beta e^2}{\varepsilon}\sum_{i=1}^{m} z_i^2\rho_i\right\}^{1/2} \tag{2.8}$$

is the *Debye inverse screening length*. The zero of potential has been chosen so that $\phi(\infty)=0$.

The potential difference across the double layer is

$$V = \phi(0) = \frac{2}{\beta ze}\sinh^{-1}\frac{b}{2} + \frac{E\sigma}{2\varepsilon} . \tag{2.9}$$

The first term in Eq. (2.9),

$$\phi(\sigma/2) = \frac{2}{\beta ze} \sinh^{-1} \frac{b}{2} , \tag{2.10}$$

is called the *diffuse potential* of the double layer. The second term in Eq. (2.9), $E\sigma/2\varepsilon$, is the potential drop across the "inner" region of thickness $\sigma/2$, where the electric field is E/ε.

The *differential* capacitance C across the interface is given by

$$C = \frac{1}{4\pi} \frac{dE}{dV}$$

$$= \frac{\varepsilon}{4\pi} \frac{\kappa\sqrt{1+b^2/4}}{1+\frac{\kappa\sigma}{2}\sqrt{1+b^2/4}} . \tag{2.11}$$

Equation (2.11) is *formally* identical to a diffuse layer capacitor with capacitance

$$C_d = \frac{\varepsilon}{4\pi} \kappa\sqrt{1+\frac{b^2}{4}} \tag{2.12}$$

in series with an "inner layer" capacitor, with capacitance

$$C_H = \frac{\varepsilon}{2\pi\sigma} . \tag{2.13}$$

It is common to interpret this formal identity in terms of physical models in which the *inner layer* is separable from the rest of the interfaced region (called the *diffuse layer*). However, we shall see that such models can be misleading.

At contact

$$g_i(\sigma/2) = \exp\{\beta ze\phi(\sigma/2)\}$$

$$= 1 + \frac{b^2}{2} \pm b\sqrt{1+\frac{b^2}{4}} . \tag{2.14}$$

Thus, in the *Gouy-Chapman-Stern* (GCS) theory

$$\sum_{i=1}^{m} \rho_i(\sigma/2) = \sum_{i=1}^{m} \rho_i + \beta E^2/8\pi\varepsilon \ . \tag{2.15}$$

This is to be compared with the primitive model version of Eq. (1.20)

$$\sum_{i=1}^{m} \rho_i(\sigma/2) = \beta\left(p + E^2/8\pi\varepsilon\right) , \tag{2.16}$$

where p is the osmotic pressure of the ions. The GCS theory satisfies the exact contact value theorem for the primitive model with exception that p is replaced by the ideal pressure

$$\rho kT = kT \sum_{i=1}^{m} \rho_i \ . \tag{2.17}$$

At low concentrations

$$p \sim \rho kT \ . \tag{2.18}$$

As the electrode is charged, the term $E^2/8\pi\varepsilon$ rapidly becomes dominant. Thus except at high concentrations, where Eq. (2.18) is not satisfied and at very low electrode charge densities, where ρkT is comparable to or bigger in magnitude than $E^2/8\pi\varepsilon$, the contact values of the GCS profiles should be very accurate.

Analytic results for the integrals of $g_i(x)-1$ can also be obtained from the GCS theory. For the symmetric salt

$$\int_{\sigma_{i/2}}^{\infty} [g_i(t)-1]dt = \frac{2}{\kappa}\left[\sqrt{1+\frac{b^2}{4}} \pm \frac{b}{2} - 1\right] . \tag{2.19}$$

From Eq. (2.19) we see that the charge neutrally condition, Eq. (1.12), is satisfied. This means that the area under the GCS profiles must be given with reasonably good accuracy. This in turn means that the GCS results for the integrals of $g_i(x)-1$ must be at least qualitatively reasonable. Moreover, since these profiles have essentially the correct contact value and essentially the correct area, the profiles themselves must also be qualitatively reasonable.

Except for preventing the ions from approaching closer to the electrode than $\sigma/2$, the diameter of the ions plays no role in the GCS theory. This means that, in principle at least, the GCS theory can allow unphysical packings in the neighborhood of the electrode. The number of ions/unit area in the neighborhood of the electrode is approximately

$$\sum_{i=1}^{m} \rho_i \int_{\sigma/2}^{\sigma} g_i(t)dt \ .$$

At a high enough charge on the electrode, this integral can exceed the two-dimensional close packed density. This is clearly unphysical. At these high electrode charges, the real profile must fall off more quickly than the GCS result. The charge neutrality result, Eq. (1.12), must still be satisfied. As a result, the real profile must be greater than the GCS result at larger distances. Since the potential is the first moment of the profile, the real potential must ultimately exceed the GCS result and the real capacitance must be less than the GCS result.

As it turns out, this packing problem occurs only at very high electrode charge densities, higher than is generally of physical interest. As a result, with few exceptions, this packing problem can be ignored.

3. COMPARISON OF GCS THEORY WITH SIMULATION RESULTS

The recent Monte Carlo (MC) computer simulations of Torrie and Valleau [8,9] permit a rigorous test of the GCS theory. Comparison with experiment does not give a rigorous test since representing the solvent by a dielectric continuum has uncertain consequences. As is seen in Fig. 2, at least for a 1:1 salt ($|z_i|=1$) the GCS profiles are in fairly good agreement with the simulation values [8]. Larger differences can be seen in Fig. 3 where the diffuse layer potential, $\phi(\sigma/2)$, is plotted. At the charge densities of physical interest, the GCS values for $\phi(\sigma/2)$ appreciately exceed the simulation values. As has been mentioned, ultimately the reverse must be true. In fact, the beginnings of this phenomena are seen in Fig. 3 at the very highest charge densities. A comparison of the MGC charge adsorption isotherms with simulation values would also be of interest. However, Torrie and Valleau did not calculate these isotherms from their profiles.

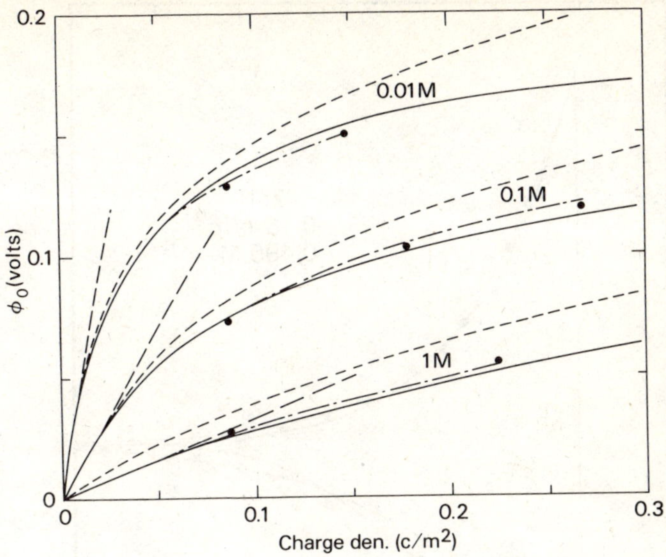

Figure 3. *Diffuse layer potential difference for the primitive model of an electrolyte near a hard charged electrode (T=298k, σ=4.25Å, $z_1 = z_2 = 1$, ϵ = 78.5). The points give the computer simulation values while the curves marked ———, —— – ——, ——— ———, and - - - give the HNC/MSA, MPB5, MSA and GSC results, respectively. Each set of curves is labelled with the value of the concentration.*

In Figs. 4 and 5, GCS profiles are compared with simulation results [9] for the unsymmetric 2:1 and 1:2 salts. Expressions for the GCS profiles were first obtained by Grahame [10]. As observed by Torrie and Valleau [9], as soon as the electrode is charged, the double layer is dominated by the counterions. Thus, the 2:1 results when the counterion is univalent, shown in Fig. 4, are similar to the 1:1 results in Fig. 2 whereas the 1:2 results, where the counterion is divalent, shown in Fig. 5, are similar to typical 2:2 results. The MGC theory is much less successful in dealing with divalent ions. This is also seen in Figs. 6 and 7 where GCS values for $\phi(\sigma/2)$ and the charge adsorption isotherms are compared with simulation values. The simulation values for the charge adsorption isotherms were calculated by Barojas *et al.* [11] from the Torrie-Valleau [9] profiles.

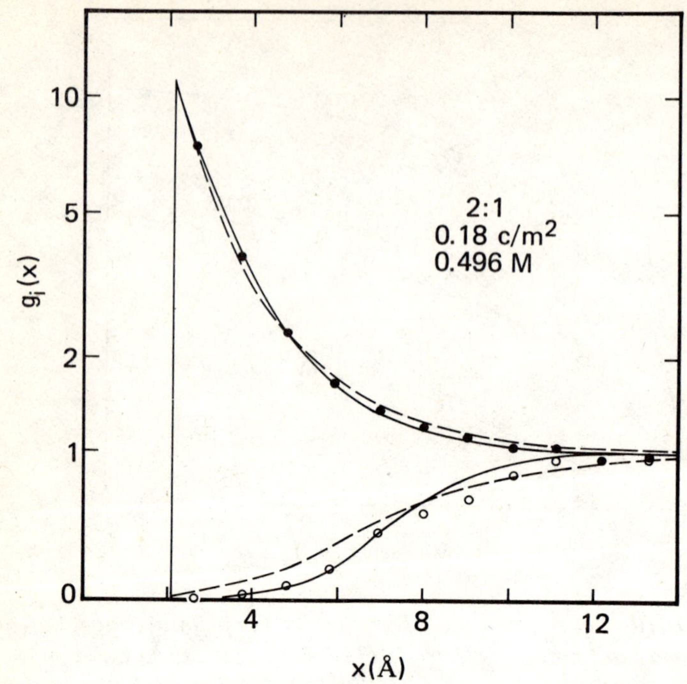

Figure 4. *Ionic profiles for the primitive model of a 0.496M electrolyte near a hard charged electrode ($T=298k$, $\sigma=4.25$Å, $z_1=-1$, $z_1=2$, $\epsilon=78.5$). The points and curves have the same meaning as in Fig. 2.*

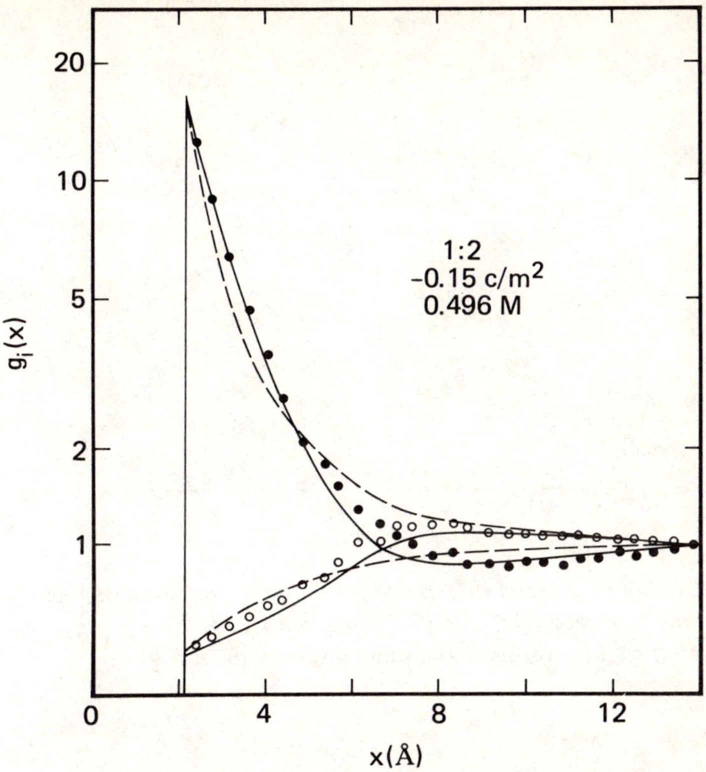

Figure 5. *Ionic profiles for the primitive model of a 0.496M electrolyte near a hard charged electrode ($T=298k$, $\sigma=4.25$Å, $z_1=-1$, $z_2=2$, $\epsilon=78.5$). The points and curves have the same meaning as in Fig. 2.*

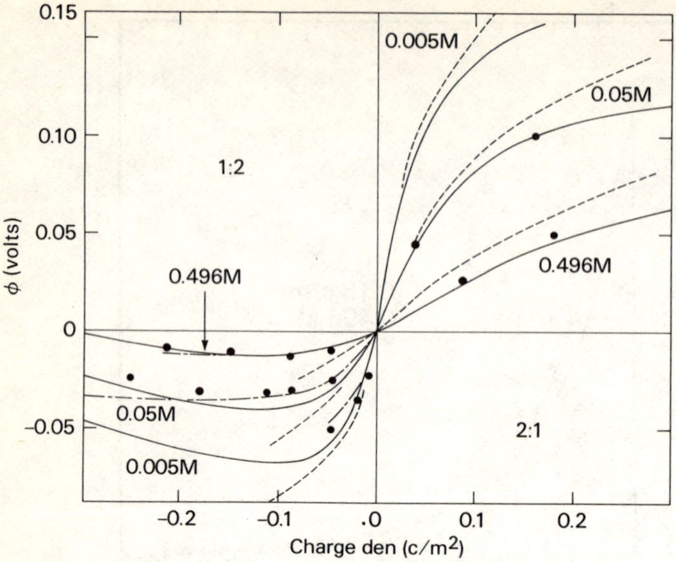

Figure 6. *Diffuse layer potential difference for the primitive model of an electrolyte near a hard charged electrode ($T=298k$, $\sigma=4.25$Å, $z_1=-1$, $z_2=2$, $\epsilon=78.5$). The points and curves have the same meaning as in Fig. 3.*

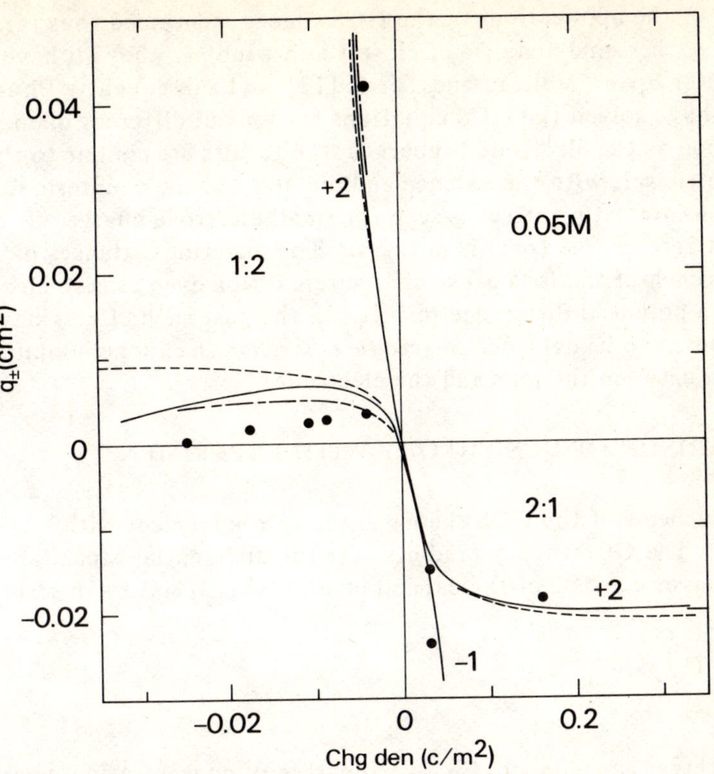

Figure 7. *Excess charge adsorption isotherms for the primitive model of a 0.05M electrolyte near a hard charged electrode. The points and curves have the same meaning as in Fig. 3.*

In all of the applications of the GCS theory mentioned thus far, all the ions have the same distance of closest approach *i.e.*, they all have the same diameter $\sigma_i=\sigma$. Valleau and Torrie [12], and subsequently Bhuiyan *et al.* [13], have solved the GCS equations for ions of different diameter. Again as soon as the electrode is charged, the results are similar to those of a symmetric salt with the valence and diameter of the counterion. However, an interesting effect is seen for small electrode charge. The potential of zero charge (pzc) is not zero. The differing distances of closest approach of the ions cause charge separation even at zero electrode charge and a potential difference results. In the past such effects have been considered to be evidence of *specific adsorption* (*i.e.*, noncoulombic interactions between the ions and the electrode).

4. COMPARISON OF GCS THEORY WITH EXPERIMENT

Some aspects of the GCS theory are in fair agreement with experiment. The GCS theory predicts that the differential capacitance of the double layer consists of the capacitance of two capacitors in series, *i.e.*,

$$C^{-1} = C_d^{-1} + C_H^{-1} . \tag{4.1}$$

The diffuse layer capacitance, C_d, is a function of concentration whereas the inner layer capacitance, C_H, is independent of concentration.

Experimentally, Eq. (4.1) seems valid and C_d is fairly well described by the GCS theory. Also the difference between the experimental value for C^{-1} and C_d^{-1} is independent of concentration, to a good approximation. However, the GCS theory would suggest that C_H is a constant independent of the electrode charge and of the order

$$C_H = \frac{\varepsilon}{2\pi\sigma} \sim 200\ \mu f/cm^2 . \tag{4.2}$$

As is seen in Fig. 8, experimentally C_H is much smaller and is not independent of electrode charge.

If it is assumed, as in Fig. 9, that the dielectric constant in the "inner layer", $0<x<\sigma/2$, is $\varepsilon^*<\varepsilon$, then C_H is given by

$$C_H = \frac{\varepsilon^*}{2\pi\sigma} , \tag{4.3}$$

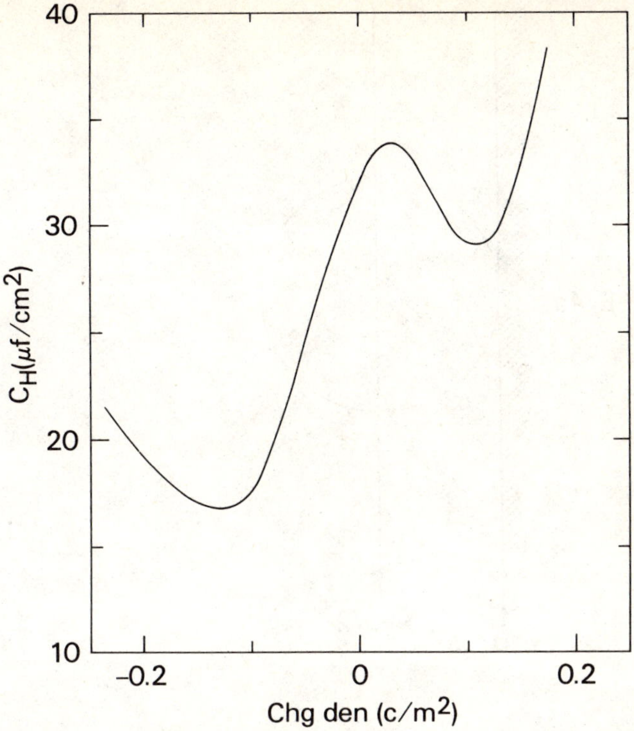

Figure 8. *"Inner" layer capacitance of an aqueous NaF solution near a mercury electrode.*

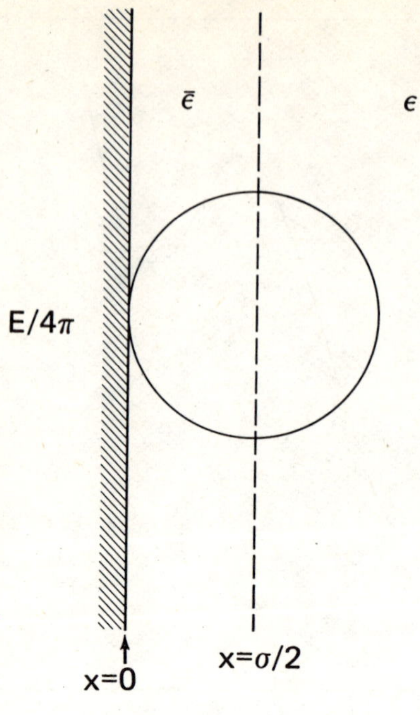

Figure 9. *Conventional model of inner and diffuse layers.*

and values of C_H of the order of the experimental values can be obtained. To explain the charge dependence of C_H from Eq. (4.3) one would have to assume that ε^* was a function of charge density. This is a bit too empirical to be satisfying.

More satisfying are the approaches based on the model of Watts-Tobin [14] which has been refined by several workers, most recently Fawcett and Guidelli. In this model the inner layer is thought of a two-dimensional layer of solvent molecules as shown in Fig. 10. Using this model with some adjustable parameters it is possible to describe the charge dependence of C_H. This procedure will not be described here since presumably it will be covered in detail by Fawcett and Guidelli in their lectures [15,16].

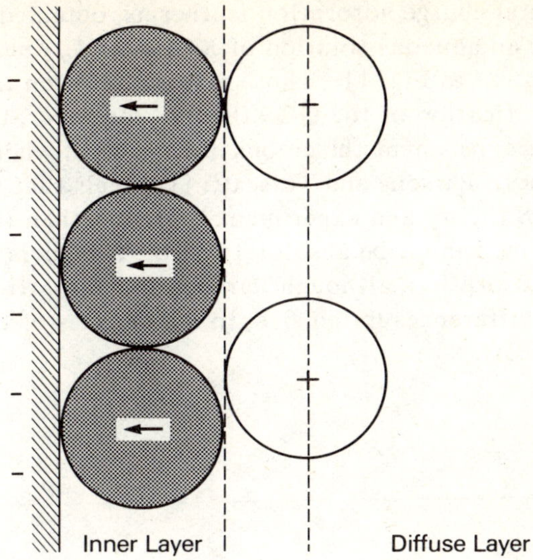

Figure 10. *Conventional model of inner and diffuse layers with the inner layer consisting of a dipole layer.*

The Watts-Tobin model and its refinements are based on the idea that C_i arises from the properties of some two-dimensional "inner layer" and that the remainder of the double layer can be described adequately by the GCS theory. This view is too extreme. As will be seen in Section 7 where a microscopic model of the solvent is considered, the interfacial region for the solvent extends over the *entire* double layer. It is not correct to think of an compact inner layer of solvent molecules and an extended diffuse layer of ions as a microscopically valid model of the inner layer.

From the discussion of Section 3, it is to be expected that deviations from the GCS predictions might be observable, especially if divalent ions are present or if there are significant differences in the ion diameters. There has been no systematic search for such deviations. They will occur mainly at high charge or high concentration where C_d is large and so C_d^{-1} is a small contribution to C^{-1}. Deviations from C_d would be smaller still. Despite these problems it is worth considering the experimental results of Parsons and Trasatti [17] and Parsons and Zobel [18].

Experimental charge adsorption isotherms, obtained by Parsons and Trasatti [17] for an aqueous solution of KCl + $MgCl_2$ near a mercury electrode are plotted in Fig. 11. These results are often taken as an experimental verification of the GCS theory. However, the agreement is only fair. In the experiment the coions are adsorbed whereas they are not in the GCS theory. Parsons and Trassatti [19] feels that the difference between the GCS theory and experiment is greater than the experimental error, although one cannot be absolutely sure. The deviations could be due to specific adsorption, although Parsons and Trasatti feel that this is not the case, to differences in ion size, to deficiencies of the GCS theory or to solvent effects.

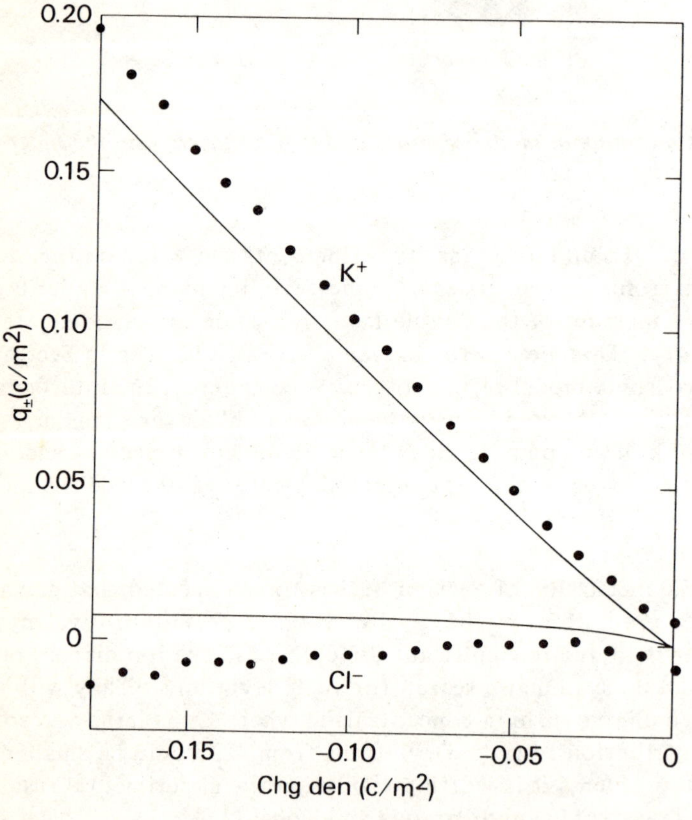

Figure 11. *Excess charge adsorption isotherms for* K^+ *and* Cl^- *in a 0.03 M aqueous solution at 25° C. The points give the experimental results and the curve gives the GCS results.*

If Eq. (4.1) is valid, then C^{-1} will be a linear function of C_d^{-1}, as calculated from the GCS theory, with the intercept being the value of C_H^{-1}. The experimental values [18] of C^{-1} for an aqueous NaH_2PO_4 solution near a mercury electrode are plotted in Fig. 12. The results are linear to a good approximation. However, deviations are seen at high densities. These deviations are unlikely to be experimental error since they are present all electrode charges (unless, of course, the error is systematic but that seems unlikely). They could be due to specific adsorption, although Parsons and Zobel feel that this is not the case, to deficiencies in the GCS theory, or to solvent effects.

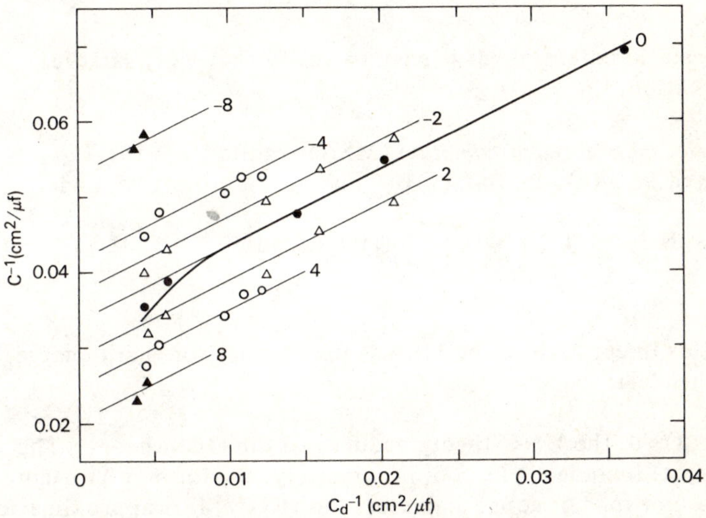

Figure 12. *Inverse capacitance of an aqueous solution of NaH_2PO_4 at 25° C near a hanging mercury electrode as a function of the inverse diffuse layer capacitance calculated from the GCS approximation. The light straight lines of unit slope give the results obtained from Eq. (4.1) and the low concentration MSA results obtained from Eq. (7.11) with $\sigma=\sigma_d=2.76$Å, $\epsilon=78.4$ and $T=25$° C. The heavy curve gives the full MSA results obtained using the same parameters. Both the low concentration and full MSA results include the electrode contribution (-0.026 cm^2/μf).*

The remainder of this article will be concerned with the development of theories which explain some of the differences between the GCS theory and computer simulations and experiment discussed in Section 3 and in this section.

5. INTEGRAL EQUATIONS: THE PRIMITIVE MODEL

Integral equations have been used with considerable success to describe the properties of bulk fluids. One of the most successful of these equations is the *hypernetted chain* (HNC) equation. Using the primitive model, the HNC equations for the double layer are

$$\ln\ g_i(x) = \beta z_i e\phi(x) + 2\pi \sum_{j=1}^{m} \rho_j \int_{-\infty}^{\infty} h_j(t)\mathscr{C}_{ij}(x,t)dt\ , \tag{5.1}$$

where

$$\phi(x) = \frac{4\pi e}{\varepsilon} \sum_{i=1}^{m} z_i\rho_i \int_{x}^{\infty} (t-x)g_i(t)dt \tag{5.2}$$

is the electrostatic potential. It is easy to verify that $\phi(x)$ satisfies Poisson's equation, Eq. (2.3).

The *mean spherical approximation* (MSA) results from the linearization of $\ln\ g_i(x)=\ln[1+h_i(x)]$. Thus, in the primitive model,

$$h_i(x) = \beta z_i e\phi(x) + 2\pi \sum_{j=1}^{m} \rho_j \int_{-\infty}^{\infty} h_j(t)\mathscr{C}_{ij}(x,t)dt\ . \tag{5.3}$$

Because of this linearization, the MSA is useful only for small charge density on the electrode.

If $\mathscr{C}_{ij}(x,t)=0$, the HNC theory reduces to the GCS theory. The function $\mathscr{C}_{ij}(x,t)$ include, at least approximately, the ion-ion hard-core interactions. For ions of equal diameter, the HNC/MSA approximation is obtained by using the MSA approximation (in the bulk) to calculate $\mathscr{C}_{ij}(x,t)$. For $\sigma_i=\sigma$, the result is

$$\begin{aligned}\mathscr{C}_{ij}(x,t) &= c_1\frac{\sigma^2-(x-t)^2}{2} + c_2\frac{\sigma^3-|x-t|^3}{3\sigma} + c_3\frac{\sigma^5-|x-t|^5}{5\sigma^3} \\ &+ \frac{\beta z_i z_j e^2}{\varepsilon}\left\{\sigma - |x-t| - \left(\frac{\Gamma}{1+\Gamma\sigma}\right)[\sigma^2-(x-t)^2]\right. \\ &\left. + \left(\frac{\Gamma}{1+\Gamma\sigma}\right)^2[\sigma^3-|x-t|^3]\right\}\end{aligned} \tag{5.4}$$

for $|x-t|<\sigma$ and $\mathscr{C}_{ij}(x,t)=0$ for $|x-t|>\sigma$. In Eq. (5.4)

$$c_1 = -\frac{(1+2\eta)^2}{(1-\eta)^4}\ , \tag{5.5}$$

$$c_2 = 6\eta \frac{(1+\eta/2)^2}{(1-\eta)^4} , \tag{5.6}$$

$$c_3 = \frac{1}{2} \eta c_1 , \tag{5.7}$$

where

$$\eta = \frac{\pi}{6} \sigma^3 \sum_{i=1}^{m} \rho_i \tag{5.8}$$

and

$$\Gamma = \frac{\sqrt{1+2\kappa\sigma}-1}{2\sigma} . \tag{5.9}$$

The MSA expressions for the general case, where the σ_i are unequal, are known but are complex.

The HNC/HNC (or more simply HNC) theory is obtained by using the HNC theory to calculate the $\mathscr{C}_{ij}(x,t)$. Since the HNC theory is a better theory for bulk electrolytes one would expect the HNC theory to give better results for the double layer than the HNC/MSA theory. However, this is not the case, at least for the systems studied to date. Evidently there is a fortuitous cancellation of error in the HNC/MSA approximation. Although a bit puzzling, this is fortunate since the HNC/MSA theory is considerably easier to use (especially in complex situations).

In the HNC approximation, for the primitive model, [20]

$$\sum_{i=1}^{m} \rho_i(\sigma_i/2) = \frac{1}{2} \left[1 + \beta \frac{\partial p}{\partial \rho} \right] \sum_{i=1}^{m} \rho_i + \beta E^2/8\pi\varepsilon , \tag{5.10}$$

where $\beta\partial p/\partial\rho$ is the inverse of the compressibility of the bulk electrolyte computed using whatever approximation is used to obtain $\mathscr{C}_{ij}(x,t)$. The HNC approximation does not satisfy the exact result, Eq. (2.16). However, for most cases the error is small because, at low concentrations,

βp is not greatly different from

$$\frac{1}{2}\left[1 + \beta\frac{\partial p}{\partial \rho}\right]\sum_{i=1}^{m} \rho_i$$

and because, as E is increased, the term $\beta E^2/8\pi\varepsilon$ rapidly becomes dominant.

The MSA/MSA (or more simply MSA) theory is obtained by combining Eq. (5.3) with the MSA expression for $\mathscr{C}_{ij}(x,t)$. In the primitive model and for equal diameters, the results are analytic [21]. For example,

$$g_i(x) = g_0(x) - \frac{\beta z_i eE}{\varepsilon\kappa} f(x-\sigma/2) , \tag{5.11}$$

where $g_0(x)$ is the profile for uncharged hard spheres near an uncharged hard wall. The function f(x) is also very complex. Expressions for both $g_0(x)$ and f(x) are known [22] but their precise form is not needed in this review. The main result which is of interest here is that at low concentrations,

$$f(x) \simeq e^{-\kappa x} \tag{5.12}$$

so that, at low concentrations, the MSA goes over to the linearized GCS theory.

The MSA satisfies charge neutrality, Eq. (1.12). The primitive model MSA differential capacitance for the equal diameter case is

$$C = \frac{\varepsilon}{4\pi} 2\Gamma . \tag{5.13}$$

That is, the MSA expression for the capacitance is just the linearized GC result with κ replaced by the *renormalized* inverse screening length, 2Γ. From Eq. (5.9)

$$2\Gamma = \kappa-\kappa^2\sigma/2 + \kappa^3\sigma^2/2 - \cdots . \tag{5.14}$$

Thus

$$C^{-1} = \frac{4\pi}{\varepsilon\kappa} + \frac{2\pi\sigma}{\varepsilon} - \frac{\pi\kappa\sigma^2}{\varepsilon} + \cdots . \tag{5.15}$$

If C^{-1} is plotted as a function of $4\pi/\varepsilon\kappa$, the GCS diffuse layer capacitance, the curve will be linear only at low concentrations. Moreover, at high concentrations the deviations will be negative just as is seen experimentally in Fig. 12.

The MSA total and diffuse layer potential are

$$V = \frac{E}{\varepsilon(2\Gamma)} \tag{5.16}$$

and

$$\phi(\sigma/2) = \frac{E}{\varepsilon\kappa}(1-\Gamma^2\sigma^2) , \tag{5.17}$$

respectively. Thus, the MSA V is a monotonically increasing function of E as required by the exact theorem of Blum *et al.* [4]. However, at high concentrations where $\Gamma\sigma>1$, the MSA result for $\phi(\sigma/2)$ is a monotonically decreasing function of E. Although such behavior is surprising it is not excluded by any exact theorems.

Recently, Khater *et al.* [23] have obtained analytic MSA results for the general primitive model case where diameter and valences can be unequal.

For large charge density on the electrode, the HNC approximation (or some other nonlinear approximation) should be used. In contrast to the MSA, the HNC equations must be solved numerically [11,20,24-27]. As is seen in Figs. 2-7, the HNC/MSA results are in much better agreement with the computer simulations than are the GCS results.

It is of interest to note that recently Marconi *et al.* [28] have solved the HNC/MSA equations for the case of ions of unequal diameter.

Calcuations have been made using the Born-Green-Yvon equation [29-31]. For the particular case where the superposition approximation,

$$g_{ij}(x_1,x_2,r_{12}) = g_{ij}(r_{12}) , \tag{5.18}$$

is used to describe the *pair* correlation function in the interfacial region, the BGY equations for the primitive model are

$$\ln\ g_i(x) = \beta z_i e\phi(x) + \frac{\pi}{3}(x^3-3x+2)[1-H(x-1)]\sum_{j=1}^{m}\rho_j g_{ij}(1)$$

$$+\ \pi\sum_{j=1}^{m}\rho_j g_{ij}(1)\int_{\max[0,x-1]}^{x+1}\left[1-(s-x)^2\right]h_j(s)ds$$

$$+\ \frac{\pi\beta z_i e^2}{\varepsilon}\sum_{j=1}^{m}z_j\rho_j\int_x^{\infty}\left[x^3-3xt+2t^3\right]\frac{h_{ij}(t)}{t^2}dt\ ,$$

$$-\ \frac{\pi\beta z_i e^2}{\varepsilon}\sum_{j=1}^{m}z_j\rho_j\int_0^{\infty}h_j(s)ds\int_{|x-t|}^{\infty}\left[t^2-(s-x)^2\right]\frac{h_{ij}(t)}{t^2}\ dt\ , \quad (5.19)$$

where H(x) is the heaviside step function

$$H(x) = \begin{cases} 0 & , x<0 \\ 1 & , x>0\ . \end{cases} \quad (5.20)$$

Equation (5.19) gives results which are somewhat inferior to the HNC/MSA results. However, if Eq. (5.18) is replaced by a more general approximation, results comparable to the HNC/MSA results are obtained. The BGY approximation is very promising because Eq. (2.16) is satisfied exactly. This means that the BGY equations may well be more satisfactory than the HNC/MSA equations for fused salts where βp is quite different from

$$\frac{1}{2}\rho\left[1+\beta\,\frac{\partial p}{\partial\rho}\right]\ .$$

While fused salts are being considered, it is worth mentioning that the GCS approximation has long been known to be a poor approximation to the double layer in a fused salt system. Experimentally, the capacitance at the point of zero charge increases with increasing temperature whereas the GCS approximation yields the opposite behavior. Recently Painter *et al.* [32] have shown the MSA results to be a considerable improvement over the GCS approximation. Even more promising results are obtained by Ballone *et al.* [32] using some rough but plausible modifications of the MSA.

6. MODIFIED POISSON-BOLTZMANN APPROXIMATION; THE PRIMITIVE MODEL

The GCS approximation results from the combination of Poisson's equation, Eq. (2.3), with the Boltzmann approximation, Eq. (2.4) and, as a result, is often referred to as a Poisson-Boltzmann approximation. Considering the simplicity of the approach, the GCS approximation gives good results. Improvements to this approximation are certainly worth considering. Outhwaite, Levine and Bhuiyan [33-37] have developed one such scheme which they refer to as the *modified Poisson-Boltzmann* (MPB) approximation.

The approach is similar in spirit to the BGY scheme considered in the preceding section. The profile $g_i(x)$ is related to the pair correlation function, $g_{ij}(x_1, x_2, x_{12})$. In the proceeding section the superposition approximation was used. The MPB expression for $g_{ij}(x_1, x_2, r_{12})$ is developed in analogy to the Debye-Hückel (DH) theory of bulk electrolytes [38].

As is well known, in the DH theory, a differential equation for the pair correlation function $g_{ij}(r)$ of the bulk electrolyte is obtained by substituting the linearized Boltzmann approximation

$$g_{ij}(r) = 1 - \beta z_i e \phi_j(r) \tag{6.1}$$

into Poisson's equation to obtain the differential equation

$$\nabla^2 \phi_j(r) = \kappa^2 \phi_j(r) , \quad r>\sigma . \tag{6.2}$$

Following Loeb [39], the pair correlation function in the interfacial region is written as

$$g_{ij}(1,2) = 1 - \beta z_i e \phi_j(1,2) \tag{6.3}$$

and ϕ_j (1,2) is obtained from

$$\nabla^2 \phi_j(1,2) = \kappa^2(x_2)\phi_j(1,2) , \quad r_{12}>\sigma,\ x_2>\sigma/2 , \tag{6.4}$$

where $\kappa(x_2)$ is a local inverse screening parameter, given by

$$\kappa(x_2) = \left\{\frac{4\pi\beta e^2}{\varepsilon}\sum_i z_i^2 \rho_i g_i(x_2)\right\}^{1/2} . \tag{6.5}$$

Although similar in spirit to the GCS approximation, the MPB is an improvement since the Boltzmann approximation is applied to the pair correlation function rather than to the profile (a singlet correlation function). This should be a much better approximation than Eq. (2.4). Equation (6.3) is also better than the superposition approximation (in principle, at least) because the superposition approximation implies that the pair correlation function is always identical to the bulk expression. The question is whether $g_{ij}(1,2)$ is better approximated by an accurate bulk $g_{ij}(r)$ or by the highly approximate result given above. The answer to this question is not obvious. However, the above expression is probably better because it does include the exclusion from the wall and the effects of correlations among pairs of ions which are both near the wall. These effects are neglected completely in the superposition approximation.

Equation (6.4) must be coupled with some equation relating $g_i(x)$ and $g_{ij}(1,2)$ and solved subject to the usual boundary conditions of continuity of ϕ and $\partial\phi/\partial n$ (n being the normal to the wall or ion). The BGY hierarchy [40], already mentioned in the preceding section for the specific case where the superposition approximation is used, could be employed. Generally, the very similar Kirkwood hierarchy [41] is used although there has been some work based on the BGY hierarchy.

An approximate solution of the resulting equations has been obtained. The result [42], which satisfies the charge neutrality condition, Eq. (1.12), is

$$g_i(x) = g_i^0(x)\exp\Big[-\frac{\beta z_i e}{2}\Big\{\frac{z_i e}{\varepsilon\sigma}[F(x)-F_0] + F(x)[\phi(x+\sigma)+\phi(x-\sigma)] - \frac{F(x)-1}{\sigma}\int_{x-\sigma}^{x+\sigma}\phi(y)dy\Big\}\Big], \tag{6.6}$$

where $\phi(x)$ is the singlet potential,

$$F(x) = \begin{cases} \dfrac{1}{1+\dfrac{\kappa(x)(\sigma+2x)}{4}} & , \sigma/2<x<\dfrac{3\sigma}{2} \\[2ex] \dfrac{1}{1+\kappa(x)\sigma} & , x>\dfrac{3\sigma}{2} \,. \end{cases} \tag{6.7}$$

The function F(x) is not smooth. This is a consequence of the fact that Eq. (6.6) only uses an approximate solution to the MPB equations for

$\phi_j(1,2)$. The parameters F_0 and κ_0 given by

$$F_0 = \lim_{x \to \infty} F(x) = \frac{1}{1 + \kappa_0 \sigma} \tag{6.8}$$

and

$$\kappa_0 = \lim_{x \to \infty} \kappa(x) \ . \tag{6.9}$$

The function $g_i^0(x)$ is the profile for an uncharged sphere i in a sea of charged ions and includes volume exclusion effects and surface charge terms. To complete the MPB approximation, this function must be determined. In the earlier versions of the theory a density expansion was employed. However, in the most recent version of the theory, called the MPB5 approximation [43], the result

$$g_i^0(x) = \begin{cases} 0 & , x<\sigma/2 \\ \exp\left[-2\pi \int_{\infty}^{x} \sum_j \rho_j \int_{\max[\frac{\sigma}{2},y-a]}^{y+a} (x-y) g_i(x) \times \exp\{-\beta z_j e \phi(1;2;e_j=0,r_{12}=0)\} \, dx dy\right] & , x>\sigma/2 \end{cases} , \tag{6.10}$$

obtained from the BGY hierarchy, is used. The MPB5 approximation results from the simultaneous numerical solution of Eqs. (6.6), (6.10) and Poisson's equation, Eq. (2.3).

It is interesting to note that although the MSA and HNC/MSA approximations reduce to the GCS approximation for point ions, the point ion limit for the MPB approximation includes a correction from the local variation of κ. The point ion result for the MPB approximation is

$$g_i(x) = \exp\left[-\beta\left\{z_i e \phi(x) + \frac{z_i^2 e^2}{2\varepsilon} [\kappa(x)-\kappa_0]\right\}\right] . \tag{6.11}$$

It is important to note that there is a difficulty associated with the presents form of the MPB approximations. For $\kappa_0\sigma>7.83$ pure oscillatory solutions are obtained and the approximation cannot be used because the profiles no longer decay to the bulk values. This is no problem for electrolyte solutions within the primitive model since for these systems

$\kappa_0\sigma>7.83$. However, it does mean that the MPB5 approximation, in contrast to the MSA , HNC/MSA and BGY approximations, cannot be applied to high density systems such as fused salts or nonprimitive models of electrolyte solutions.

On the positive side, the MPB approximation is more physically motivated than the MSA and HNC approximations and is thus very appealing. Furthermore, both the BGY and MPB approximations involve the pair distribution function in the interfacial region and can give results when image forces are present. For simplicity, the image terms in the MPB5 have not been exhibited in this discussion. Including image terms is certainly a formal advantage. Whether it is a practical advantage is difficult to determine as real metals are quantum solids where the classical electrostatic image expression is probably not a very good approximation.

Some results obtained from the MPB5 approximation are compared with the simulation, GCS and HNC/MSA results in Figs. 2-7. The MPB5 results are very good. There has been some discussion [37,44] about whether the MPB or the HNC/MSA approximation is more accurate. This is largely a matter of taste since one approximation is slightly better for one property and slightly worse for another. For example, as seen in Fig. 6 the MPB5 gives slightly better values for $\phi_0=\phi(\sigma/2)$ whereas the HNC/MSA gives slightly better values for the slope of ϕ_0 as a function of the charge. In any case, such discussions are not very profitable since both approximations are very good. The agreement of the two approximations is far more relevant than any differences between them. Both theories agree that the GCS theory overestimates $\phi(\sigma/2)$ and this underestimate the differential capacitance. This is not only in agreement with the simulation studies but also is in qualitative agreement with the Parsons-Zobel experimental results in Fig. 12.

Further it is premature to be overly concerned with the relative accuracy of the HNC/MSA and MPB approximations for the primitive model since this model is mainly of theoretical interest (except for fused salt double layers). Real electrolyte solutions contain solvent molecules. It is of far more interest to consider the application of these theories to more realistic models. A first step in this direction is considered in the next section.

7. THE NONPRIMITIVE ION-DIPOLE MODEL

Rather than represent the solvent as a dielectric continuum, as is done in the primitive model, it is better to represent the solvent molecules as a set of hard spheres of diameter σ_s with embedded point dipoles of dipole moment μ. For this system the interaction among the ions is

$$u_{ij}(r) = \begin{cases} \infty & , r<\sigma_{ij} \\ \dfrac{z_i z_j e^2}{r} & , r>\sigma_{ij} . \end{cases} \tag{7.1}$$

Notice that ε does not appear in the denominator, as was the case in Eq. (2.1). The interactions are considerably stronger than those in the primitive model. The nonprimitive model is a more severe test for a theory.

Carnie and Chan [45] and Blum and Henderson [1] have solved the MSA equations (useful for small electrode charge) for the ion-dipole system near a hard charged wall. The solution is analytic but is, in general, hopelessly implicit. However, in the limit $\kappa \to 0$ the results become explicit. In this low concentration limit

$$g_i(x) = g_0(x) - \frac{\beta z_i e E}{\varepsilon \kappa} e^{-\kappa(x-\sigma/2)} , \tag{7.2}$$

where σ is the diameter of the ions and $g_0(x)$ is the profile for hard spheres near a hard wall [22]. The profile of the dipoles is

$$g_s(x,\Omega) = g_0(x) + \sqrt{3}\, \Delta h_s(x) \cos \theta , \tag{7.3}$$

where $g_0(x)$ is again the profile for hard spheres near a hard wall. The functions $g_0(x)$ and $\Delta h_s(x)$ satisfy Eqs. (1.14) and (1.17). Comparing Eqs. (7.3) and (1.16) it is seen that only one spherical harmonic appears in the MSA $g_s(x,\Omega)$.

Equation (7.2) is similar to Eqs. (5.11) and (5.12) and to the linearized GCS result. However, one difference must be borne in mind, in the primitive model the density of hard spheres is just the density of ions and is small. As a result, for the primitive model

$$g_0(x) \simeq 1 , \tag{7.4}$$

whereas for the ion-dipole model the density of hard spheres is the density of ions and dipoles and is large. As a result, $g_0(x)$ in Eq. (7.2), for example, is quite different from 1 near the wall.

Despite this real difference, most of the properties calculated from Eq. (7.2) will be the similar as those obtained from the GCS theory.

Thus, the MSA gives insight into why the GCS theory is useful even though it is based on an implausible model of the solvent. The function $g_0(x)$ does not contribute to the potential because of charge neutrality. Thus, contribution of the ions is the same as in the primitive model except that ε no longer appears in the denominator. Hence, at low concentrations,

$$4\pi e \sum_{i=1}^{m} z_i \rho_i \int_{\sigma/2}^{\infty} t g_i(t) dt = \frac{E}{\kappa}\left[1 + \frac{\kappa\sigma}{2}\right] + O(\kappa) . \tag{7.5}$$

The charge adsorption isotherms are defined as excesses relative to the solvent profile. Thus

$$q_i = e z_i \rho_i \int_{\sigma/2}^{\infty} [g_i(t) - g_s^0(t)] dt . \tag{7.6}$$

Since, in the MSA $g_s^0(x)=g_0(x)$ we see that, if the ions and dipoles all have the same diameters, the charge adsorption isotherms are identical to those obtained from the MSA using the primitive model.

The solvent contribution to the potential difference is, at low concentrations,

$$\frac{4\pi}{\sqrt{3}} \rho_s \mu \int_{\sigma/2}^{\infty} \Delta h_s(t) dt = \frac{1-\varepsilon}{\varepsilon}\left[1 + \frac{\kappa\sigma}{2}\left(1 - \frac{1}{\lambda}\right)\right]\frac{E}{\kappa} + O(\kappa) , \tag{7.7}$$

where λ is a parameter related to ε through the relation

$$\lambda^2(1 + \lambda)^4 = 16\varepsilon . \tag{7.8}$$

For $\varepsilon=1$ $\lambda=1$ and for $\varepsilon\sim 78$ $\lambda\sim 2.65$. Thus, λ is a weak function of ε. Ultimately λ is proportional to the sixth root of ε.

Combining Eqs. (7.5) and (7.7) gives the MSA expression for the potential difference at low concentrations

$$V = \frac{E}{\varepsilon\kappa}\left[1 + \frac{\kappa\sigma}{2}\left(1 + \frac{\varepsilon-1}{\lambda}\right)\right] . \tag{7.9}$$

It is to be noted that there has been a cancellation between terms of order 1 and $(1-\varepsilon)/\varepsilon$ to produce a term of order $1/\varepsilon$. This has been no problem with the MSA since all the terms are analytic. However, in a theory where the analogue of (7.5) and (7.7) are obtained numerically and where

$\varepsilon \sim 80$, a one percent error in either term may produce a hundred percent error in V.

For the case where $\sigma_s \neq \sigma$ (the ion diameter), Carnie and Chan [45] have shown that Eq. (7.9) becomes

$$V = \frac{E}{\varepsilon\kappa}\left[1 + \frac{\kappa}{2}\left(\sigma + \frac{\varepsilon-1}{\lambda}\,\sigma_s\right)\right]. \tag{7.10}$$

The resultant differential capacitance is

$$C = \frac{\varepsilon}{4\pi}\,\frac{\kappa}{1 + \frac{\kappa\sigma}{2} + \left(\frac{\varepsilon-1}{\lambda}\right)\frac{\kappa\sigma_s}{2}}. \tag{7.11}$$

Setting $\varepsilon=1$ or $\sigma_s=0$ gives the primitive model result.

Equation (7.11) is of the form of Eq. (4.1) with

$$C_H^{-1} = \frac{2\pi}{\varepsilon}\left[\sigma + \frac{\varepsilon-1}{\lambda}\,\sigma_s\right]. \tag{7.12}$$

Thus, the MSA correctly predicts that at low concentrations the differential capacitance is equivalent to the capacitance of two capacitors in series and that C_H is independent of concentration. If fact, Eq. (7.12) is of the same form as (4.3) with

$$\varepsilon^* = \frac{\varepsilon}{1 + \frac{\varepsilon-1}{\lambda}\left(\frac{\sigma_s}{\sigma}\right)}. \tag{7.13}$$

However, this does not mean that C_H is the capacitance of some "inner" layer or specific region of space. Rather, as is seen in Fig. 13, C_H comes from an integration overall space. The point is that the solvent molecules will be oriented by the charge on the electrode at all distances where there is a field and where the screening is incomplete, *i.e.*, over the entire double layer. The interfacial regions for the ions and the solvent molecules is of the same thickness. It is a gross oversimplification to picture all the interfacial solvent effects as taking place in some "inner" region. In addition, the dielectric constant is a macroscopic concept and it is not meaningful to talk of an effective dielectric constant in some microscopic region.

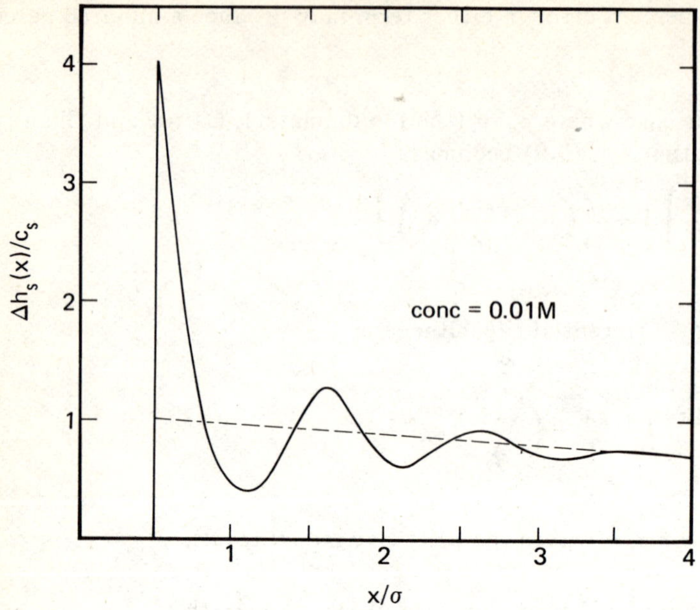

Figure 13. *$\Delta h_s(x)$ as a function of x in the MSA for a 0.01M model electrolyte solution ($|z_i| = 1$, $\sigma = \sigma_d = 2.76$Å, $\epsilon = 78.4$ and $T = 25°C$) near a hard charged wall. The parameters c_s is a normalization parameter whose value is not significant for the present discussion.*

For water $\varepsilon \sim 78$, $\lambda \sim 2.65$ and $(\varepsilon - 1)/\lambda \sim 29$. Thus, the contribution of the ions to C_H is very small compared to that of the solvent molecules in agreement with experiment. Using the above values,

$$C_H \sim 17\ \mu f/cm^2 , \tag{7.14}$$

which is of the correct order of magnitude and a considerable improvement over Eq. (4.2). Rosinberg [46] has found that Eq. (7.12) gives an good description of the solvent contribution to the capacitance for a wide range of solvents.

Equation (7.14) is, however, smaller than the experimental values of C_H for the potential of zero charge (pzc). The difference between Eq. (7.14) and experiment can be ascribed to electronic effects in the metal. These effects will be described in the next section. For the moment this will be regarded as an empirical fact. For a mercury electrode/aqueous solution interface at the pzc

$$C^{-1} = C_{MSA}^{-1} - 0.026\ cm^2/\mu f , \tag{7.15}$$

which, at low concentrations becomes

$$C^{-1} = \frac{4\pi}{\varepsilon\kappa} + \frac{2\pi}{\varepsilon}\left[\sigma + \frac{\varepsilon-1}{\lambda}\,\sigma_d\right] - 0.026\ \mathrm{cm}^2/\mu\mathrm{f}\ . \tag{7.16}$$

It is has been pointed out that, both experimentally and within the MSA, Eq. (4.1) is valid only at low concentrations (see Fig. 8). It is of interest to see if the full MSA accounts for the experimentally observed deviations from Eq. (4.1). If Eq. (4.1) is satisfied then C^{-1} will be a linear function of C_d^{-1} (as calculated from the GCS theory). As is seen in Fig. 12, C^{-1}, calculated from Eq. (7.15), is a linear function of

$$C_d^{-1} = \frac{4\pi}{\varepsilon\kappa} \tag{7.17}$$

at low concentrations and that at these low concentrations the results are in close agreement with the experimental results at the pzc of Parsons and Zobel. However both experimentally and in the MSA, deviations occur at the higher densities. It is very pleasing to see that the full MSA results [47] are in excellent agreement with experiment.

The main defect of results reported above is the fact that they are limited to the pzc because of the linearization where not in the MSA. Various simple nonlinear extensions are possible. Schmickler will outline one in his lecture [48]. Another possible extension would be to write

$$V = \frac{E\sigma}{2} + \phi(\sigma/2) - 2\pi\rho_s\mu\sigma\mathscr{L}\left[\beta E\mu\frac{9}{\lambda^2(\lambda+2)^2}\left(1 - \frac{\sigma_s}{\lambda\sigma}\right)\right], \tag{7.18}$$

where $\phi(\sigma/2)$ is the diffuse potential given by whichever approximation is used for the ions, ρ_s is the density of the solvent, which in the MSA is related to the dipole moment, μ and ε by

$$\frac{4\pi\beta\mu^2}{3}\,\rho_s\,\frac{9}{\lambda^2(\lambda+2)^2} = \frac{\varepsilon-1}{\varepsilon}\,, \tag{7.19}$$

and

$$\mathscr{L}(x) = \coth x - \frac{1}{x} \tag{7.20}$$

is the Langevin function which gives the thermally averaged orientation of freely orientable dipoles [49].

A more satisfactory procedure would be to use a nonlinear approximation, such as the HNC approximation. Some progress along this line has been made and it is to be hoped that some results will be reported in the next few years.

The MPB approximation has also been applied to the ion-dipole model of the double layer [50]. Some analytic results, very similar to Eqs. (7.10) and (7.11) have been obtained. Unfortunately, as has been mentioned already the MPB has technical problems at the high densities characteristic of the ion-dipole model of an electrolyte. These will have to be overcome before the MPB can be as useful for this model as it was for the primitive model.

Another approach which has been applied to the ion-dipole model is perturbation theory which also yields Eqs. (7.10) and (7.11). This approach has been surveyed recently [51].

8. THE CONTRIBUTION OF THE METAL; JELLIUM

The simplest model of a metal is the jellium model where the electrons are treated as a free fermion gas in the positively charged background of the smeared out metallic ions. Ideally, the electronic profile would be obtained by a solution of the schrodinger equation in a manner similar to that of Lang and Kohn [52]. The approach here is simpler. A family of trial functions containing some undetermined parameter will be employed. The values of these parameters which will used are those for which the surface energy is minimal written this family.

The trial function of Smith [53] and Partenskii and Smorodinskii [54]

$$n_-(x) = \begin{cases} n - \frac{1}{2}\left(n + \frac{\alpha E}{8\pi e}\right)e^{\alpha x} & , x<0 \\ \frac{1}{2}\left(n - \frac{\alpha E}{8\pi e}\right)e^{-\alpha x} & , x>0 , \end{cases} \tag{8.1}$$

is particularly simple and will be the one used here. This family has only one parameter, α. A more sophisticated choice would be that given by Eq. (1.5). The oscillations in Eq. (1.5) are an approximation to the Friedel oscillations in a metal. Functions similar to Eq. (1.5) have been used by Schmickler and Henderson [55] and will be considered by Schmickler in his lecture [48].

In principle, the electron and electrolyte profiles affect each other and the whole problem becomes highly complex. In practice, α is of the order 2Å^{-1} and the spill over of the electrons into the fluid, although large enough to affect the capacitance, is not large enough to make either profile dependent on the detailed nature of the other. As a result, the two problems can be solved separately.

The surface energy consists of coulomb kinetic, exchange and correlation energies and inhomogentity terms as well as the terms coupling the electrons and the electrolyte. The first terms have been given by Smith [53]. The kinetic energy term will be derived in detail. The others follow analogously. It is well known [49] that the kinetic energy per particle of a fermion gas is

$$E/N = \frac{3}{5}\,\epsilon_F\,, \tag{8.2}$$

where the fermi energy, ϵ_F, is given by

$$\epsilon_F = \frac{h^2}{2m}\left(\frac{3n}{8\pi}\right)^{2/3}. \tag{8.3}$$

Writing Eq. (8.2) in terms of n, rather than ϵ_F, gives

$$E/N = \frac{3}{5}\left(\frac{3}{8\pi}\right)^{2/3}\frac{h^2}{2m}\,n^{2/3}\,, \tag{8.4}$$

so that

$$\epsilon(n) = E/V = \frac{3}{5}\left(\frac{3}{8\pi}\right)^{2/3}\frac{h^2}{2m}\,n^{5/3}\,. \tag{8.5}$$

If atomic units, where energies are measured in units of

$$E_0 = \frac{4\pi^2 me^4}{h^2} = 27.2\text{ eV} \tag{8.6}$$

and distances in units of the Bohr radius,

$$a_0 = \frac{h^2}{4\pi^2 me^2} = 0.529\text{Å}\,, \tag{8.7}$$

then Eq. (8.5) becomes

$$\epsilon(n) = \frac{3}{10}(3\pi^2)^{2/3} n^{5/3} . \tag{8.8}$$

Including the exchange, correlation and inhomogenity terms gives

$$\epsilon(n) = \frac{3}{10}(3\pi^2)^{2/3} n^{5/3} - \frac{3}{4}\left(\frac{3}{\pi}\right)^{1/3} n^{4/3} - 0.056 \frac{n^{4/3}}{0.079+n^{1/3}} + \frac{1}{72}\frac{(\nabla n)^2}{n} . \tag{8.9}$$

The energy is then

$$E = \int \epsilon(n) d\underset{\sim}{r} \tag{8.10}$$

and the surface energy is

$$E_s = \int_{-\infty}^{\infty} [\epsilon(n) - \epsilon(nH(-x))] dx , \tag{8.11}$$

where H(z) is again the heaviside step function. To illustrate the procedure, the kinetic energy contribution to E_s will be evaluated. Using Eq. (8.1) with E=0 (*i.e.*, vanishing charge on the electrode),

$$E_s = \frac{3}{10}(3\pi^2)^{2/3} n^{5/3} \left\{ \int_{-\infty}^{0} \left[\left(1 - \frac{1}{2}e^{\alpha x}\right)^{5/3} - 1 \right] dx + \int_0^{\infty} \left(\frac{1}{2}e^{-\alpha x}\right)^{5/3} dx \right\} + \cdots .$$

$$= \frac{3}{10}(3\pi^2)^{5/3} \frac{n^{5/3}}{\alpha} \left\{ \sum_{n=1}^{\infty} (-1)^n \frac{1}{2^n} \frac{(5/3)!}{n!(5/3-n)!} \int_0^{\infty} e^{-nt} dt + \left(\frac{1}{2}\right)^{5/3} \frac{3}{5} \int_0^{\infty} e^{-t} dt \right\} + \cdots$$

$$= \frac{3}{10}(3\pi^2)^{5/3} \frac{n^{5/3}}{\alpha} \left\{ -\frac{5}{6} + \frac{5}{72} + \frac{1}{648} + \cdots + \left(\frac{1}{2}\right)^{5/3} \frac{3}{5} \right\} + \cdots$$

$$= -\frac{3}{10}(3\pi^2)^{5/3} \frac{n^{5/3}}{\alpha}(0.572) + \cdots . \tag{8.12}$$

The coulomb contribution to the surface energy is

$$E_s^{coul} = \frac{1}{8\pi}\int_{-\infty}^{\infty} \{E(x)\}^2 dx . \tag{8.13}$$

For vanishing electrode charge, it is easy to show that

$$E(x) = \begin{cases} \frac{2\pi ne}{\alpha}\, e^{\alpha x} & , x<0 \\ \frac{2\pi ne}{\alpha}\, e^{-\alpha x} & , x>0 . \end{cases} \tag{8.14}$$

which gives

$$E_s^{coul} = \frac{\pi n^2 e^2}{2\alpha^3} . \tag{8.15}$$

Putting everything together, the zero charge jellium expression (in atomic units) for the surface energy is

$$\begin{aligned} E_s = {} & \frac{\pi n^2}{2\alpha^3} - \frac{3}{10}(3\pi^2)^{2/3}\frac{n^{5/3}}{\alpha}\,(0.572) \\ & + \frac{3}{4}\left(\frac{3}{\pi}\right)^{1/3}\frac{n^{4/3}}{\alpha}\,(0.339) + \frac{I(n)}{\alpha} \\ & - \frac{0.084n}{\alpha}\left(a^2 - \frac{1}{2}a + \frac{1}{3} + a^3 \ln \frac{a}{a+1}\right) + \frac{\alpha n}{72} \ln 2 , \end{aligned} \tag{8.16}$$

where

$$I(n) = \frac{0.168}{1+k}\, n \int_{2^{1/3}}^{1} \frac{(1+k)x^4 - (x+k)}{(x+k)(1-x^3)}\, x^2 dx , \tag{8.17}$$

$$k = 0.079/n^{1/3} , \tag{8.18}$$

and

$$a = 2^{1/3}k . \tag{8.19}$$

The integral in I(n) can be evaluated analytically. The result is

$$\int \frac{(1+k)x^4-(x+k)}{(x+k)(1-x^3)}dx = \frac{1}{k^2-k+1}$$

$$\left\{k^6 \ln(x+k) + \sum_{n=1}^{6}(-1)^n \frac{k^{6-n}x^n}{n}\right\}$$

$$-\frac{1}{3}\sum_{n=1}^{6}\frac{x^n}{n} + \frac{(1+k)(k-2)}{3(k^2-k+1)}$$

$$x\left\{\frac{1}{2}\ln(1+x+x^2) - \frac{1}{\sqrt{3}}\tan^{-1}\frac{2x+1}{\sqrt{3}}\right.$$

$$\left. + \frac{x^6}{6} - \frac{x^5}{5} + \frac{x^3}{3} - \frac{x^2}{2}\right\}$$

$$+ \frac{(1+k)(2k-1)}{3(k^2-k+1)}\left\{\frac{1}{3}\ln(1+x+x^2) + \frac{2}{\sqrt{3}}\tan^{-1}\frac{2x+1}{\sqrt{3}}\right.$$

$$\left. + \frac{x^5}{5} - \frac{x^4}{4} + \frac{x^2}{2} - x\right\}. \quad (8.20)$$

although Eq. (8.16) has been derived from Eq. (8.1) using E=0, it can be shown that (8.16) is correct to order E.

There are three jellium-electrolyte coupling terms. The first is the jellium-ion interaction energy. From Eq. (8.14), the potential, due to the jellium, is

$$\phi(x) = -\frac{4\pi ne}{\alpha^2} + \frac{2\pi ne}{\alpha^2}e^{-\alpha x}, \quad x>\sigma/2 . \quad (8.21)$$

As has been mentioned, $\alpha \sim 2\text{Å}^{-1}$. As a result, $e^{-\alpha x}$ is negligible for $x > \sigma/2$ and only the first term need be considered. Thus, the jellium-ion interaction energy is

$$
\begin{aligned}
E_s^{ion} &= \int_{\sigma/2}^{\infty} g(x)\phi(x)dx \\
&= -\frac{4\pi ne}{\alpha^2}\int_{\sigma/2}^{\infty} g(x)dx \qquad (8.22) \\
&= \frac{neE}{\alpha^2} .
\end{aligned}
$$

It is to be noted that because $e^{-\alpha\sigma/2} \sim 0$, E_s^{ion} does not depend upon the actual ion profiles but only upon the total charge in the double layer.

The second term is the jellium-dipole interaction energy. This term involves $e^{-\alpha\sigma_s/2}$ and can safely be neglected. If only these two terms and those in Eq. (8.16) are included in the surface energy, the result is that of Partenskii and Smorodinskii and the model does not distinguish between a metal-vacuum and a metal-electrolyte interface. Clearly there is at least one additional term. Schmickler and Henderson [56] have pointed out that there is a repulsion of the electrons from the solution and have represented this repulsion semiempirically by an energy barrier of constant height V_b, in the region $x>0$. Thus,

$$
\begin{aligned}
E_s^{rep} &= V_b\int_0^{\infty} n_-(x)dx \\
&= \frac{V^b}{\alpha}\left(\frac{n}{2} - \frac{\alpha E}{8\pi e}\right) . \qquad (8.23)
\end{aligned}
$$

The parameter V_b is regarded as a property of the solvent and is chosen so as to give results for an interface involving that solvent and a mercury electrode. For an aqueous solution, V_b=3eV is used. For a vacuum, V_b=0.

The total surface energy, E_s is obtained by summing Eqs. (8.16), (8.22) and (8.23). The parameter α is their determined by minimization of E_s. Since this jellium model is not sensitive to the microscopic structure of the electrolyte it can be combined with *any* model of the electrolyte. Following Schmickler and Henderson the MSA ion-dipole results for the electrolyte will be used. The jellium contribution to the potential difference is

$$V_m = -\,4\pi \int_{-\infty}^{\infty} t\, q(t)dt$$

$$= 4\pi ne \int_0^{\infty} te^{-\alpha t}dt \tag{8.24}$$

$$= \frac{4\pi ne}{\alpha^2}\ .$$

Thus, the total potential difference, for small electrode charge, is

$$V = \frac{E}{\varepsilon\kappa} + \frac{E}{2\varepsilon}\left(\sigma + \frac{\varepsilon-1}{\lambda}\,\sigma_s\right) + \frac{4\pi ne}{\alpha^2} \tag{8.25}$$

and the differential capacitance (at low concentration and low electrode charge) is

$$C^{-1} = \frac{4\pi}{\varepsilon k} + \frac{2\pi}{\varepsilon}\left(\sigma + \frac{\varepsilon-1}{\lambda}\,\sigma_s\right) - \frac{8\pi ne}{\alpha^3}\,\frac{d\alpha}{dE}\ . \tag{8.26}$$

Because the α is not sensitive to the microscope structure of the electrolyte at the interface, α and $d\alpha/dE$ are independent of concentration and, therefore,

$$C_H^{-1} = \frac{2\pi}{\varepsilon}\left(\sigma + \frac{\varepsilon-1}{\lambda}\,\sigma_s\right) - \frac{8\pi ne}{\alpha^3}\,\frac{d\alpha}{de}\ . \tag{8.27}$$

It is to be noticed that the jellium contribution to C_H^{-1} in negative as suggested by Eq. (7.16).

Badiali *et al.* [57] have constructed a similar model coupling the ion-dipole mixture to a jellium model of a metal electrode. The two models differ in some technical details, as pointed out by Schmickler and Henderson [56]. The two models lead to differing physical interpretations. In the above Schmickler-Henderson model the capacitance of the double layer differs for different metals (immersed in the same electrolyte) because of differences in the electronic density n. In the Badiali *et al.* model, the capacitance of different metal-electrolyte interfaces is strongly influenced by a "distance of closest approach" d, which is of the order of the metal ion radius. The results given here are those obtained from the Schmickler-Henderson model.

For an aqueous electrolyte (V_b=3eV), the values of α and $d\alpha/dE$ at E=0, determined by minimization of E_s, are given by good accuracy by

$$\alpha(E=0) = 904\ n^2 - 61.118n + 2.73174 \tag{8.28}$$

and

$$\frac{e}{\sigma^2}\left(\frac{d\alpha}{dE}\right)_{E=0} = -\ 606.784(n-0.0125)^2 + 0.55359\ . \tag{8.29}$$

These values of α and $d\alpha/dE$ differ slightly from those of Schmickler and Henderson. The above results are obtained by minimizing the surface energy as obtained from Eqs. (8.16), (8.22) and (8.23) whereas in the calculation of Schmickler and Henderson the small jellium-dipole term, the term of order $e^{-\alpha\sigma/2}$ missing from Eq. (8.22), are included.

The values of C_H at the pzc which result from the values of α and $d\alpha/dE$, obtained from Eqs. (8.28) and (8.29), are plotted as a function of n and compared with experiment [58] in Fig. 14. The agreement with experiment is very good.

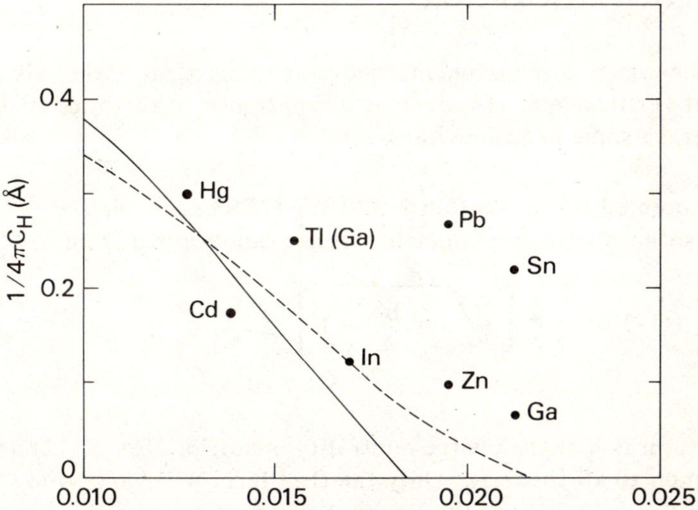

Figure 14. *Inverse "inner" layer capacitance as a function of the electronic density of the metal electrode. The points give the experimental results for several sp metals taken from Trasatti. The solid curve gives the results of the simple theory outlined here. The broken curve gives the results of the somewhat more sophisticated theory of Schmicker and Henderson.*

Schmickler and Henderson [56] have also calculated C_H for two other solvents, DSMO and acetonetrile and have obtained good agreement with experiment for these systems also.

The jellium model discussed above is deficient in two respects. Firstly, C_H diverges for $n \gtrsim 0.0185$ ($n > 0.0215$ in the slightly more refined calculations reported by Schmickler and Henderson). The experimental results do not exhibit this behavior. Secondly, the model outlined above

$$\left(\frac{d^2\alpha}{dE^2}\right)_{E=0} < 0 . \tag{8.30}$$

If correct, this would imply that the maximum in a plot of C_H would occur for negative charge. This is the opposite of what is observed experimentally in most systems. Both defects can be overcome by parameterizations which are more sophisticated than Eq. (8.1). These more sophisticated results will be covered by Schmickler [48].

9. ANOMALOUS ADSORPTION

This phenomenon is just mentioned in passing because there is no experimental verification. However, it is mentioned in the hope that it might encourage some experimental work.

It was pointed out in Section 2 that the GCS excess charge adsorption isotherms for a symmetric salt can be obtained from

$$\int_{\sigma_{i/2}}^{\infty} [g_i(t)-1]dt = \frac{2}{\kappa}\left[\sqrt{1+\frac{b^2}{4}} - 1\right] \pm \frac{b}{\kappa} . \tag{2.19}$$

The second term is just the charge neutrality condition, Eq. (1.12) and will be common to all theories. Only the first term will vary from theory to theory.

Henderson *et al.* [59] have shown that, in the primitive model at least, there should be a factor $\beta(\partial p/\partial\rho)_T$, where p is the osmotic pressure, in the denominator. Actually this term is present in the GCS result. It is just that the GCS theory treats the hard sphere part of the fluid as a perfect gas where $\beta(\partial p/\partial\rho)_T=1$. This suggests that Eq. (2.19) might be generalized in an *ad hoc* fashion to

$$\int_{\sigma_{i/2}}^{\infty} [g_i(t)-1]dt = \frac{2kT}{\kappa}\left(\frac{\partial\rho}{\partial p}\right)_T \left[\sqrt{1+\frac{b^2}{4}} - 1\right] \pm \frac{b}{\kappa}\,. \tag{9.1}$$

Phase separation is known to occur in selected electrolytes and it has been suggested [60] that the analogue in the primitive model of this phenomenon in real electrolytes is the vanishing of $(\beta\partial p/\partial\rho)_T$. This suggests that the excess charge adsorption isotherms in electrolytes exhibiting phase separation may be anomalously high when the electrolyte is close to the transition point.

Conceivably the adsorption of the coions seen by Parsons and Trasatti [17] might be an incipient form of this anomalous adsorption. However, this is unlikely to be the case since it is seen for both KCl and $MgCl_2$. Presumably, the 1:1 system would be much further from a transition point than would the 1:2 system and unlikely to exhibit such effects.

10. SUMMARY

Until recently, models of the double layer were based on the primitive model in which the solvent is regarded as a dielectric continuum and on the notion that the electrode could be regarded as a charged hard wall.

The Gouy-Chapman-Stern (GCS) approximation gives a reasonable approximation to most of the properties of the primitive model double layer. However, computer simulations and all of the more modern theories (MSA, HNC, BGY and MPB) predict potential difference and inverse capacitances which are less than those of the GCS theory.

To obtain agreement with experiment from the GCS theory, adjustable parameters must be used. This is not only unappealing but leads to unphysical molecular models which assume the interfacial solvent layer to be a monolayer. Recently, progress has been made towards replacing this primitive model by a microscopic or molecular model of the solvent and a electronic model of the metal. Assuming, for simplicity, that the solvent molecules are dipolar hard spheres and that the metal can be modelled as jellium leads to results which are nearly free of empirical parameters and which are in close agreement with experiment. Moreover, this microscopic model gives information unavailable from the primitive model, such as the thickness of interfacial solvent layer, the breakdown of the additivity of concentration dependent and independent inverse capacitances and the possibility of anomalously large adsorption in some systems.

11. ACKNOWLEDGMENTS

The author is grateful to Drs. G. Torrie and J. P. Valleau for sending the results of their computer simulations and to Dr. C. W. Outhwaite for sending his MPB5 results. He is also grateful to Drs. L. Blum, M. Lozada-Cassou, C. W. Outhwaite, R. Parsons and W. Schmickler for many fruitful discussions. The financial support of the National Science Foundation (Grant No. CHE80-01969) is also gratefully acknowledged. Finally thanks are due to Dr. A. F. Silva for organizing this institute and the NATO Science Committee whose financial support made the attendance of the author possible.

REFERENCES

1. Blum, L. and Henderson, D. 1981, J. Chem. Phys. 74, pp. 1902-1910.
2. Blum, L. and Henderson, D. 1978, J. Chem. Phys. 69, pp. 5441-5449.
3. Henderson, D., Blum, L., Lebowitz, J. L. 1979, J. Electroanal. Chem. 102, pp. 315-319.
4. Blum. L., Lebowitz, J. L. and Henderson, D. 1980, J. Chem. Phys. 72, pp. 4249-4250.
5. Gouy, G. 1910, J. Phys. 9, pp. 457-467.
6. Chapman, D. L. 1913, Phil. Mag. 25, p. 475.
7. Stern, O. 1924, Z. Eleutro. Chem. 30, pp 508-516.
8. Torrie, G. M. and Valleau, J. P. 1979, Chem. Phys. Letters 65, pp. 343-346; 1980, J. Chem. Phys. 73, pp. 5807-5816.
9. Torrie, G. M. and Valleau, J. P. 1982, J. Phys. Chem. 86, pp. 3251-3257.
10. Grahame, D.C. 1953, J. Chem. Phys. 21, pp. 1054-1060.
11. Barojas, J., Henderson, D. and Lozada-Cassou, M. 1983, J. Phys. Chem. 87, pp. 4547-4548.
12. Valleau, J. P. and Torrie, G. M. 1982, J. Chem. Phys. 76, pp. 4623-4630.
13. Bhuiyan, L. B., Blum, L. and Henderson, D. 1983, J. Chem. Phys. 78, pp. 442-445.
14. Watts-Tobin, J. 1961, Phil. Mag. 6, pp. 133-153.
15. Fawcett, W. R. 1986, these proceedings.
16. Guidelli, R. 1986, these proceedings.
17. Parsons, R. and Trasatti, S. 1969, Trans. Faraday Soc. 65, pp. 3314-3323.
18. Parsons, R. and Zobel, F. G. R. 1965, J. Electroanal. Chem. 9, pp. 333-348.
19. Parsons, R. and Trasatti, S. 1983, private communication.
20. Carnie, S. L., Chem, D. Y. C., Mitchell, D. J. and Ninham, B. 1981, J. Chem. Phys. 74, pp. 1472-1478.
21. Blum, L. 1977, J. Phys. Chem. 81, pp 136-147.
22. Henderson, D. and Smith, W. R. 1978, J. Stat Phys. 19, pp. 191-200.
23. Khater, A. F., Henderson, D., Blum, L. and Bhuiyan, L.B. 1984, J. Phys. 88, pp. 3682-3684.
24. Henderson, D., Blum L. and Smith, W. R. 1979, Chem. Phys. Letters 63, pp. 381-383.
25. Henderson, D. and Blum, L. 1980, J. Electroanal. Chem. 111, pp. 217-222.
26. Lozada-Cassou, M., Saavedra-Barrera, R. and Henderson, D. 1982, J. Chem. Phys. 77, pp. 5150-5155.
27. Lozada-Cassou, M. and Henderson, D. 1983, J. Phys. Chem. 87, pp. 2821-2824.
28. Marconi, U. M. B., Wiechen, J. and Forstmann, F. 1984, preprint.

29. Croxton, T. L. and McQuarrie, D. A. 1981, Mol. Phys. 42, pp. 141-151.
30. Henderson, D., Blum, L. and Bhuiyan, L. B. 1981, Mol. Phys. 43, pp. 1185-1188.
31. Blum, L., Hernando, J. and Lebowitz, J. L. 1983, J. Phys. Chem. 87, pp. 2825-2832.
32. Painter, K. R., Ballone, P., Tosi, M. P., Grout, P. J., and March, N. H. 1983, Surface Sci. 133, pp. 89-100; Ballone P., Pastore, G. and Tosi, M. P. 1984, Phys. Chem. Liq. 13, pp. 269-278.
33. Outhwaite, C. W. 1974, Mol. Phys. 27, pp. 561-575.
34. Outhwaite, C. W. and Thomlinson, M. M. 1974, Chem. Phys. Letters 25, pp. 375-377.
35. Outhwaite, C. W. 1978, J. Chem. Faraday II 74, pp. 1214-1221.
36. Levine, S. and Outhwaite, C. W. 1978, J. Chem. Soc. Faraday II 74, pp. 1670-1689.
37. Levine, S., Outhwaite, C. W. and Bhuiyan, L. B., 1981, J. Electroanal. Chem. 123, pp. 105-119.
38. Debye, P. and Hückel, E. 1923, Phys. Z. 24, pp. 185-206.
39. Loeb, A. L., 1951, J. Colloid Sci. 6, pp. 75-91.
40. Yvon, J. 1935, *La Theorie Statistique des Fluides et l' Equation d' Etat* (Hermann et Cie) ; Born M. and Green, H. S. 1946 Proc. Roy. Soc. (London) A188, pp.10-18.
41. Kirkwood, J. G. 1934, J. Chem. Phys. 2, pp. 767-781; 1935 *ibid* 3, pp. 300-313.
42. Outhwaite, C. W. 1970, Chem. Phys. Letters 7, pp. 636-638; Bell, G. M. and Rangecroft, 1972, Mol. Phys. 24, pp. 255-267.
43. Outhwaite, C. W. and Bhuiyan, L. B. 1983, J. Chem. Soc. Faraday II 79, pp. 707-718.
44. Torrie, G. M. and Carnie, S. L. 1984, Adv. Chem. Phys., 56, pp. 141-253.
45. Carnie, S. L. and Chan, D. Y. C. 1980, J. Chem. Phys. 73, pp. 2949-2957.
46. Rosinberg, M. L. 1983, Ph. D. Thesis, Université M & P. Curie, Paris, p. 149.
47. Blum, L., Henderson, D. and Parsons, R. 1984, J. Electroanal. Chem. 161, pp. 389-392.
48 Schmickler, W. 1986, these proceedings.
49 Eyring, H., Henderson, D., Stover B. J. and Eyring, E. M. 1982, *Statistical Mechanics and Dynamics* 2nd Ed. (Wiley-Interscience) pp. 283-287.
50. Outhwaite, C. W. 1980, Chem. Phys. Letters 76, pp. 619-621.
51. Henderson, D., Barojas J. and Blum, L. 1984, Revista Mexicana de Fisica 30, pp. 139-216.
52. Lang, N. D. and Kohn, W. 1970, Phys. Rev. B1, pp. 4555-4568.
53. Smith, J. R. 1969, Phys. Rev. 181, pp. 522-529.

54. Partenskii, M. B. and Smorodinskii, Ya. G. 1974, Sov. Phys. Solid State 16, pp. 423-424.
55. Schmickler, W. and Henderson, D. 1984, Phys. Rev. B 30, pp. 3081-3803.
56. Schmickler, W. and Henderson, D. 1984, J. Chem. Phys. 80, pp. 3381-3386.
57. Badiali, J. P., Rosinberg, M. L., Vericat, F. and Blum, L. 1983, J. Electroanal. Chem. 158, pp. 253-267.
58. Trasatti, S. 1981, J. Electroanal. Chem. 123, pp. 121-139.
59. Henderson, D., Barojas, J. and Blum, L. 1983, J. Phys. Chem. 87, pp. 4544-4547.
60. Friedman, H. L. 1981, Ann. Rev. Phys. Chem. 32, pp. 179-204.

THE SEMICONDUCTOR-ELECTROLYTE INTERFACE

L.M. Peter

Department of Chemistry,
The University, Southampton SO9 5NH
Great Britain

ABSTRACT This lecture considers the distribution of charge and potential at the semiconductor-electrolyte interface and examines the influence of different types of surface sites on the electrochemical behaviour of semiconductors. A brief analysis of the illuminated semiconductor-electrolyte interface follows, and it is shown that non-steady state photocurrents can be related to the density and energy level of surface states. The lecture concludes with some experimental results which illustrate the effects of surface recombination at p-GaP and suggests some new experimental approaches to the study of surface states and photochemical kinetics.

1. INTRODUCTION

The central emphasis of this NATO ASI has been placed on the properties and, more particularly, on the theoretical description of the metal-solution interface. Such an emphasis is a natural consequence of the fact that precise experimental data for the mercury-solution interface have been available for a number of years, but a more important stimulus for theoretical progress has been the steady accumulation of a body of reliable information obtained with oriented single crystal metal electrodes. Recent attempts to describe the metal-electrolyte interface have shifted attention away from the solvent towards a more realistic consideration of the electronic properties of the metal, which is no longer considered as a simple dielectric mirror but rather as a free electron gas associated in the 'jellium' model with the periodic array of core charges (1). This degree of theoretical sophistication is restricted to the study of metal electrodes,

A. F. Silva (ed.), Trends in Interfacial Electrochemistry, 523–546.

and by contrast, understanding of the properties of the interface between non-metallic solids and electrolyte solutions is limited essentially to an electrostatic point of view. The boundary between the solid and the solvent is considered to be abrupt in the case of semiconductor electrodes, and little attention has been paid to the properties of the 'inner layer' of solvent. Semiconductor electrodes are, in any case, less ideal than many metal electrodes. They are 'ideally polarisable' only in the sense that the current flow under depletion conditions is sufficiently small that equilibrium is not perturbed, but in general the thermodynamic stability of compound or elemental semiconductors is limited. Nevertheless, semiconducting materials are of considerable importance in electrochemistry, not only as single crystal electrodes but also as phases formed during the anodic oxidation of metals, and it is therefore worthwhile to try to achieve a realistic view of their interfacial properties. Since detailed discussions of the semiconductor-electrolyte interface are available elsewhere (2-4), this lecture attempts to focus attention on the *surface* properties of the semiconductor, and in particular on the way in which interfacial states influence both equilibrium and non-equilibrium behaviour.

2. THE DISTRIBUTION OF CHARGE AT THE SEMICONDUCTOR-ELECTROLYTE INTERFACE IN THE ABSENCE OF SURFACE STATES.

The density of free charge carriers in a semiconductor is generally much smaller than in a metal. For semiconductors with band-gaps greater than 1 eV, the contribution of thermally generated carriers to the room-temperature conductivity is negligible, and the conductivity is determined by the density of donor or acceptor impurities in the crystal, giving rise to n- or p-type conductivity. Typical donor and acceptor densities lie in the range 10^{15} - 10^{18} cm^{-3}, although values outside this range may also be encountered. By comparison, the ionic density in a 0.1 mol dm^{-3} electrolyte solution is around 10^{20} cm^{-3}, at least two orders of magnitude higher, and these carrier densities in turn are considerably smaller than the free electron density of simple metals ($\sim 10^{22}$ - 10^{23} cm^{-3}). Therefore, whereas in the case of metals it may be an acceptable approximation to consider that excess charge in the solid is located essentially in a surface plane, the lower charge carrier densities in the electrolyte and semiconductor phases give rise to regions of *diffuse* charge, and the electrostatic parts of the problem are virtually identical for both phases.

If electrons are withdrawn from the surface region of an n-type semiconductor, the immobile ionised donor states give rise to a *space-charge*. The distribution of charge and potential across the semiconductor-electrolyte interface follows from

Poisson's equation and Fermi-Dirac (or Boltzmann) statistics. Fig. 1 illustrates the profiles of charge and potential expected

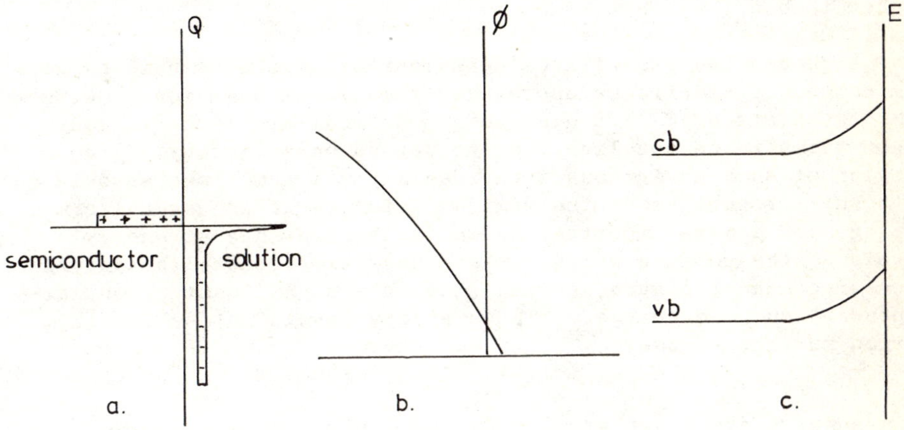

Figure 1 Distribution of (a) charge and (b) potential at the semiconductor electrolyte interface under depletion conditions. The corresponding band diagram is shown in (c).

under these *depletion conditions* and contrasts them with the commonly accepted picture of the electron energy profile (the 'band-bending'). The validity of this diagram depends on a number of assumptions. The charge in the semiconductor is supposed to be composed only of the electron density and the ionised donor density, and the surface itself has no excess charge. Fortunately for the progress of semiconductor electrochemistry, these simplifying assumptions are satisfactory for a number of stable wide band-gap semiconductors. The work of Dewald (5), for example, demonstrated that ZnO, unlike the elemental semiconductors Ge and Si, appears to be free of *surface states* which can store charge. The measurement of the *space-charge capacitance*, C_{sc}, of materials such as ZnO and CdS under depletion conditions gives data which fit closely to the *Mott-Schottky* relationship

$$C_{sc}^{-2} = \left(\frac{2}{eN\varepsilon\varepsilon_o}\right)\left(U - U_{fb} - \frac{k_BT}{e}\right) \qquad 1.$$

where U is the electrode potential, U_{fb} is the flat-band potential, ε is the relative permittivity of the semiconductor and N is the donor density (in the case of n-type semiconductors). The measured electrode capacitance corresponds in the simplest case to a series connection of the space-charge capacitance and the double-layer capacitance, but for semiconductors with average doping levels, the latter is at least two orders of magnitude higher than C_{sc}.

Although the capacitance behaviour of several single crystal semiconductor electrodes approaches this ideal behaviour, Mott-Schottky plots of C^{-2} vs electrode potential are found in many other cases to be non-linear or to depend on frequency. The origins of such deviations from ideality are sometimes associated with bulk properties of the semiconductor, such as non-uniform doping, but a more important cause is the presence of energy levels at the surface which can exchange electrons with the valence or conduction band, giving rise to a significant potential-dependent surface charge. These energy levels may be loosely termed 'surface states'.

3. SURFACE SITES AND SURFACE STATES; DEFINITIONS AND DISTINCTIONS.

The ability of the surface of semiconductors to store charge has been recognised for some time in gas-phase and vacuum studies and the concept of *surface states* is not new. The formation of surface states at the junction between two solid phases has been widely proposed as the origin of deviations from ideal barrier heights at semiconductor-metal contacts and at semiconductor heterojunctions. In the case of the semiconductor-electrolyte junction, the presence of surface states has been invoked to explain the non-ideal behaviour of materials such as Ge (6,7), but it is clear that the original meaning of the term 'surface state' has been changed to allow a more general explanation of non-ideality at two phase junctions. It is therefore sensible to re-examine the definition of surface states in the special context of the semiconductor-electrolyte junction.

The surface of a covalent or ionic crystal corresponds to an abrupt discontinuity in the periodic lattice structure. As a consequence, the environment of the outermost layer of atoms differs substantially from the interior of the crystal, and it is clear that the electronic energy levels associated with the surface will not correspond to those in the energy bands of the solid. If the solid is covalent, the surface is considered to have 'dangling bonds' or orbitals which are directed out from the solid. The high surface energy which results from such a discontinuity may lead to surface reconstruction, and the surface orbitals are able to interact strongly with adsorbed materials or adjacent phases. These states are referred to as *Shockley surface states* (8). The surfaces of ionic materials are also expected to exhibit surface states that arise from the asymmetric electrostatic environment; these are the *Tamm surface states* (9). Quantum calculations of both types of surface state have provided valuable insight into the solid-gas interface (10), but no equivalent calculations have been made for the solid-electrolyte interface.

The surface states discussed above are associated with electron energy levels which arise from the solid. It is also important to consider the Lewis acid/base properties of surfaces, and it is possible to define *Lewis sites* which either donate or accept electron pairs from adsorbates. Such sites are particularly important at solid-solution interfaces, since they can exchange H^+ and OH^- ions with the electrolyte. For example, a Lewis base site such as a lattice oxide ion can accept a proton from solution as follows

$$O_L^{2-} + H^+ \rightleftharpoons O_L{:}H^2 \qquad 2.$$

or a Lewis acid site such as a metal ion can interact with hydroxyl ions or water;

$$M_L + OH^- \rightleftharpoons M_L{:}OH^- \qquad 3.$$

An important consequence of ion exchange processes of this kind is that the surface dipole potential depends on the proton activity of the solution phase. A simple picture of the oxide-solution interface is shown in fig. 2. The surface is com-

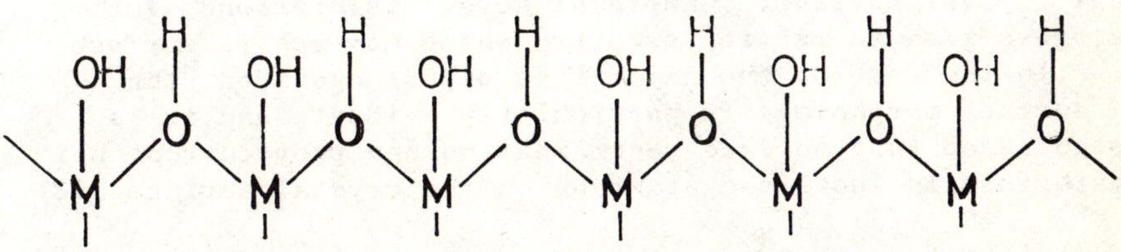

Figure 2 Illustration of 'bound water' at the surface of a metal oxide in contact with an aqueous solution.

pletely covered in this case with bound water, and it is

immediately clear that the interface is quite different from that formed at a metal electrode. The water is strongly chemisorbed on the surface of the oxide, and the acid/base equilibrium corresponding to eq. 2 and 3 will determine, at least in part, the surface charge of the solid. The pH at which the surface densities of OH^- and H^+ groups are equal is the *point of zero charge*. The most obvious manifestation of these Lewis sites is the pH dependence of the flat-band potential which is observed for oxide semiconductors (and for many other compound semiconductors) in aqueous solution.

The interaction of the surface of the semiconductor with other components of the electrolyte phase can also change the potential distribution. The adsorption of anions or cations can occur by ion exchange, particularly on oxide semiconductors, e.g.

$$M{:}OH^- + Na^+ \rightleftharpoons M{:}ONa + H^+ \qquad 4.$$

$$M{:}OH^- + Br^- \rightleftharpoons M{:}Br^- + OH^- \qquad 5.$$

Ionic adsorption is not restricted to oxides; particularly striking effects of anion adsorption are observed for S^{2-} on CdS (11) and for I_3^- on the layer type transition metal dichalcogenides (12).

The interaction of the surface of the semiconductor with redox species in solution may, in principle, involve electron exchange, and in the case of metal centres which have more than one stable valence state, it is possible to define a *redox surface state*, although the distinction between such a state and the more general concept of electronic surface states is not well-defined.

The distribution of surface sites is probably never uniform at a real crystal surface. Emergent screw dislocations, for example, give rise to particular sites which may act as surface states. In the case of the layer-type dichalcogenides, the role of surface morphology is particularly evident, and it is well established that surface recombination and photocorrosion in these materials is localised at steps on the crystal surface (13).

In most real systems, the interface is less ideal than the preceding discussion suggests. We need to distinguish between true intrinsic or extrinsic surface states and, for example, intermediates formed during dark or photoreactions which lead to lattice decomposition. The capture of a hole by a surface bond may be an essentially irreversible process if the rupture of the bond leads to an atom leaving the surface of the crystal. The density of surface intermediates of this kind is a complex function of the kinetic parameters, and is likely to be different in the dark and under illumination. In some cases, intermediates such as H• or small atoms or ions may diffuse into the surface region, giving rise to *near-surface states*. These are levels which interact preferentially with the electronic energy levels of the semiconductor since they are in rather than on the surface. It has been shown, for example, that the near-surface states produced by illumination of p-GaP in acid solution are due to the presence of hydrogen atoms which have diffused into a thin layer close to the surface of the semiconductor (14). The surface states and surface sites which have been discussed above are listed in Table 1.

Table 1. Surface sites and surface states

Type	Exchange Mechanism
intrinsic (Shockley, Tamm) states	electronic (with semiconductor and/or solution)
Lewis acid/base sites	ionic (H^+, OH^- exchange with solution)
redox sites	electronic and/or ionic
special lattice sites (dislocations, defects)	ionic (e.g. anion adsorption) or electronic
surface intermediates (products of dark or photoreactions)	electronic or ionic
'near surface' states	preferential electron exchange with semiconductor

In many cases, lattice decomposition can result in the formation of bulk phases, for example oxides, on the surface of a semiconductor electrode. If the oxide layer is continuous and does not store appreciable charge, it may be more appropriate to describe the system as a semiconductor-insulator-solution junction. In other cases, porous oxide films are formed, and ion and electron exchange processes involving the surface phase profoundly modify the behaviour of the junction. The wide range of possible complications in electrode behaviour is indicated schematically in fig. 3.

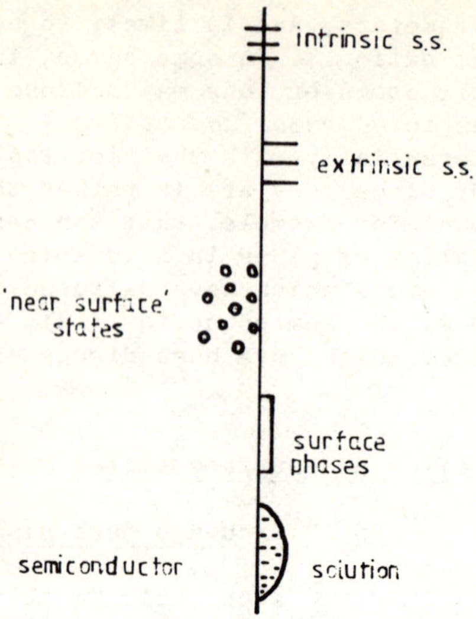

Figure 3 Surface states and surface phases at the semiconductor-electrolyte interface.

4. THE INFLUENCE OF SURFACE STATES ON THE DISTRIBUTION OF CHARGE AND POTENTIAL.

The presence of an appreciable density of surface states at the semiconductor-solution interface leads to changes in the potential and charge distribution that arise from the ability of the surface to store charge. If the density of surface states approaches the surface atom density, the electrode will behave as a metal. In this case, changes in electrode potential appear primarily across the electrical double layer in the solution. Consider, for example, a set of surface states located at a discrete energy level as shown in fig. 4. The *occupancy* of this surface energy level is determined by electron exchange with the solution and with the semiconductor. If we assume for the moment that the electrolyte solution does not contain any redox species which can exchange electrons with surface states (the 'ideally polarisable' case), then the occupancy in the dark is given by

$$f_s = \left[1 + \exp\left\{ \frac{E_F^o - E_t - |e|\Delta\phi_{sc}}{k_B T} \right\} \right]^{-1} \qquad 6.$$

where $\Delta\phi_{sc}$ is the potential drop in the semiconductor and $E_F^o - E_t$ is defined in fig. 4. The positive charge stored in the surface states can be written as

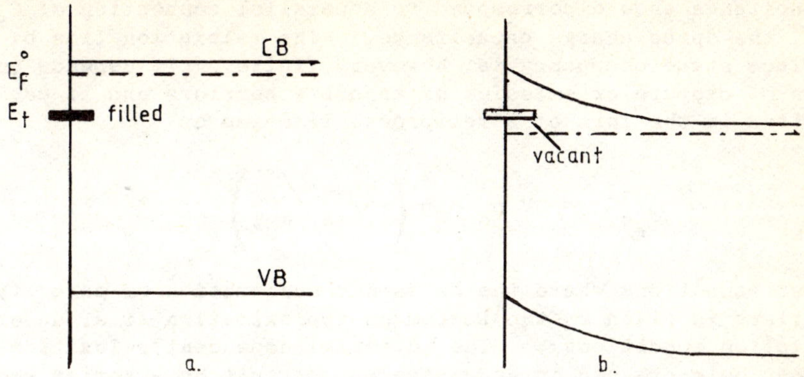

Figure 4 Simple model for surface states located at a discrete energy level E_t. The occupation of the surface state is determined by $E_F - E_t$. The flat band situation is shown in (a) and the effect of band bending is illustrated in (b).

$$Q_s = N_s |e| \left(1 - f_s\right) \tag{7.}$$

and it is clear that this charge will vary rapidly as the Fermi level passes through the surface energy level. The potential drop in the solution phase can be expressed in terms of the double-layer capacitance C_H in the form

$$V_H = \frac{Q_s}{C_H} + \text{const} = \frac{N_s |e| \left(1 - f_s\right)}{C_H} + \text{const} \tag{8.}$$

(It is assumed in the derivation of eq. 8 that the dipole contribution to V_H remains unchanged). It is now possible to obtain the *surface state capacitance* C_s by differentiation of eq. 7;

$$C_s = \frac{dQ_s}{d\Delta\phi_{sc}} = \frac{e^2}{k_B T} \quad \frac{N_s \exp\left(\dfrac{E_F^o - E_t + e\Delta\phi_{sc}}{k_B T}\right)}{\left\{1 + \exp\left(\dfrac{E_T^o - E_t + e\Delta\phi_{sc}}{k_B T}\right)\right\}} \tag{9.}$$

If the density of surface states is not too high ($N_s \lesssim 10^{12}$ cm^{-2}), the change in V_H can be neglected, and the measured electrode capacitance should correspond to a parallel connection of C_s with C_{sc}, the space charge capacitance. The relaxation time of the surface state occupancy is, however, finite. It depends on the rate of capture or emission of majority carriers and it can be written in the form of a reciprocal lifetime as

$$\tau_s^{-1} = \sigma V n_o \exp\left(-\Delta\phi_{sc}/k_B T\right) \qquad 10$$

under conditions where the surface concentration of majority carriers is given by the Boltzmann approximation (i.e. under depletion conditions). The potential-dependent relaxation time τ_s can be expressed in an equivalent circuit by a series combination of a resistance R_s with the surface-state capacitance C_s. These elements appear in the more general equivalent circuit shown in fig. 5. In practice, the relaxation time of surface states may

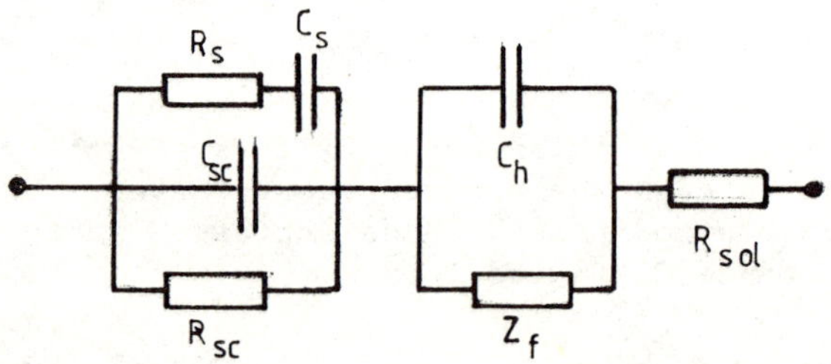

Figure 5 Equivalent circuit for the semiconductor-electrolyte interface under depletion conditions in the presence of surface states with a discrete relaxation frequency given by eq. 10. R_{sc} is the space-charge leakage resistance, Z_f is the Faradaic impedance and R_{sol} is the solution resistance. The other elements are defined in the text (note that $\tau_s = R_s C_s$).

be too slow for them to contribute to the capacity measured at audio-frequencies. Fig. 6 illustrates the surface state capacitance calculated from eq. 9 and shows how the relaxation time increases rapidly with band-bending.

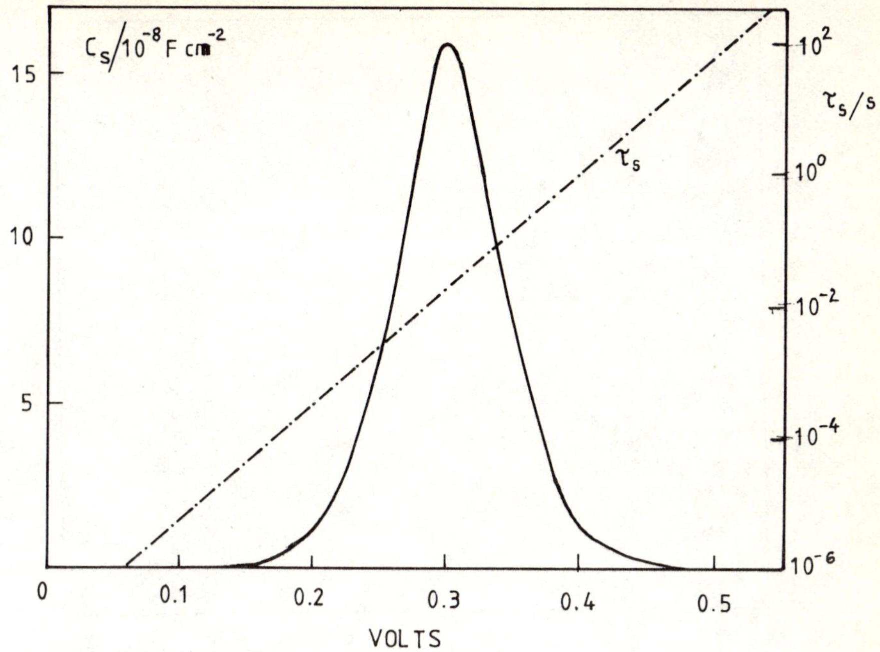

Figure 6 Potential dependence of C_s calculated from eq. 9 for $E_F^o - E_t = 0.3$ eV, $N_s = 10^{11}$ cm^{-2}. Note that the relaxation time exceeds 1s for potentials above 0.4V. For larger values of $E_F^o - E_t$, the relaxation time will be even longer, so that the surface state capacitance is not seen in a.c. measurements at audio-frequencies.

The total capacitance obtained by proper analysis of the electrode impedance is expected to show a maximum when the Fermi level passes through the surface level, and fig. 7 shows the result calculated for a typical semiconductor. Surface state capacitance measurements have been made, for example on Ge electrodes (15), but fewer reliable data are available for compound semiconductors, although Dare-Edwards et al (16) have analysed the impedance of p-GaP and shown that deviations from ideal behaviour can be attributed to a surface state located near the valence band. One of the most important effects of surface states is that they can lead to a phenomenon known as *Fermi-level pinning*. As the Fermi level moves into the surface level, the accumulation of charge results in the appearance of an additional potential drop across the Helmholtz layer rather than

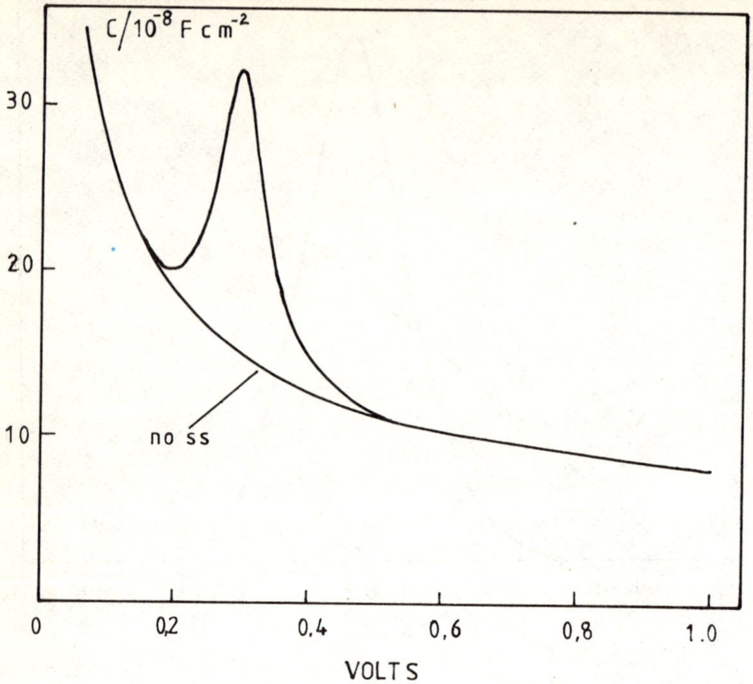

Figure 7 Total capacitance calculated for a semiconductor with surface states. $N_s = 10^{11}$ cm^{-2}, $N = 10^{17}$ cm^{-3}, $\varepsilon = 10$. $E_F^o - E_t = 0.3$ eV.

across the space-charge layer in the semiconductor. The Fermi level therefore becomes 'pinned' at the surface level until the occupation approaches zero (or unity). The pinning of the Fermi level therefore corresponds to an 'unpinning' of the band edges with respect to the solution as shown in fig. 8. Fermi-level pinning will only be important if the surface state density is high ($N_s > 10^{13}$ cm^{-2}); this is made clear by fig. 9 which shows how the potential drop in the space charge and Helmholtz layers varies with electrode potential for different values of the surface state density N_s (17).

The purpose of this discussion has been to show for a simple case how the presence of surface states which communicate preferentially with the semiconductor modifies the potential distribution. The extension of these ideas needs to take into account that surface states are likely to be distributed in energy. At the same time, the interaction of the surface state with solvent fluctuations has been neglected, and electron exchange with redox species has been specifically excluded. It is clear that substantial improvements in the model are possible, but the paucity of experimental impedance data has discouraged

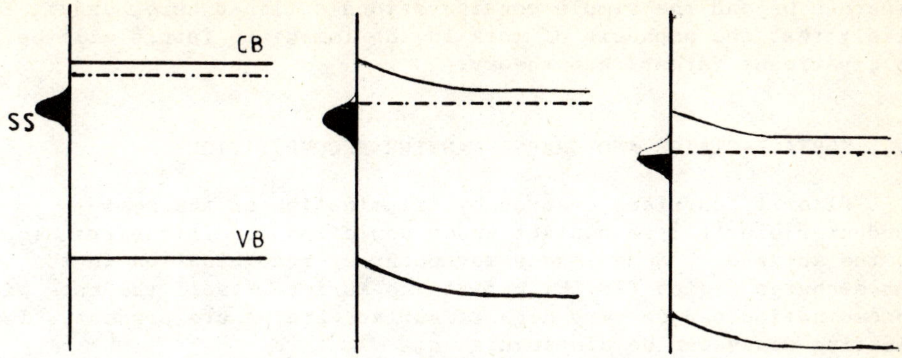

Figure 8 Fermi level pinning. Note that the band edges move with respect to the solution due to the accumulation of surface charge while the band-bending remains almost constant. The electrode potential is increased from left to right.

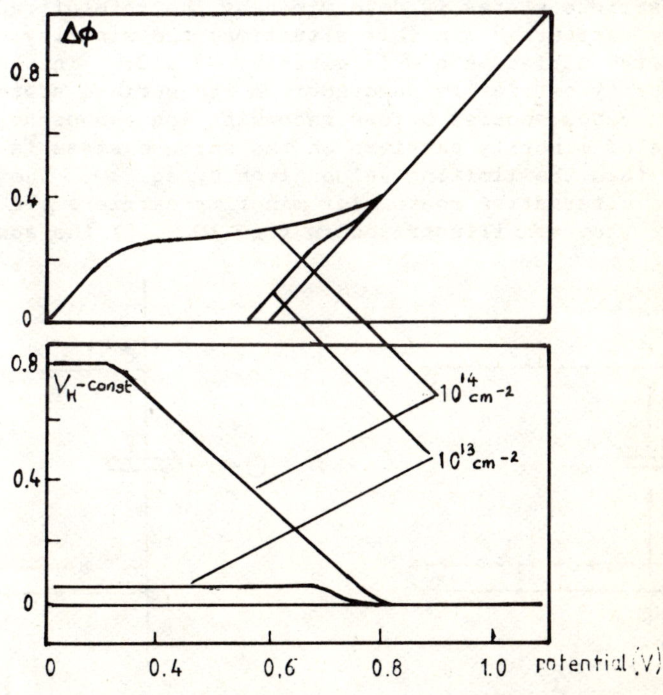

Figure 9 Distribution of potential across the semiconductor ($\Delta\phi_{sc}$) and double-layer (V_H) as a function of electrode potential in the presence of surface states. (N_s as shown).

progress beyond the simple considerations outlined here, and it is likely that the emphasis of work in the immediate future will be on experiment rather than theory.

5. SURFACE-STATES AND CHARGE-CARRIER RECOMBINATION

Minority carriers created by illumination of the semiconductor-electrolyte contact under depletion conditions are swept to the surface. Some losses may occur by recombination in the space-charge region (18,19) but at the surface itself the rate of recombination can be very high if surface states are present. Two limiting cases can be distinguished;

(i) the surface state interacts preferentially with the solid

(ii) the surface state interacts preferentially with redox species in solution.

In the first case, the lifetime of minority carriers trapped at vacant surface states is determined by the rate of capture of majority carriers. In this situation, the minority carrier lifetime on the surface state is given by eq. 10. In the second case, the minority carrier is removed from the surface state by a solution redox species before recombination can occur, so that the lifetime of minority carriers on the surface state is considerably shorter than the limiting value given by eq. 10. There are therefore two alternative routes for minority carriers via surface states; they are illustrated in fig. 10. If the energy level of

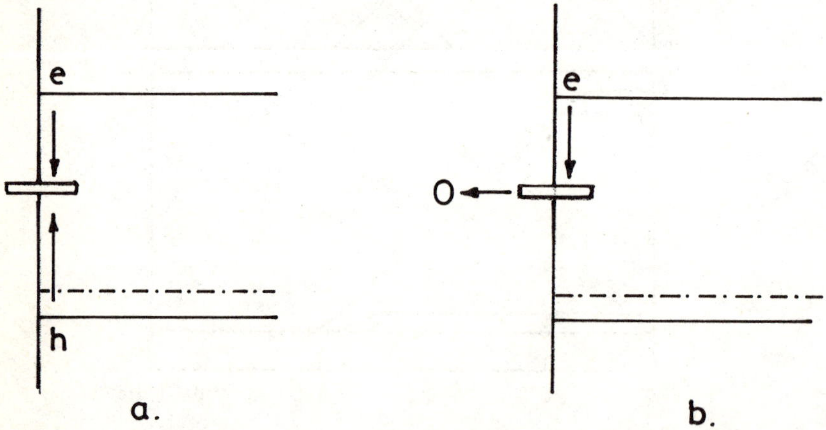

Figure 10 Limiting behaviour of surface states. In (a), the surface state promotes carrier recombination, whereas in (b) the minority carrier is passed to a redox species before recombination has time to occur

a surface state overlaps with the solution redox species, we might expect electron transfer to be sufficiently rapid in most cases that surface recombination will not occur. In this limit, the occupancy of the surface state is determined by the redox potential (or *redox Fermi level)*, rather than by the Fermi level in the solid. It is therefore surprising that recombination is seen at all illuminated semiconductor electrodes. There are several reasons, however, why electron exchange with the solution may not be able to compete favourably with recombination. In many experimental situations, the surface of a semiconductor electrode may be partly covered with an oxide layer or other insoluble decomposition product. The surface states in this case may be located at the semiconductor-oxide junction, where they can act as efficient recombination centres according to the first scheme in fig. 10. A second possibility is that the surface states are not on the surface in the true sense, but simply *near* the surface. If this is the case, they will communicate preferentially with the semiconductor and act as efficient recombination centres. These are the 'near surface states' that have already been mentioned.

Several treatments have been made of the influence of surface recombination on the steady-state photocurrent response of the semiconductor-electrolyte interface (18-20). The discussion which follows is based on a more detailed examination of steady state and transient photocurrents (17). In the absence of space-charge recombination, the flux of excess minority carriers into the surface is given by the *Gärtner equation* (21);

$$g = I_o \left[1 - \exp(-\alpha W) \Big/ (1 + \alpha L) \right] \qquad 11.$$

where I_o is the incident photon flux, α is the absorption coefficient, L is the diffusion length of minority carriers and W is the width of the space charge region, given in the Boltzmann approximation by

$$W = \left(2\Delta\phi_{sc}\varepsilon\varepsilon_o \Big/ |e|N \right)^{\frac{1}{2}} \qquad 12.$$

The time dependent *excess* minority carrier density in surface states is given by

$$\Delta n_s(t) = N_s \Delta f_s(t) \qquad 13.$$

where

$$\Delta f_s(t) = f_s(t) - f_s(o) \qquad 14.$$

and $f_s(o)$ is given by eq. 6 (here it is assumed that we are dealing with a p-type semiconductor, where the minority carriers are electrons). The *excess hole flux* due to the reaction of valence band holes with excess electrons in surface states is then

$$j_p(t) = k_p \rho_s N_s \Delta f_s(t) \tag{15.}$$

where

$$\rho_s = \rho_o \exp\left(-|e|\Delta\phi_{sc} \Big/ k_B T \right) \tag{16.}$$

The net time dependent photocurrent is therefore

$$j_{photo} = g - j_p(t) \tag{17.}$$

It is important to note that the time-dependent fluxes of minority and majority carriers into the surface state are not equal except under steady-state conditions ($t \to \infty$). The solution of this problem (17) gives expressions for the steady-state photocurrent which have been used to generate the photocurrent-voltage curves shown in fig. 11.

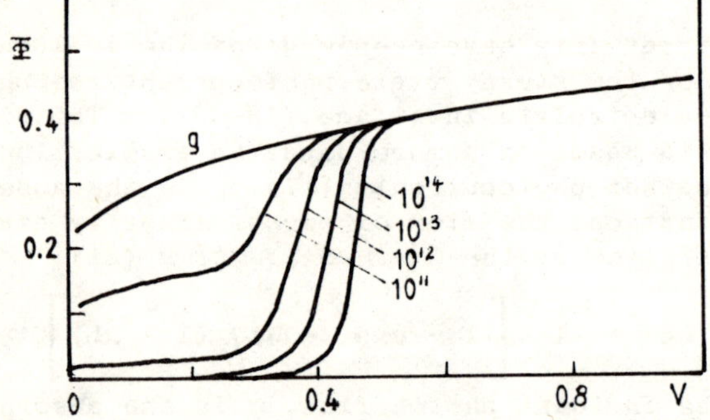

Figure 11 Steady state photocurrent-voltage curves calculated for a surface state situated 0.3 eV above the bulk Fermi level of a p-type semiconductor. $I_o = 10^{14}$ cm^{-2} s^{-1}, $\alpha = 3.3 \times 10^4$ cm^{-1}, N_s as shown (for further details see (17)). Note the delayed onset of the photocurrent which results from surface recombination. The curve for $N_s = 0$ corresponds to the Gärtner equation.

Perhaps more interesting from a diagnostic point of view are the photocurrent transients predicted in the presence of surface recombination. In the limit of low illumination levels or high surface-state densities, the photocurrent response to an illumination step is given by the expression

$$\frac{j_{photo}(t) - j_{photo}(\infty)}{j_{photo}(o) - j_{photo}(\infty)} = \exp(-t/\tau_s) \qquad 18.$$

where τ_s is the minority carrier lifetime in surface states as defined by eq. 10. When the light is interrupted, the reaction of excess trapped electrons with holes gives rise to a transient current of opposite sign which decays with a lifetime τ_s. A typical photocurrent transient (calculated for conditions similar to those used to compute the curves in fig. 11) is shown in fig. 12.

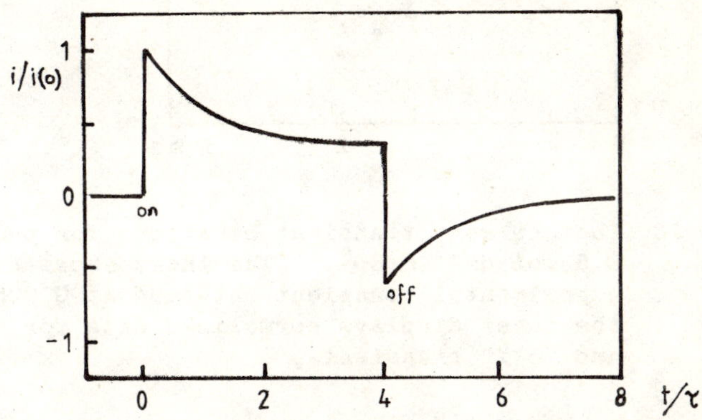

Figure 12 Photocurrent transient calculated for the same conditions as those in fig. 11. Potential 0.4V, $N_s = 10^{13}\ cm^{-2}$.

6. NEAR SURFACE STATES IN p-GaP

The preceding discussion has shown that the effects of surface recombination are likely to be most obvious if the surface states communicate predominantly with the semiconductor. Particularly striking effects have been observed in the case of p-GaP electrodes where the onset of the photocurrent is delayed by up to 0.5V with respect to the flat-band potential. Photocurrent transients obtained in the rising portion of the photocurrent-voltage characteristic show pronounced relaxation effects; an example is shown in fig. 13 together with a semilogarithmic plot according to eq. 18. A detailed study of the near surface states in p-GaP has shown that they are produced by the diffusion of photogenerated hydrogen atoms into the semiconductor lattice (14).

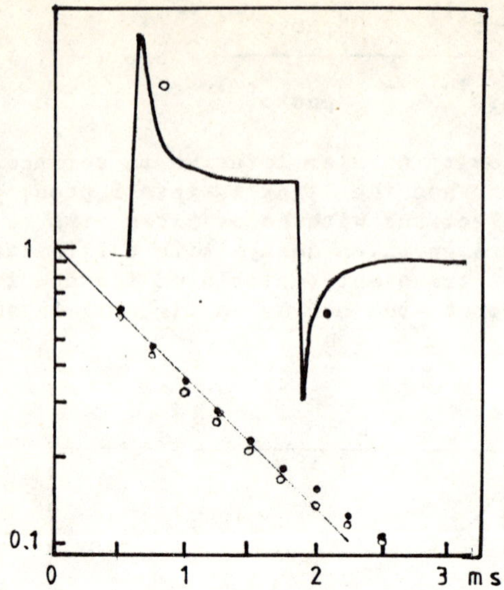

Figure 13 Photocurrent transient behaviour for p-GaP in 0.5 mol dm^{-3} H_2SO^4. The inset shows an experimental transient obtained at 0.225V, and the inset displays normalised data for the "on" and "off" transients.

The vacant near surface state consists of a proton which is formed when the Fermi level in the solid passes through the energy level of the hydrogen atoms incorporated into the 'near surface' region of the p-GaP (here 'near surface' refers to a depth of 30-50 Å). The recombination process in this case is simply

$$H^+_{ns} + e \longrightarrow H^{\cdot}_{ns} \qquad \text{19a}$$

$$H^{\cdot}_{ns} + h \longrightarrow H^+_{ns} \qquad \text{19b}$$

The photoinduced near surface states in p-GaP can be removed by polarising the electrode at the flat-band potential since the proton space-charge creates an electrical field which expels the ions from the solid. It is therefore possible to begin measurements with a reproducible electrode surface free of surface states. The build up of hydrogen can then be followed as a function of the period of illumination and the total amount of hydrogen incorporated at any time can then be determined by linear sweep voltammetry. It has been shown that the hydrogen atoms diffuse at least 30 Å into the crystal surface, reaching surface densities of up to 10^{14} cm^{-2} (14).

The formation of surface states can also be followed by *sub-bandgap spectroscopy*. In this technique sub-band gap light is used to excite electrons from surface states into the conduction band, and the photocurrent conversion efficiency in this region gives a measure of the density of surface states. A set of sub-bandgap spectra is shown in fig. 14. In principle the technique

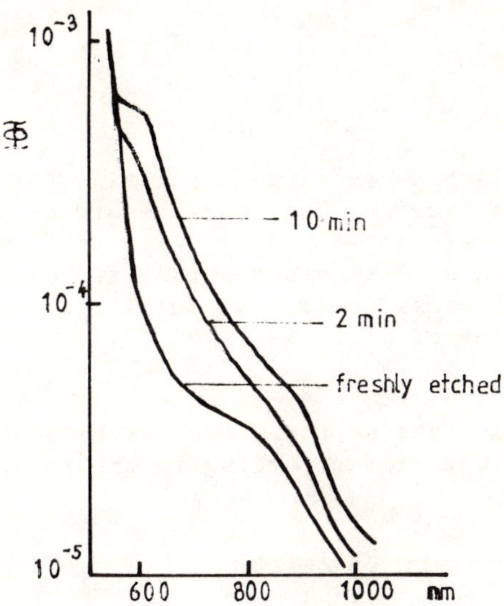

Figure 14 Sub-bandgap response of p-GaP in 0.5 mol dm^{-3} H_2SO_4 at -0.5 V <u>vs</u> SCE showing increase brought about by illumination with white light for different periods.

can be used to examine the distribution of surface states in energy, but no detailed quantitative work has been carried out with this promising technique.

Surface states can also be introduced by chemical pretreatment of the p-GaP electrode with Ru(III) (14,16). These surface states are evidently able to communicate preferentially with the electrolyte since they catalyse the hydrogen evolution process and reduce the rate of surface recombination. Since a high density of hydrogen near surface states is still present in the treated electrodes, the reduction in recombination rate must be interpreted in terms of the competition mechanism illustrated in fig. 15.

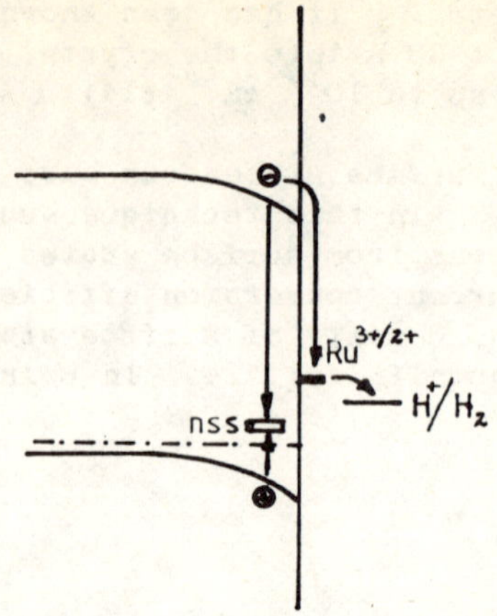

Figure 15 Energy scheme showing competition for electrons between hydrogen near surface states and Ru(III) surface states at p-GaP. The efficiency of hydrogen evolution shows that the electron capture cross section of Ru(III) is much greater than that of H^+.

The ruthenium surface states therefore act as efficient mediators for electron transfer according to the following scheme;

Ru catalysed hydrogen evolution:

$$Ru(III) + e \longrightarrow Ru(II) \quad 20.$$

$$Ru(II) + H^+ \longrightarrow Ru(III) + H^{\cdot}_{ads} \quad 21.$$

$$2H^{\cdot}_{ads} \longrightarrow H_2 \quad 22.$$

or

$$H^{\cdot}_{ads} + H^+ + Ru(II) \longrightarrow H_2 + Ru(III) \quad 23.$$

recombination:

$$H^+_{ns} + e \longrightarrow H^{\cdot}_{ns} \quad 24.$$

$$H^{\cdot}_{ns} + h \longrightarrow H^+_{ns} \quad 25.$$

The different pathways for photoexcited electrons illustrate very well the two limiting types of behaviour of surface states. The improved hydrogen evolution characteristics achieved with Ru(III) treated surfaces suggest that the electron capture cross section of Ru(III) is higher than that of H^+; this would be the case if the Ru centres carry a multiple positive charge, i.e. if they are effectively Ru^{3+}.

7. AC PHOTOCURRENT METHODS

Semiconductor electrochemistry has been well served by a number of basic concepts and a limited range of techniques. The reexamination of some of these concepts is now in progress and the use of more sophisticated experimental techniques is likely to become widespread. The discussion of surface recombination given in this lecture has shown the advantage of non-steady-state techniques, and it is clear that pulsed light experiments can give more information than steady-state photocurrent-voltage curves. We have recently developed an A.C. method for the measurement of non-steady-state photocurrent behaviour, and it appears to offer a promising new approach to the study of photoelectrochemical kinetics and surface processes at illuminated semiconductor electrodes. An acousto-optic modulator is used to produce a sinusoidally modulated laser beam which illuminates the semiconductor electrode, and the magnitude and phase of the resulting signal are measured as a function of frequency with a frequency response analyser. The response of the electrode can be considered in terms of the equivalent circuit shown in fig. 16. The following limiting cases can be considered

(i) recombination dominant

(ii) mixing recombination and charge transfer

(iii) charge-transfer dominant.

The photocurrent response can be conveniently defined in terms of the *complex quantum efficiency* $\tilde{\Phi}$ as

$$\tilde{\Phi} = j_{photo}/g$$

The quantity $\tilde{\Phi}$ is conveniently displayed in the complex plane as shown in fig. 17 for the general case (ii).

Preliminary experimental results have already been obtained for n-GaAs with this technique, and an example is shown in fig. 18.

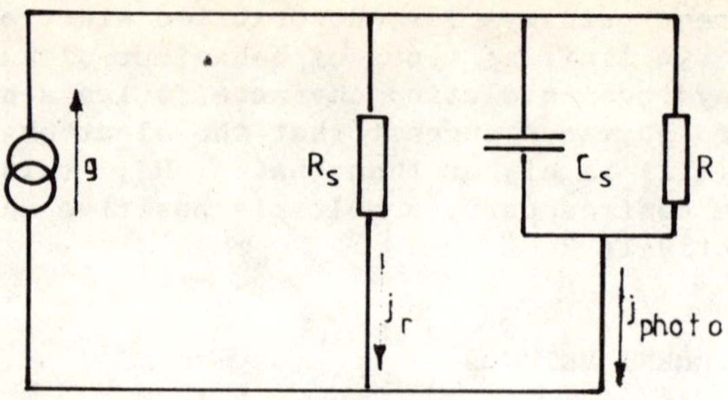

Figure 16 Simple equivalent circuit for AC photocurrent response in the presence of surface recombination. The current generator corresponds to the flux of minority carriers given by the Gärtner equation (see eq. 11). R_s and C_s have been defined above and R_o is related to the rate constant for interfacial electron transfer to redox species in solution. The recombination flux is $\tilde{j}_r$ and the measured photocurrent is $\tilde{j}_{photo} = \tilde{g} - \tilde{j}_r$.

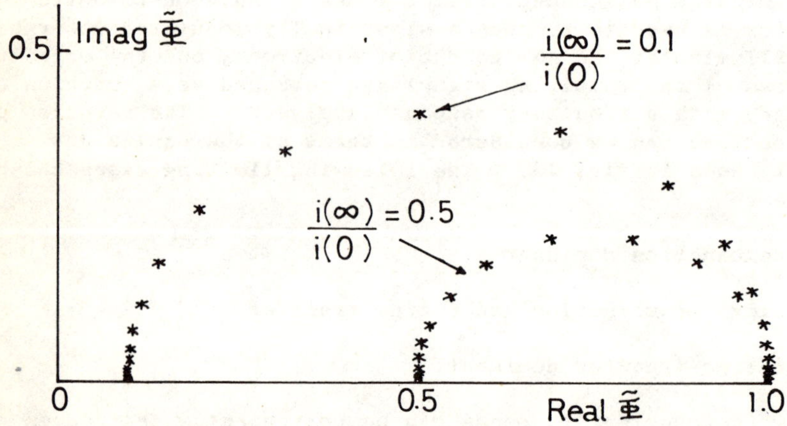

Figure 17 Complex quantum efficiency as a function of frequency (case (ii)). Note that $\tilde{\Phi} \to 1$ in the high frequency limit.

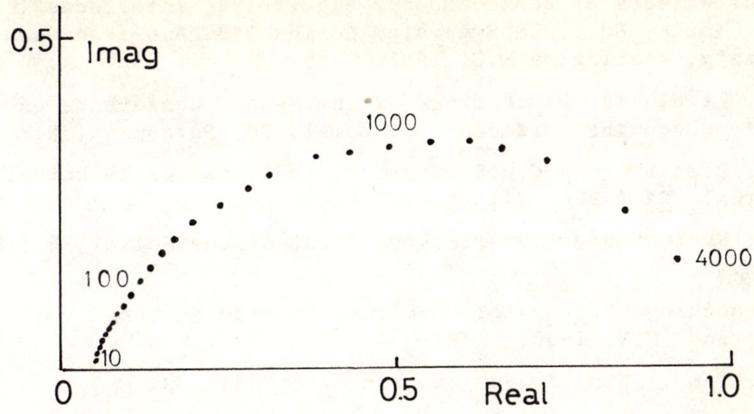

Fig. 18 Plot of complex photocurrent conversion efficiency obtained with n-GaAs in 1.0 mol $dm^{-3}Na_2SO_4$ at 0.5V vs SCE (illuminations He-Ne).

The potential of the method for the study of more complex systems is now being explored.

8. Outlook

The development of semiconductor electrochemistry during the past ten years or so has been rapid. Experimental progress has resulted largely from the search for new and more stable materials, and less attention has been paid to fundamental aspects of the inferfacial properties of the semiconductor-electrolyte junction. The next few years will see the application of non-stationary techniques as well as a range of *in-situ* spectroscopic techniques to the study of semiconductor electrodes. As more reliable fundamental information becomes available, it is to be hoped that theoretical progress will follow. The semiconductor-electrolyte interface still remains as an exciting challenge to experimentalist and theoretician alike.

REFERENCES

1. see, for example, the lecture by Henderson and Schmickler in this volume.

2. V.A. Myamlin and Y.V. Pleskov, "Electrochemistry of Semiconductors", Plenum, N.Y. 1967.

3. S.R. Morrison, "Electrochemistry of Semiconductor and Oxidised Metal Electrodes", Plenum, N.Y. 1980.

4. "Photoeffects at semiconductor-electrolyte interfaces", A.J. Nozik, Ed., ACS Symposium Series 146. American Chemical Society, Washington D.C. 1981.

5. J.F. Dewald in "Proceedings of the Second Conference on Semiconducuting Surfaces". J. Zemel, Ed. Pergammon, N.Y. 1960.

6. W.H. Brattain and C.G.B. Garrett, Bell Systems Technical Journal, 34 (1963) 2119.

7. M.D. Krotova and Y.V. Pleskov, Phys. Status Solidi, 3 (1963) 2119.

8. W. Shockley, "Theory of Electronic Semiconductors" van Nostrand, N.Y. 1950.

9. I.E. Tamm, Phys. Z. Sowjetunion, 1 (1932) 733.

10. S.R. Morrison, "The Chemical Physics of Surfaces" Plenum, N.Y. 1977.

11. T. Inoue, T. Watanabe, A. Fujishima and K. Honda, Bull. Chem. Soc. Japan, 52 (1979) 1243.

12. H. Tributsch, J. Electrochem. Soc., 125 (1978) 1086.

13. H.J. Lewerenz, A. Heller, H.J. Leamy and S.D. Ferris, in ref. 4, p.17.

14. J. Li, R. Peat and L.M. Peter, 165 (1984) 41.

15. U. Plitt, H. Schulte and H. Seliger, Phys. Lett. 30A (1969), 266.

16. M.P. Dare-Edwards, A. Hamnett and J.B. Goodenough, J. Electroanal. Chem., 119 (1981) 109.

17. L.M. Peter, J. Li and R. Peat, J. Electroanal. Chem., 165 (1984) 29.

18. J. Reichmann, Appl. Phys. Lett., 36 (1980) 574.

19. W.J. Albery, P.N. Bartlett, A. Hamnett and M.P. Dare-Edwards, J. Electrochem. Soc., 128 (1981) 1492.

20. J.J. Kelly and R. Memming, J. Electrochem. Soc., 129 (1982) 730.

21. W.W. Gärtner, Phys. Rev. 116 (1959) 84.

LIST OF PARTICIPANTS

Abrantes, L.: R. Escola Politécnica, Lisboa, Portugal
Amadeo, M.: Instituto di Scienze Chimica, Piazza Renascimento, 6 61029 Urbino, Itália
Andreu, R.: Dep. Química-Física c/Palos da Fronteira, 1 Sevilha, Spain
Barbosa, R.: Dep. Química Pr. Gomes Teixeira, 4000 Porto, Portugal
Barros, A.: Dep. Química Pr. Gomes Teixeira, 4000 Porto, Portugal
Battisti, A.: Universidade di Ferrara, Via Borsari, 46, 44100 Ferrara, Itália
Baumgartel, H.: Universitat Berlin, Takustasse 3, 1000 Berlin 33, W. Germany
Benedetti, L.: Universita di Modena, Via Campi 183, 41100 Modena, Itália
Berlouis, L.; Dep. of Chemistry, University Road, Southampton SO9 5NH, England
Bewick, A.: Dep. of Chemistry, University Road, Southampton SO9 5NH, England
Bijsterbosch, B.: Lab. Physical Colloid Chemistry, de Dreijen 6, 6703 Wageningen, The Netherlands
Brett, A.: Dep. de Química, Faculdade de Ciências e Tecn., 3000 Coimbra, Portugal
Brett, M.: Dep. de Química, Faculdade de Ciências e Tecn., 3000 Coimbra, Portugal
Buck, R.: Dep. of Chemistry - Venable and Kenan Labs 045 A Chapel Hill L.N.C., 27514 U.S.A.
Buess-Herman, C.: Universite Libre Bruxelles, Av. Franklin Roosevelt, Bruxelas, Belgique
Camões, M.: Faculdade Ciências, R. Escola Politécnica 1200 Lisboa Portugal
Eskikaya, T.: Technical University of Istambul, Abidei Hurriet Cad, No 79/81 D7, Turkey
Fawcett, W.R.: University of Carolina, Davis, California 95616, U.S.A.
Ferreira, M.: Instituto Superior Técnico, Av. Rovisco Pais, 1200 Lisboa, Portugal

Fonseca, I.: Faculdade Ciências Lisboa, R. Escola Politécnica 1200 Lisboa, Portugal

Girault, H.: University of Southampton, University Road, Southampton SO9 5NH, England

Guidelli, R.: Universitat di Firenze, Via Gino Capponi 9 50121 Firenze, Itália

Gutierrez, C.: Consejo Superior Inv. Cient. Serrano 119, Madrid 6 Spain

Halley, J.W.: University of California, Santa Barbara, California U.S.A.

Hamelin, A.: CNRS., Bellevue 1, Place Aristide-Briand, 92190 Meudon, France

Henderson, D.: 5600 Cottle Road, San Jose, California U.S.A.

Hillman, A.R.: University of Bristol, Cantock's Close, Bristol BS8 1TS, England

Jesus, J.: Universidade de Aveiro 3800 Aveiro, Portugal

Koklü, U.: IST. Technishni, 11 Blok. Daire B, Istambul, Turkey

Kolb, D.M.: Fritz, Haber Institut, Faradayweg 4-6, 1000 Berlin 33/Dahlen, W. Germany

Lopes, M.H.: Univ. do Porto, Pr. Gomes Teixeira, 4000 Porto, Portugal

Montenegro, I.: Largo do Paço 4719 Braga, Portugal

Moura, C.: Faculdade de Ciências, Pr. Gomes Teixeira, 4000 Porto, Portugal

Nelson, L.A.: Prosprect place Plymouth Pl 1 eDH, England

Neto, M.M.: C.E.C.U.L., Calçada Bento Rocha Cabral, 14, 1200 Lisboa, Portugal

Newmark, A.: Columbia University New York 10027, U.S.A.

Parsonage, N.G.: Imperial College London SW7 ZAY, England

Parsons, R.: CNRS, Bellevue 1, Place Aristide Briand, 92190 Meudon, France

Pereira, M.I.: Faculdade de Ciências, R. Escola Politécnica, 1200 Lisboa, Portugal

Peter, L.: University of Southampton, Southampton SO9 5NH, England

Rangel, C.M.: Est. Paço Lumiar, 22, 1600 Lisboa, Portugal

Rehbach, M.S.: Rijksunifresiteit Utrecht, Padualaan 8 Postbus 80.051, 3508, TB Utrecht, The Netherlands

Rocha, M.I.: Faculdade Ciências, Pr. Gomes Teixeira, 4000 Porto, Portugal

Sanz, F.: Universidade de Barcelona, Av. Diagonal, 647, Barcelona 28, Spain

Sass, J.K.: Fritz-Haber Institut, Faradayweg 4-6, 1000 Berlin 33/ /Dahlem, W. Germany

Schmickler, W.: Universitaet Dusseldorf D, 4000 Dusseldorf, W. Germany

Sertel, M.: Ondokuzmaysis Universitesi 19 Mayis, Universitesi Atakum/Samsum, Turkey

Silva, A.F.: Departamento Química, Pr. Gomes Teixeira, 4000 Porto Portugal

Silva, C.J.: Departamento Química, Pr. Gomes Teixeira, 4000 Porto Portugal

Silva, M.D.: Departamento Química, Pr. Gomes Teixeira, 4000 Porto Portugal

Simão, J.E.: Universidade do Minho, Largo do Paço, 4719 Braga, Portugal

Simões, A.: Inst. Superior Técnico, Av. Rovisco Pais, 1096 Lisboa Codex-Portugal

Sottomayor, M.J.: Faculdade Ciências, Pr. Gomes Teixeira, 4000 Porto, Portugal

Tascon, J.: Instituto de Catalisis, Serrano 119, Madrid 6, Spain

Tieje, Y.: University of Southampton, Southampton SO9 5HN, England

Trasatti, S.: University Milan, Via Venezian 21, 20133 Milan, Italy

Tronel-Peyroz, E.: LA 330-CNRS Route de Mende, B.P. 5051, 34033 Montpellier Cedex, France

Valdes, J.L.: Columbia University, Terrace Building NCW York 10027 U.S.A.

Van der Eeden, A.L.G.: Rijksuniversiteit, Utrecht, Padualaan 8, Postbus 80.051 350 8 TB, Utrecht, The Netherlands

Van der Linden, J.G.: University of Nijmegen, Toernooveld, 6525 ED Nijmegen, The Netherlands

Velasco, J.: Universidade de Madrid Cantoblanco Madrid, 34 Spain

Viana, C.: Faculdade de Ciências, R. Escola Politécnica, 1200 Lisboa, Portugal

Weaver, M.: Chemistry Department, West Lafayette, Indiana 47907, U.S.A.

Yeter, B.: Gasy University/Ank Enevis SOK.41/10 Ankara, Turkey

Yildiz, A.: Hacettepe University, Ankara, Turkey

LIST OF CONTRIBUTORS

Bewick, Alan: Department of Chemistry, University Road, Southampton SO9 5NH, England.

Bijsterbosch, B. H.: Lab. Physical Colloid Chemistry, de Dreijen 6, 6703 Wageningen, The Netherlands.

Buck, Richard: Department of Chemistry, Venable and Kenan Labs 045 A, Chapel Hill, N.C. 27514, U.S.A.

Buess-Herman, C.: University Libre Bruxelles, Av. Franklin Roosevelt, Bruxelles, Belgique.

Guidelli, Rolando: Universitat di Firenze, Via Gino Capponi 9, 50121 Firenze, Italia.

Hamelin, Melle. A.: CNRS-Bellevue 1, Place Aristide-Briand, 92190 Meudon, France.

Henderson, Douglas: 5600 Cottle Road, San Jose, California, U.S.A.

Kolb, D.M.: Fritz-Haber-Institut, Faradayweg 4-6, 1000 Berlin 33/Dahlem, W. Germany.

Parsonage, N.G.: Imperial College, London SW7 2AY, England.

Parsons, Roger: CNRS-Bellevue 1, Place Aristide-Briand, 92190 Meudon, France.

Peter, L.: University of Southampton, Southampton SO9 5NH, England.

Sass, J.K.: Fritz-Haber-Institut, Faradayweg 4-6, 1000 Berlin 33/Dahlem, W. Germany.

Schmickler, W.: Universitaet Dusseldorf, D 4000 Dusseldorf, W. Germany.

Silva, A.F.: Departamento Química, Pr. Gomes Teixeira, 4000 Porto, Portugal.

Trasatti, Sergio: University Milan, Via Venezian 21, 20133 Milan, Italia.

Weaver, Michael: Chemistry Department, West Lafayette, Indiana 47907, U.S.A.

SUBJECT INDEX